# 33rd Ethylene Producers Conference 2021

Topical Conference at the 2021 AIChE Spring Meeting and 17th Global Congress on Process Safety

Online
18 - 23 April 2021

ISBN: 978-1-7138-5329-9

**Printed from e-media with permission by:**

Curran Associates, Inc.
57 Morehouse Lane
Red Hook, NY 12571

**Some format issues inherent in the e-media version may also appear in this print version.**



Printed with permission by Curran Associates, Inc. (2022)

For permission requests, please contact AIChE
at the address below.

AIChE
120 Wall Street, FL 23
New York, NY 10005-4020

Phone: (800) 242-4363
Fax: (203) 775-5177

www.aiche.org

**Additional copies of this publication are available from:**

Curran Associates, Inc.
57 Morehouse Lane
Red Hook, NY 12571 USA
Phone: 845-758-0400
Fax: 845-758-2633
Email: curran@proceedings.com
Web: www.proceedings.com

# TABLE OF CONTENTS

# AIChE Paper Number 8a

# How to Reduce $CO_2$ Emissions of Steam Cracking Furnaces: The Million Dollar Question

***Ismaël Amghizar***
*Modeling Researcher*
*AVGI BV, Ghent, Belgium*

***David J Brown***
*Senior R&D Manager*
*AVGI BV, Ghent, Belgium*

***Guy B Marin***
*Emeritus Professor*
*Laboratory for Chemical Technology (LCT), Ghent University, Ghent, Belgium*

***Oleksii Mynko***
*R&D Engineer*
*AVGI BV, Ghent, Belgium*

***Kevin M. Van Geem***
*Professor*
*Laboratory for Chemical Technology (LCT), Ghent University, Ghent, Belgium*

Prepared for Presentation at the 2021 Spring National Meeting
Dallas, TX, April 18 – April 23, 2021

# How to Reduce $CO_2$ Emissions of Steam Cracking Furnaces: The Million Dollar Question

***Ismaël Amghizar***
*Modeling Researcher*
*AVGI BV, Ghent, Belgium*
***David J Brown***
*Senior R&D Manager*
*AVGI BV, Ghent, Belgium*
***Guy B Marin***
*Emeritus Professor*
*Laboratory for Chemical Technology (LCT), Ghent University, Ghent, Belgium*
***Oleksii Mynko***
*R&D Engineer*
*AVGI BV, Ghent, Belgium*
***Kevin Van Geem***
*Professor*
*Laboratory for Chemical Technology (LCT), Ghent University, Ghent, Belgium*

Abstract:

In state-of-the-art steam cracking plants more than 90% of the $CO_2$ emissions can be directly related to the high energy consumption of the endothermic conversion in the cracking furnaces. Steam cracking accounts for a global emission of more than 300 million tonne of $CO_2$ per annum. Enhancing heat transfer in the radiation section, using green energy, and reducing coke formation are key to substantially reduce $CO_2$ emissions. Heat transfer can be increased by implementing three-dimensional (3D) coil technologies. These reactor technologies also reduces coke formation because of the lower wall temperatures that are consequently obtained. Advanced manufacturing techniques and better computational abilities have opened the doors to novel and improved 3D reactor technologies that are designed to increase the heat transfer while minimizing the pressure drop penalty. At the same time applying high emissivity coatings on the furnace refractory and reactor tubes can further reduce $CO_2$ emissions. Substantial fuel savings can also be obtained by a novel furnace design, where the heat recovery scheme is substantially modified. Combining all these technologies could result in reducing emissions by 30%. Shifting completely to green electricity, which is practically infeasible today, is another alternative but the technologies that would potentially allow this are still in their infancy. These new technologies, combined with advanced process innovations and $CO_2$ capture, will help the industry to meet future emissions targets. The potential benefit of small and larger modifications have been quantified by accounting for the additional annual production of high value chemicals, extended run length as well as reduction in fuel consumption and hence direct $CO_2$ emissions, compared to a base case of a conventional operating steam cracking furnace. COILSIM1D is used to carry out the evaluation.

## Introduction

Climate change, of which greenhouse gas emission is the main driver, is one of the most urgent challenges humanity is currently facing. The atmospheric CO2 concentration has been rising rapidly since the start of the measurements in March 1958, with an average increase of approximately 2 ppm per year over the past decade. In 2016, the atmospheric CO2 concentration stayed above the symbolic 400 ppm mark all year round for the first time, corresponding to a 30% increase compared to the pre-industrial (before circa 1750) levels of 270 ppm. The Paris Agreement under the United Nations Framework Convention on Climate Change adopted in 2015 illustrates the worldwide commitment to reduce greenhouse gas emissions to mitigate global warming, but this will require an order-of-magnitude increase in public and private investments in research and development between 2019 and 2030.

Strategies to reduce CO2 emissions can be divided in four categories that focus on either avoidance of CO2 emissions or binding the emitted CO2 in a natural or non-natural sink. The first category is improving energy efficiency, which currently provides the greatest return on investment and has already been successfully applied in many industrial contexts. Although this approach still has potential, annual improvements of 1-2% will not be sufficient to meet the climate targets. The second category, using non- or low-carbon energy sources (e.g. solar, wind, geothermal), is at large scale still challenging due to the fluctuating nature of the energy supply and the slow rate at which the electricity production is becoming more renewable. Carbon Capture and Sequestration (CCS), i.e. a series of technologies combining CO2 capture from large point sources such as power plants, transportation to a storage site, and sequestration into a (natural) sink, is the third category, but its potential is currently rather limited due to technical and economic hurdles. The fourth category is Carbon Capture and Utilization (CCU), in which CO2 is converted to (high-value) products. This category can be considered as a special case of the third category with the utilization part acting as a non-natural sink.

Also in the recently completed project 'IMPROOF' [ '**I**ntegrated **M**odel guided **PRO**cess **O**ptimization of steam cracking **F**urnaces'] [1], [2], [3] the objective was to reduce CO2 emissions.

Outside IMPROOF other interesting technologies are being developed such as 'RDR'. RDR is a new technology under development that uses mechanical work to provide the energy needed for the endothermic cracking reactions. 'RDR' has been presented already at this meeting [4], and an update is expected in 2022.

Table 1 below lists the 11 cases that were evaluated in this work. A more detailed description and explanation of each case and each technology is given in the paper. The evaluation of RDR is considered separately towards the end.

**Table 1: List of studied technologies and simulation cases**

| Case | Bare coil | 3D coils | High-emissivity coating | Alumina forming alloy | Biogas | Oxyfuel | Novel furnace arrangement | RDR reactor |
|---|---|---|---|---|---|---|---|---|
| BASE | ✓ | | | | | | | |
| HEC | ✓ | | ✓ | | | | | |
| ALU | ✓ | | | ✓ | | | | |
| 3D | | ✓ | | ✓ | | | | |
| 3D+HEC | | ✓ | ✓ | ✓ | | | | |
| BIO | ✓ | | | | ✓ | | | |
| OXY | ✓ | | | | | ✓ | | |
| 'L-E 20%' | ✓ | | | | | | ✓ | |
| 'L-E 30%' | ✓ | | | | | | ✓ | |
| RDR | | | | | | | | ✓ |

'L-E' and 'Novel furnace arrangement' are alternative names given to the 'Low-Emission Furnace' concept developed and patented by one of IMPROOF partners [5]. Percentage stands for relative reduction in overall firing rate compared to baseline scenario.

For biofuel, oxyfuel, and for the Low-Emission Furnace, sensitivity studies have been performed in order to quantify (a) the impact of the biogas and combustion oxygen source; (b) specific embodiments of the Low-Emission Furnace; and (c) the carbon footprint of the electrical grid.

## Modelling approach and technologies evaluated

### *Base case*

Reactor coils have been simulated with a custom geometry in line with what was presented at this meeting [6], [7]. For this work we have concentrated on the performance when cracking propane.

### *High emissivity refractory coating ('HEC')*

Emissivity is defined as the ratio of total radiative output from a surface per unit time per unit area at a specific temperature and wavelength to that of a black surface under the same conditions.

The efficiency of furnaces can be improved if the emissivity of the refractory linings and/or the external coil walls is increased. The mechanism for radiant (thermal) efficiency increase, and hence fuel saving, is quite complex and it is outside the scope of this paper to delve into it. An excellent and concise description is given by Adams et al [27].

The emissivity values is in the range of 0.85 -0.95, so in this work an average value of 0.9 is taken. In order to quantify the impact of radiant heat transfer COISLIM1D uses the concept of overall effective exchange factor F which is multiplier for incident heat flux calculated by Stefan–Boltzmann law that account geometrical and operational factors that affect radiant heat transfer to reactor coils.

$$q_{radiant} = \sigma \boldsymbol{F}\left(T_{gas}^{4} - T_{coil}^{4}\right) \quad (1)$$

In order to estimate exchange factor a series of CFD simulation were carried out for various conditions (flue gas temperature and composition, tube metal temperature, radiant box geometry) for coated, uncoated and oxyfuel furnaces.

The CFD simulations use the weighted sum of grey gases model (WSGGM) built into the ANSYS Fluent software package. Obtained effective exchange factors have been introduced to simulation as correction factors for radiant heat transfer. Enhanced emissivity for radiant walls is applied to *HEC and 3D+HEC* cases. In the current version of COILSIM1D (v. 3.9.1) a large matrix of CFD simulations has been fitted using an artificial neural network ('ANN') so that the user can simply input the emissivity of either the refractory lining or the coil and the software will accurately and directly calculate the value on 'F'.

### *Special Reactor Alloy ('ALU' scenario)*

An advanced and proven method for improving the coil material to make it less prone to coke formation is the addition of special elements into the coil alloy [8], [9], [10]. An important trend for the new generation of special coils is the addition of aluminum into the formulation. Aluminum-containing alloys provide excellent coke resistant properties by forming an alumina ($Al_2O_3$) layer that can resist high temperatures [11], [12], [13].

### *3D Reactor Geometry ('3D', '3D+HEC' scenarios)*

The use of 3D geometries for the manufacture of advanced reactor coils has been subject of significant interest in recent years in the steam cracking industry. Typically 3D geometries enhance heat transfer by increasing the effective surface area; perturbing the gas boundary layer; or increasing the local gas velocity by adding a tangential component to the normal, predominantly axial flow. So, compared to a bare tube, the 3-D-enhanced tube will have a lower tube metal temperature for the same net heat flux. This lower 'TMT' reduces the rate of coke formation and increases run-length. Since most coking occurs in the last pass, where the process gas is hottest, and where the severity/conversion is highest, 3D geometry is only applied to the last pass of the coil.

### *Oxyfuel combustion ('OXY')*

One of the more radical but promising ways to reduce carbon emissions of cracking furnace is fuel combustion with pure oxygen instead of air. It is associated with carbon capture and sequestration system (CCS) to remove $CO_2$ from the flue gas. The absence of nitrogen in the flue gas significantly simplifies CCS and reduces its energy consumption. '.... after cooling and condensation of water a stream which consists of a high concentration of carbon dioxide is received. This stream can be cleaned comparatively easy to a high purity stream for sequestration in geological formations....' [14]. Moreover, application of oxyfuel combustion allows for increased efficiency in the radiant section compared to air/fuel combustion and therefore gives additional benefit for the use of this type of technology. Unfortunately, a high oxygen fraction in the fuel-oxidizer mixture inevitably leads to increased adiabatic flame temperature that might damage reactor coils. In order to reduce the adiabatic flame temperature to allow classic furnace design to be operated with oxygen as oxidizer, fresh oxygen is mixed with recycled flue gas captured at stack before being introduced into firebox. Assumed oxidant composition is given in Table 2 just below.

**Table 2: Oxidant – recycle mixture composition**

| Component | Volume fraction, % |
|---|---|
| $O_2$ | 23 |
| $CO_2$ | 44.2 |
| $H_2O$ | 32.8 |

Thanks to recycle, a significant fraction of heat that is normally lost at stack with flue gas is recovered, resulting in an oxidant mixture temperature of 129 °C.

***Biogas fuel ('BIO')***

One of studied pathways to reduce $CO_2$ emissions of the furnace is biogas combustion. While it cannot improve efficiency of the furnace, the fuel itself is biogenic and therefore, all carbon it contains has been previously extracted from atmosphere via photosynthesis: this means that net $CO_2$ emission of any biofuel combustion is zero. Two approaches for biogas combustion are possible: purification biogas to biomethane with ~ 96 vol% fraction of $CH_4$ and direct combustion of biogas that contains a significant (up to 50 vol%) amount of $CO_2$. The latter has been assumed for simulations. Simulation presented in this work uses the following fuel composition on a wet basis (shown in Table 3).

**Table 3: Biogas composition on wet basis**

| Specie | | Vol. fraction, % |
|---|---|---|
| $CH_4$ | | 60 |
| $CO_2$ | | 33.5 |
| $N_2$ | | 1 |
| $O_2$ | | 0.5 |
| $H_2O$ | | 5 |

***Low-Emission Furnace***

The Low-Emission Furnace has significantly lower convection section duty than a conventional furnace, therefore the heat recovery scheme has been adjusted in favor of the so-called 'process duty'. The process duty is the duty required for:

- preheating, evaporation and superheating of the naphtha
- superheating of dilution steam

The total process duty cannot be reduced, but it is possible to partly superheat the naphtha/ dilution steam mixture using cracked gas instead of flue gas. In the new scheme, the BFW is no longer preheated. It is directly fed to the steam drum. The saturated VHP steam is still superheated in the convection section, but its quantity is significantly reduced. Since overall steam output is significantly reduced and an additional driving force, such as electric motors, is required for separation stage operation. In this work two possible embodiments of the Low-Emission Furnace are studied: ordinary air fired furnace with novel heat recovery arrangement with firing rate reduced by 20% ('L-E 20%') and similar furnace but with combustion air preheater that allows to increase radiant box efficiency even more resulting in 30% reduction in overall firing rate ('L-E 30%').

### *Rotodynamic Reactor ('RDR')*

It is not the purpose of this paper to described in detail Rotodynamic Reactor ('RDR'). As mentioned above, RDR has already been presented at this meeting [4], and an update is expected in 2022. Yet more information and background can be found from the following references: [15],[16], [17], [18].

## Global analysis

### *Scope*

The purpose of the work described in this part of the paper is to benchmark the environmental performance of the studied technologies. In order to avoid burden-shift, and to fulfil major guidelines, the system boundary has been expanded to include the separation stage (a.k.a. 'Recovery Section' or 'cold train') and the steam system within the plant utilities. Besides process operating parameters, LCA requires upstream and downstream data obtained from a secondary source: 'ecoinvent 3.5' [19], one of the world's leading Life Cycle Inventory (LCI) commercial database. A schematic of the selected system boundary is shown on Figure 1, below.

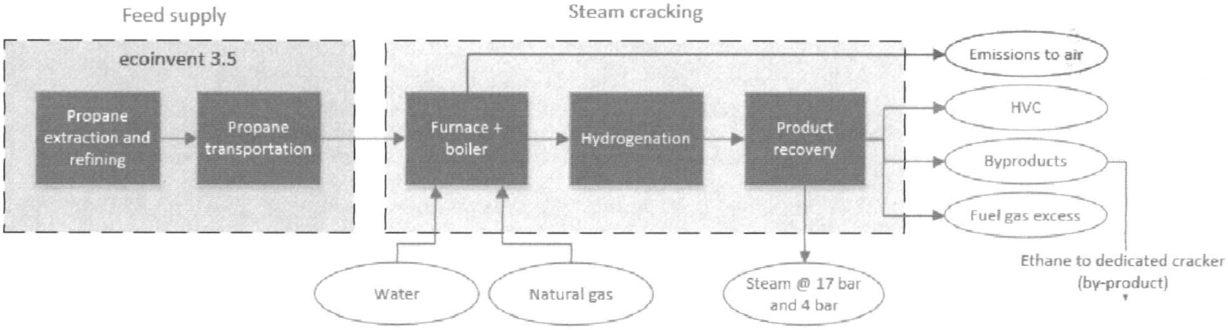


**Figure 1: Overview of system boundaries (dotted line) considered in the study**

This study uses mass allocation of environmental burden among high value chemicals (HVC). The hydrogen fraction of hydrogen-rich off-gas is treated as a co-product. The remaining methane rich fuel gas is assumed to be partially used for on-site energy needs or exported and a fuel credit applied based on its energy content (LHV). The production of 1000 kg of HVCs has been selected as functional unit (FU).

Cracking products treatment operations that are usually considered to be the part of downstream value chain, but are not part of every ethylene plant are omitted, as follows:

- Hydrogen valorization (pressure swing adsorption)
- C4 extraction
- Aromatics extraction
- Hydro-treating of pygas

## Methodology

As mentioned above, a complete LCA study includes the separation stage of the plant in order to avoid burden-shift within the facility and to get the correct environmental burden allocation among HVCs. Therefore, a separation section model is needed to account for cracked gas recovery rates as well as steam demand required by the separation stage to drive main (CG) and cooling cycles (refrigerant, C2- and C3-) compressors and heat rectification column reboilers. A generic demethanizer-first cold train with front end hydrogenation has been selected[1] and modelled with process simulation software. Separation process has been modelled for a whole ethylene plant assuming a total combined cracked gas furnace effluent flow of 150 metric tons per hour. Heat integration has been carried out in order to estimate best possible case. A grand composite curve can be found below in Figure 2.

---

1 We recognize that there are many different possible configurations for the separation train but since the technological changes are in the furnace section and we are dealing with relatively small perturbations to the recovery and utilities areas of the plant the LCA results should not be sensitive to the 'design' chosen.

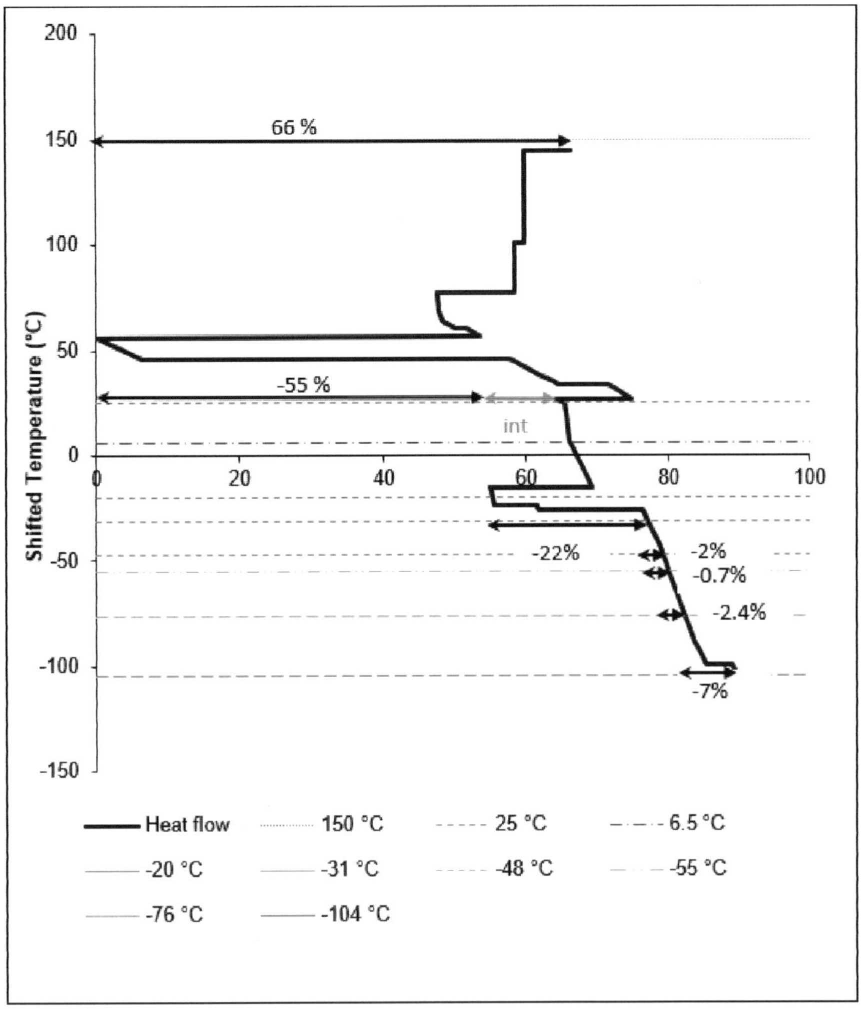


**Figure 2: Grand composite curve for heat integration on ethylene plant cold train**

The assumed configuration of the steam system is given in Table 4 below.

**Table 4: Steam levels used in simulation study**

| Name | Pressure, bar(g) | Temperature, °C | Description |
|---|---|---|---|
| **VHP** | 106.39 | 315 | Very high-pressure steam (saturated) |
| **HPSS** | 106.39 | 510 | High pressure superheated steam |
| **HP** | 40 | 383 | High pressure steam |
| **MP** | 17 | 290 | Medium pressure steam (exported) |
| **LP** | 5 | 185 | Low pressure steam (exported) |
| **DS** | 12 | 200 | Decoke steam |
| **BFW** | 136.78 | 120 | Boiler feedwater |
| **Water** | 1 | 15 | Used to produce decoke steam |

Hydrogenation has been modelled with the typical set of reactions presented in Table 5 below.

**Table 5: Hydrogenation reactions and their selectivity**

| Reaction | Selectivity, % |
|---|---|
| **C2H2 + H2 = C2H4** | 50 |
| **C2H2 + 2H2 = C2H6** | 50 |
| **MA + H2 = C3H6** | 81.465 |
| **MA + 2H2 = C3H8** | 15.265 |
| **PD + H2 = C3H6** | 81.465 |
| **PD + 2H2 = C3H8** | 15.265 |
| **BUTD + H2 = 1-BUTENE** | 7.4 |

Where 'MA' stands for methylacetylene and 'PD' for propadiene, as usual. Aforementioned models describe all relevant mass and energy flows in the ethylene plant considered for LCA and LCC.

### *Life cycle impact assessment (LCIA) method*

The Environmental Footprint (EF) method v 2.0 (as available in the software Simapro 9.0) has been applied in this study. It is a method developed in 2013 with a specific Recommendation (2013/179/EU), within the framework of the 'Single Market for Green Products' communication (COM/2013/0196) [6]. It has been selected because it includes all impact categories suggested by World Business Council for Sustainable Development (WBCSD) in applicable guidelines [20], quantified with most up-to-date models (among available). For more information and details refer the extensive list of references at the end of this paper.

### *Life-cycle inventory*

Ecoinvent 3.5 life cycle inventory database has been used to quantify the environmental impact of material streams present at the system boundary. For steam flows custom inventory consists of two unit processes: heat flow, quantified as the amount of heat required to produce 1 mass unit of the given steam level in a boiler with thermal efficiency of 93%, and deionized water to make up for blowdown losses (or to be imported for dilution steam production that is being purged to atmosphere instead of condensing). For fuel gas export, natural gas equivalent (based on lower heating value) has been used to calculate the avoided burden. Table 6 is an overview of all unit processes for accounted mass and energy flows.

**Table 6: Overview of background data used**

| Flow description | Unit process (ecoinvent 3.5) |
|---|---|
| **Propane feed** | Propane extraction, from liquefied petroleum gas {global} |
| **HPSS production** | Heat, district, or industrial, natural gas {Europe}; Water, deionized, from tap water, at user {Europe} |
| **HP production** | Heat, district, or industrial, natural gas {Europe}; Water, deionized, from tap water, at user {Europe} |

| | |
|---|---|
| **Decoke steam** | Heat, district, or industrial, natural gas {Europe}; Water, deionized, from tap water, at user {Europe} |
| **Superheating duty** | Heat production, natural gas, at industrial furnace >100kW |
| **Electricity (CCGT[2]) for decoke air blower in CUP furnaces** | Electricity production, natural gas, combined cycle power plant {NL} |
| **Steam export (MP and LP levels)** | Heat, district, or industrial, natural gas {Europe}; Water, deionized, from tap water, at user {Europe} |
| **Fuel gas export** | Natural gas, from high pressure network (as energy equivalent). |
| **Biogas** | Unit processes for market and by-product biogas |
| **Oxygen flow for oxyfuel** | Unit process for liquid oxygen and energy equivalent for gaseous flow |
| **Grid electricity for decoke air blower in Low-Emission furnaces** | Electricity, medium voltage market group {Europe without Switzerland} |
| **Grid electricity for Low-Emission cold train compressors and RDR reactor** | Various national grid unit processes for sensitivity study |

### *Data sources*

Due to model limitations, data used for life cycle inventory step has been taken from different simulations involving various software packages. Overview of used models and their level of detail is presented on following list:

- Propane, mass flow: since furnace capacity is fixed, the mass flow of propane imported to studied production system boundaries depends only on mass flow of propane, recovered from cracking effluent which is defined by the reliable and validated COILSIM1D simulation.
- Utility steam production by auxiliary boiler, mass flow: the amount of steam required for separation stage is determined from the process simulation model and the amount of steam produced by the furnace itself estimated by COILSIM1D model.
- Superheating duty, energy flow depends solely on the amount of steam produced by the furnace derived from the COILSIM1D model .
- Decoke steam, mass flow: mass flow and steam parameters are provided by industrial partners .
- Electricity (from on-site generation or grid) for decoke air blower, energy flow is defined by process simulation based on fixed air flow and pressure (industrial data)
- Steam export, mass flow is fixed for all considered scenarios based on the generic ethylene plant process simulation.
- Fuel gas export, mass flow is defined by cracking effluent composition (validated COISLIM1D model) and its impact is accounted as avoided burden.

And some flows are only present in specific cases:

- Oxygen, mass flow is defined by process simulation of flue gas recycle loop and COILSIM1D furnace simulation.
- Biogas, mass flow: defined by COILSIM1D furnace simulation, environmental impact is quantified with two different unit processes for sensitivity study.

### *Greenhouse gas (GHG) protocol*

---

[2] Combined cycle gas turbine power plant

The GHG Protocol Corporate Standard provides standards and guidance for companies and other types of organizations preparing a GHG emissions inventory. It covers the accounting and reporting of the six greenhouse gases covered by the Kyoto Protocol - carbon dioxide ($CO_2$), methane ($CH_4$), nitrous oxide ($N_2O$), hydrofluorocarbons (HFCs), perfluorocarbons (PFCs), and sulfur hexafluoride ($SF_6$). Emissions are divided into 3 categories, or so-called Scopes [21]:

1. Scope 1. Direct GHG emissions. Direct GHG emissions are from sources that are owned or controlled by the company, for example, emissions from combustion in owned or controlled boilers, furnaces, vehicles, etc., emissions from chemical production in owned or controlled process equipment. Direct $CO_2$ emissions from the combustion of biomass is not included in Scope 1 but is reported separately.
2. Scope 2. Electricity indirect GHG emissions. Scope 2 accounts for GHG emissions from the generation of purchased electricity consumed by the company. Purchased electricity is defined as electricity that is purchased or otherwise brought into the organizational boundary of the company. Scope 2 emissions physically occur at the facility where electricity is generated.
3. Scope 3. Other indirect GHG emissions. Scope 3 is an optional reporting category that allows for the treatment of all other indirect emissions. Scope 3 emissions are a consequence of the activities but from sources not owned or controlled by the company. Some examples of Scope 3 activities are extraction and production of purchased materials; transportation of purchased fuels; and use of sold products and services.

## LCA Results and Discussion

### *Overall comparison*

The impact on climate change or radiative forcing as Global Warming Potential (GWP100) measured in kg of $CO_2$ equivalent per 1 kg of HVC is presented on

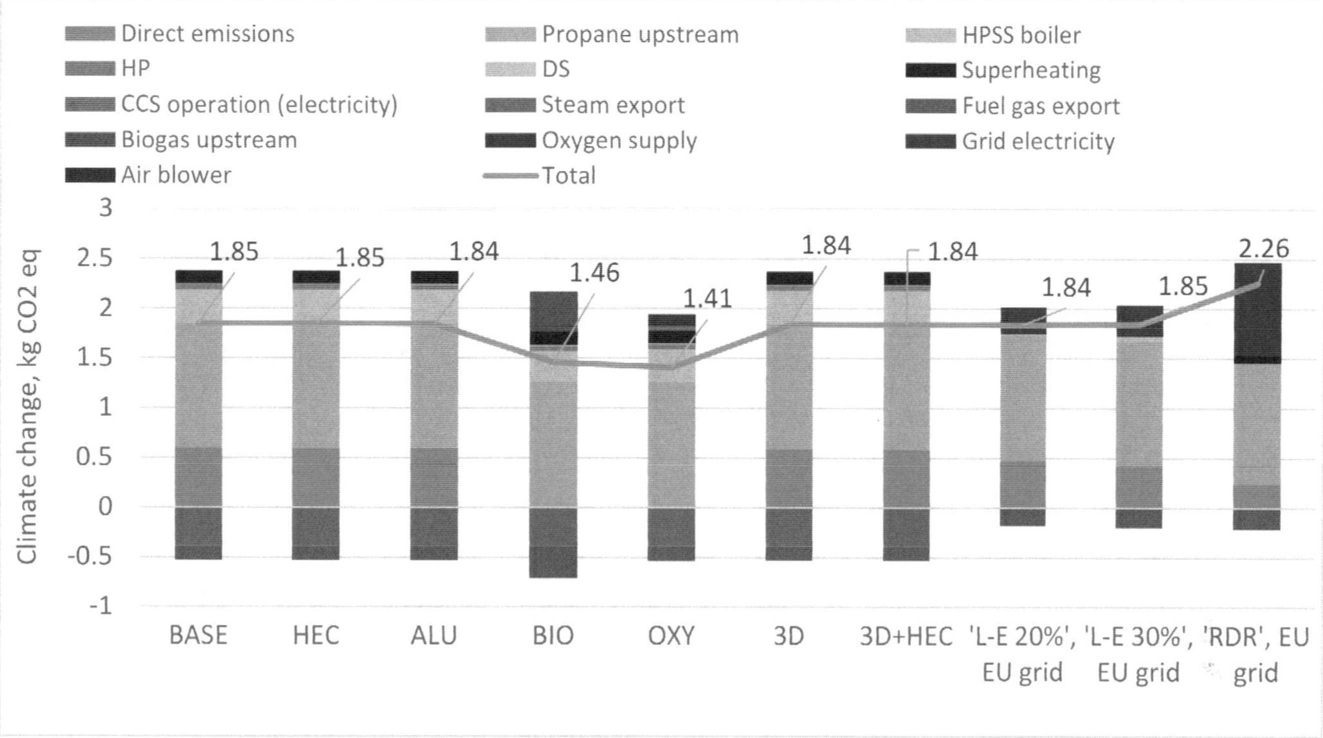


Figure 3. The propane upstream system causes the major fraction of the overall indicator score (~65% for the base case). The high footprint of feedstock supply can be attributed to emissions that occur during production and refining of liquid petroleum gas that is further separated in order to obtain propane. Its contribution remains constant among all considered cases since the only factor that impacts it is the cracking severity that has been fixed within simulation setup as 85% $C_3H_8$ conversion. Only our 'RDR' scenario allows to reduce the feedstock contribution by 3.5% due to significantly higher selectivity towards high value chemicals. Negative contributions represent avoided burden due to excessive fuel gas and medium/low pressure steam exports. Due to the carbon neutrality of biogas combustion and the avoided burden taken for whole fuel gas flow being exported it is the best case from a climate change perspective. Also traditional air fired furnace improvements, in combination with 3D coils, alumina rich coil alloy and high emissivity radiant wall coating (scenario '3D+HEC') shows the largest reduction of carbon footprint (0.5%). Oxyfuel combustion with carbon capture and sequestration shows a total absence of direct $CO_2$ emissions from the stack, and 4.5% reduction of the firing rate thanks to the reduced stack losses and increased radiant box efficiency. However, the carbon footprint of oxygen upstream slightly increases the overall carbon equivalent of HVC production. Another significant benefit of oxyfuel furnace is zero $NO_x$ emission due to the absence of nitrogen in the oxidant. The consequence is zero impact on terrestrial and marine eutrophication and other impact categories influenced by $NO_x$ emissions.

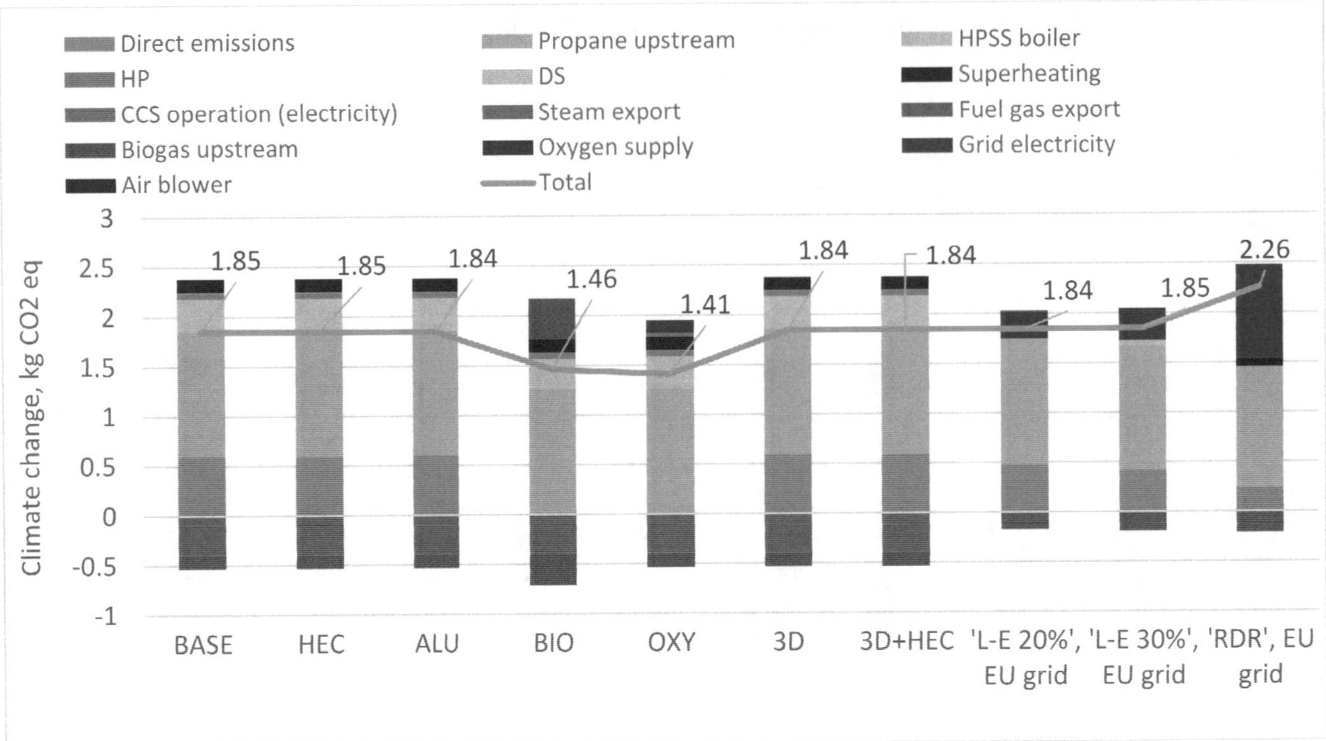


**Figure 3: Global Warming Potential (GWP100) as Radiative forcing impact of 1 kg of HVC production**

*Legend:*

- *Direct emissions – impact, attributed to flue gas leaving the furnace at stack*
- *HP – high pressure steam production in auxiliary boiler*
- *Air blower – electricity consumption by air blower that feeds decoke air into reactor coils during decoking*
- *Biogas upstream –biogas production (supply chain)*
- *CCS operation (electricity) – carbon capture and sequestration unit; only electricity consumption is accounted for*
- *Propane upstream –propane feed upstream*
- *DS – decoke steam production in natural gas fired boiler*
- *Steam export – negative impact (credits) for low and medium pressure export*
- *Oxygen supply – supply of oxygen to be used as oxidizer in oxyfuel furnace*
- *HPSS boiler – HPSS production with natural gas fired auxiliary boiler*
- *Superheating – superheating VHP saturated steam produced in the steam drum in natural gas fired auxiliary boiler to HPSS level*
- *Fuel gas export – avoided burden (of gaseous fuel production) attributed to exported excess fuel gas (quantified by calorific value)*
- *Grid electricity – electricity supply for electric motors that drive compressors of Low-Emission furnace separation stage and RDR reactor*

The Low-Emission Furnace shows a significant reduction of stack emissions (20% and 30% for base embodiment and for arrangement with air preheater respectively) but, unlike conventional arrangements, it is not taking credits for steam export. The overall performance of such a design depends heavily on the nature of the electricity grid used to drive the cold train compressors. Important to notice that due to low steam output scenario 'L-E 30%' requires additional steam import from utility boiler in order to supply required heating duty to separation train.

Another solution with a significant reduction of the carbon footprint is combustion of biogas instead of methane recovered from cracked gas. Due to biogenic nature, carbon present in the gas is a part of natural carbon loop and hence, emissions of such furnace can be considered net $CO_2$ neutral. Moreover, since byproduct fuel gas is not combusted it's exported from system boundary and such production system takes credit for avoided burden. However, introduction of inert $CO_2$ stream along with fuel leads to reduced thermal efficiency of the furnace. Moreover, combustion of raw biogas emits gas with corrosive properties, mostly due to high sulfur oxides content which can damage steel parts of the furnace so biogas pre-treatment is required.

***Electrified furnaces***

As mentioned earlier, the Low-Emission Furnace and Rotodynamic reactor require extra imported electric power compared to the base case and therefore, the environmental footprint (of a modified ethylene plant based on the Low-Emission Furnace or RDR) depends very much on the carbon intensity of the electric grid itself. Results with different sources of electricity are presented in Figure 4. Thanks to its high selectivity towards HVC's and zero stack emissions from reactor itself only $CO_2$ emissions that occur at stack are caused by steam superheater that contributes for up to 6.5% of overall carbon footprint and feedstock/dilution steam mix preheater (up to 17.5% of overall footprint), RDR performs clearly better than any of the IMPROOF technologies in case if carbon neutral electricity is available. .

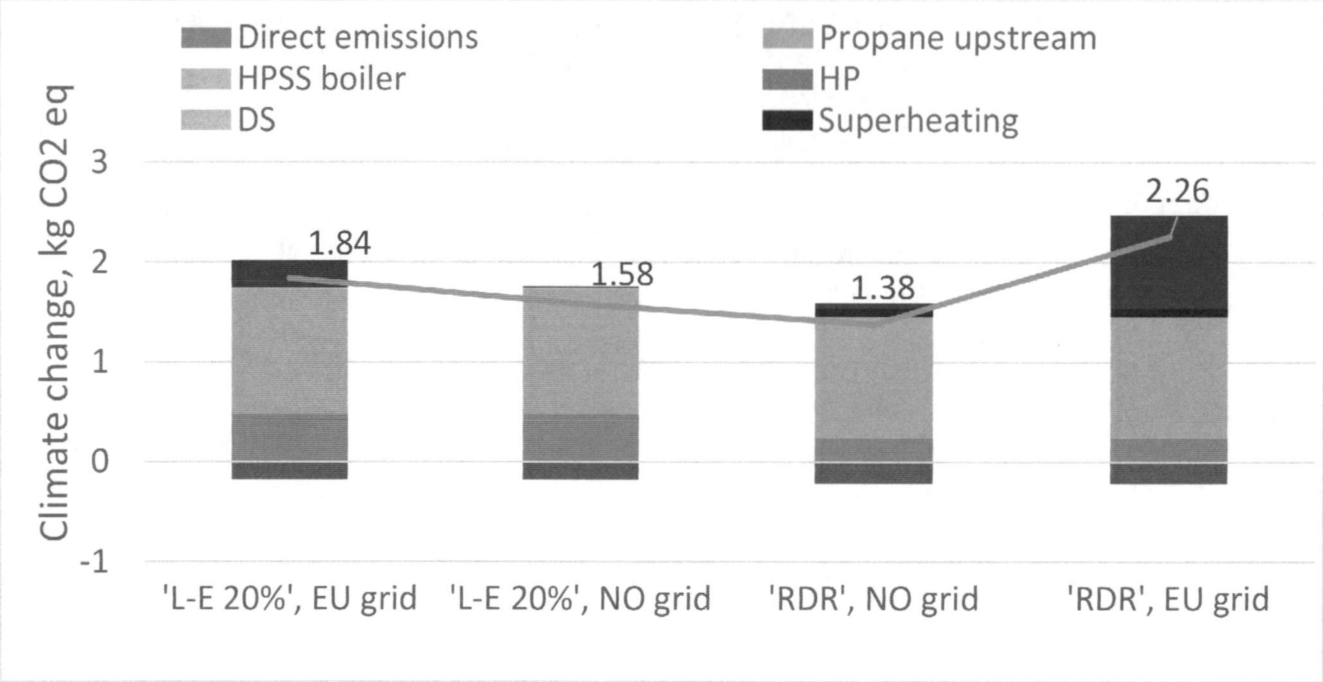


**Figure 4: Carbon footprint of Low-Emission furnace, dependence on national grid carbon intensity (NO= Norway grid [95% green electricity]; EU = EU grid [20% renewable] )**

***Greenhouse gas protocol***

Scope 1+2 emissions are shown in Figure 5 and Scope 3 in Figure 6. As can be seen on Figure 5, the use of the Low-Emission Furnace significantly reduces Scope 1 emissions. Moreover, since production systems do not take credits or avoided burden under the GHG reporting protocol, the LE furnace has an even bigger advantage than what was shown in the LCA procedure. As can be seen in Figure 6, by far the largest Scope 3 emissions come from propane upstream. Standard GHG reporting procedure include emissions caused by company-specific factors such as logistics, employee commute and so on. This report shows estimates based on mathematical models and so excludes company-specific factors.

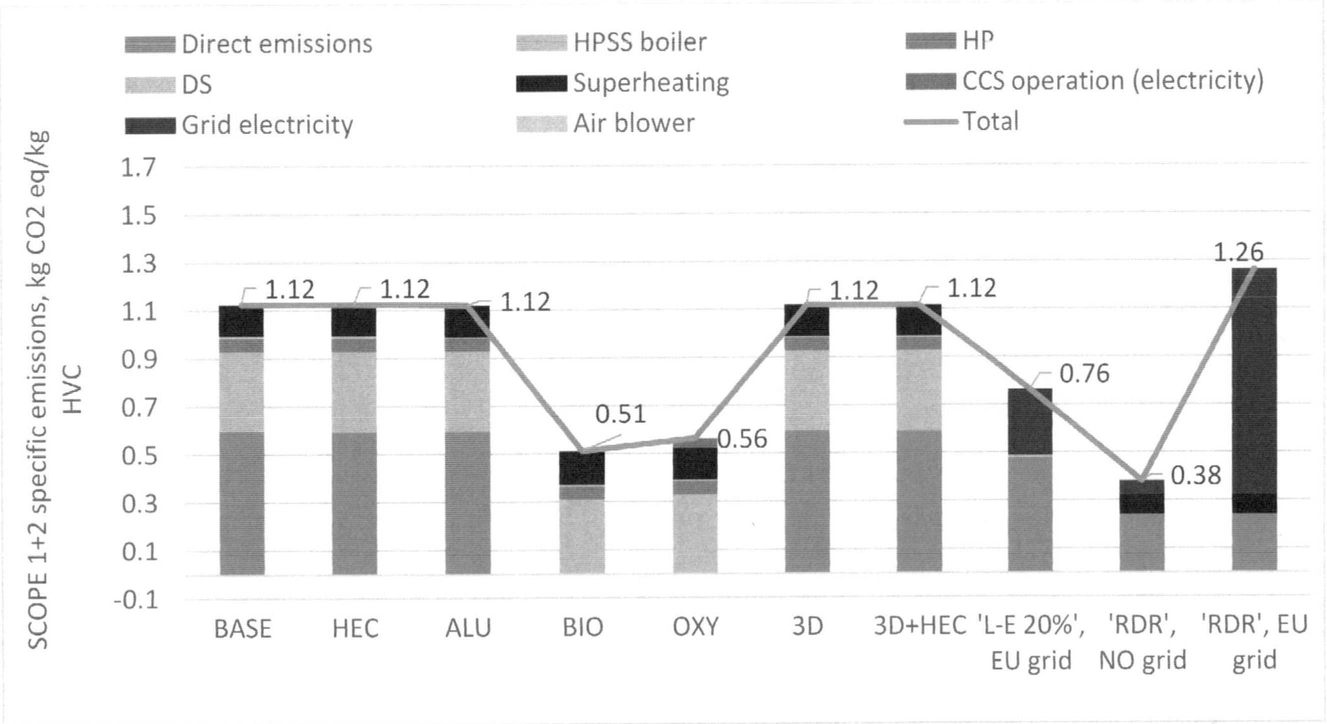


**Figure 5: Scope 1+2 greenhouse gas emissions of various cases; Scope 2 emissions are limited to grid electricity import for Low-Emission Furnace**

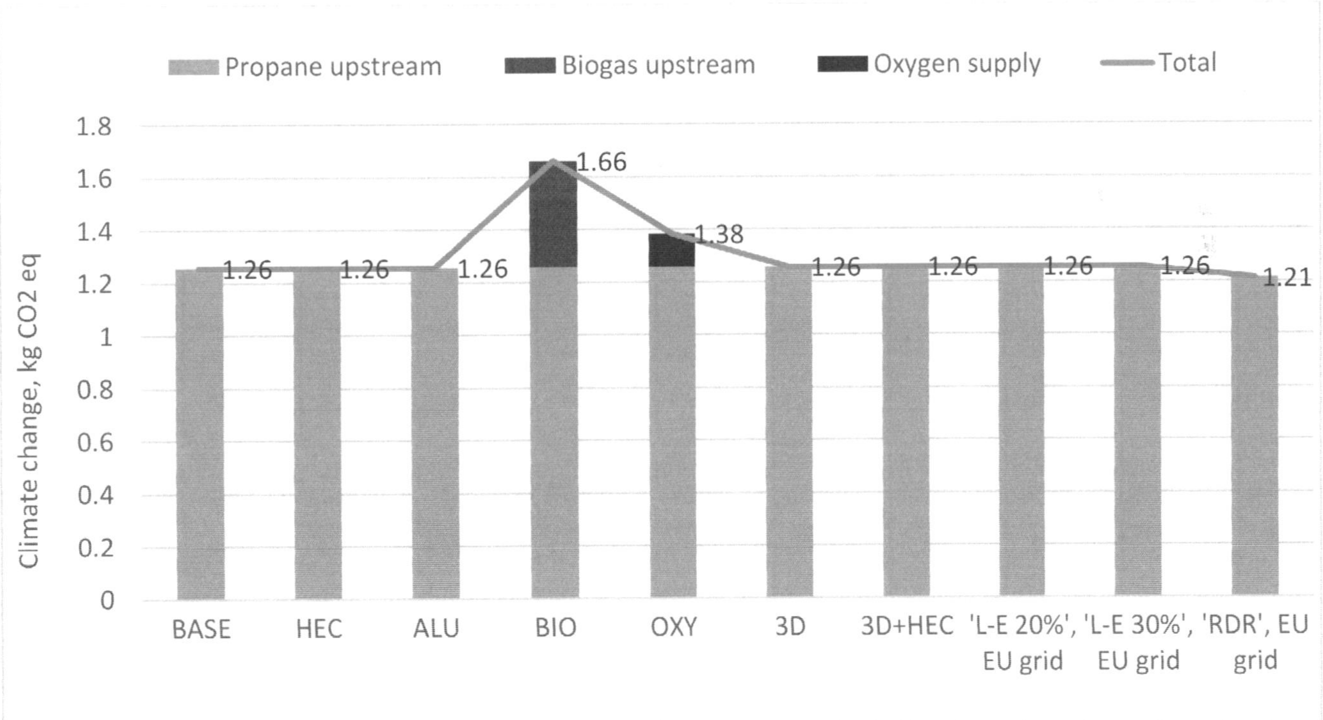


**Figure 6: Scope 3 greenhouse gas emissions**

## Conclusion

The environmental and economic benefit expected for a propane cracking furnace has been estimated by means of life cycle assessment and greenhouse gas reporting protocols.

Among the more conventional steam cracking solutions, the combination of alumina rich coil alloy, 3D reactor geometry and high emissivity radiant wall coating is a clear optimum leading to 0.5% reduction in carbon footprint of the studied process.

In case that electrified steam cracking is considered, the Low-Emission Furnace has the biggest potential for reducing greenhouse gas emissions. However, in order to realize the full benefits, the plant must be tied into an electrical grid with a high proportion of generation from renewable, green sources such as wind and solar. With a full renewable grid such solution can offer up to 16.5% reduction is greenhouse gas emission per mass unit of HVC.

The use of biogas and oxyfuel also have the potential to reduce greenhouse gases but the application of the full range of lifecycle tools has helped us identify negative effects that might outweigh the benefits.

RDR has an even greater upside than the Low-Emission Furnace if green electricity would be used and available, of course, it is at a much earlier stage of development and represents a much bigger step change in technology.

## Acknowledgement

Funding from the European Union Horizon H2020 Programme (H2020-SPIRE-04-2016) under grant agreement n°723706 is acknowledged

**References:**


1. Van Geem, K.M.; Battin-Leclerc, F; et al. IMPROOF: INTEGRATED MODEL GUIDED PROCESS OPTIMIZATION OF STEAM CRACKING FURNACES, Paper 124b, AIChE Spring National Meeting, 2017.
2. Symoens, S.H.; Djokic, M.R.; Zhang, J; et al IMPROOF: INTEGRATED MODEL GUIDED PROCESS OPTIMIZATION OF STEAM CRACKING FURNACES, Paper 129f, AIChE Spring National Meeting, 2018.
3. Vangaever, S; Zhang, Y; Djokic, M.R.; et al Reducing greenhouse gas emissions of steam cracking furnaces by oxy-fuel combustion, renewable fuels and high emissivity coatings, Paper 124d, AIChE Spring National Meeting, 2020.
4. Johannesdahl, H; RDR – revolution in ethylene yield, Paper 129a, AIChE Spring National Meeting, 2018.
5. Oud, P. Cracking Furnace system and method for cracking hydrocarbon feedstock therein, US Patent Application US 2020/0172814 A1, 2020.
6. Sinke, D.J.; Barandregt, S. Cracking Furnace Revamp Experience As Part of a Major Expansion At Dow Benelux B.V., Terneuzen. Paper 91c, AIChE Spring National Meeting, 2002.
7. Sinke, D.J.; Pitcher, M.; Martin, M. Low-$NO_x$ burner retrofit experiences in cracking furnaces at Dow, Terneuzen. Paper 52b, AIChE Spring National Meeting, 2013.
8. Fedorova, E.; Monceau, D.; Oquab, D. Quantification of Growth Kinetics and Adherence of Oxide Scales Formed on Ni-Based Superalloys at High Temperature. Corros. Sci. 2010, 52 (12), 3932–3942. https://doi.org/10.1016/j.corsci.2010.08.013.
9. A., H.; Lindvaag, O. A.; Trimm, D. L. Surface Effects on the Steam Cracking of Propane. In Coke Formation on Metal Surfaces; ACS Symposium Series, 1983; pp 45–58. https://doi.org/10.1021/bk-1983-0202.ch003.
10. Holmen, A.; Lindvaag, O. A.; Trimm, D. L. Coke Formation during Steam Cracking of Hydrocarbons. Part 2. Effect of Preoxidation and Prereduction of the Reactor Surface. J. Chem. Technol. Biotechnol. Chem. Technol. 1985, 35 (7), 358–364.
11. Munoz Gandarillas, A. E.; Van Geem, K. M.; Reyniers, M.-F.; Marin, G. B. Influence of the Reactor Material Composition on Coke Formation during Ethane Steam Cracking. Ind. Eng. Chem. Res. 2014, 53 (15), 6358–6371. https://doi.org/10.1021/ie500391b.
12. Prescott, R.; Graham, M. J. The Formation of Aluminum Oxide Scales on High-Temperature Alloys. Oxid. Met. 38 (3–4), 233–254. https://doi.org/10.1007/bf00666913.
13. Munoz Gandarillas, A. E.; Van Geem, K. M.; Reyniers, M. F.; Marin, G. B. Coking Resistance of Specialized Coil Materials during Steam Cracking of Sulfur-Free Naphtha. Ind. Eng. Chem. Res. 2014, 53 (35), 13644–13655. https://doi.org/10.1021/ie502277e.
14. Weikl, M.C.; Schmidt, G. Carbon capture in cracking furnaces, Paper 12c, AIChE Spring National Meeting, 2010.
15. Piloting of a 'revolutionary' approach to olefin cracking, Chemical Engineering, May 2017
16. Bushuev, V.A.; Process for producing low-molecular olefins by pyrolysis of hydrocarbons, US Patent 7,232,937, 2007
17. Seppala, J; Hiltunen, J; Purola, V-M. Process and rotary machine type reactor, US Patent 9,234,140, 2016
18. Bushuev, V.A., Bladed reactor for the pyrolysis of hydrocarbons, US Patent 9,494,038, 2016.

19. Wernet, G., Bauer, C., Steubing, B., Reinhard, J., Moreno-Ruiz, E., and Weidema, B. Ecoinvent Version 3. Int. J. Life Cycle Assess. 2016, 21 (9), 1218–1230.
20. WBCSD. Life Cycle Metrics for Chemical Products; 2014.
21. WBCSD; WRI. A Corporate Accounting and Reporting Standard. Greenh. Gas Protoc. 2004.

**AIChE Paper Number: 617345**

# LOW-EMISSION CRACKING FURNACE

*Presenter:* ***Tijn van Boekel***
*Senior Process Engineer*
*Technip Energies, Zoetermeer, The Netherlands*

*Author:* ***Peter Oud***
*Product Development Manager Ethylene*
*Technip Energies, Zoetermeer, The Netherlands*

Prepared for Presentation at the 2021 Spring National Meeting
Dallas, Texas, April 18, 2021

# LOW-EMISSION CRACKING FURNACE

*Presenter:* ***Tijn van Boekel***
*Senior Process Engineer*
*Technip Energies, Zoetermeer, The Netherlands*

*Author:* ***Peter Oud***
*Product Development Manager Ethylene*
*Technip Energies, Zoetermeer, The Netherlands*

Abstract:

As the global temperature is rising due to the increase of greenhouse gases such as carbon dioxide in the atmosphere, governments are struggling to find cost-effective ways to meet the 2020 emission targets. In light of these developments, the European Commission is investigating the possibility to reduce $CO_2$ emissions from steam cracking furnaces by more than 25 percent. As compared to the power industry, the contribution of these cracking furnaces is limited, but in the framework of the chemical and petrochemical industry, these are major emission sources. Steam cracking furnaces are an integral part of ethylene plants in which the bulk chemicals for the plastic industry are produced. Due to the large capacity of the bulk chemicals produced along with the high energy demand of the cracking process itself, ethylene plants are significant contributors to $CO_2$ emissions. The new cracking furnace design proposed by Technip Energies can meet the target of the European committee. By modifying the heat recovery scheme, fuel consumption can be reduced by 30 percent and so can the associated $CO_2$ emissions.

The furnace modification can be done cost-effectively with positive profit margins and requires only short-term investments.

A revolutionary new design of ethylene furnaces is presented in this paper, which relies on a patent filed by Technip Energies. This concept leads to major changes in the ethylene plant energy balance as well as the flowsheet. The presentation analyzes the impact of lower generation of high pressure steam on electric power and sheds light on the $CO_2$ emissions balance for the ethylene plant as well as the power plant. Overall economics look promising.

## Introduction

IMPROOF is a European project aiming at improving the energy efficiency of steam cracking furnaces, while reducing emissions of greenhouse gases and NOx.

Within the framework of the IMPROOF project Technip Energies has been requested to design the future low-emission cracking furnace. This furnace had to meet the following objectives:

- Improve the energy efficiency of steam cracking furnaces by at least 20 percent, and in a cost-effective way, while simultaneously reducing emissions of greenhouse gases and NOx per ton ethylene produced by at least 25 percent.

This analysis had to include an evaluation of the impact in terms of CAPEX and OPEX that these objectives have on two to three high efficiency furnace design scenarios relative to a base case conventional furnace design.

To be able to meet this specific KPI, Technip Energies did extensive research, and filed a patent for the furnace design in June 2017, which covered a number of furnace configurations. This patent was awarded last year.

## Conventional furnace definition

As a template for the Base Case conventional furnace design, Technip Energies' largest liquid furnace was selected. This state-of-the-art, world-scale furnace based on Technip Energies' two-lane GK6 technology was designed for processing light hydrocarbon feedstock.

### *Process description of conventional cracking furnace design*

Figure 1 below shows the furnace lay-out of the Base Case furnace. The cracking furnace will be described on the basis of this layout.

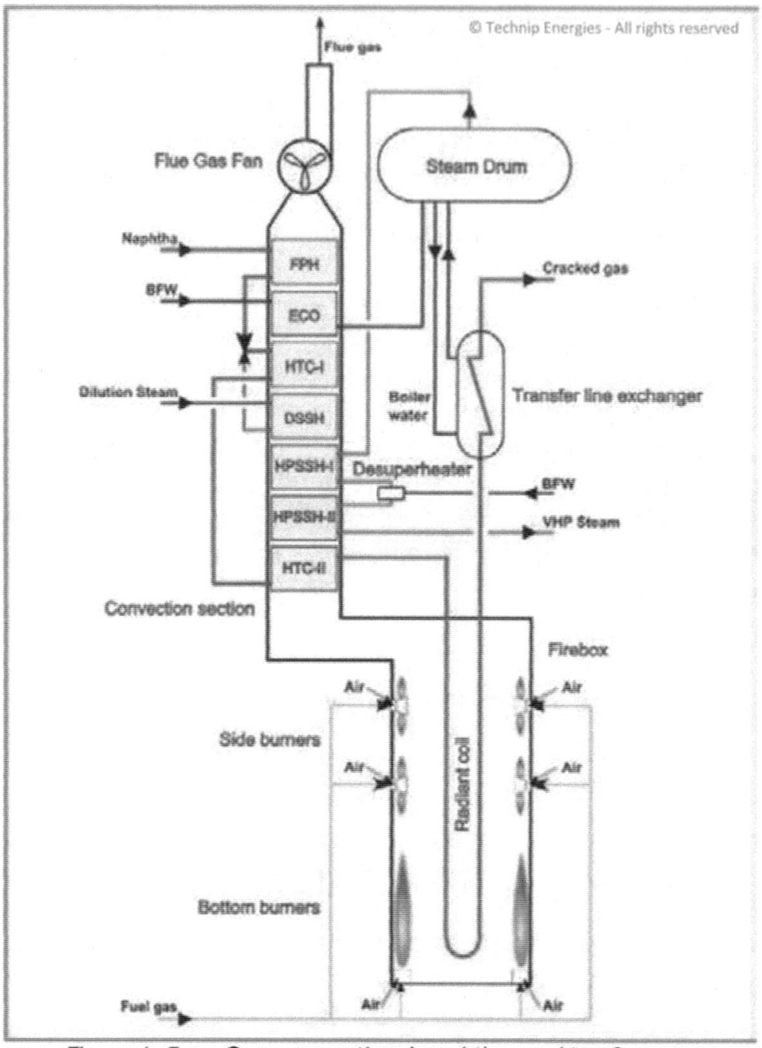

Figure 1: Base Case conventional naphtha cracking furnace

In general, hydrocarbon feedstock enters the feed preheater (FPH). This is one of the many convection banks in the convection section of a cracking furnace. This feedstock can be any kind of hydrocarbon, preferably paraffinic or naphthenic in nature, but small quantities of aromatics and olefins can also be present. Examples of such feedstock are: ethane, propane, butane, natural gasoline, naphtha, kerosene, natural gas condensate, gas oil, vacuum gas oil, hydrotreated or desulphurized or hydrodesulphurized (vacuum) gas oils or combinations thereof.

Depending on the state of the feedstock, the feed is preheated and/or partly or fully evaporated in the feed preheater before being mixed with dilution steam. Dilution steam can be injected directly or it is superheated first before being mixed with the feedstock. There can also be multiple steam injection points, for the heavier feedstock, but this goes beyond the scope of this paper. In this case, the feedstock is naphtha, and further descriptions will be based on this particular feedstock.

The feedstock is partly vaporized in the FPH, while the dilution steam is further superheated in the dilution steam superheater (DSSH).

The mixed feedstock/dilution steam mixture is heated in the high temperature coils (HTC-I and HTC-II) to reach an optimum temperature for introduction in the radiant coil. The radiant coil is positioned in the firebox. In the radiant coil, the hydrocarbon feedstock is quickly heated up to the point where the pyrolysis reaction starts and the hydrocarbon feedstock is converted into products and by-products. Such products include hydrogen, ethylene, propylene, butadiene, benzene, toluene, styrene and xylenes, amongst others. By-products primarily include methane and fuel oil. The resulting mixture of dilution steam, unconverted and converted feedstock, is reactor effluent called "cracked gas". This cracked gas is cooled quickly in the transfer line exchanger, also called TLE, to freeze the equilibrium of the reactions in favor of the products. The waste heat in the cracked gas is recovered in the transfer line exchanger and is used to generate saturated very high pressure (VHP) steam. This steam is generated from boiler water coming from the steam drum. In the TLE, the boiler water is partly vaporized. This partly vaporized boiler water is flowing back to the steam drum by natural circulation. In the steam drum, the generated saturated steam is separated from the boiler water and sent to the convection section to be superheated.

The heat of reaction for the highly endothermic pyrolysis reaction is supplied by the combustion of fuel (gas) with combustion air in the furnace firebox, also called radiant section. This is accomplished by means of bottom burners and/or side wall burners. Other firing options are possible, but this is beyond the scope of this paper. It is essential for the furnace run length to distribute the heat as evenly as possible in the firebox. Generally speaking, bottom burners have relatively long combustion zones (flames), as the number of burners is restricted. Side wall burners are usually higher in number, but their combustion zones are smaller. Furnace efficiency is determined by the heat release pattern of the bottom burners and the physical location of the side wall burners. In the example drawing of the furnace shown in Figure 1 above, the more the bottom burners release heat in the lower regions of the furnace and the lower the elevation of the side wall burners, the better the efficiency. In this case, there are both bottom and side burners.

In the combustion zones, fuel and air are converted to combustion products such as water and $CO_2$, the so-called "flue gas". The waste heat from the flue gas is recovered in the convection section using various types of convection banks. Part of the heat is used for the process side, the preheat and/or evaporation of the hydrocarbon feed and the dilution steam, and the rest is used for the non-process side, such as the generation of VHP steam.

In the drawing of the Base Case cracking furnace, these are the sections from bottom to top:

- The second high temperature coil (HTC-II), used to heat the hydrocarbon/dilution steam mixture
- The second high pressure steam superheater (HPSSH-II), used to superheat very high pressure steam
- The first high pressure steam superheater (HPSSH-I), used to superheat saturated very high pressure steam
- The dilution steam superheater (DSSH), used to superheat dilution steam

- The first high temperature coil (HTC-I), used to heat the hydrocarbon/dilution steam mixture
- The economizer (ECO), used to preheat boiler feed water
- The feed preheater (FPH), used to preheat and partly vaporize feedstock

The flue gas is evacuated to the atmosphere through a stack using a flue gas fan.

To recover excess waste heat from the flue gas, as described above by generating VHP steam, boiler feed water is heated up in an economizer in the convection section before being delivered to the steam drum. In the steam drum, it is mixed with the boiler water present. As described, the boiler water is used in the TLE to generate steam. Saturated steam is sent to the convection section, where it is superheated by the high pressure steam superheaters. To control the VHP steam temperature, boiler feed water is injected in an intermediate desuperheater.

## Low-emission cracking furnace configurations

### *Case scenarios*

Three furnace configurations have been investigated:

- Air preheat
- Full oxyfuel combustion
- Partial oxyfuel combustion

Each of these cases meets the required project objectives:

- Improve the energy efficiency of steam cracking furnaces by at least 20 percent in a cost-effective way, while simultaneously reducing emissions of greenhouse gases and NOx per ton ethylene produced by at least 25 percent.

One of these cases, the case with air preheat, will be described below.

## Process description of Low-emission cracking furnace design with air preheat

Figure 2 below shows the furnace lay-out of the Low-emission cracking furnace with air preheat.

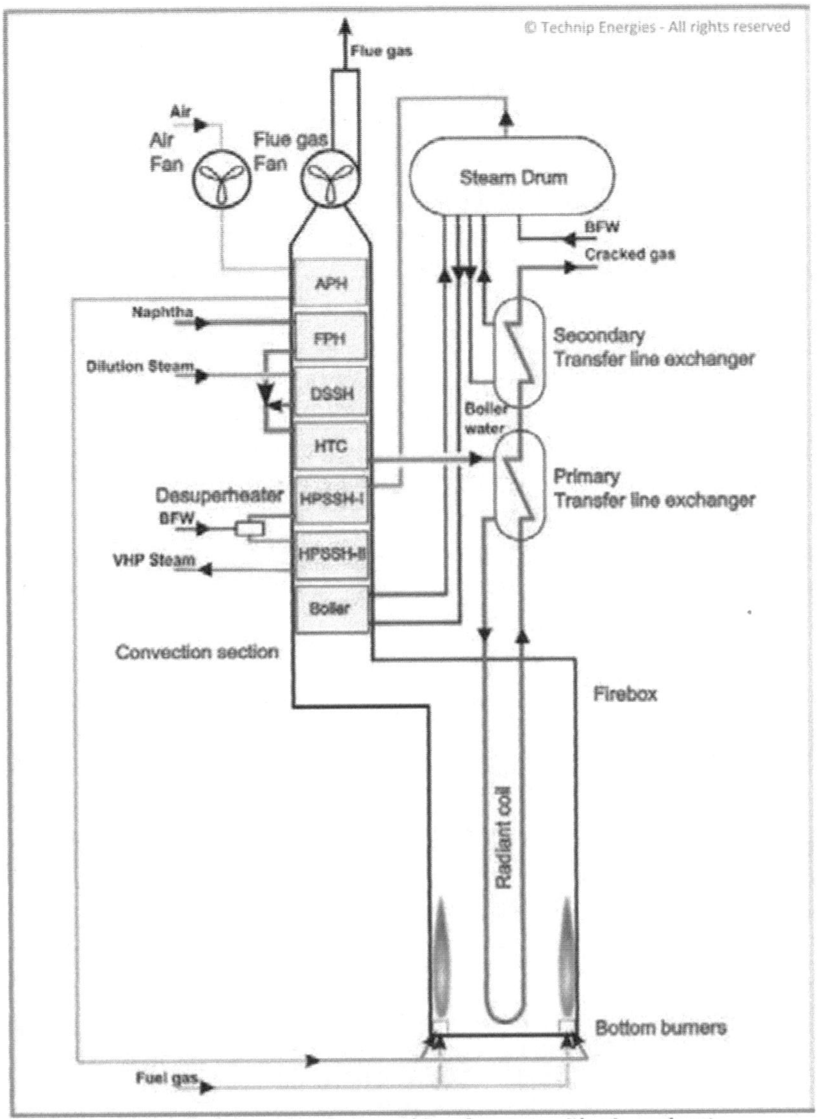


Figure 2: Low-emission cracking furnace with air preheat

Just as in the Base Case, naphtha is partly vaporized in the FPH, while the dilution steam is further superheated in the dilution steam superheater (DSSH).

The mixed feedstock/dilution steam is further heated in a single high temperature coil (HTC), but contrary to the Base Case, it is not superheated in the HTC-II using flue gas, but instead, in the primary TLE using cracked gas. The primary TLE is designed such that the hydrocarbon/ steam mixture reaches the optimum temperature for introduction in the radiant coil.

In the radiant coil, the hydrocarbon feedstock is quickly heated up to the point where the pyrolysis reaction starts and the hydrocarbon feedstock is converted into products and byproducts. This is the same as in the Base Case.

What differs from the Base Case is the two-stage cooling of the cracked gas:

- Stage 1: Primary TLE
- Stage 2: Secondary TLE

The primary TLE quickly cools the cracked gas to freeze the equilibrium of the reactions in favor of the products. This new design is a typical feed-effluent exchanger. The waste heat from the radiant coil outlet stream, the cracked gas, is used to preheat the hydrocarbon/steam mixture coming from the HTC bank in the convection section to the radiant coil inlet conditions.

The secondary TLE further cools down the cracked gas to the required battery limit temperature by generating saturated VHP steam.

The heat of reaction for the highly endothermic pyrolysis reaction in the radiant coils is supplied by the combustion of fuel (gas) with combustion air in the furnace firebox, also called the radiant section. To increase firebox efficiency, firing should be done in the lower regions of the firebox. This is achieved by selecting a design with bottom burners only.

Further increase of the firebox efficiency is achieved by raising the temperature in the combustion zone. This is done by preheating the combustion air using a combustion air preheater (APH). This APH is situated in the top of the convection section. Combustion air is supplied by means of an air fan, forcing the air flow through the air preheater, where it is heated up using the waste heat recovered from the flue gas. The preheated air is sent through ducting to the burners.

In the combustion zones, fuel and air are converted to flue gas. The waste heat from the flue gas is recovered in the convection section using various types of convection banks. Part of the heat is used for the process side, the preheat and/or evaporation of the hydrocarbon feed and the dilution steam, and the rest is used for the non-process side, such as the generation of VHP steam.

In the drawing of the Low-emission cracking furnace with air preheat, these are the sections from bottom to top:

- The boiler coil, used to generate very high pressure steam
- The second high pressure steam superheater (HPSSH-II), used to superheat very high pressure steam
- The first high pressure steam superheater (HPSSH-I), used to superheat saturated very high pressure steam
- The high temperature coil (HTC), used to heat the hydrocarbon/dilution steam mixture
- The dilution steam superheater (DSSH), used to superheat dilution steam
- The feed preheater (FPH), used to preheat and partly evaporate feedstock
- The air preheater (APH), used to preheat combustion air

The flue gas is evacuated to the atmosphere through a stack using a flue gas fan.

In the new design BFW is no longer preheated in the convection section, but is delivered to the steam drum directly. In the steam drum, it is mixed with the boiler water present. This boiler water, in this new scheme, is not only sent to the secondary TLE, but also to the boiler coil to generate steam.

As in the Base Case, saturated steam is sent to the convection section, where it is superheated by the high pressure steam superheaters. To control the VHP steam temperature, boiler feed water is injected in an intermediate desuperheater.

***DeNOx unit requirement***

Unless the low NOx burner technology is improved considerably, it is expected that the NOx emissions will increase with air preheat. By increasing the combustion air temperature, the adiabatic flame temperature which is the highest temperature in the flame, is increased. Since the thermal NOx production is highly sensitive to flame temperature, the NOx emissions will increase, given the same burner technology.

To meet the targets of the EU with respect to NOx emissions, the high efficiency furnace with air preheat requires a DeNOx unit. This has been considered in Figure 3 below, which shows the same scheme as discussed above in Figure 2, but with a DeNOx unit included.

The DeNOx unit is situated in the convection section in between the DSSH and the FPH banks. An ammonia injection grid is provided upstream in the flue gas and is located in between the HTC and the DSSH banks. The ammonia injection grid is used to equally disperse ammonia in the flue gas. The ammonia and the NOx react selectively with each other on the surface of the DeNOx catalyst to be converted to nitrogen and water.

The relevant reaction equations are the following:

- $4\ NO + 4\ NH3 + O2 \rightarrow 4\ N2 + 6\ H2O$
- $NO + NO2 + 2\ NH3 \rightarrow 2N2 + 3\ H2O$
- $6\ NO2 + 8\ NH3 \rightarrow 7\ N2 + 12\ H2O$

The DeNOx unit consists of a bed of multiple catalyst modules arranged in parallel to each other; there can be multiple of these beds in a series.

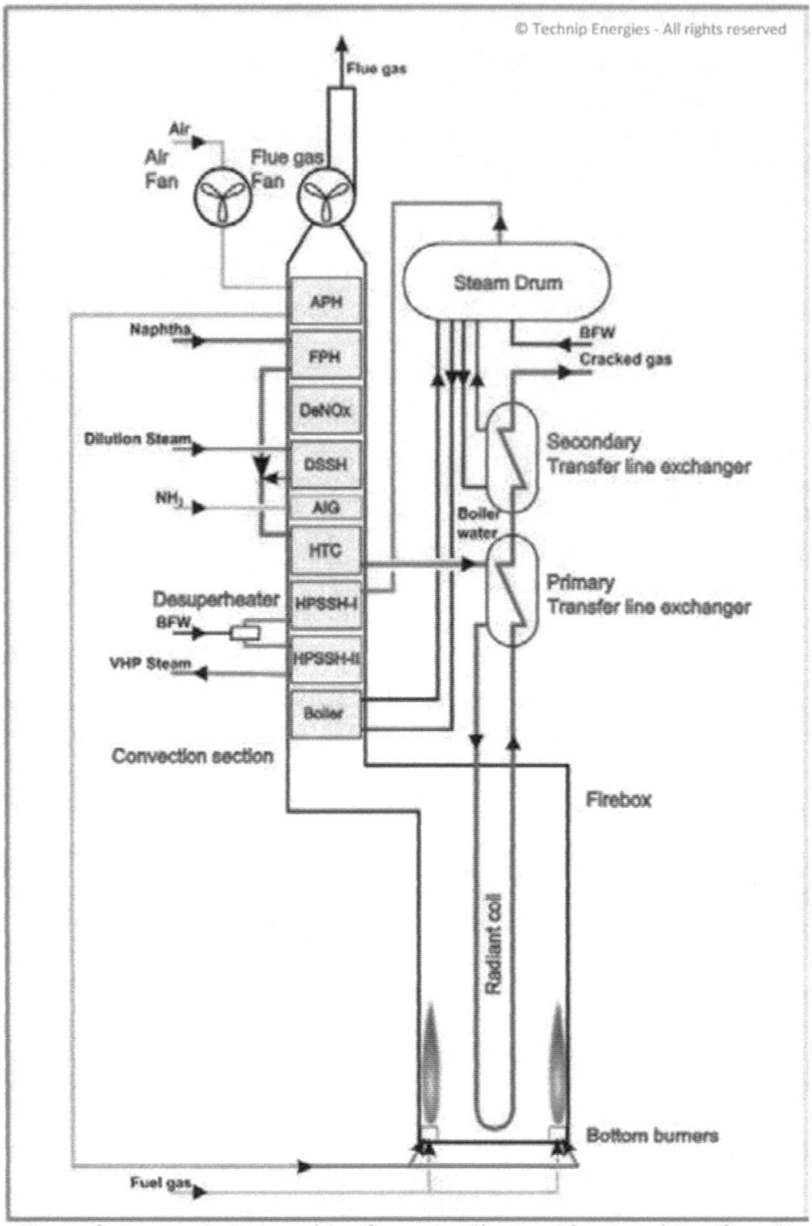


Figure 3: Low-emission cracking furnace with air preheat and DeNOx unit

***New cracking heat recovery scheme***

It was found during the investigations, that the heat recovery scheme of the conventional furnace configuration, as described above, was limiting the extent to which the fuel efficiency can be improved. A new heat recovery scheme has been developed to overcome this limitation. Technip Energies filed a patent for this new configuration to protect its intellectual property. This patent was subsequently awarded.

Basically, fuel efficiency can be increased by raising the firebox efficiency. However, this reduces the convection section duty. In the Base Case, the convection section duty is used for the following purposes:

- preheating, vaporizing and superheating of the naphtha
- superheating of dilution steam
- preheating of BFW
- superheating of saturated VHP steam that is generated in the TLE

In the modified scheme, the convection section duty is significantly lower and therefore, the heat recovery scheme has been adjusted in favor of the so-called "process duty". The process duty is the duty required for the:

- preheating, vaporizing and superheating of the naphtha
- superheating of dilution steam

The total process duty cannot be reduced, but it is possible to partly superheat the naphtha/ dilution steam mixture using cracked gas instead of flue gas.

What can be reduced is the so-called "non-process duty". This is the duty related to the production of VHP steam:

- preheating of BFW
- superheating of saturated VHP steam generated in the TLE

In the new scheme, the BFW is not preheated; it is directly fed to the steam drum. The saturated VHP steam is still superheated in the convection section, but its quantity is significantly reduced due to the resulting heat balance.

To reduce the quantity of VHP steam generated, the duty recovered from the cracked gas for VHP steam generation in the TLE has to be reduced. This is achieved by splitting the TLE is two parts:

- Primary TLE
- Secondary TLE

In the new configuration, the primary TLE is used to superheat the naphtha/dilution steam mixture, while the secondary TLE is used to generate VHP steam.

In addition to generating saturated VHP steam in the secondary TLE, the new scheme also generates VHP steam in a so-called "boiler coil". This coil is situated in the convection section upstream of the HPSSH-II bank and generates saturated VHP steam using flue gas instead of cracked gas. The purpose of this additional convection section bank is to recover excess heat from the flue gas and to protect the HPSSH-II bank against excessively high tube metal temperatures. The duty of this convection bank is considerably lower than that of the primary TLE, thus enabling a net duty shift from the non-process side to the process side. This finally results in a reduction of the generated VHP steam, while still being able to superheat the naphtha/ dilution steam mixture to the optimum radiant coil inlet temperature. This is illustrated in Figure 4 below.

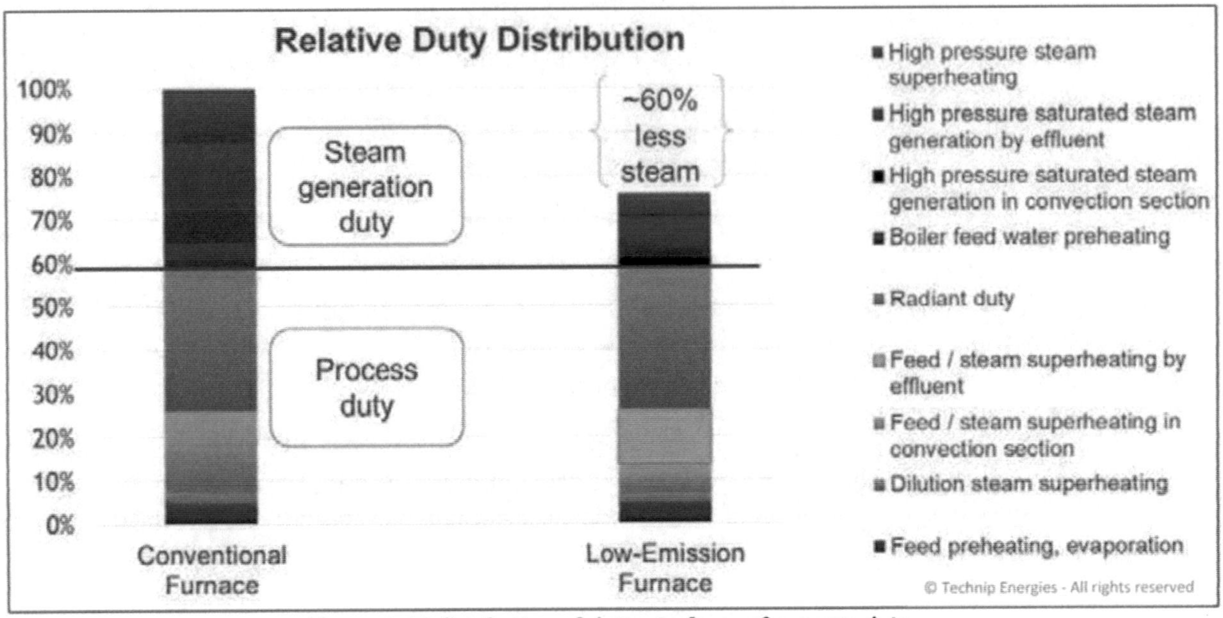


Figure 4: Redistribution of duties in favor of process duty

By maintaining a constant process duty, the VHP steam production decreases by approximately 60 percent. This has an impact on the downstream units and this issue will be addressed in more detail below.

This special furnace heat recovery scheme allows the furnace fuel efficiency to be increased.

***Cost-effectiveness***

Any plant's operating margins are defined by the emerging prices of feedstock, fuel and products. Pay-back times depend on each plant's individual market situation and the selected furnace configuration.

A detailed evaluation of the OPEX and the CAPEX shows that ethylene plant's economics are dominated by the operating costs. The most attractive furnace configuration is the one presented in this paper with an air preheater. Even though this configuration requires the highest investment from furnaces standpoint, it shows a pay-back time of roughly one year, while operating under normal market conditions, due to the more favorable operating costs. This scheme can raise the operating margins, especially under difficult market conditions.

## Road map towards a low-emission ethylene plant

***Impact of the Low-emission cracking furnace on the downstream ethylene plant.***

As already indicated above, the Low-emission cracking furnace has two main issues with respect to the operation of the ethylene plant:

- ~30 percent less fuel gas consumption in cracking furnaces
- ~60 percent less VHP steam production

The reduced fuel consumption poses an opportunity to increase power production from excess fuel gas by raising the heat-to-power efficiency, while the reduced VHP steam production presents an opportunity to introduce electricity from renewable power sources, as the reduced VHP steam production is insufficient to drive all machines.

Ethylene plants operate 24/7, 365 days a year for as long as five years in a row. The availability of a reliable power source is of paramount importance to be able to achieve these kind of on-stream factors. Conventional ethylene plants rely on their own fuel gas production to generate the power they require in the rest of the plant. This is done in the cracking furnaces.

The introduction of the low-emission furnace concept changes that traditional ethylene plant power generation scheme. The Low-emission cracking furnace is able to reduce the $CO_2$ emission in excess of the 25 percent target of the IMPROOF program. It can even reduce the $CO_2$ emission of the cracking furnace beyond 30 percent. However, it is doing this at the expense of the plant's own power production. The fuel reduction is cutting the power for running the plant by as much as 60 percent. Due to the fluctuating nature of renewable power sources, such as solar power and wind power, direct application of renewable power in the ethylene power is not practically feasible. This does not mean that renewable power can't be introduced in the long run, but as a first step in the energy transition, it is necessary to upgrade the heat-to-power conversion level of the excess fuel gas that is now available in the ethylene plant because of the higher fuel efficiency of the low-emission cracking furnaces.

This will be explained in more detail in the next section by using an example. Using the Low-emission cracking furnace production figures as the basis, two flow sheets can be presented:

- Conventional ethylene plant flow sheet
- An integrated low-emission ethylene plant/power plant flow sheet

### *Conventional ethylene plant flow sheet*

The conventional ethylene plant flow sheet is shown in Figure 5 below.

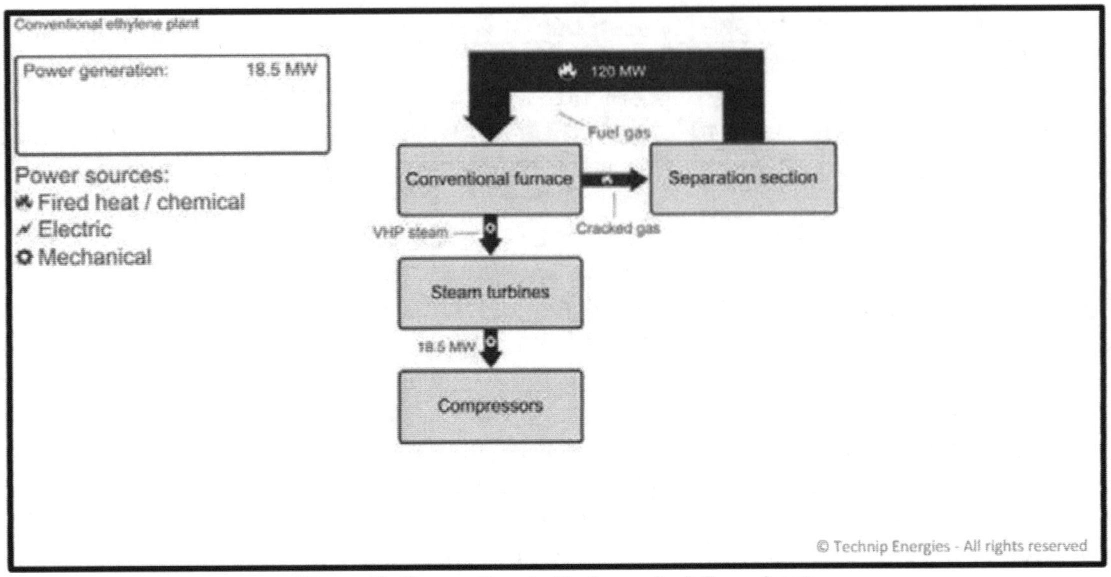

Figure 5: Conventional ethylene plant flow sheet

The conventional furnace, having a production capacity of approximately 140 kilo tonnes of ethylene per year, receives from the separation section the 120 MW of fired heat that is required to drive the furnace by means of combustion of the fuel gas produced in the separation section. In the furnace, this fired heat is used to produce the products, amongst others 17.4 metric tonnes of ethylene, from the hydrocarbon feedstock, which in this case is naphtha. In addition, part of the fired heat is used to generate VHP steam. Approximately 18.5 MW of shaft power is produced by the steam turbines to drive the ethylene plant machines, mainly compressors and a few pumps.

### *Integrated low-emission ethylene plant/power plant flow sheet*

An integrated low-emission ethylene plant/power plant flow sheet is shown in Figure 6 below. A patent has been filed by Technip Energies for this flow sheet.

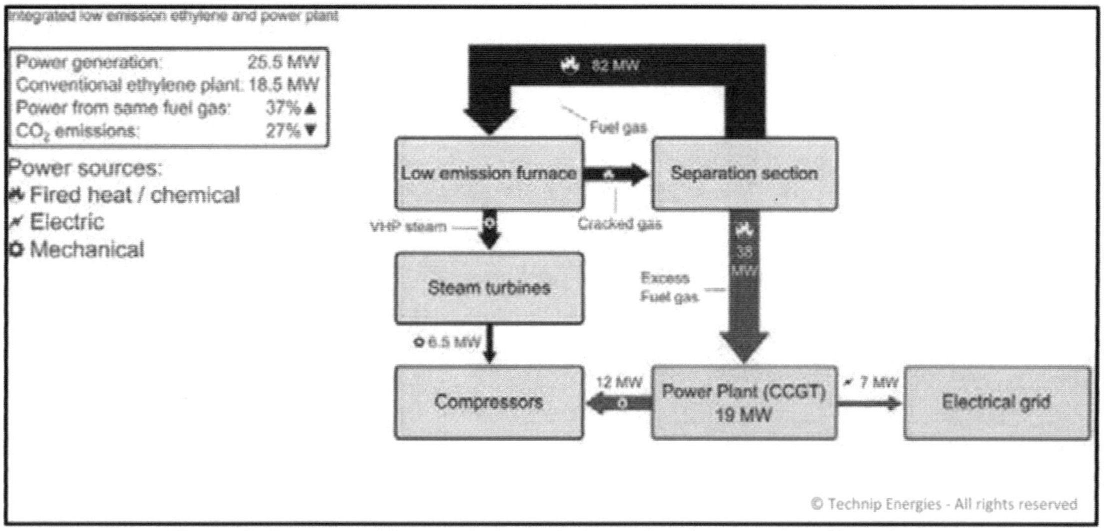


Figure 6: Integrated low-emission ethylene plant/power plant flow sheet

An ethylene plant arrangement having a Low-emission cracking furnace, adapted to include an integrated power plant, is provided and compared to an ethylene plant arrangement with a conventional cracking furnace (i.e. not low-emission).

By using this arrangement, having a Low emission cracking furnace at the heart of the plant, the fired heat is reduced by 32 percent from 120 MW (for a conventional furnace, see Figure 5) to 82 MW and the power generated within the ethylene plant is reduced from 18.5 MW to 6.5 MW. This is shown by comparing the generation and consumption figures of fuel and power of Figure 5 for the conventional ethylene plant circuit with those of Figure 6, for an ethylene plant arrangement having a Low emission cracking furnace. This is achieved by raising the firebox efficiency of the cracking furnace from 40 to 53 percent. This means that the ethylene plant has a shortage of 12 MW power and an excess fuel gas production of 32 percent, as this fuel gas was previously fired on the cracking furnace.

By the integration of the cracking furnace with the power plant, it is possible to convert this excess fuel gas to power in the combined cycle gas turbine (CCGT) power plant at 50 percent heat-to-power efficiency instead of 32 percent, which is the heat-to-power efficiency in the ethylene plant as explained below.

The saved fuel amounts to 38 MW heat (120-82 MW), as compared to a conventional ethylene plant system, where the cracking furnace firebox efficiency is 40 percent. In the latter situation, this converts to 12 MW power. This corresponds to a heat-to power conversion of a conventional ethylene plant system of 12/38 = 32 percent. It is possible to convert this excess fuel gas in the CCGT at 50 percent efficiency, resulting in 0.5*38 MW heat= 19 MW power, of which 12 MW is delivered to the ethylene plant to compensate for the shortage discussed above and 7 MW power is available for export. This 12 MW shaft power can be delivered to the ethylene plant as electric power directly to an electro motor in order to drive a machine, but it can also be provided as high pressure steam, in order to drive a machine via a steam turbine, and supply this energy in the form of shaft power. Alternatively, the gas turbine of the CCGT can be used to supply the shaft power directly, thus avoiding the investment in both a power generator to supply the power to an electro motor, as well as the electro motor itself. The VHP steam generated in the waste heat boiler of the corresponding gas turbine can be used to drive new or existing steam turbines, or can be used to drive a generator in case the rest of the machines are driven by electro motors or a combination thereof.

As the power generated is raised from 18.5 MW to 25.5 MW by the extra 7 MW raised in the CCGT, an extra 37 percent more power generation (duty) for the same amount of $CO_2$ produced results in a reduction of the specific $CO_2$ emission of 27 percent. This is 27 percent less $CO_2$ per MW of power produced. These extra 7 MW of power can be delivered as a continuous stream to the external grid. At the same time the ethylene plant is provided with a highly reliable power source.

This flow sheet example shows that it is possible to make better use of the excess fuel gas that is made available because of the higher fuel efficiency of the low-emission furnace and that upgrading the loop efficiency from 32 to 50 percent amounts to a 27 percent reduction of specific $CO_2$ emissions from the same fuel gas. The specific emission being expressed in terms of $CO_2$ emissions per MWh of fuel fired.

### *Ethylene plant specific emission, excluding the power plant*

The ethylene plant specific emissions from the furnace stack, expressed as emissions per metric ton of ethylene produced, are shown in Figure 7 for the state-of-the-art conventional furnace and for the low emission furnace.

Due to the reduction of the fired heat from 120 MW to 82 MW the $CO_2$ emissions per ton of ethylene produced reduce from 1.31 MT $CO_2$ per MT to 0.91 MT $CO_2$ per MT of ethylene produced. This is a reduction of 31 percent.

### *Ethylene plant specific emissions, including the power plant*

The ethylene plant specific emissions, including the emissions from the power plant firing the excess fuel gas that is made available in the ethylene plant because of the higher fuel efficiency of the low-emission furnace and the emissions from a 7 MW virtual coal firing power plant, are shown in Figure 8 for the state-of-the-art conventional furnace and for the low emission furnace.

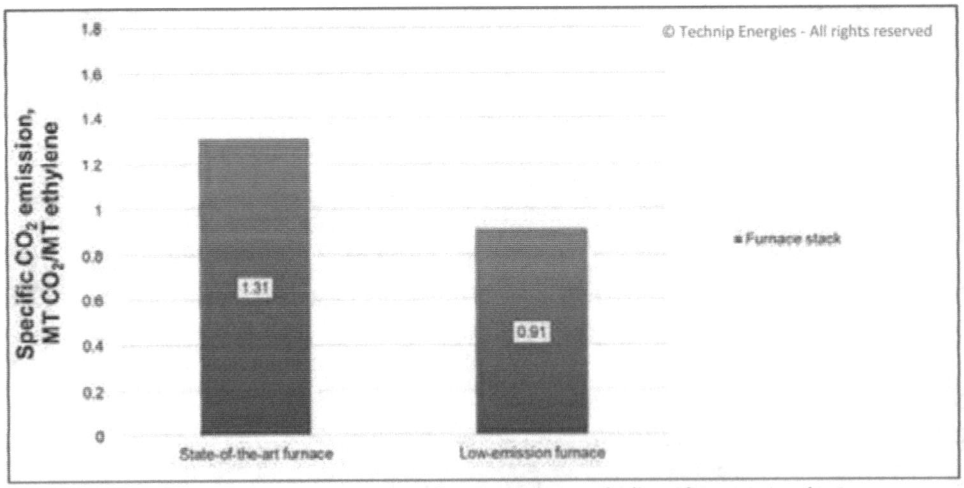


Figure 7: Ethylene plant specific emission, excluding the power plant

The reduction within the ethylene plant from 1.31 to 0.91 MT $CO_2$ per MT of ethylene produced (indicated in blue) is the emission reduction due to introduction of the low-emission furnace itself. On top of that, we see the emission of 0.25 MT $CO_2$ per MT of ethylene produced associated with the supply of the 12 MW from the CCGT for the in-plant power demand of the ethylene plant (indicated as green) and the emission 0.15 MT $CO_2$ per MT of ethylene produced associated with the delivery of the extra 7 MW produced by the CCGT that's exported to the external power grid (indicated as white). The total emission of the CCGT is 0.40 MT $CO_2$ per MT of ethylene produced, together with the emission of the ethylene plant of 0.91 MT $CO_2$ per MT of ethylene produced, brings the total back to 1.31 MT $CO_2$ per MT of ethylene produced. This result is as it should be, as we are firing the same amount of fuel gas in both cases.

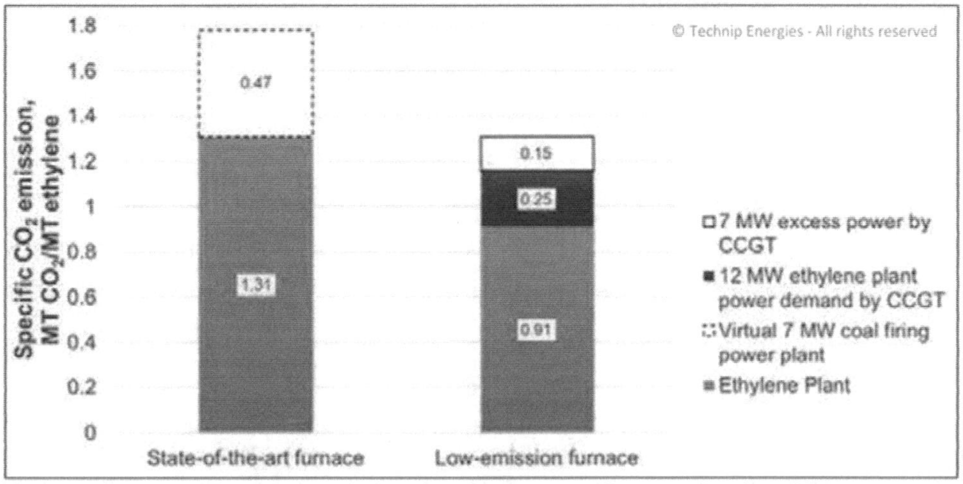


Figure 8: Ethylene plant specific emission during the phasing out of conventional coal firing stage, including the power plant

At this point, the emission reduction becomes complicated. How do we account for the avoided emission by the 7 MW power that we are producing from this same fuel gas? Obviously, this power would replace power generated by another source, for example, a conventional coal firing power plant or a CCGT. These avoided emissions depend on the energy transition stage.

Two stages can be distinguished:

- Phasing out conventional coal firing
- Phasing out fossil gas fuels firing

Figure 8 reflects the situation for the first stage, in which the coal is being phased out. To visualize the avoided emissions, a virtual conventional coal firing power plant has been added to the graph on the left side, producing 7 MW, the same amount as is being sent to the grid as excess power from the CCGT. When comparing the emission from the coal fired plant with that of the CCGT for the same power, it is immediately clear that the conventional coal firing power plant produces significantly more $CO_2$ than the integrated CCGT. The carbon footprint of the coal firing plant is 1.2 MT $CO_2$/MWh, while that of the CCGT is 0.38 MT $CO_2$/MWh. That is more than three times higher, which is also reflected in the specific $CO_2$ emissions of 0.47 MT $CO_2$/MT ethylene produced for the coal firing plant, which is more than three times higher than the 0.15 MT $CO_2$/MT ethylene produced by the CCGT. This 0.47 MT $CO_2$/MT ethylene produced is the avoided emission level in this first stage of the energy transition. This represents a $CO_2$ emission reduction of 26 percent.

Figure 9 reflects the situation for the second stage, the stage in which the fossil gas fuels firing is being phased out.

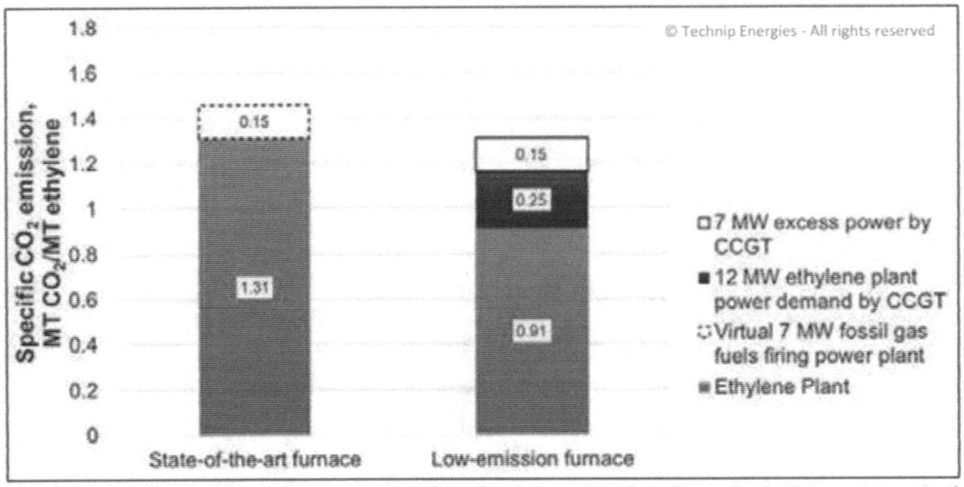


Figure 9: Ethylene plant specific emission during the phasing out of fossil gas fuels firing stage, including the power plant

In this case, the virtual conventional coal firing power plant has been replaced by a fossil gas fuels firing plant to the graph on the left side, also producing 7 MW. The graph shows the contribution of the avoided emissions if the source being replaced would be a fossil gas fuels firing power plant. In this case, this contribution is the same as the 0.15 MT $CO_2$/MT ethylene produced by the integrated CCGT on the right side of the graph. This makes sense, as they fire the same fuel gas with the same technology, a CCGT. Nevertheless, there is a reduction of the $CO_2$ emissions of 0.15 MT $CO_2$/MT ethylene produced. This avoided emission represents a $CO_2$ emission reduction of 10 percent in this next stage of the transition.

Which level can be applied depends on the local situation. The figure below shows the situation for the power generation in the Netherlands in 2019.

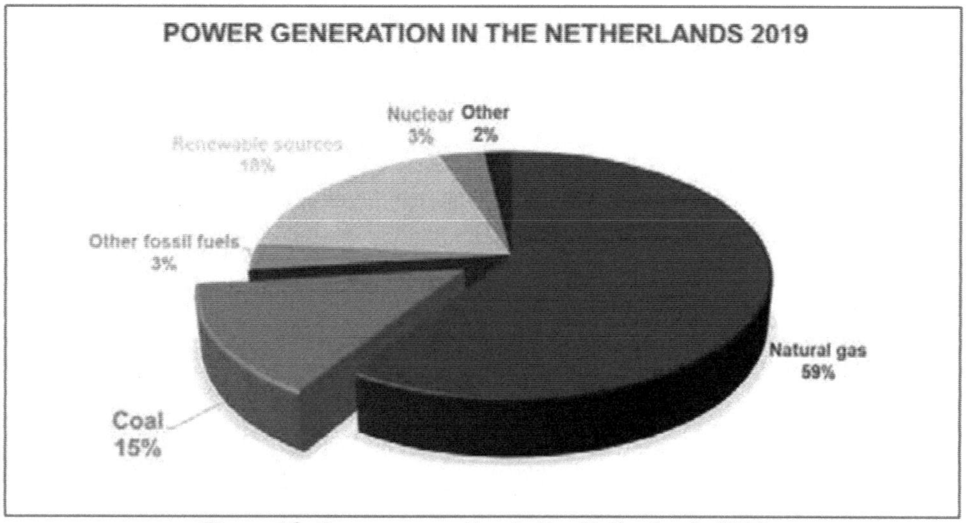


Figure 10: Power generation in the Netherlands 2019
* Source: Dutch Central Office of Statistics (CBS) 2019

Figure 10 shows that the contribution of fossil fuels in the generation of power is substantial. Fifteen percent of the countries' power generation is from coal, while 59 percent is from natural gas and 3 percent is from other fossil fuel sources. Any power produced from the excess fuel gas produced by the ethylene plant would add to the natural gas part of the slice and would enable the phasing out of the coal firing. This would justify using the avoided emission level of 0.47 MT $CO_2$ per MT ethylene produced during the period that conventional coal firing is being phased out.

In the next phase, during the further introduction of renewables and the further reduction of the fossil fuel gas firing, the excess fuel gas produced by the ethylene plant would be a base load in the grid. It would be the one of the last sources to be phased out as its production, being a by-product of the plastics production, is not voluntarily but compulsory. Considering that its carbon footprint over the complete life cycle is probably less than that of natural gas, which also has a considerable well-to-tank contribution, and that this source is a highly reliable power source in the grid, this is not necessarily a bad thing. Natural gas firing is already a huge improvement over coal firing, as it has a three times lower carbon footprint. As long as natural gas firing is not fully replaced by renewable sources, there is an incentive to combust the excess fuel gas from the plastics production in a CCGT at a high heat-to-power efficiency.

Whether these proposed avoided emission levels will be recognized and will be ground for carbon credits will be subject to discussions with local governments.

## Economic evaluation

### *Low-emission ethylene plant economics*

An economic evaluation has been performed to evaluate the impact of the low-emission ethylene furnace on plant overall economics. The price-set as shown in Table 1 below has been used for this purpose. This table shows the 2019 price level of the $CO_2$ credits. In addition, it shows the power and natural gas prices on European level (Eur28) for large industrial consumers.

Table 1: Price-set utilities and CO2 credits

| Price-set | Units | 2019 |
|---|---|---|
| $CO_2$ Emissions * | €/MT $CO_2$ | 25.0 |
| Power ** | €/MWh | 50.8 |
| Natural Gas ** | €/MWh | 16.2 |

* Source: https://markets.businessinsider.com
**Data source: Eurostat – Data Explorer.
Electricity: Band IG Eur28 2019S2
Natural gas: Band I6 Eur28 2019S2

Figure 11 presents the results of this economic evaluation of the low-emission ethylene plant as compared to a conventional ethylene plant. This evaluation is for the ethylene plant only, excluding the power plant.

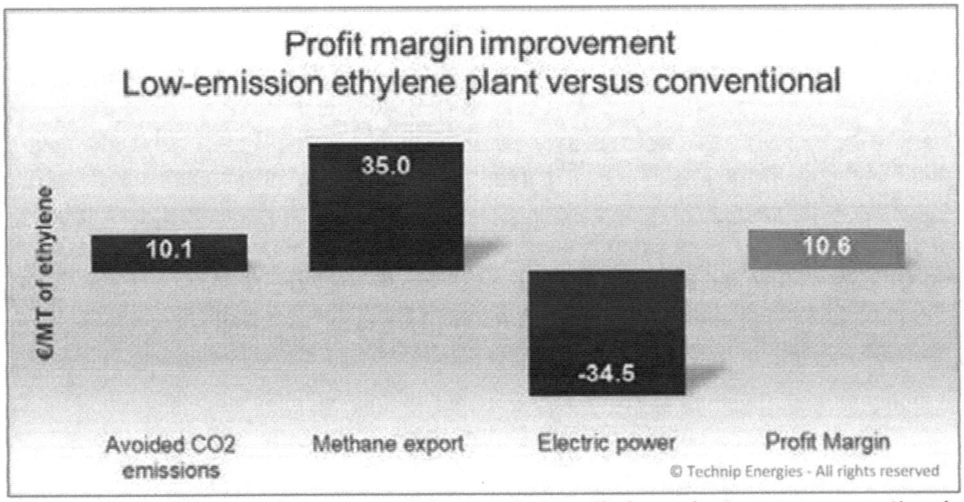


Figure 11: Profit margin improvement: Low-emission ethylene plant versus conventional

As the ratio between the price of the natural gas and that of the power, 0.32, is practically the same as the heat-to-power efficiency of the ethylene plant, the value of the exported excess fuel gas to the CCGT, corresponding to 19 MW of fired heat, is closely matching that of the imported 12 MW power from the CCGT. The main source of the additional profit is from the carbon credits. The more than 30 percent reduction of the fuel gas consumption in the cracking furnace gives rise to a credit of 10.1 € of MT of ethylene produced. The total additional profit is 10.6 € per MT of ethylene produced.

### *Integrated low-emission ethylene plant/power plant economics*

The same evaluation has been performed for the integrated ethylene plant/power plant from the example. This time, the power plant is included in the evaluation. The result is presented in Figure 12.

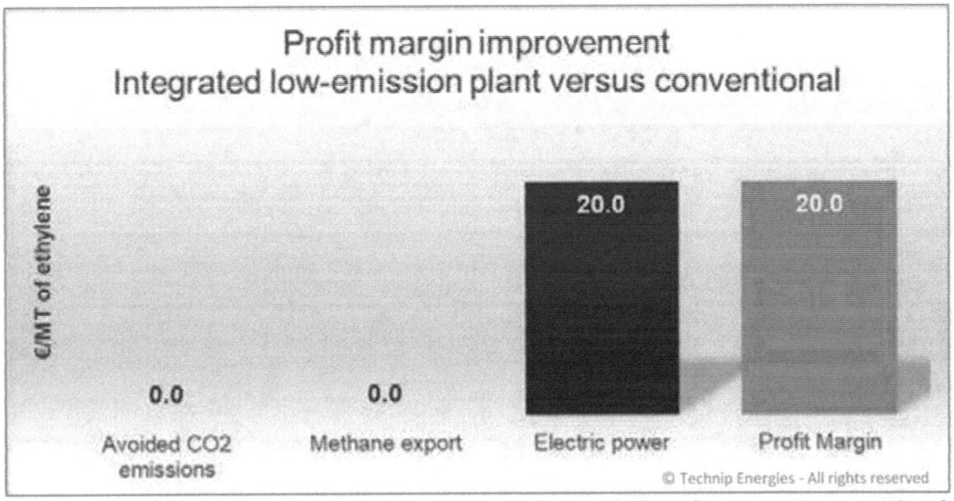


Figure 12: Profit margin improvement: Integrated low-emission plant versus conventional

As the CCGT is fully using the excess fuel gas produced, there is no net export of fuel gas. As the fuel gas fired in the integrated ethylene plant is the same as fired in the conventional ethylene plant, corresponding to 120 MW fired heat, the carbon credits from lowering fuel gas consumption are also zero.

The only source of additional profit is the export of the 7MW excess power generated by the CCGT from the excess fuel gas. This amounts to 20.0 € per MT of ethylene produced. This shows that integration nearly doubles the additional profit margin and that improving fuel efficiency directly improves the profit margins even without carbon credits.

### *Potential additional credits from avoided emissions*

As discussed above, there is a potential for carbon credits during the period that fossil fuels are being phased out. Table 2 shows the situation for the two steps in the energy transition:

- Phasing out of conventional coal firing power plants
- Phasing out of fossil fuel gas firing power plants

The table is based on a 2019 price level of 25 €/MT $CO_2$ for the carbon credits. Based on this level the potential additional profit margin from avoided emissions can initially be as high as 11.8 € per MT of ethylene produced during the phasing out of the coal, and will probably drop to 3.8 € per MT of ethylene produced once the coal has been phased out.

Table 2: Credit levels and credit value for avoided specific emissions

Table 2: Credit levels and credit value for avoided specific emissions

| | Avoided specific emission, MT $CO_2$/MT ethylene | Corresponding $CO_2$ credits, €/MT ethylene |
|---|---|---|
| Phasing out coal | 0.47 | 11.8 |
| Phasing out fossil gas fuel | 0.15 | 3.8 |

As previously stated, whether these carbon credits will be recognized and awarded will be subject to discussion with local governments.

***Profit margin improvement and CO2 reduction***

Using the numbers in the example above for a 1,400 kta ethylene plant the profit margins are as follows:

- During the phasing out conventional coal firing: ~45 Million €/year
- During the phasing out fossil gas fuels firing: ~33 Million €/year

For a period of 10 years this would amount to: ~333 - 445 Million €

The main investment, associated with the 190 MW CCGT required for this case, would cost approximately half of this.

The corresponding annual CO2 reduction is as follows:

- During the phasing out conventional coal firing: ~1770 kta
- During the phasing out fossil gas fuels firing: ~565 kta

Any future price developments of natural gas, electricity and carbon credits are not taken into account in these numbers, as these numbers are highly dependent on the local economic environment. They are merely shown to indicate that with this flow scheme it is possible to reduce $CO_2$ emissions, while at the same raise the plant's overall profit margins. With the example that is being presented in this paper, individual operating companies can assess the economic viability of the low-emission ethylene plant flow sheet for their own specific economic environment.

We want to specifically mention here, that the OPEX evaluation of the low-emission ethylene plant as being depicted in Figure 6 does not include an evaluation of the CAPEX. It should be fully understood that an investment in a power plant is not a short term investment, contrary to the pay-back time of the low-emission cracking furnace itself as presented above. The payback time still needs to be determined, but is considered to be outside of the scope of the original project, which was aimed exclusively at cracking furnace emission reduction and not at ethylene plant reduction. Yet, we felt an obligation to also present a practically feasible road map to reduce the $CO_2$ level for the ethylene plant as well as its power supply, as these are being affected by the low-emission cracking furnace concept.

***Global market outlook***

It is beyond the scope of this paper to present a full market outlook. The situation can be different for each and every region in the world and for each and every nation, depending on its own national resources. For a global market outlook, reference is made to the IEA World Energy Outlook 2020 [1] and the BP Energy Outlook 2020 (https://www.bp.com/en/global/corporate/energy-economics/energy-outlook.html). Both outlooks present similar trends.

In its World Energy Outlook 2020, the IEA sketches the development for the global energy market over the coming 20 years for a number of scenarios, taking into account the impact of the COVID-19 pandemic.

Based in this report, in particular the outlook for fuel supply section, coal is the first fossil fuel to be phased out gradually over the coming 30 years, followed by oil [2]. The natural gas market is least affected, as this fuel source is expected to play a dominant role in the transition phase, during the phasing out of coal and oil, as it is seen as the lesser evil in terms of $CO_2$ emissions. Especially in the emerging economies, India, Southeast Asia, Africa and China, the energy demand is expected to grow rapidly over the coming 10 years [3] . In Asia, which is expected to account for half of the world's energy consumption by 2050 [4], electricity demand is expected to double by 2050. Natural gas and renewables will be the main drivers to meet this strong increase in demand. The natural gas price is expected to increase in these regions as is illustrated in Figure 12 below [4]. This figure shows the expected evolution of the natural gas price in the region for the period 2019 through 2050 period. It shows an natural gas price annual growth rate of 3.9 through 8.1 percent in China and slightly lower growth rates in the neighboring countries identified as the Other Non-OECD Asia countries [5].

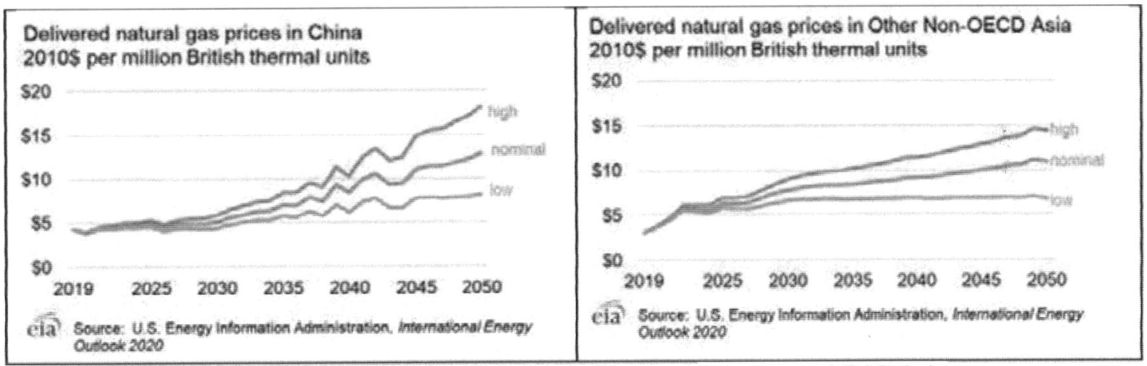


Figure 12: Expected evolution of Natural gas price in China and Other Non-OECD Asia

The low-emission plant concept can play a major role in this energy transition by supplying not only the fuel gas, but also generating the electricity to meet this increasing energy demand. This would replace part of the demand for LNG. At the same time, the ethylene plant operator would have a reliable power source, independent from the external grid.

## Conclusions

- The low-emission cracking furnace is able to reduce the $CO_2$ emission in excess of the 25 percent target of the IMPROOF program. It can reduce the $CO_2$ emission of the cracking furnace beyond 30 percent.
- By improving the heat-to-power conversion of the excess fuel gas from approximately 31 to ~50 percent, using a CCGT, the power generated from the same fuel can be increased by 37 percent, thus reducing the specific emission level, expressed in MT $CO_2$/MWh of power produced, by 27 percent.
- For the ethylene plant, the only source of emissions is from the cracking furnace stack. The specific ethylene plant emission, expressed as emissions per metric ton of ethylene produced, reduces by approximately 30 percent, in line with the emission reduction of the low-emission cracking furnace itself.
- For the emissions beyond the ethylene plant, such as the power generation system running on excess fuel gas, also expressed as emissions per metric ton of ethylene produced, the emissions reduce by approximately 26 percent initially during the phasing out of the conventional coal firing power plants. In the next stage, this value reduces to 10 percent.
- The impact of the low-emission cracking furnace on the ethylene plant profit margin is mainly from exported carbon credits and amounts to approximately 10.6 € per MT of ethylene produced at the 2019 price-set used.
- Integrating the power plant with the ethylene plant can be beneficial. It provides the ethylene plant with a reliable power source, and by using the gas turbine to provide shaft power to the ethylene plant machines directly, equipment cost associated with power generators and electro motors, can be reduced. At the same time, profit margin can be nearly doubled to a level of approximately 20.0 € per MT of ethylene produced at the 2019 price-set used.
- Potential additional profit from avoided emissions by phasing out fossil fuels is ~4-12 €/MTA ethylene at 2019 $CO_2$ emission prices. Whether these carbon credits will be recognized and awarded will be subject to discussion with local governments.
- An integrated low-emission ethylene plant is not only feasible, it is also profitable. We expect this can serve as the foundation for additional modifications of the ethylene plant that will facilitate the introduction of renewable power sources in the power grid to further reduce $CO_2$ emissions and reach the European Committee's target of 55 percent $CO_2$ reduction by 2030.
- Globally, the low-emission ethylene plant concept can be used to meet the rising energy demand, especially in the emerging economies, reducing the dependence on LNG.

## Acknowledgment

The work leading to this invention has received funding from the European Union Horizon H2020 Programme (H2020-SPIRE-04-2016) under grant agreement n°723706.

## References

1. IEA (2020), World Energy Outlook 2020, IEA, Paris https://www.iea.org/reports/world-energy-outlook-2020
2. IEA, Global fuel supply by scenario, 2010-2040, IEA, Paris https://www.iea.org/data-and-statistics/charts/global-fuel-supply-by-scenario-2010-2040
3. IEA, Electricity demand outlook in selected regions/countries in the Stated Policies Scenario, 2019-2030, IEA, Paris https://www.iea.org/data-and-statistics/charts/electricity-demand-outlook-in-selected-regions-countries-in-the-stated-policies-scenario-2019-2030
4. U.S. Energy Information Administration (2020), Effects of Changes in Natural Gas Prices and Renewable Capital Costs on the Electric Sector in Asia https://www.eia.gov/outlooks/ieo/section_issue_Asia.php
5. See https://www.eia.gov/outlooks/ieo/pdf/appm.pdf) for the region definitions.

SPRING21
+17TH GCPS
A Joint AIChE and CCPS Meeting

# Process Safety in Transit:
# A Fatal Explosion of a Truckload of Ammonium Nitrate

**Michael S. Schmidt**
**Engineering Management and Chemical Engineering Departments**
**Missouri University of Science and Technology (MS&T)**
and
**Bluefield Process Safety, LLC**
**3101 Cherokee St.**
**St. Louis, MO 63118**
**MSchmidt@BluefieldSafety.com**

Prepared for Presentation at
American Institute of Chemical Engineers
2020 Spring Meeting and 17th Global Congress on Process Safety
Virtual
April 18 - 22, 2021

# Process Safety in Transit: A Fatal Explosion of a Truckload of Ammonium Nitrate

**Michael S. Schmidt**
**Engineering Management and Chemical Engineering Departments**
**Missouri University of Science and Technology (MS&T)**
and
**Bluefield Process Safety, LLC**
**3101 Cherokee St.**
**St. Louis, MO 63118**
**MSchmidt@BluefieldSafety.com**

**Keywords:** Hazard Identification and Risk Analysis, Transportation Safety

## Abstract

On March 27, 2019, a truck loaded with ammonium nitrate exploded on an Arkansas highway, killing the driver. While recent ammonium nitrate explosions, like the Beirut explosion in August 2020, the Port of Tianjin explosion in August 2015, and the West Fertilizer explosion in April 2013, are widely reported in the global press, explosions of ammonium nitrate while in transit have been fatal as well but often go underreported, escaping the attention of process safety specialists.

Process safety responsibilities do not end at the fence-line. Process safety specialists have a professional duty to protect the public and environment as hazardous materials move through and between communities.

Why did the truck in Arkansas explode? A tire catching fire is an inadequate explanation. And given the warning that the driver had, why was he killed? Understanding the answers to these questions will help all process safety specialists to better protect the public and the environment, as well as workers, from the consequences of process safety incidents, whether they involve ammonium nitrate or other hazardous chemicals.

## 1 A Brief History of Ammonium Nitrate Disasters

Catastrophic ammonium nitrate disasters have been happening for as long as we have been producing ammonium nitrate. In 2014, Han et al. compiled a list of 74 ammonium nitrate incidents going back to 1896. **[1]** A common source of information, Wikipedia, has a list of 35 disasters **[2]**, of which all but the most recent three are also in the Han compilation.

The Wikipedia list includes the most well-known ammonium nitrate disasters, the disasters with which all students of process safety are familiar:

September 21, 1921 Oppau, Germany 561 dead
An attempt at disaggregation with blasting elements at the BASF plant in Oppau resulted in the explosion of about 450 tons of ammonium nitrate.

April 16, 1947 Texas City, Texas, United States 581 dead
The cargo ship *Grandcamp* caught fire and subsequently exploded, involving about 2,100 tons of ammonium nitrate.

December 13, 1994 Sergeant Bluff, Iowa, United States 4 dead
An explosion in a neutralizer at the Terra Industries Port Neal Plant, which reacts nitric acid with ammonia to make ammonium nitrate.

April 17, 2013 West, Texas, United States 15 dead
Arson at the West Fertilizer Company warehouse where about 240 tons of ammonium nitrate was stored, followed by an explosion as firefighters attempted to put out the fire.

August 12, 2015 Port Tianjin, China 173 dead
A fire at a hazardous goods warehouse set off an explosion of about 800 tons of ammonium nitrate stored there.

August 4, 2020 Beirut, Lebanon 204 dead
A fire at a Port of Beirut warehouse triggered the explosion of about 2,750 tons of impounded ammonium nitrate.

Of the 35 disasters on the Wikipedia list, over a third involved the transportation of ammonium nitrate. Four were explosions of cargo ships loaded with ammonium nitrate. Three were explosions of freight trains pulling cars loaded with ammonium nitrate. Six were explosions of trucks hauling loads of ammonium nitrate.

They follow a common pattern: An incident—sometimes a wreck, sometimes an onboard failure—ignites a fire. The fire burns and attempts are made to bring the fire under control. Then, without warning, the ammonium nitrate explodes. Later, in a typical news account, a reporter will find a witness that says that 'it was like a war zone.'

An incident that is not on any list is the explosion of a truck hauling a load of ammonium nitrate across southwest Arkansas in 2019 that killed the driver of the truck. This case study looks at that incident and shares some lessons that should benefit every enterprise involved in the shipment of hazardous materials by truck.

## 2 Background of the Incident

On Wednesday morning, March 27, 2019, a truck pulling a hopper trailer loaded with 25 tons of ammonium nitrate prill left El Dorado, in southeast Arkansas, for Texarkana, Texas, approximately 100 miles to the west. About 40 miles into the trip, the driver pulled the truck onto the shoulder of the highway to address a fire at the back of the truck. While there, the truck exploded, killing the driver.

### *2.1 The driver*

The driver was 63-year-old Randall McDougal. He drove for Blann Tractor Company, headquartered in Hampton, Arkansas, also in southeast Arkansas. Nicknamed, "Billy Goat", Mr. McDougal had been a truck driver for over 40 years, his entire adult life. He was an extremely experienced driver and was quite familiar with the region. [3]

### *2.2 The Load*

The load was ammonium nitrate prill. The trailer was a hopper trailer, similar to that shown in Figure 1.

**Figure 1. Hopper trailer typically used for hauling ammonium nitrate. (IME 2020)**

A typical load is about 25 tons.

### *2.3 The Route*

The driver picked up the load at the El Dorado Chemical Company facility on West Avenue, north of the city of El Dorado, Arkansas. The destination for the load was Texarkana, Texas, approximately 100 miles to the west. The route involved heading northwest from the plant on Arkansas Highway 7, through Smackover and toward Camden.

Before reaching Camden, the likely route involved a left turn onto westbound Arkansas Highway 376, bypassing the town of Camden, and then a left turn onto westbound U.S. Highway 278. The remainder of the route consisted of about 40 miles west on U.S. Highway 278 toward Hope, Arkansas, and then about 30 miles southwest on Interstate 30 to Texarkana at the Texas-Arkansas state line. The intended route is shown in Figure 2.

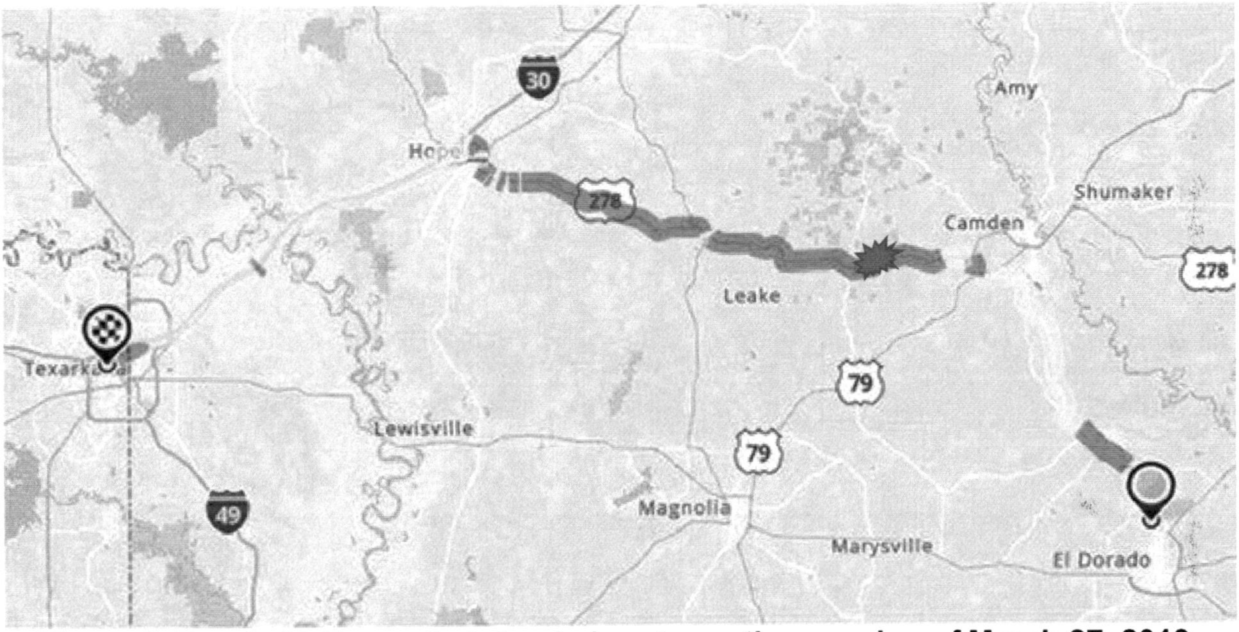


**Figure 2. Randy McDougal's intended route on the morning of March 27, 2019.**

The trip ended at about 6:40 am on the shoulder of U.S. Highway 278, just a couple of miles east of the intersection of U.S. Highway 278 with Arkansas Highway 57.

# 3 The Incident

The account of the incident that follows was compiled from the news release issued by the Arkansas State Police, the reports published in the *Arkansas Democrat-Gazette* [4], the *Texarkana Gazette* [5], *KSLA – Shreveport* [6], *5News Online – Fayetteville* [7], and an interview with David Archer, District Engineer for District 7 of the Arkansas Department of Transportation [8].

### *3.1 The call to 911*

At 6:40 am, Mr. McDougal contacted 911 to notify emergency services that he had smoke coming from the rear wheels of his rig. He told them that he was hauling ammonium nitrate and that he intended to stop when he got to a remote, unpopulated area and to put the fire out with his on-board fire extinguishers. Operators at the 911 center dispatched police and firefighters to the scene. David Archer, District 7 Engineer for the Arkansas Department of Transportation said that from the spot on the shoulder where Mr. McDougal parked, the closest residence was about a quarter mile away.

### *3.2 Arrival of first responders*

The Camden Fire Department and the Stephens Fire Department responded to the call, as did the Ouachita County Sherriff's department and the Arkansas State Police. Aware of the hazard of ammonium nitrate and its potential to explode, first responders parked their vehicles about a quarter mile from the burning truck.

Their first action at the scene was to begin evacuating nearby residents to a distance of one mile from the burning truck.

### *3.3 Driver's actions at the scene*

Mr. McDougal stopped on the shoulder of westbound U.S. Highway 278, at spot where there were no nearby houses. He got out of the cab of the tractor and went back to the burning wheel and attempted, unsuccessfully, to put the fire out with his on-board fire extinguisher.

His next immediate concern was the safety of nearby residents. He began efforts to evacuate them, but when he understood that emergency responders were attending to that, he returned to his truck with the intention of unhooking the tractor from the trailer and driving the tractor away from the burning trailer.

As Mr. McDougal was in the process of dropping the trailer, the load of ammonium nitrate exploded.

### *3.4 Explosion*

The explosion occurred at 7:02 am, as recorded seismically at the White Oak Lake, Arkansas station of the Arkansas Geological Survey and shown in Figure 3. This was around 20 minutes after the initial call to 911.

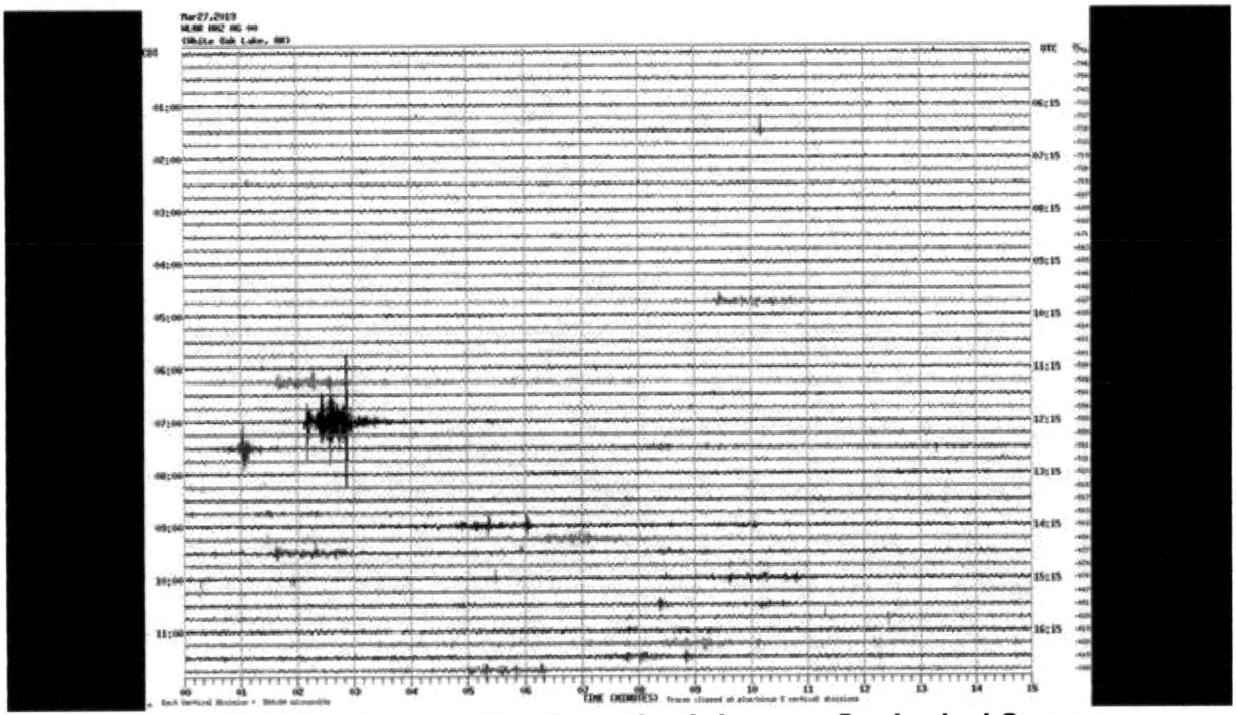

**Figure 3. Seismic recording from the Arkansas Geological Survey, showing the explosion at 7:02 am. (Lewis 2019)**

David Archer described it this way: "Where there had been an 18-wheeler, there was just a crater. The truck was just gone. The biggest piece I found was a foot-long piece of radiator hose." He also noted that the explosion was heard as far away as Little Rock, Arkansas, almost 100 miles to the north [8].

The photograph in Figure 4 was taken by Mike Rowe, a paramedic with ProMed Ambulance who happened to be on the highway at the time of the explosion. The photograph was featured prominently in the reports from several news outlets. It shows the red smoke characteristic of ammonium nitrate decomposition.

**Figure 4. Immediately following the explosion, showing the characteristic red cloud of ammonium nitrate (photo by Mike Rowe, ProMed Ambulance)**

After the explosion, firefighters withdrew to a distance of about one mile from the explosion. After they were confident that there would be no more explosions, firefighters worked to put out the fire. The *Arkansas Democrat-Gazette* reported that Capt. Adam LaDuke with the Ouachita County sheriff's office indicated that crews had extinguished most of the fire by mid-morning. [4]

# 4 The Impact of the Explosion

## *4.1 Safety Impacts*

The driver, Randal McDougal, was killed instantly at the scene.

In addition, three firefighters were injured during the explosion and brought to a hospital. Two were released soon after, and the third was released later.

There was also a school bus from the Camden Fairview School District on the road when the explosion occurred. "The bus driver tried turning the vehicle around when the explosion happened," District Superintendent Mark Keith said. "No children were on board, and the driver was uninjured." [4]

## *4.2 Environmental Impacts*

The environmental impact of the explosion was primarily one of downed trees. The aerial photograph shown in Figure 5 shows downed trees radiating out from the crater. [4]

**Figure 5. Aerial photograph of crater and associated damage to surrounding forest. (photo by Thomas Metthe, *Arkansas Democrat-Gazette*)**

In addition to knocking down trees, the force of the explosion stripped needles from trees. David Archer commented, "Even though it was barely spring, everything was green for a 200-yard radius because the ground was blanketed with all the needles blown off the pine trees." [8]

### *4.3 Impacts on assets*

The immediate asset impact was to the truck and its cargo. The tractor and trailer were obliterated, and the ammonium nitrate went up in a cloud of red smoke. The debris field extended several hundred yards in all directions, as suggested in Figure 6.

**Figure 6. Crews with heavy equipment clean up the debris field and begin repairs on U.S. Highway 278. (photo by J.D.Bailey, Banner News)**

Another immediate impact of the explosion was broken windows. A fire truck and a school bus that were around a quarter mile from the explosion suffered broken glass. The shattered windscreen of the fire truck is shown in Figure 7. Nearby residences also suffered broken windows, the closest also being about a quarter mile away from the explosion.

**Figure 7. Damaged windscreen of Camden Fire Department engine parked about a quarter mile from the explosion.**

Additionally, the highway required repairs. The crater was 15 feet deep and stretched all the way across the highway, as shown in Figure 8. It required that a 500 foot stretch of highway be rebuilt. In an amazing community effort, District 7 Maintenance Engineer Drew Hoggard assembled the resources needed to completely rebuild the highway and open it to traffic in less than 48 hours, an effort that was awarded the 2019 Director's Diamond Award. David Archer noted that most of the time was spent waiting for layers of asphalt to cool enough to allow the next layer to be applied.

**Figure 8. The smoldering crater was 15 feet deep and stretched all the way across the highway. (photo by David Archer, Arkansas Department of Transportation)**

# 5 An Analysis of the Explosion

## *5.1 Stability of ammonium nitrate*

At this point in the history of process safety, it should come as no surprise that ammonium nitrate has the potential to be a powerful explosive. Knowing that, we sometimes lose sight of how stable ammonium nitrate is.

Pure ammonium nitrate is not combustible. It is an oxidizer, however, and will accelerate the combustion of flammable and combustible materials. It is stable at ambient temperatures and pressures, and as a pure compound it is reported as having very low shock sensitivity. (It is important to recall that prior to the 1921 explosion at Oppau, BASF reportedly used small charges of dynamite to disaggregate solid masses of ammonium nitrate/ammonium sulfate mixtures as many as 20,000 times without incident.) [9]

### 5.1.1 Impurities

However, ammonium nitrate is sensitized when contaminated with organic materials. In fact, ANFO, a bulk industrial explosive consists of a 94% ammonium nitrate/6% fuel oil mixture.

(Sometime, reporters or the public will sieze on terms like "fertilizer grade" vs. "explosive grade" ammonium nitrate to explain different stabilities or energies of explosion between various inventories of ammonium nitrate.The difference is in the porosity of the prill. Agricultural grade is not very porous, so it ships a little denser. Explosive grade is porous, so it can absorb fuel oil more readily to make ANFO. Without fuel oil, though, explosive grade AN is just AN, and no more or less likely to explode.)

Spilled diesel fuel at truck accidents has been suspected as a sensitizer in some cases of ammonium nitrate explosions. The Institute of Makers of Explosives (IME, Washington, D.C.) recommends that trailers be dedicated to hauling ammonium nitrate, and if not, that they be washed out and dried before using to haul ammonium nitrate. Contaminants from previous loads that they particularly warn about include, but are not limited to

- Animal fats
- Baled rags, burlap, or cotton bags
- Charcoal
- Foam rubber
- Hay
- Metal powders
- Sawdust
- Straw
- Vegetable oil
- Wood, wood chips, or wood shavings

Other impurities are known as sensitizers for ammonium nitrate, including chlorates, mineral acids, and metal sulfides.

### 5.1.2 Elevated Temperatures

The temperature at which ammonium nitrate decomposes is widely reported as 210 C (410°F). [10]. It is the sudden and catastrophic decomposition of ammonium nitrate that results in explosions.

There are two decomposition reactions. The first dominates at lower temperatures, in the range of 200 ~ 300 C (390 ~ 570°F):

$$NH_4NO_3 \rightarrow N_2O + 2\ H_2O(g)$$

This means that for every pound of ammonium nitrate, the decomposition produces 14.5 standard cubic feet of gas, or at 410°F, it releases 23.8 cubic feet of gas to the open atmosphere. On the other hand, if the one pound of ammonium nitrate were in a contained volume, say by being encased in a fused mass of ammonium nitrate, the volume of one pound of ammonium nitrate prill — 0.0186 cubic feet, given a specific gravity for ammonium nitrate prill of 0.86 –

would remain constant and the pressure after decomposition would increase to almost 19,000 psig. An explosive release would happen long before reaching 19,000 psig.

Then at around 300 C (~ 570°F), the reaction shifts:

$$2\ NH_4NO_3 \rightarrow 2\ N_2 + O_2 + 4\ H_2O(g)$$

At this higher temperature, the decomposition of a pound of ammonium nitrate produces 16.9 standard cubic feet of gas, or at 570°F, it releases 32.9 cubic feet of gas to the open atmosphere. Again, if the pound of ammonium nitrate prill, occupying 0.0186 cubic feet, decomposes without venting, it will build 26,000 psi of pressure.

Ammonium nitrate will decompose whether air is present or not, so as long as the temperature is high enough, smothering it has no impact.

### *5.2 Truck fires*

There are about 6,000 ~ 7,000 truck fires each year involving large, over-the-road trucks [11] [12] That is out of about 8.5 million trucks registered, meaning that there is a little less than a 0.1% probability that any truck in any given year will catch fire. That probability is per truck, not per trip. Obviously, the probability of a specific truck catching fire during a specific trip depends on the age of the truck, its maintenance history, and the experience and skill of the driver.

When trucks do catch fire, however, trailer fires are most common. As shared at a 2018 conference on truck safety, [13] the causes of trailer fires are distributed between four causes:

| | |
|---|---|
| 32% | Brake fires |
| 32% | Wheel bearing fires |
| 24% | Air leaks |
| 12% | Tire fires |

The two causes that get the most attention are tire fires, because the other causes are likely to set tires on fire, and brake fires, perhaps because they are so difficult to put out.

#### 5.2.1 Tire fires

Tire fires are typically not the initiating cause of truck fires but are a source of fuel for fires once they have an ignition source in the form of brake fires or wheel bearing fires. The normal operating temperature for truck tires is around 100 ~ 150°F. At 250°F, tires begin rapid deterioration, which can lead to blowouts. At 500 ~ 550°F, tire decomposition produces flammable vapors, which can ignite if there is an ignition source. The autoignition temperature for the decomposition products of tire deterioration is around 850 ~ 900°F. [14]

While brake fires and wheel bearings are typically the ignition sources of tire fires, it is possible for truck tires to catch fire even when not started by overheated brakes or bearings. The most likely cause then is a flat tire. Flat tires can overheat and catch fire by spinning on the rim, rubbing against the other tire or frame rail, or becoming stuck and dragging on the road. [15]

The temperatures from a tire fire impinging on a load of ammonium nitrate prill are well above those necessary to cause the material to melt and then to decompose. Heat transfer from a tire fire to the AN load will depend on the configuration and geometry of the trailer and tires, the rate at which the heat generated by the burning tire is carried away from the trailer as opposed to be directed at the trailer, and weather conditions (cold and wet will impede heat transfer, warm dry will not.) While all these factors will influence the outcome, none are so strong that credit should be taken for them, however.

#### 5.2.2 Brake fires

Brake fires are the most common cause of truck fires, and most of these fires occur within 30 to 50 miles of the start of the trip. [13]

Brakes can catch fire if the driver rides the brakes. This is especially a problem with inexperienced drivers on highways with steep grades, as are found in the mountains.

More often, though, brake fires result when the brake drags. Brakes can drag because the brakes do not release after the brake is applied, because the truck is travelling while the parking brake is still engaged, corrosion that interferes with the operation of the brake, or a failure of the any of the moving parts of the brake system. [14]

The problem with dragging brakes is that the driver will often not notice that brakes are dragging. Eight of the wheels on a typical 18-wheeler are on the trailer, and each has a brake. The hopper truck shown in Figure 1 has sixteen wheels on the trailer. If the driver feels the drag at all, they may attribute it to a heavier than normal load or to headwinds.

Brett Aquila, truck driver and editor of *TruckingTruth*, shared this note about brake fires in a blog about another truck brake fire:

> "In case you're wondering why the fire wasn't put out with two fire extinguishers, it's because brake fires are incredibly difficult to put out. The brake shoes and drums can reach very high temperatures and take a long time to cool off! Once they've reached high enough temperatures to ignite the shoes themselves and any grease that may be in the area, the fire continues to re-ignite even after you've tried putting it out several times. The fire won't stay out until the brakes can be cooled down. So, it isn't uncommon to hear about brake fires that end up burning the trailer to the ground because there wasn't enough water or fire extinguishers around to put the fire completely out. If you ever happen to see a tractor trailer with a brake fire, make sure the fire department is called. A fire extinguisher will not often be able to put it out." [16]

### *5.3 Explosive energy of load*

The energy of explosion for ammonium nitrate is around 1,445 kJ/kg. [17], [18]

The energy of explosion for TNT is generally reported as 4,184 kJ/kg, although the U.S. EPA uses a value of 4680 kJ/kg. [19]

A typical load of ammonium nitrate prill is about 25 tons, or 50,000 lb, or 22,700 kg. That means that the explosive energy of a typical load of ammonium nitrate prill is about the same as that of 8.6 tons, or 17,300 lb, or 7,800 kg, of TNT.

## *5.4 Consistency with impacts*

### 5.4.1 Shattered glass

The overpressure that results in shattering single pane glass is 0.5 to 1.0 psid. [20] The EPA stipulates that an endpoint overpressure for explosions of 1.0 psid, which corresponds to the overpressure that shatters single pane glass, be used in worst case analyses under the RMP Rule, 40 CFR 60. [19] The EPA also requires that a TNT-equivalency model be used, with the following scaled distance equation:

$$D_{mi} = 0.0081\ \text{mi/lb}^{1/3}\ (m_{TNT})^{1/3}$$

where the equivalent mass of TNT, $m_{TNT}$, is given in pounds.

In the case of a load of 25 tons of ammonium nitrate prill, the equivalent mass of TNT is 17,300 lb. So, the distance to a 1 psid over pressure is 0.21 miles. The distance to 0.5 psid is even further. This is consistent with windows breaking at nearby residences, given that the nearest residence was about a quarter mile away. It is also consistent with the broken windshield of the fire engine, which firefighters initially parked about a quarter mile away from the burning truck.

### 5.4.2 Crater size

Weapons experts are quick to point out that the size of a crater produced by an explosion depends on a number of factors; the amount and type of explosive, the height above or depth below the surface of the explosion, and the type of soil or rock being cratered are the primary factors. The calculations for estimating the depth and diameter of a crater can be quite detailed. [21], [22]

On the other hand, a rough estimate of the radius of a crater can be made using the overpressure associated with a crater. *Lees' Loss Prevention* gives the overpressure at the lip of a crater as 300 psid. [20] With this, the scaled distance equation gives an estimate of the crater radius:

$$r = z_e(m_{TNT})^{1/3}$$

where the equivalent mass of TNT, $m_{TNT}$, the scaled distance at a specific overpressure, $z_e$, and the radius of the blast circle, r, are given in consistent units. All that is required is the scaled distance for an overpressure of 300 psid.

A commonly used equation for estimating the scaled distance, $z_e$, as a function of the overpressure, $P_O$, is found in the Automated Resource for Chemical Hazard Incident Evaluation (ARCHIE), a program developed for use by FEMA, DoT, and the EPA. [23]

$$z_e = \exp\left(3.5031 - 0.7241\, \mathit{ln} P_O + 0.0398 (\mathit{ln} P_O)^2\right)$$

where $z_e$ is scaled distance, ft/lb$^{1/3}$
$P_O$ is peak overpressure, psid

For a peak overpressure of 300 psid, $z_e$ is 1.95 ft/lb$^{1/3}$, meaning that the radius of a crater created by an explosion with an equivalent mass of TNT, $m_{TNT}$, of 17,300 lb is 50 ft, or a diameter of 100 ft. The Arkansas Department of Transportation reported that the crater was about 48 feet across, while photographs of the crater suggest that it was about 70 feet across. An estimate of 100 feet across, based on a 300 psid overpressure and a 100% yield of explosive energy is not unreasonable, given that such estimates tend to be conservative.

# 6 Recommendations to Prevent Similar Incidents in the Future

Legal liability for shipments is a question for lawyers and the courts. But we also operate in court of public opinion. When a headline screams, 'Chemical Spill on I-70 Leads to Fire and Evacuations: Highway Shut Down for 8 Hours,' the public is not thinking, *The trucking industry just can't seem to operate safely.* No, the public is thinking, *The chemical industry is sending these dangerous chemicals through our community.* So, legal liability aside, like it or not, we have a shared responsibility to not only make our plants safer, but to make the roads safer as well. Here are some recommendations for doing that.

## *6.1 Only ship in well-maintained trailers*

Trucks hauling ammonium nitrate do not catch fire because they are hauling ammonium nitrate. They catch fire because they are trucks. Tire fires and brake fires are not more likely because of the load. If the load is ammonium nitrate, however, there is a potential for the load to explode, something that will not happen if the load is furniture or potatoes.

That means that there is special responsibility to assure that the trailers loaded with ammonium nitrate are in good repair. The tires need to be properly inflated. The brakes need to be properly maintained.

When a trailer first arrives on site for loading, someone should have a look at it. Does it appear to be in good repair? Do the brakes smell hot? Does the lubricant in the hubs smell burnt? Is the trailer clean and dry? While quality issues are typically the only consequence of cross-contamination between loads, cross contamination may also result in incidents on the road.

Before the truck leaves, have another look. Are valves securely closed? Are caps in place and secure? Are there fire extinguishers on board? The driver is responsible for the load, but a second set eyes can help.

## *6.2 Confirm that the driver understands the hazards of the material they are hauling*

Someone who understands the hazards of the material being shipped should talk to the drivers about the hazards of the material. It is important that drivers understand those hazards as well, and that will rarely be accomplished by simply including a Safety Data Sheet in the packet of paperwork going with the shipment. It is unreasonable to expect a truck driver, someone hired for their skill in operating a rig, to also be skilled in deciphering a Safety Data Sheet, although they certainly may be. In case they're not, though, talk to them.

Specifically, make sure that they have thought about and know what they will do in the event of a brake fire, a jackknife, or a collision that results in a leak or spill.

In the case of ammonium nitrate, note that the explosive energy of a 25-ton load is equivalent to that of 8.6 tons of TNT. That if it explodes, the detonation of a 25-ton load of ammonium nitrate is enough to put a crater in the ground that stretches all across the highway and to shatter windows a quarter mile away. Remind them that fire extinguishers are just for incipient fires and that a tire or brake fire is hot enough to cause the catastrophic decomposition of ammonium nitrate, whether it is contaminated or not. Repeat that the Department of Transportation's Emergency Response Guide (ERG) recommends isolation and evacuation (including emergency responders) for a mile in all directions when a tank, rail car, or truck loaded with ammonium nitrate are involved in a fire. [24]

### *6.3 Emergency response*

#### 6.3.1 Only use fire extinguishers for incipient fires

The IME recommends that a truck loaded with ammonium nitrate have "at least two fire extinguishers, each with a rating of 2A-10B:C or higher." [25] Several guidance documents on truck fires review the PASS method of using a fire extinguisher: Pull pin, Aim at base of fire, Squeeze trigger, Sweep across base of fire. Implicit in all of this is that the fire is early and small enough that a fire extinguisher can put it out. Fire extinguishers are only for incipient fires, and a fire that requires two or more fire extinguishers has advanced beyond the incipient stage.

Successfully extinguishing an advanced fire requires copious amount of water, and that will require a fire department with access to copious amounts of water.

#### 6.3.2 Assure that the load is placarded with the correct placards

Before a truck leaves any facility, it should be properly placarded. Those placards are not for the facility and they are not for the driver. No, the placard is for any emergency responder that must react to an incident involving the truck.

Emergency responders are trained to approach truck incidents from upwind and uphill, and then scan the truck for placarding before deciding how to approach the incident. [26] The few seconds they take to do this can make the difference between life and death. All too often, when there are fatalities or other injuries during an incident, it is the first responders that suffer them.

They don't have a chance of doing the right thing if the placards are wrong.

#### 6.3.3 Call 911 with the information they need to respond correctly

As soon as a call can be made safely, call 911 and report the fire. Identify the location as best as possible. Make sure that the dispatcher understands the hazard to which the fire department will be responding. Explain that by the time the fire department arrives, they will probably not be able to get close enough to fight the fire.

The ERG states that small ammonium nitrate fires should be extinguished with water, not dry chemicals or foams. [24] Water works, not by smothering the fire, but by cooling the material to less than its decomposition temperature. Liquid water has a large heat capacity; water also provides cooling due to its large heat of vaporization when used as an extinguishing agent for fires. Because ammonium nitrate decomposes whether in contact with air or not, extinguishing media that simply smother the ammonium nitrate are of no use. So, while vaporizing water to steam will help, steam itself will not, as the captain of the *Grandcamp* learned.

For fires involving tanks, railcars, or truck loads—fires bigger than "small fires"—the ERG urges the use of unmanned master stream devices or monitor nozzles, aimed to cool the trailer. In no uncertain terms, the ERG warns emergency responders to stay away from trucks engulfed in fire. If it is not possible to use unmanned master stream devices or monitor nozzles, the ERG urges first responders to withdraw from the area and let the fire burn.

Typically, the most important role for first responders will be to shut down the highway to keep others out of harm's way.

### *6.4 Telemetrics*

In a plant, any safety critical process condition is typically instrumented so that a Safety Instrumented System continuously monitors it and shuts down the process automatically when conditions are unsafe or alarms to prompt an operator to shut down the process. Increasingly, it is also possible to continuously monitor conditions in a load that is on the road. Fleet tire pressure monitoring systems allow for both drivers and fleets to stay aware of tire pressures.

There are also fleet temperature monitoring systems. They are currently used primarily to assure that food in cold chain logistics does not spoil, but are also being used to monitor the logistics of Covid-19 vaccines. Monitoring the temperature of ammonium nitrate loads, especially directly above the rear axles could also be useful in preventing future incidents. Temperature monitoring should also reduce the number of truck fires in general, regardless of the load. It may be time to apply what we have learned about Safety Instrumented Systems in fixed locations to transportation.

## 7 A Final Note

We do not know exactly what caused the fire that resulted in the explosion of a truckload of ammonium nitrate on a rural Arkansas highway early on Wednesday morning, March 27, 2019. Randy McDougal reported it as a brake fire. Given his forty years of experience as a truck driver and the preponderance of truck fires ignited by brake fires, there is little reason to doubt his assessment. As for the cause of the brake fire, the terrain he was driving through was not especially hilly, and even if it were, it is unlikely that the brake fire started because Mr. McDougal was riding the brakes.

His only action that warrants any criticism is that he went back to the burning truck to drop the trailer and drive his tractor away. We do not know why he did that. If he had not, he probably could have escaped the explosion with his life, perhaps without injury.

If he had managed to uncouple his tractor and drive away, that would be incorporated into the lore of truck fires. Perhaps, if it had been a load of furniture or potatoes, he could have. Ammonium nitrate fires do not always result in explosions. We have learned a lot about the factors that contribute to explosions—often contamination—but not so much that we could predict with confidence which AN truck fires will result in explosions and which will not. But we do know that when ammonium nitrate in a fire explodes, it explodes without warning.

When Mr. McDougal noticed the fire, he contacted 911, fully explained the situation, and found a safe place to park his burning truck. He attempted to extinguish the fire and then satisfied himself that neighbors were out of harm's way.

His family, friends, and fellow drivers call him a hero. He was.

## 8 References

[1] Z. Han, S. Sachdeva, M. Papadaki and M. S. Mannan, "Study of Ammonium Nitrate Thermal Hazards with Additives," in *Proceedings of the 17th Annual International Symposium on Process Safety*, College Station, Texas, 2014.

[2] Wikipedia, "List of Ammonium Nitrate Disasters," 22 January 2021. [Online]. Available: https://en.wikipedia.org/wiki/List_of_ammonium_nitrate_disasters . [Accessed 17 February 2021].

[3] Perry's Funeral Chapel, "Randy's Obituary," 5 April 2019. [Online]. Available: https://www.perrysfuneralchapels.com/obituary/randy-mcdougal. [Accessed 17 February 2021].

[4] Y. Rddad and J. Snyder, "Driver killed after fertilizer truck explodes in south Arkansas; Area evacuated after blast that was heard miles away," 27 March 2019. [Online]. Available: https://www.arkansasonline.com/news/2019/mar/27/fertilizer-truck-explosion-spurs-evacuations-south/. [Accessed 12 February 2021].

[5] J. Bailey, "Texarkana-bound truck explodes near Camden," 27 March 2019. [Online]. Available: https://www.texarkanagazette.com/news/texarkana/story/2019/mar/27/fertilizer-truck-bound-texarkana-explodes-ouachita-county-killing-driver/771719/. [Accessed 12 February 2021].

[6] D. Scruggs and A. Onken, "Driver killed when truck explodes on Arkansas highway," 27 March 2019. [Online]. Available: https://www.ksla.com/2019/03/27/arkansas-highway-closed-following-truck-explosion/. [Accessed 12 February 2021].

[7] G. Lewis, "Camden Explosion Measured Seismically, Heard For Miles," 27 March 2019. [Online]. Available: https://www.5newsonline.com/article/weather/camden-explosion-measured-seismically-heard-for-miles/527-38f9974e-c911-4bd7-b886-9e11b11e86d6. [Accessed 12 February 2021].

[8] D. Archer, Interviewee, *District Engineer, Arkansas Department of Transportation District 7.* [Interview]. 8 February 2021.

[9] U. Hörcher, "Oppau 1921: Old Facts Revisited," 2016. [Online]. Available: https://www.aidic.it/cet/16/48/125.pdf. [Accessed 17 February 2021].

[10] National Center for Biotechnology Information, "Ammonium nitrate," [Online]. Available: https://pubchem.ncbi.nlm.nih.gov/compound/Ammonium-nitrate. [Accessed 17 February 2021].

[11] J. Pearlman and N. Meltzer, "Risk of Commercial Truck Fires in the United States: An Exploratory Data Analysis," 2011. [Online]. Available: file:///C:/Users/nbres/Downloads/724611.pdf. [Accessed 16 February 2021].

[12] M. Ahrens, "U.S. Vehicle Fire Problem," January 2013. [Online]. Available: https://www.nfpa.org/-/media/Files/News-and-Research/Fire-statistics-and-reports/US-Fire-Problem/ostypeofvehicle.ashx. [Accessed 16 February 2021].

[13] P. J. Fisher, "Many trailer, truck fires can be prevented," 22 October 2019. [Online]. Available: https://www.tirebusiness.com/opinion/fisher-many-trailer-truck-fires-can-be-prevented. [Accessed 17 February 2021].

[14] J. Park, "How to Prevent Wheel-End Fires," 9 September 2019. [Online]. Available: https://www.truckinginfo.com/340034/how-to-prevent-wheel-end-fires. [Accessed 16 February 2021].

[15] R. Harris, "Truck Fires - Top 10 Causes," 8 September 2016. [Online]. Available: https://www.linkedin.com/pulse/truck-fires-top-10-causes-randolph-harris/. [Accessed 17 February 2021].

[16] B. Aquila, "Tractor Trailer Brakes on Fire!," 2015. [Online]. Available: https://www.truckingtruth.com/trucking_blogs/Article-30/tractor-trailer-brakes-on-fire. [Accessed 2 February 2021].

[17] R. Meyer, J. Köhler and A. Homburg, "Ammonium Nitrate," in *Explosives*, Sixth Edition ed., Weinheim, Germany, Wiley-VCH Verlag, 2007, p. 15.

[18] S. Chaturvedi and P. N. Dave, "Review on Thermal Decomposition of Ammonium Nitrate," 2013. [Online]. Available:

https://www.researchgate.net/publication/263578419_Review_on_Thermal_Decomposition_of_Ammonium_Nitrate. [Accessed 16 February 2021].

[19] EPA, "RMP Guidance for Offsite Consequence Analysis," March 2009. [Online]. Available: https://www.epa.gov/rmp/rmp-guidance-offsite-consequence-analysis. [Accessed 1 December 2020].

[20] S. Mannan, "Explosions," in *Lees' Loss Prevention in the Process Industries*, vol. 2, S. Mannan, Ed., Burlington, MA, Elsevier Butterworth-Heinemann, 2005, p. 17.

[21] M. M. Swisdak, "Explosion Effects and Properties Part I - Explosion Effects in Air," 6 October 1975. [Online]. Available: https://apps.dtic.mil/dtic/tr/fulltext/u2/a018544.pdf. [Accessed 15 February 2021].

[22] K. E. Gould and K. Tempo, "High Explosive Field Tests: Explosion Phenomena and Environmental Impacts," 1 October 1981. [Online]. Available: https://apps.dtic.mil/dtic/tr/fulltext/u2/a135737.pdf. [Accessed 15 February 2021].

[23] EPA, "Commercial Explosives And Their Hazards," February 1994. [Online]. Available: https://nepis.epa.gov/Exe/ZyPDF.cgi/9100CHJC.PDF?Dockey=9100CHJC.PDF. [Accessed 18 February 2021].

[24] Pipeline and Hazardous Materials Safety Administration, "Guide 140 - Oxidizers," in *2020 Emergency Response Guidebook*, Washington, D.C., U.S.Department of Transportation, 2020, pp. 216-217.

[25] IME, "IME Guideline for the Responsible Transportation of Solid Ammonium Nitrate by Motor Vehicle," September 2020. [Online]. Available: https://www.ime.org/uploads/public/COVID19/Transportation%20of%20TGAN%20Guidance%20FORMATTED%2011_2_2020.pdf. [Accessed 15 February 2021].

[26] B. Stoothoff, "Lesson Learned at a Tractor Trailer Fire," 7 July 2010. [Online]. Available: https://www.fireengineering.com/firefighting/lessons-learned-tractor-trailer-fire/#gref. [Accessed 18 February 2021].

# 619243 SIL2 Torquemeter for Functional Safety Process Control of Steam Turbine Driven Compressor

**Abstract:**

Use of a torquemeter to accurately measure shaft horsepower independent of fuel, pressure, temperature, and flow variations is known to many oil and gas plant turbomachinery users worldwide. Used to inform the end user of turbomachinery train location specific degredation, the basic condition monitoring torquemeter system is a useful plant tool for performance trending and diagnostics. The large fuel use of steam turbines in ethylene manufacture makes plant efficiency extremely important to bottom line considerations.

For end users of torquemeter devices, it has always been possible for them to be used as process control devices, since accurate delivered power measurement helps run close to performance limits, maximising plant throughput and therefore revenue. The single measurement aspect of the torquemeter system does not provide for a functionally safe system in process control terms.

This paper outlines the successful development of a process system, using novel technology, which can be applied as a retrofit or new build system.

Summary

For one ethylene plant operated by Total, their existing torquemeter was successfully monitoring shaft horsepower and was set to react to an increase in demand from the compressor by ensuring the steam turbine power delivery was within acceptable limits for the compressor rotor rating. Any potential over run in power from the steam turbine was prevented by linking the torquemeter reading to the steam demand valve, which closed on the first sign of potential over torque. Since this process loop was safety critical, after a customer functional safety review a requirement was identified for a two out of three (2oo3) redundant voting system of torque measurement, so that in the unlikely event of a torquemeter failure, the manufacturing process could continue safely. The Safety Intergity Level required for plant compliance from the customer was SIL 2. Accuracy during normal running was to be maintained at +/-1% of reading at normal torque and operating speed.

Single failure mode (plant still running) had to maintain accuracy at +/-5% of reading at normal torque and speed. To move from a single reading torquemeter to a 2oo3 voting functional safety torquemeter required the integration of three pairs of measuring coils, detecting three independent values, which could be processed in three independent SIL Modules to provide 3 completely independent values for comparison by the customers data acquisition system and associated Emergency Shutdown system (ESD). This paper will explain how proof of concept study was carried out using a small R&D developed torquemeter; how the recorded data showing optimal alignment, seperation and signal strengths led to the final design; how the end-user was involved in all stages of the development; and how the final design was installed. It will also summarise some of the SIL concepts and work carried out to achieve the required level 2 product.

# Early warning solution of gas leaks

*R. Braun*

*CEO – Grandperspective GmbH*

*Berlin, Germany*

Automated early detection of gas leaks and tracking of moving gas clouds in case of an incident – this combination of risk prevention and incident mitigation was the design task for the early warning solution that we installed in an ammonia complex. The site in close vicinity to a densely populated campus requires the warning of a release incident at any point within minutes. Once ammonia is in the air, the propagation of the moving cloud has to be tracked live until it reaches the fence. A remote monitoring solution is the logical answer to the task. Remote monitoring covers large areas and can locate a gas at any point because the sensor doesn't have to be within the gas cloud.

Warning for ammonia releases becomes an urgent problem in light of increasing demands for green fuels. Hydrogen fuel is safely stored and transported in form of ammonia. Large scale storage sites and loading stations for ships, trucks or trains demand reliable monitoring. As of today, there is no such technology available. Gas sensors pick up only local concentration values without information on the total amount of gas that is released.

While the detection and visualization of an ammonia cloud was the primary target of the installation, the early warning includes other process gases, both flammable as well as toxic gases from the surrounding production sites. The early warning solution therefore is based on spectroscopic analysis (FTIR spectroscopy). This technology allows the secure identification of hundreds of different chemicals from large distances.

We present the installation of a sensing solution that detects gas leaks at any point of the plant from the distance. The gas cloud is assessed in total so the location, dimension and propagation can be monitored live.

## AICHE PAPER NUMBER

# A SUCCESSFUL PROGRAM IN PROCESS WATER CORROSION CONTROL

***Emerentino Quadro***
*Olefins Process Engineer*
*Braskem*
***Joana Arnaldo Ribeiro***
*Olefins Plant Engineer*
*Braskem*
***Henrique Silva***
*Account Manager*
*SUEZ Water Technologies & Solutions*
***Tatiana Barbosa***
*Product Application Engineer*
*SUEZ Water Technologies & Solutions*

Prepared for Presentation at the 2021 Spring National Meeting
Virtual, April 18-22, 2021

# A SUCCESSFUL PROGRAM IN PROCESS WATER CORROSION CONTROL

***Emerentino Quadro***
*Olefins Process Engineer*
*Braskem*
***Joana Arnaldo Ribeiro***
*Olefins Plant Engineer*
*Braskem*
***Henrique Silva***
*Account Manager*
*SUEZ Water Technologies & Solutions*
***Tatiana Barbosa***
*Product Application Engineer*
*SUEZ Water Technologies & Solutions*

**Abstract:** Controlling corrosion in a Process Water System is key for reliable operation of an Olefins Plant. This paper discusses a Multidisciplinary Root Cause Analysis conducted at the Braskem Olefins plant at the Camacari Site together with SUEZ Water Technologies (responsible for the chemical treatment in this section of the plant). The resulting actions significantly reduced the corrosion rate and improved control of process water quality to avoid reoccurrence of past failures that caused emergency shutdowns.

This Process Water System is susceptible to corrosion by many mechanisms, including acid, and/or caustic corrosion, pitting, flow accelerated corrosion (FAC), corrosion under deposit, etc. This paper will describe the process conditions and water chemistry conducive to corrosion, the strategy adopted, and the results achieved. This strategy was based on the selection of neutralizers, adjustment of pH range, adoption of a chemical responsible to generate an anti-corrosive film inside the pipe and equipment, efficient corrosion monitoring and use of a chemical treatment in the blow down of this system, in order to achieve the required specification by the liquid effluent treatment unit.

## Introduction

Braskem is a Brazilian petrochemicals manufacturer, producing monomers, resins and chemicals specialties. In 2019, Braskem supplied 64% of the Brazilian resin demand.

Braskem's ethylene capacity is currently 5,000 kty, distributed across seven Olefins plants, located on five different sites, three of them based on naphtha steam cracking and two sites based on ethane steam cracking, including 200 kty from ethanol to produce green polymer, from renewable resources. Four of these sites are in Brazil and one of them is in the port city of Coatzacoalcos (Mexico), built together with the Mexican group Idesa.

This paper will focus on one of the oldest Braskem Olefins plants (*UO-1 Q1 BA)*, started up in 1978 and located in northeastern Brazil. Its ethylene capacity increased from 360 kty to 620 kty after several revamps. The last revamp of this unit was performed in 2002. The total ethylene capacity of this site is complemented by another Olefins plant, which produces 660 Kty for a total of 1,280 kty for the whole site.

One of the main concerns related to this plant's reliability is the corrosion rate in the Process Water System. A major Process Water reboiler failure resulted in an emergency plant shutdown. Several alternatives to monitor and keep the corrosion rate in control have been tried through the years. Despite some design differences, this system is present in all ethylene units and it is a reliability concern for many plants. The aim of this paper is to share our experience and explain the program currently implemented and responsible to successfully keep this corrosion rate in control.

## Brief Process Description

The UO-1 Q1 BA has a typical naphtha steam cracker design. The dilution steam fed to the furnaces is condensed by using cooling water (CW) and separated from the Pygas in the bottom of the Quench Water Tower, and the stream from the top of this tower is sent to the Charge Gas Compressor (CGC). The resulted quench water is stripped in a tower by using low-pressure (LP) steam in two reboilers. Alternatively, LP steam is directly injected in this stripper. The bottom stream from this stripper, called now Process Water (PW), is sent, by pump, to the dilution steam generator (DSG) vessel. Thermosiphon reboilers supply the required heat duty to this vessel by using quench oil (QO) and medium-pressure (MP) steam. The dilution steam is generated at 5.2 $kgf/cm^2g$. The schematic presented as Figure 1 illustrates the process in UO-1 Q1 BA, where are indicated the locations of corrosion monitoring and of the highest corrosion rates (in red, based on inspection) and highest frequency of failures.

The Process Water corrosion rate in this plant is monitored by a corrosion coupon and a probe. An electrical resistance (ER) probe provides results every few days when exposed to a corrosive environment from the process. The ER probe is retractable with a wire loop element of 40 or 80 mil thickness, depending on corrosion severity. In comparison to the ER probe, the corrosion coupon has the advantage to show the corrosion morphology, giving more details about the type of corrosion. On the other hand, the ER probe enables to measure corrosion more frequently in a defined period which allow the process deviation to be identified and corrected earlier. These points of corrosion monitoring are localized in the inlet of the Process Water reboiler where heat is recovered from Quench Oil.

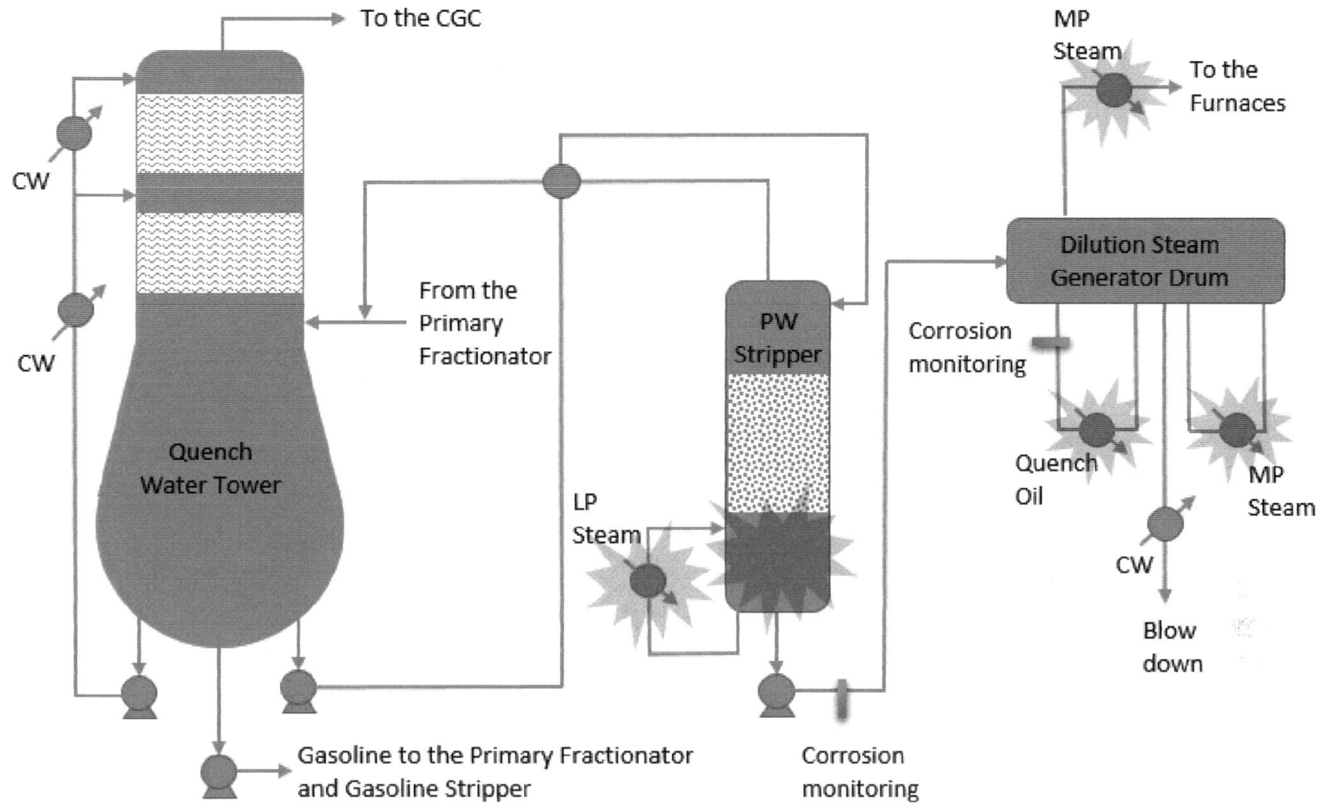


Figure 1. Simplified sketch of the dilution steam generation system.

The failure rate was typically high in the (i) PW stripper reboiler, (ii) bottom region of the PW stripper and (iii) the thermosiphon reboilers to dilution steam generation by using quench oil and, mainly, MP steam.

While the Quench Water pH was typically controlled in low ranges (usually 6,0 to 7,5) in order to minimize emulsion and reduce polymerizable hydrocarbon in water sent downstream, the Process Water pH needed to be controlled in a higher range (8,5 to 9,5) in order to minimize corrosion rate at the PW stripper (particularly at the bottoms) and DSG, including its reboilers.

Although low range pH has been adopted in the quench water circuit for several years, it has never been clear that it was a benefit for better phases separation and reduced hydrocarbon content in the water, particularly in this plant and probably due to its design.

The neutralizer used to adjust the Process Water pH was injected just downstream of the PW stripper, instead of in the feed as typically desired, due to the risk of significant entrainment from this stripper overhead to the Quench Water Tower. As a consequence, the pH in the bottom of this stripper was lower than desired.

The Process Water pH in the Dilution Steam Generation drum is monitored by an online pH meter, in addition to the lab analysis. However, the neutralizer's injection flowrate is adjusted manually. Other Process Water parameters are routinely analyzed in the laboratory: iron, chloride, weak and strong acids, and sodium, as well as turbidity.

Failures in the MP and LP steam reboilers are not so critical, since they result in direct injection of steam in the Process Water stripper and the consequent increase of the blowdown flowrate. However, a leak in the bottom section of the Process Water stripper, in the PW reboiler with Quench Oil, or in most of the piping in this system could result in an immediate plant shutdown.

Another important concern for this system, not discussed in this paper, is the fouling rate usually resulted from poor phase separation in the Quench Water Tower. A high fouling rate can increase the corrosion rate by creating an environment that is conducive to corrosion.

## Types of Corrosion that Could Be Present in this System

The Process Water is a complex mix of ions, organic molecules, dissolved gases, and inorganic compounds in an aqueous solution. As a result, many corrosion mechanisms could take place in this system [1]. Some on the most relevant types of corrosion present in Process Water system are listed below.

Acid corrosion

Acidic corrosion is a familiar degradation of materials contacting acidic media. Usually, without any kind of chemical treatment, the Process Water has acid characteristics (pH < 7) due the presence of acid compounds, like chlorides ions, acetic acid, carbon dioxide, oxygenates and sulfide elements. The presence of these elements is due to the naphtha's composition and pyrolysis furnaces process conditions. In order to prevent this corrosion, a neutralizer is frequently injected into this system.

Caustic corrosion

A poor pH control results in a variation of fluid chemistry and an eventual pH close to 10, causing caustic corrosion. Caustic corrosion occurs when caustic is concentrated and dissolves the protective magnetite ($Fe_3O_4$) layer, causing a loss of base metal and eventual failure. This mechanism is particularly problematic when the metal is stressed and/or silica is present.

Additionally, frequent fluctuation of pH in the Process Water results in the alternation between the aforementioned acid and caustic corrosion. This phenomenon could cause corrosion fatigue in the material [2].

Pitting

According to the National Association of Corrosion Engineers (NACE), pitting corrosion is a localized type of corrosion by which cavities are produced in the material in a small area or point. It is generally more dangerous than uniform corrosion damage because it is more difficult to detect, predict and design against. Typically, the corrosion products cover the pits. The presence of chloride ions and oxygen can increase the rate of this kind of corrosion [3].

Corrosion under deposit

Under deposit corrosion is defined as localized corrosion that occurs underneath or around deposits present on a steel surface. Accelerated corrosion can be found under solid deposits on the bottom of horizontal lines, mainly close to the vaporization area. Under deposit corrosion is usually caused by acid and base neutralization salt products. Under deposit corrosion has been more frequently cited as being responsible for different failures within the oil and gas industries over the past few years, in Process Water type systems [3].

Flow Accelerated Corrosion (FAC)

FAC refers to the removal of the magnetite ($Fe_3O_4$) layer by fluid flow due the certain conditions of shear stress and water chemistry followed by an increase in the corrosion of the exposed metal, leading to wall thinning (metal loss) to the extent of perforation. Susceptible components include carbon steel and low alloy steel piping (with low chromium concentration) and tubing and vessels exposed to flowing water (single-phase or double phase). This mechanism is aggressive particularly at bends/elbows, Tees, orifices, downstream of distributors and valves and any location where turbulent flow exists. Despite being overlooked by many olefins units, this mechanism can contribute to failures in the dilution steam generation system. The temperature at which the FAC rate reaches its maximum is around 150°C, which is close to the usual process conditions in the dilution steam generation system. FAC is strongly dependent on the pH; thus, the FAC rate is significantly reduced when the pH is higher than 9.3 [4]. Braskem has identified the presence of this mechanism in this system. As a reminder, the literature includes several cases of fatalities due to pipe failures caused by FAC, including in an Olefins unit. It occurs because the failure mode for FAC results in an instantaneous liberation of a significate quantity of energy, like an explosion.

## Historical Data – Timeline of Modifications

Until the early part of last decade, the chemical treatment of Process Water of this unit was restricted to caustic solution injection in order to achieve the desired window for the Process Water pH (between 8.5 and 9.0) and also in the Quench Water. In fact, the pH control was very poor. Caustic solution is a too strong to get a stable control of the water chemistry. It is necessary to keep in mind the numerous factors affecting the process water pH, including naphtha contaminants, pyrolysis severity, recycle streams, etc. In addition, the caustic solution increases the tendency for emulsion and foam formation in the quench water, due to the presence of the remaining pygas.

In 2013, a new approach was adopted to keep the Process Water pH in control by using an all amine treatment. The desired window of Process Water pH (between 8.5 and 9.0) was not modified; however, but the pH controllability was improved significantly. It was observed, on the other hand, that there was a increased frequency of under deposit corrosion. These failures were studied and we concluded the root cause was related to the process conditions in this region, which were conducive to salt reversion. This is to be expected when there is a high temperature (under deposit and close the tube wall of heat exchanger) and a salt formed by a weak basic substance (amine). At this condition, the amine recovered leaves the region under deposit and the acid stays there, reducing the pH locally. In order to prevent this phenomenon, we started a fixed and small injection of caustic solution in the feed of dilution steam generator drum; however, the pH was kept on control by the amine blending dosage.

Despite these improvements, the corrosion control was still based solely on pH control and, in fact, this strategy was not effective: the uniform Process Water corrosion rate had an impressive variation of 1 to 30 mpy. Per NACE guidelines for corrosion rate classification, the corrosion measured in the system was most often severe (above 10 mpy), while the target was moderate to low corrosion (below 3 mpy). Several alternatives were studied. One of the alternatives was to inject an oxygen scavenger in this system. We decided against this approach because the oxygen concentration is very low, given that the PW stripper works as a deaerator. We were therefore not convinced the eventual passivation effect of the oxygen scavenger could change the corrosion rate. In 2017, we decided to inject a chemical treatment product provided by SUEZ Water Technologies (responsible for the chemical treatment in this section of the plant). This product, named Petroflo WS-55, contains a long-chain semi-polar organic molecule designed to provide a physical barrier between the metal wall and the fluid. It thus works as a filmer corrosion inhibitor. After that, we observed a significantly reduced uniform corrosion rate, to values between 1 and 3 mpy as presented in Figure 2.

However, the pH in the bottom of the Process Water stripper was still far from the desired range. Due to the entrainment from the overhead of this stripper to the Quench Water tower, the necessary neutralizer to adjust the pH was injected downstream to this stripper. Once the region localized in the bottom of this stripper and its reboilers are known points of high corrosion rate, again, it was necessary discuss this chemical treatment strategy. Figure 3 presents a Histogram of pH in the bottom and reboilers of the Process Water stripper at that moment of the plant campaign. In addition to this significantly variation, the mean value is lower than 7.5 (Figure 3 represents the histogram of this pH in 2018). As consequence, failures in the Process Water reboilers were frequent and the loss of material from the Process Water stripper wall was critical, resulting in a relevant risk of leakage towards the end of the plant campaign. During the plant turnaround (2019), part of the shell thickness of this stripper was reconstituted. The monitored iron content shown in Figure 4 was much higher than desired (< 0.5 ppm) in the process water that was sent to DSG, which confirmed the issue.

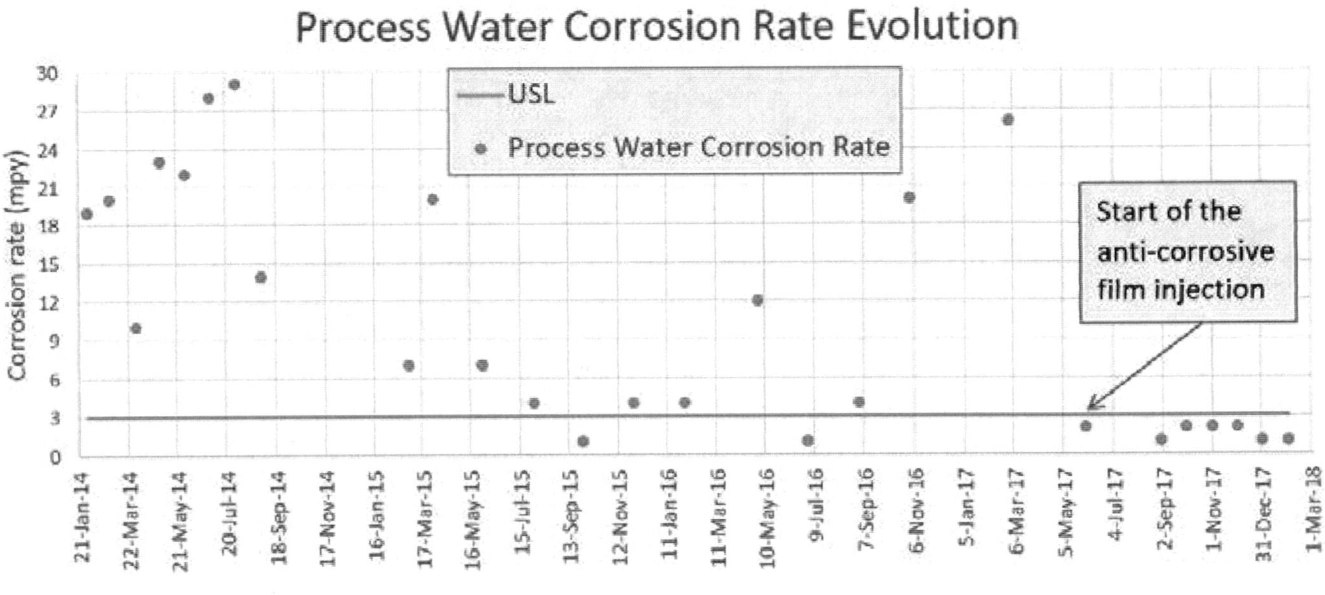

Figure 2. Effect of anti-corrosive film on Process Water corrosion rate.

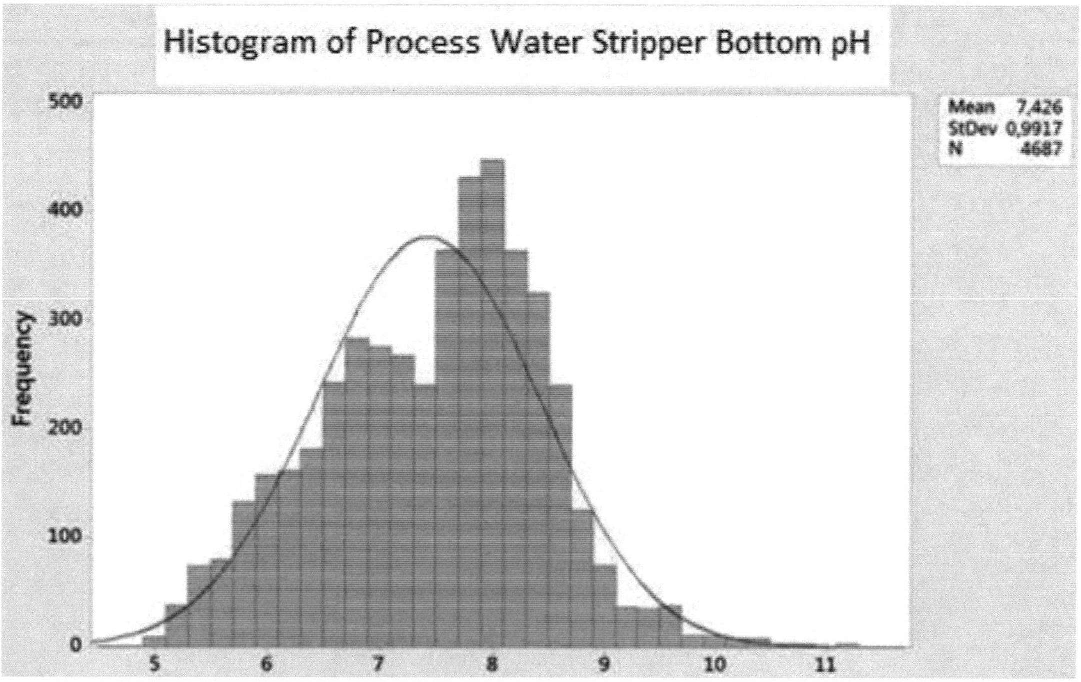


Figure 3. Histogram of pH in the bottom and reboilers of the Process Water stripper (2018).

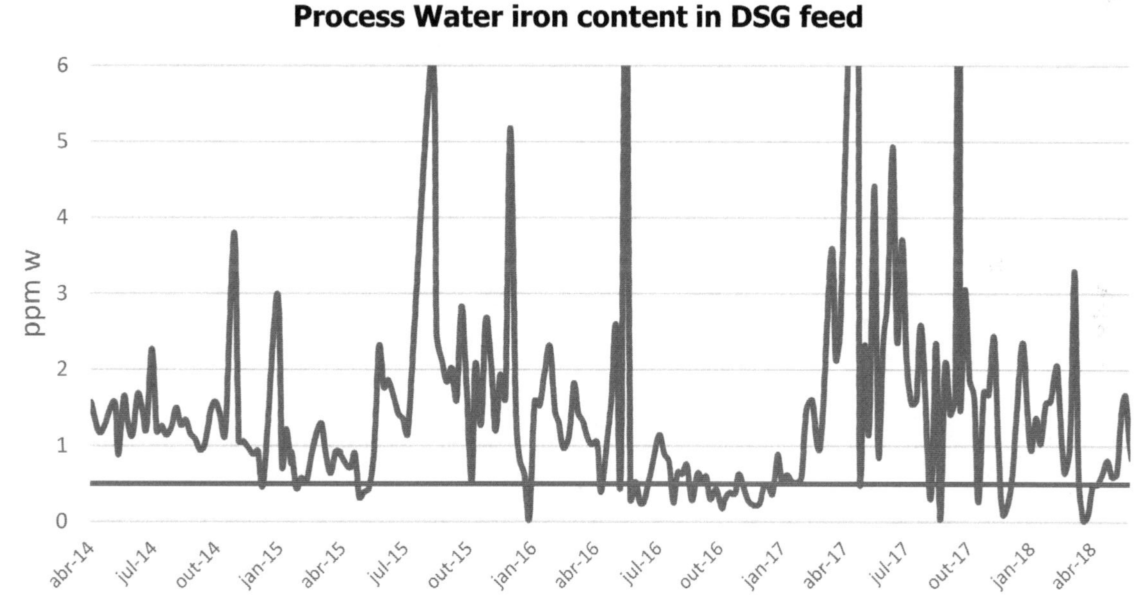


Figure 4. Trend of process water iron content in DSG feed before implemented solutions

In order to solve this issue, modifications were implemented in 2019 (during the plant's major turnaround) to extend chemical treatment improvements to the bottom section of the Process Water Stripper and to prevent FAC in this system. The main modifications are as follow:

- The target for the Process Water pH was changed from 8.5-9.0 to 9.0-9.5. This modification has required the installation of a facility to inject a weak acid in the Dilution

Steam Generation Drum Blowdown. The water blowdown treatment unit requires a pH lower than 9.0. Additionally, the effluent channel, made of cement, through where flows this Process Water blowdown, could be dissolved with pH higher than 9.0. This modification makes control of the pH much easier, since the operation point is far from the equivalence point (pH 7.0). Additionally, the new pH window significantly reduces the FAC rate;

- The injections for the neutralizers and anti-corrosive film chemical were relocated from the feed of Dilution Steam Generation drum to the bottom region of the Process Water Stripper. These new points are shown in Figure 5. As a result, pH control in this region was stabilized within an adequate operational range. As presented in Figure 6, in 2020, the mean value of the pH in the bottom region of Process Water stripper was 9.4 and the standard deviation was 0.26, which was much lower than in the past (0.99).

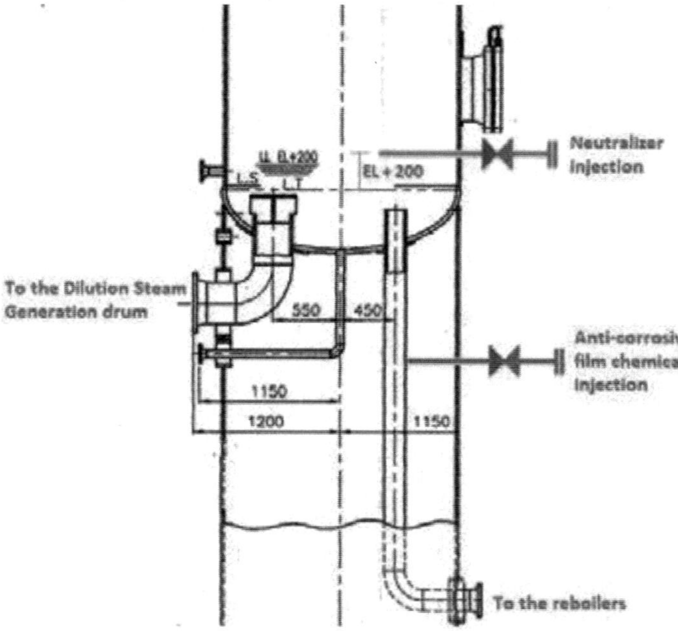


Figure 5. New chemical injection points in the Process Water stripper.

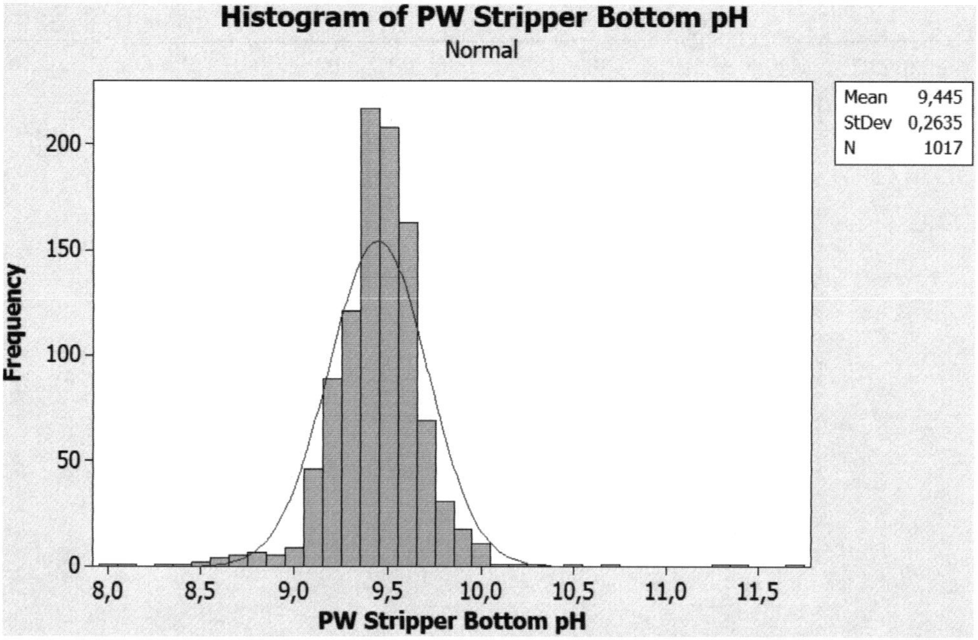


Figure 6. Histogram of pH in the bottom and reboilers of the Process Water stripper (2020).

Currently, after these modifications in the plant facilities to optimize the chemical treatment performance and the implemented adjustments to the chemical treatment program itself, we have observed very low corrosion rate, as shown in Figure 7, and no pitting in the corrosion coupons. Additionally, these corrosion rate data have been confirmed by the Corrater® corrosion monitor.

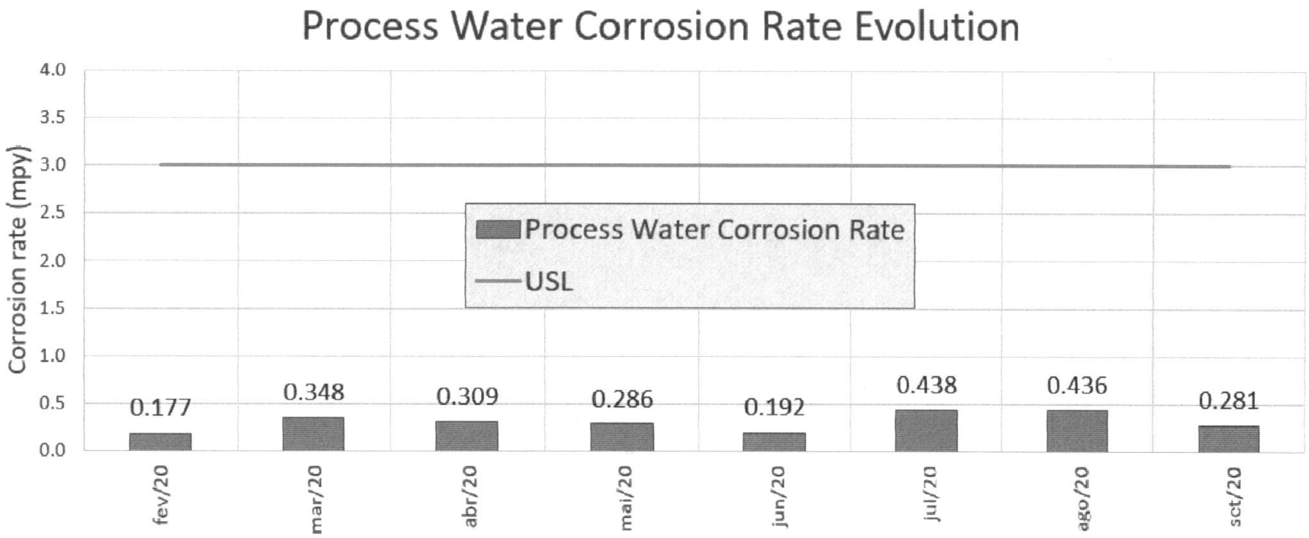


Figure 7. Process Water corrosion rate after chemical treatment improvements, including an anti-corrosive film injection and modification of pH control range

Additionally, these low corrosion rate values are consistent with the concentration of iron present in the PWS bottoms (Figure 8) and in the blowdown of the DSG drum (Figure 9). Currently, these iron concentrations (between 0 and 0.2 ppm) are much lower than the values obtained in the last plant campaign, when around half of the data were above 0.5 ppm and sometimes values higher than 5 ppm could be found.

After one year of operation in the current campaign, no failures have been detected. No reduction in the heat exchangers tube wall thickness has been reported by the inspection team during this period.

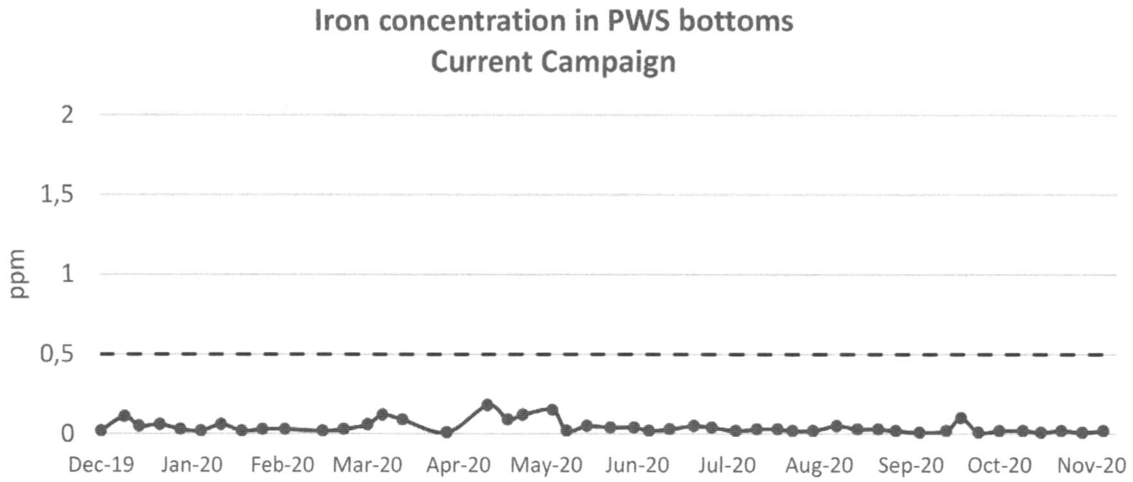


Figure 8. Historical data of iron concentration in the Process Water Stripper bottoms during the current campaign.

Despite these improvements, it is necessary to keep in mind the monitoring and chemical treatment of this system is much more time-consuming than other areas of the Olefins plant. Thus, there must be a clear definition of the key issues, operating discipline, and effective and fast communication between the operational, chemical treatment and plant laboratory team to guarantee the success of the Process Water corrosion controlling program. The dynamics of this system are very fast and the process conditions could change dramatically as function of production plan, naphtha quality, recycle streams or pyrolysis severity.

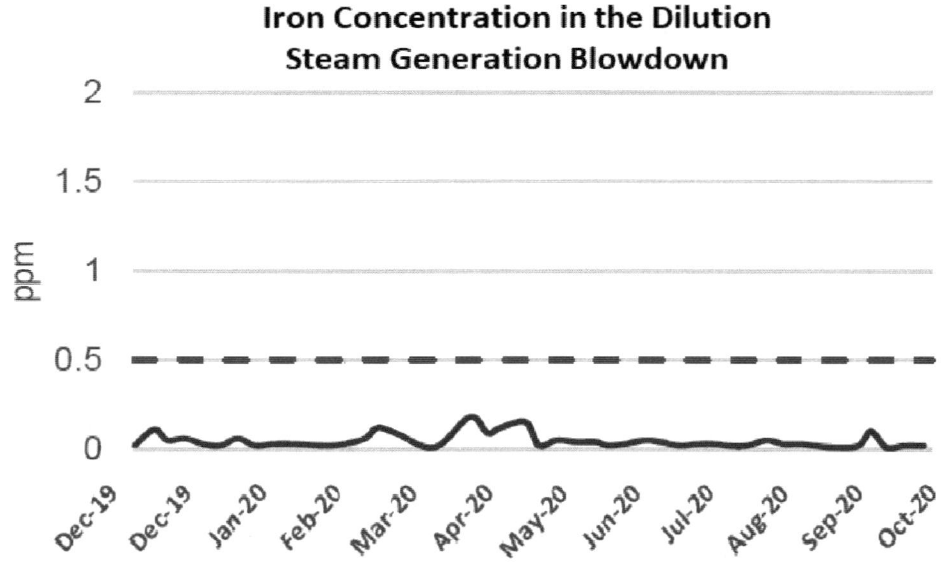


Figure 9. Historical data of iron concentration in the Dilution Steam Generation Blowdown during the current campaign.

This troubleshooting has taught us that: (i) an effective corrosion control program for Process Water should not be based uniquely on pH control, (ii) in order to prevent different types of corrosion mechanisms, the Process Water pH operational window should be carefully determined (in our case, good results have been achieved with pH between 9.0 and 9.5) and (iii) Process Water is a complex environment where many corrosion mechanism could take place and therefore the mix of neutralizers, an approach to protect the iron oxide protective layer, the adjustment of water chemistry in consonance to the process conditions in each part of the plant and the implementation of a monitoring strategy reliably and consistent with the system dynamics are key issues for a successful corrosion control program.

## Conclusions

Several improvements were implemented in a Process Water system of a naphtha-cracking Olefins plant to keep the corrosion rate in control. In one decade, a very poor corrosion control strategy has been modified and the high water corrosivity and the consequent failures (one of which resulted in a plant emergency shutdown) are now a solved issue. This statement is based on the corrosion rate (lower than 1 mpy), iron concentration in the water (lower than 0.3 ppm), and the absence of pitting (per observation of the coupons and the feedback from the inspection team).

The current implemented corrosion control program was designed by considering the fast dynamics of this system and the high complexity of the Process Water, where many corrosion mechanisms could take place. This is the key of the success of this program.

## References


1. Kolmetz, K., Nolidin, C.D., & Mustaffa, Z. (2003) "Improve the Reliability of High Flux Reboilers A Treatment Approach for Dilution Steam Generator (DSG) Reboilers in Ethylene Plants".
2. Gong, Yi & Yang, Chao & Yao, C. & Yang, Zhen-Guo. (2011) "Acidic/caustic alternating corrosion on carbon steel pipes in heat exchanger of ethylene plant. Materials and Corrosion". 62. 967 - 978. 10.1002/maco.201005741.
3. Huang, J. (2013) "Mechanistic Study of Under Deposit Corrosion of Milld Steel in Aqueous Carbon Dioxide Solution" PHD Thesis. Ohio University.
4. Dooley, B., Lister, D. (2018) "Flow Accelerated Corrosion in Steam Generating Plants", PowerPlant Chemistry, 20 (4).

**ZPC 1400KTA Ethylene Plant Startup**

**AIChE Paper Number: 23C**

*Yong Wang*

*Technip Energies, Houston, Texas, USA*

*Rui Chen*

*Technip Energies, Houston, Texas, USA*

*Astrid Han*

*Technip Energies, Houston, Texas, USA*

*Joel Guillaume*

*Technip Energies, Houston, Texas, USA*

## Abstract

Zhejiang Petroleum Chemical Co., Ltd. (ZPC) is located at the Xiaoyushan island in Zhoushan City, Zhejiang Province, in China. ZPC is developing several mega-scale grassroots integrated refinery and petrochemical complexes on the island. It is quickly becoming one of the world's largest petrochemical production bases.

Technip Energies was selected to provide the proprietary ethylene technology and process design for ZPC. Technip Energies provided key proprietary technology components including a Heat Integrated Rectifier System (HRS®), Ripple Trays™ and Wet Air Oxidation process. Technip Energies' Ultra Selective Conversion (USC®) U-coil ethylene technology was selected for its high energy efficiency and high yields. The Phase 1 ethylene plant can produce 1400KTA ethylene. The plant can process heavy feeds, such as hydrogenation cracked tail oil; liquid feeds, such as refinery naphtha and raffinate; and vapor feeds, such as LPG and ethane rich gas. The project only took 42 months from the start of PDP design to plant startup. On spec. product was achieved within one day after feed in. This paper will discuss the ethylene plant design features and the startup experiences.

Zhejiang Petroleum Chemical Co., Ltd. (ZPC) is China's first and largest private led joint venture in the energy sector. ZPC planned several phases of mega-scale refinery and petrochemical complexes on the island of Xiaoyushan in Zhoushan City, Zhejiang Province. Technip Stone & Webster Process Technology, Inc. (a subsidiary of Technip Energies) was awarded the ethylene plant PDP contracts for Phases 1, 2 and 3. Each ethylene plant produces 1400KTA ethylene.

## Ethylene Cracking Furnaces

Technip Energies' Ultra Selective Conversion (USC®) U-coil and W-coil were used for the cracking furnaces. Fresh feeds are supplied from the nearby refinery which include heavy liquid feed, such as hydrogenation cracked tail oil; light liquid feeds, such as naphtha, raffinate and rich NC5; vapor feeds, such as mixed C4 and mixed C2. As shown in Figure 1., the USC® U-coil is a U-shaped serpentine coil with inlet leg and outlet leg connected with a return bend. The simple structure ensures durable mechanical performance. Because of the short residence time, the U-coil is perfect for liquid feeds to gain high yields. Mixed C4 LPG is also recommended to be processed in the U-coil to achieve high yields. The USC® W-coil is a W-shaped, four leg serpentine coil as shown in Figure 2. The residence time of W-coil is longer than U-coil, but the loading capacity per radiant coil is higher. It is preferable to process vapor feed in the W-coil, as the yield is not too sensitive to the residence time. Mixed C2 is mainly processed in W-coil. Mixed C4 can be processed in W-coil as well. The diameter of the USC® radiant coil increases or is "swaged" approximately midway through the coil, to enhance the yield and reduce the risk of spalled coke blocking the coil.

**Figure 1.** USC® U-coil

**Figure 2.** USC® W-coil

To meet the production for ZPC Phase 1, seven 224U furnaces and two 64W furnaces were installed as shown in Figure 3. This installation of the 224U and 64W is the largest Technip Energies U-coil furnace and W-coil furnaces in operation. Both can produce approximately 200KTA ethylene per furnace. They are twin cell furnaces with eight flow passes. The 224U furnace has a total of two hundred and twenty-four swaged U-coils, arranged into two rows of one hundred and twelve coils, one row in each radiant cell. Similarly, the 64W furnace has a total of sixty-four swaged W-coils in two radiant cells. The twin cell furnace has a common convection section offset with the two radiant cells. A shared convection section reduces the convection section equipment, such as ID fan and steam drum, when compared to two single cell furnaces. The convection section is designed with high heat recovery. Thermal efficiency of about 95 percent can be achieved. Even though the liquid and gas furnaces are large, the operating flexibility is built into the design. Different feeds can be fed into different furnace passes. Thus quadra cracking (one feed in one half of the firebox and another feed in the half of the firebox), independent cell cracking (one feed in one firebox and another feed in the other firebox) and independent cell decoke (one firebox under normal cracking and one firebox under decoke) are possible. The furnace design allows for the maximum operating flexibility.

F-1190 F-1180 F-1170 F-1160 F-1150 F-1140 F-1130 F-1120 F-1110

**Figure 3.** ZPC Furnace Island

There are sixty-four ultra-low NOx burners in each furnace. Because of the tight local NOx emission regulations, Selective Catalytic Reduction (SCR) is required and is installed in the furnace convection sections. This is the first SCR installation in ethylene cracking furnaces in China. Anhydrous ammonia is injected to control the stack NOx. A combustion air preheater on each furnace is used to preheat the burner combustion air to 50°C, to reduce fuel gas consumption and improve the furnace efficiency. As shown in Figure 4, the combustion air preheater is the individual heat exchanger unit installed under the burner. Quench water flows through the exchanger to provide the heat to the combustion air. This design can make use of the low value heat from the quench water to reduce furnace fuel consumption and reduce the cooling water usage for cooling the quench water. Additionally, the combustion air preheater draws the air in with the induced draft provided from the furnace ID fan. The pressure drop across the combustion air preheater is very little. Thus no forced draft fan or complicated air duct are needed with this installation.

**Figure 4.** Combustion Air Preheater

Furnaces F-1110 and F-1120 are the heavy liquid 224U furnaces. The furnaces can crack heavy liquid feed and light liquid feed. The convection section is equipped with dual steam injection, cold dilution steam and preheated dilution steam, to fully vaporize the liquid feed before entering the radiant section. At the outlet of the radiant section, the cracking effluent is immediately quenched with double pipe primary quench exchangers, Selective Linear Exchangers (SLEs). Two radiant coils are connected to one SLE exchanger element. Downstream of the SLEs, the cracking effluent is collected and sent to the secondary quench - quench fitting. Quench oil is directly injected into the cracking effluent to lower the temperature and the effluent is routed to the quench oil tower via the wet transfer line.

Furnaces F-1130 to F-1170 are the light liquid 224U furnaces. These furnaces can crack light liquid feed and mixed C4/C2 feeds. The light liquid furnace design is similar to the heavy feed furnace. Only one dilution steam injection point is used in the convection section. The effluent quench also has two stages. The primary quench is one SLE double pipe quench exchanger for each radiant coil. Instead of the direct quench oil quenching as the secondary quench, a shell and tube transfer line exchanger (TLX-D) is used. The uniqueness of the TLX-D design is the dual cooling medium services on the shell side. When the furnace feed is liquid, saturated boiler feed water from the furnace steam drum is introduced to the shell side of TLX-D as the cooling medium. When the furnace feed is vapor, cold boiler feed water from the boiler feed water supply is introduced to the shell side of TLX-D as the cooling medium. The cracking effluent from different feeds can be cooled down to different temperatures in the furnace. Based on the feed type, different cooling media are utilized in the TLX-D to maximize the heat recovery from the cracking effluent. This allows the operating flexibility with the optimized heat integration. Figure 5 shows the picture of TLX-D exchanger. The cracking effluent from the light liquid furnace is routed to the quench oil tower via the dry transfer line.

**Figure 5.** TLX-D Transfer Line Exchanger

Furnaces F-1180 and F-1190 are the vapor feed 64W furnaces. These furnaces do not have a dilution steam preheat bank in the convection section. The cracking effluent coming out of the radiant section has three stages of quenching. The primary quench is a double pipe one-to-one Ultra Selective Exchanger (USX). The secondary quench is a shell and tube transfer line exchanger (TLX-1), which utilizes saturated boiler feed water from furnace steam drum as the cooling medium. The tertiary quench is a shell and tube transfer line exchanger (TLX-2) which utilizes cold boiler feed water as the cooling medium. The cracking effluent from the 64W furnaces is routed to the quench oil tower via a dry transfer line.

## Ethylene Recovery Section

Technip Energies implements several innovative designs in the ZPC ethylene plant recovery section. The features increase operational flexibility and reliability, and reduce the plant operation cost and/or capital cost. These features are described in the following sections.

**Low Temperature Quench System**

Technip Energies' pioneering high temperature quench system has been widely adopted in many of the Technip Energies plants in China and all over the world. The system is often referred to as the "quench oil viscosity control system" and includes a heavy fuel oil stripper to stabilize the quench oil viscosity and maintain quench oil quality, especially when the plants are designed for heavy feed cracking.

With high temperature quench systems (Quench Oil Tower bottom temperature greater than 180°C), polymerization in the quench oil is more aggressive and the rate of make-up and purge must be sufficient to maintain the quench oil quality and viscosity. In general, the heavier the feed, the more fuel oil produced. In the case of the ZPC feedstock, there is insufficient production of fuel oil with the light feeds such as Naphtha, Raffinate, C4/C5s, and ethane rich gas, and a significant amount of continuous flux oil import would have been required to allow for a good high temperature quench design. Therefore, a low temperature quench oil tower configuration is adopted.

The low temperature quench system includes the following components as opposed to a high temperature quench system. The low temperature quench tower configuration is shown in Figure 6.

a) 2-section Quench Oil (QO) Tower (quench, rectification)
b) Single loop heat recovery system
c) Single fuel oil stripper

In the low temperature quench design, the QO temperature from the QO tower is too low to be used for dilution steam generation. The heat in the QO is recovered by generating very low pressure steam and is also being used as heating medium in several process heaters in the feed/quench area. Although the design requires more medium pressure (MP) steam to produce dilution steam, there is more production of higher value super high pressure (SHP) steam in the furnaces. In addition, the import of flux oil is not necessary for the quench system operation because the quench oil quality is well maintained in the low temperature quench system. Overall plant reliability is improved and the operating cost is reduced.

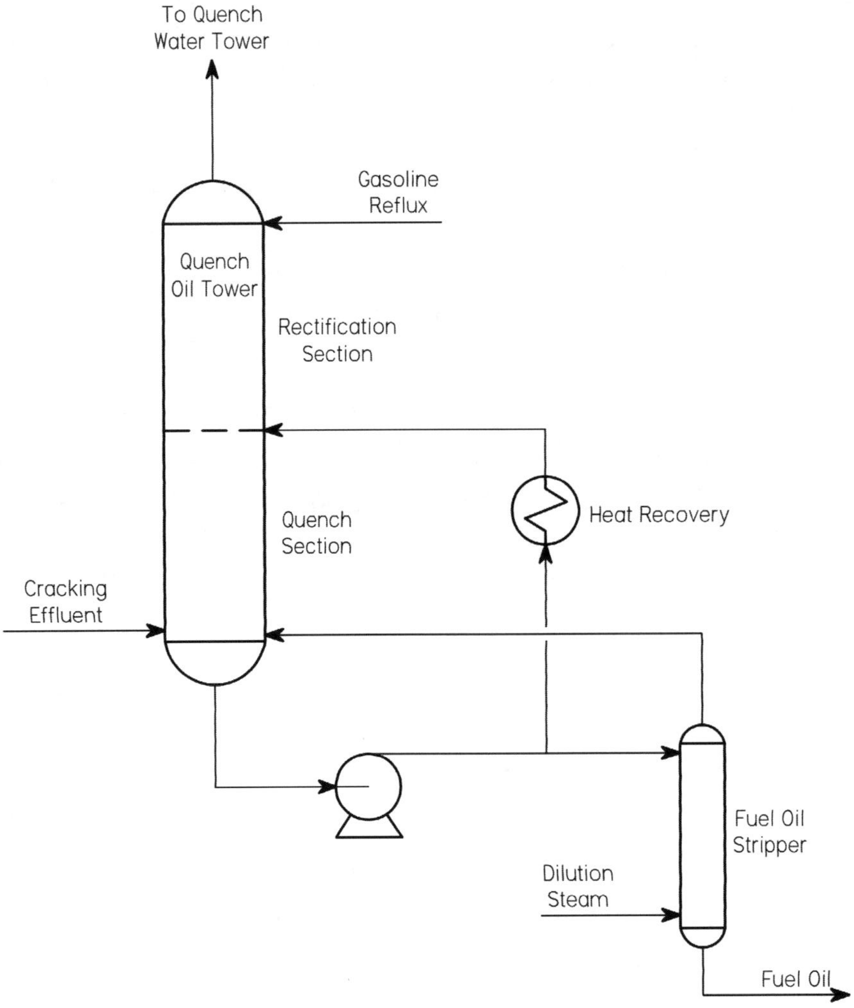


**Figure 6.** Low Temperature Quench Oil Tower Configuration

In summary, the benefits of the low temperature quench system for ZPC are:

1) Lower operating cost (no flux oil import during the normal operation and more SHP steam production)
2) Lower capital cost (no Heavy Fuel Oil Stripper and the associate equipment, savings on the furnace transfer lines)
3) Better overall plant reliability (improved performance of the quench system and the downstream pyrolysis gasoline (Pygas system)

**Maximization of Hydrogen Recovery**

Hydrogen concentrated residue gas is separated from the cracked gas as the overhead of the Technip Energies' patented HRS® (Heat Integrated Rectifier System) and goes on to pass through hydrogen purification and residue gas expansion. Normally the hydrogen produced from the hydrogen purification system is used internally in the ethylene complex for hydrogenation services such as MAPD (Methyl Acetylene/Propadiene) hydrogenation, Pygas hydrogenation, or

C4 hydrogenation. Hydrogen is also a high value by-product which can be sold to external users. However, extra production of hydrogen is at the cost of higher energy consumption within the ethylene plant. To maximize hydrogen production, the HRS® overhead will mostly be routed to the hydrogen purification system and the residue gas to expansion is reduced. As a result, ethylene and propylene refrigerant consumption has to be increased to meet the demand in the chilling train. In the case of ZPC ethylene plant design, an option of maximizing hydrogen production is provided so that ZPC can choose to produce extra hydrogen for export at the additional operating cost. The simplified hydrogen production scheme is provided in Figure 7.

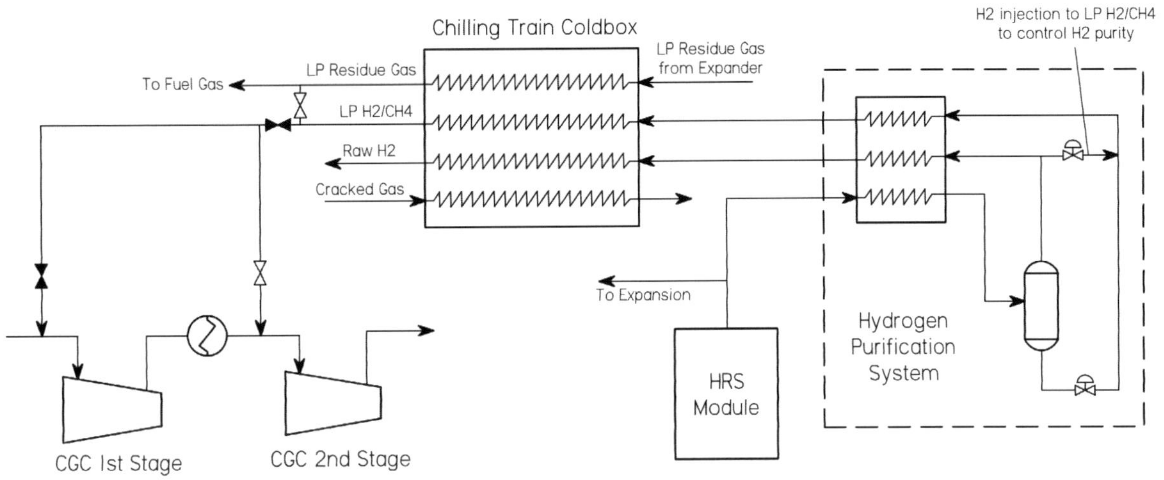


**Figure 7.** Hydrogen Recovery Scheme

A portion of the overhead of the HRS® module is routed to the hydrogen purification system to produce raw hydrogen sufficient for the hydrogenation services in the ethylene complex. The remaining overhead from the HRS® module is routed to the expander along with the demethanizer overhead to provide -140°C cooling to the chilling train cold box.

The hydrogen purification system is a two-stage, auto-refrigerated hydrogen recovery system inside the hydrogen cold box (Figure 7 only shows one stage for the sake of simplicity). The vapor from the HRS® module is sequentially cooled in hydrogen core exchangers number 1 and 2 and fed to the second hydrogen drum. Condensed liquid from the second hydrogen drum is flashed to low pressure to reach a very low temperature and provide the refrigeration to the hydrogen core exchangers. The vapor from the second drum is raw hydrogen. The purity of the hydrogen stream is controlled by the second hydrogen drum overhead temperature, which is achieved by controlling the amount of raw hydrogen injected into the LP H2/CH4 stream. Injecting raw hydrogen into the LP H2/CH4 stream reduces the methane/ethylene partial pressure, and thus achieves a lower temperature.

The normal route of LP H2/CH4 is to combine with LP residue gas after going through the chilling train cold box and go to the fuel gas header. The injected raw hydrogen goes to the fuel gas and cannot be recovered. In case hydrogen production needs to be increased, the LP H2/CH4 can be recycled back to the cracked gas compressor (CGC) to significantly reduce the raw hydrogen injection to the LP H2/CH4. The increased recovery of hydrogen in this manner is at the cost of the increased power of CGC compressor. In addition, more overhead vapor from the HRS® module can be routed to the hydrogen purification system to maximize the hydrogen recovery. This is at a price of more ethylene and propylene compressor power consumption which is required to compensate the refrigeration loss from the expander. With all of the hydrogen maximizing options implemented, more than 85 percent hydrogen recovery can be achieved.

**Dual Pressure C3 Splitter System**

Technip Energies' traditional C3 Splitter is a two-tower system (C3 Rectifier and C3 Stripper) in a series arrangement. The C3 Rectifier bottom is pumped and fed to the top tray of the C3 Stripper, and the overhead vapor of C3 Stripper is fed to the bottom of the C3 Rectifier. The reboilers of the two towers are heated with quench water (QW) in series to maximize the utilization of the QW heat.

However, in ZPC ethylene plant, there is a shortage of QW heat for C3 Splitter system reboilers due to very light mixed feed to the furnaces. If the traditional C3 Splitter design is used, there will not be enough quench water or high enough QW temperature available for a reasonable reboiler design. A modified scheme, dual pressure C3 Splitter System, is implemented in the ZPC plant design, as shown in Figure 8.

The major change of the modified system from the traditional C3 Splitter System is shown in red in Figure 8. Instead of routing the C3 Stripper overhead vapor to the C3 Rectifier bottom (as shown in the dashed line), the overhead of the C3 Splitter 2 is fully condensed against cooling water in the condenser. A portion of the condensed liquid is drawn as propylene product and the remaining is returned back to the tower as reflux. The feed from the C3 Splitter 1 bottom to the C3 Splitter 2 is not fed to the top tray, but to the middle of the tower.

The new scheme of the C3 Splitter System enables the C3 Splitter 2 to operate at a lower pressure compared with the traditional two-tower-in-series scheme. Therefore, the operating temperature of the tower is lower than a traditional C3 Stripper, which allows the QW at a lower temperature to be used as the heating medium of the reboiler.

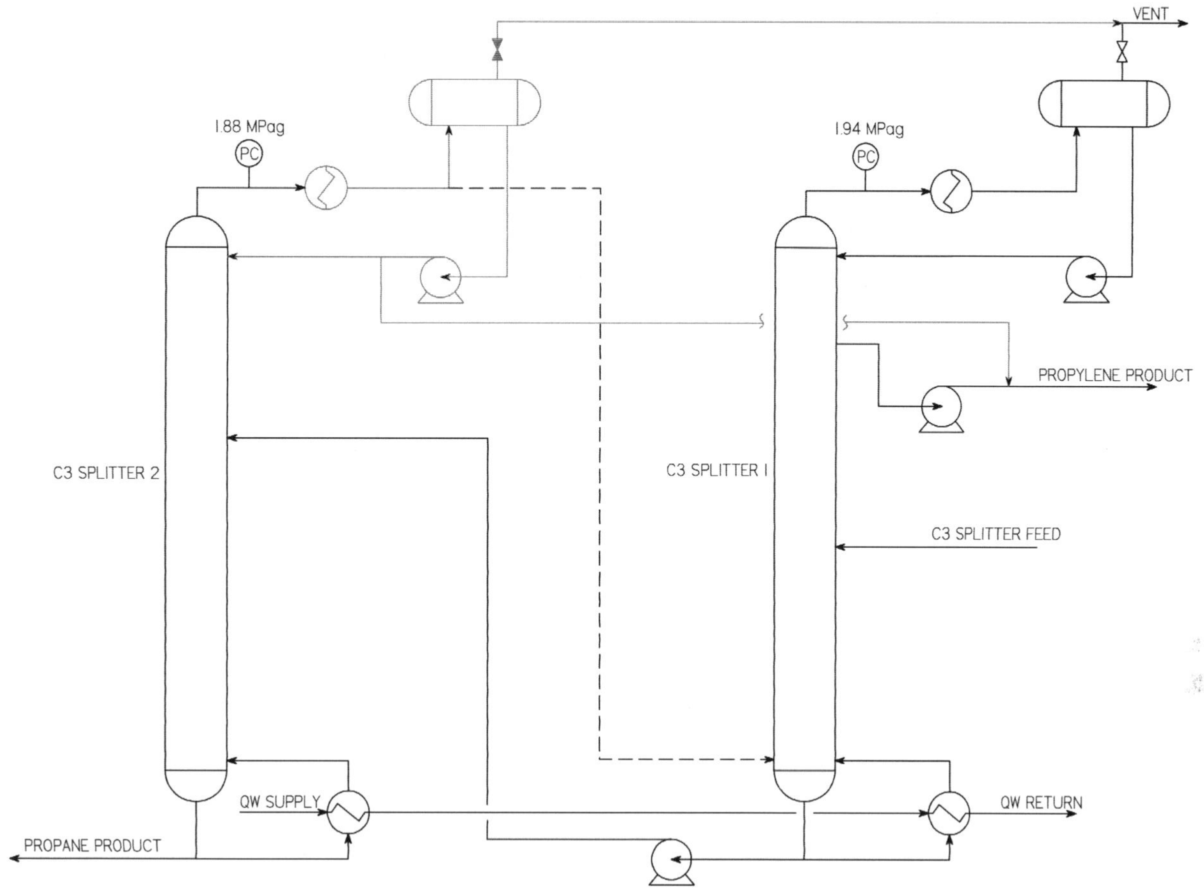


**Figure 8.** Dual Pressure C3 Splitter System

**Improvement of Ethylene Refrigeration System**

The ethylene refrigeration system of the ZPC ethylene plant is driven by a six-stage Ethylene Refrigerant/Heat Pump Compressor integrated with the C2 Splitter. The first three stages function as the compressor in the compression refrigeration cycle for the ethylene refrigerant (C2R) users at the different temperature levels. The third stage, along with the fourth stage, also function as the integrated heat pump compressor with the C2 Splitter to provide the reflux and the reboiling medium to the ethylene tower. The fifth and sixth stages are the product stages to send out the low pressure (LP) and the high pressure (HP) ethylene vapor products.

The following three ethylene product scenarios are considered in the plant design.

**Table 1.** Ethylene Product Scenarios for ZPC

| Phase | | Vapor (HP) | Vapor (LP) | Liquid | Cryo Liquid |
|---|---|---|---|---|---|
| **Temperature** | °C | 38 | 38 | -35 | -98 |
| **Pressure** | MPa(G) | 5.2 | 3.75 | 2.0 | 0.45 |
| **Flow (Normal)** | KTA | 300 | 1100 | - | - |
| **Flow (Alternate 1)** | KTA | 300 | 400 | 700 | - |
| **Flow (Alternate 2)** | KTA | - | - | - | 980 |

The Alternate 1 operation requires more propylene refrigerant (C3R) duty than normal operation to desuperheat and condense the ethylene vapor to produce ethylene liquid. The propylene refrigerant compressor is rated based on certain liquid ethylene product rate – 700 KTA or 50 percent of the plant load, so that there is no minimum flow kickback during the normal operation. Similarly, the cryogenic liquid product is limited to 980 KTA (70 percent of plant load) in the alternate scenario 2, where the -101°C C2R is used to produce cryogenic ethylene product, to avoid C2R compressor minimum flow recycle during normal operation.

The sketch of the C2R compressor discharge side is provided in Figure 9. The improvement implemented in the ZPC design is represented by the red lines and symbols added to the traditional system configuration. For the sake of simplicity, the control scheme is not shown in the sketch.

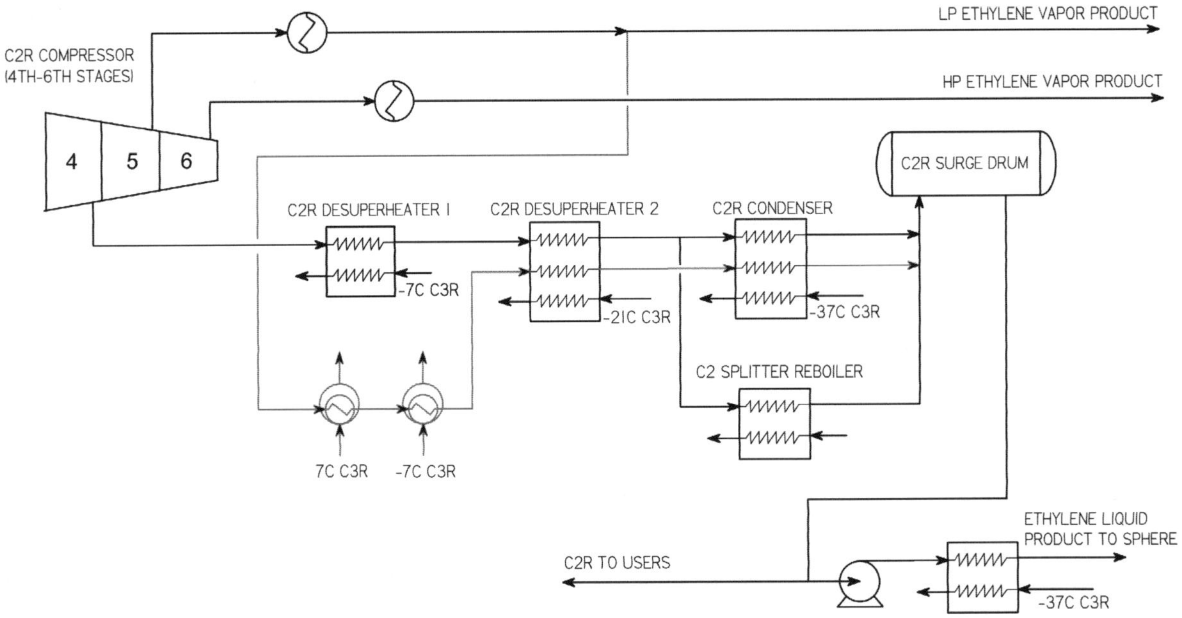


**Figure 9.** Ethylene Refrigerant Compressor Product Discharge

During the normal operation, there is only vapor ethylene product. If the downstream user(s) is/are not available, the extra LP vapor product needs to rundown to the storage sphere so that the downstream upset does not impact the ethylene plant operation. In the traditional scheme, the vapor ethylene product combines with the 4$^{th}$ stage discharge of the C2R compressor and is desuperheated and condensed in a series of plate-fin exchangers. It adds an extra duty requirement on the propylene refrigerant system, especially at the -37°C level, because the ethylene vapor is condensed at this temperature. The propylene refrigerant compressor is designed for the normal operation case in which there is no liquid ethylene product. The ethylene rundown case is rated for the C3R compressor, but the liquid ethylene rundown quantity has to be carefully selected so that the compressor is not in minimum recycle on any stage during the normal operation. With the improved ethylene refrigerant scheme, the duty at -37°C C3R is significantly reduced therefore the plant can achieve 70 percent liquid ethylene product and 30 percent vapor ethylene product at 100 percent plant load. It provides the plant more operational resilience during the downstream user upset, with the lower energy cost.

**Feed/Effluent Exchanger in Dryer Regen System**

Residue gas from the cold recovery section is used as the regeneration gas for the ethylene plant dryers, before routing to the fuel gas system. The residue gas is around 65°C - 70°C. A portion of it is cooled against cooling water and the balance is heated to 240°C. The cold and hot regeneration gases are mixed in a certain ratio based on the regeneration procedure and the temperature requirement. The regeneration gas feed/effluent exchanger (red in Figure 10) is used to significantly reduce the high pressure steam and cooling water consumption. Though this concept is used in the design of other Technip Energies licensed plants, this is the first time it is implemented in China, in the ZPC ethylene plant. Considering the mega capacity of the ZPC plant, the reduction of utility demand and energy saving is significant.

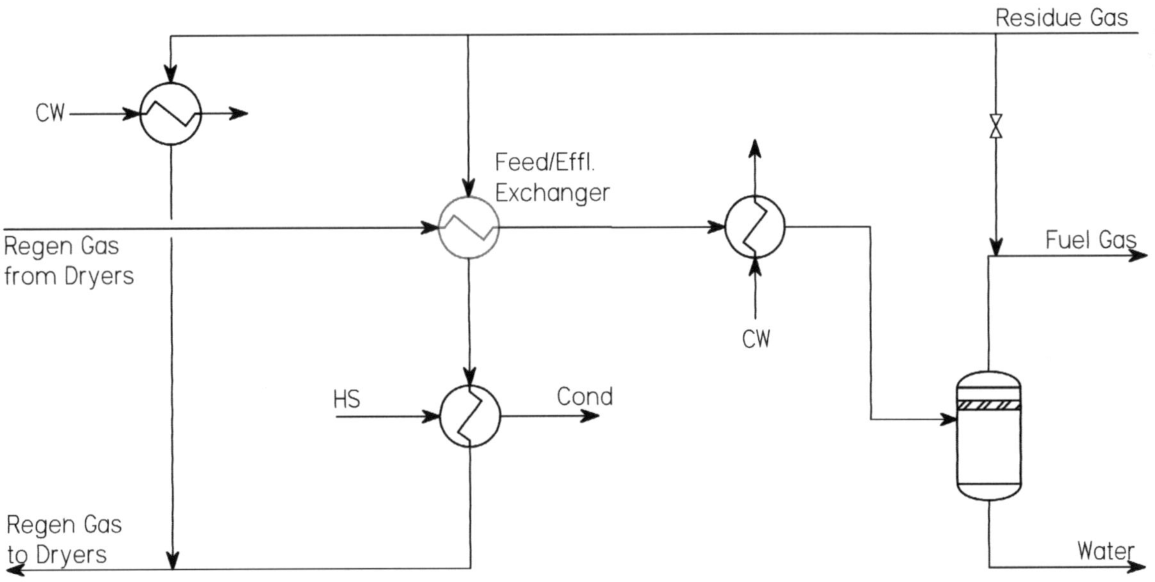


**Figure 10.** Regeneration Gas System

## Ethylene Plant Startup

The ZPC PDP contract was awarded in June 2016. The PDP design was finished in about five months. In December 2019, the plant was commissioned and started up. It was a really fast track project.

**Construction, Pre-commissioning and Commissioning**

Considering the poor infrastructure on the island, the ZPC project team decided to maximize the modularization as much as possible. This includes all nine cracking furnaces and the major towers.

Modularization helps to reduce the manpower need on the island and accelerates the schedule. For example, all nine furnaces were assembled and shipped to site for additional piping connections. This is the largest Technip Energies' cracking furnace being fully modularized and shipped to site.

After the mechanical completion, the ZPC team started the furnace dry-out one by one. Test runs were conducted on all major rotating equipment. During the pre-commissioning work, it was found that the quench water pumps vibrated when the speed was ramped up. Noise could be heard from the pumps as well. It was believed the quench water pump cavitation occurred. Pump data sheets and piping layout were checked. Pump inlet piping was modified to allow a longer straight run before entering the pump. The piping change minimized the flow turbulence and solved the pump cavitation problem. No other major issues were observed during the pre-commissioning.

**Startup**

In December 2019, the ZPC team was ready to start up the ethylene plant. The major milestones are listed below.

- December 17, 2019
  - LPG was brought into the fuel gas system.
  - Furnaces F-1110 and F-1190 pilots were lit to raise the temperature.
  - Flux oil was introduced into the quench oil tower.
- December 18, 2019
  - Boiler feed water was introduced in the F-1110.
  - Nitrogen was introduced to pressurize the quench water tower to 0.3 MPag to prepare for cracked gas compressor (CGC) startup.
- December 22, 2019
  - The CGC was started up with nitrogen.
  - More furnaces were brought up to hot steam standby conditions.
  - Started exporting super high-pressure steam to the steam header.
- December 24, 2019
  - Propylene was inventoried in the propylene refrigeration system and the propylene compressor was started up.
  - Started chilling of cold fractionation section with nitrogen recycle.
  - Established total reflux operation in the propylene tower.
- December 28, 2019
  - Ethylene was inventoried in ethylene refrigeration system and ethylene compressor was started up.
  - The cold box temperature was chilled to -120°C.
  - Introduced ethane rich gas to the F-1190 to purge the nitrogen out of the system.
  - Established total reflux operation in the ethylene tower.

- December 29, 2019
  - Started feeding furnaces one-by-one. F-1130, F-1140 and F-1160 were fed in with light liquid and F-1110 was fed in hydrogenated tail oil.
  - After feed was established to four liquid furnaces, the ethane feed to furnace F-1190 was stopped, and the furnace was decoked, in preparation to receive recycle ethane.
  - Started feeding the acetylene reactor and raising reactor temperature after the butadiene was on spec at the overhead of the HP Depropanizer.
- December 30, 2019
  - While waiting for the acetylene reactor being on spec, started forwarding the cracked gas to the cold fractionation. Recovered the off-spec C2s from the Deethanizer overhead to the off-spec storage.
  - The acetylene was on spec at the reactor outlet after three and a half hours. (It took only 19 hours from the furnace feed in for the acetylene reactor to achieve on spec.)
  - Started feeding the ethylene tower when its feed was on spec. It was only 23 hours from furnace feed in for the ethylene product to be on spec.
- January 4, 2020
  - Propylene product was on spec.

The entire startup procedure minimized flaring. There is only limited flaring during the following short periods of time:

- Between HP Depropanizer was fed with cracked gas and its overhead butadiene was on spec
- When C2 splitter and C3 splitter were establishing total reflux

**Troubleshooting During Startup**

Before the furnace feed in, some issues were identified in different sections of the plant and resolved by the startup team. During the furnace warm up to hot steam standby, hot bolting was conducted. Some leaks were found on phosphate injection flanges and steam drum man holes. Some plugging problems were found at the furnace attemperators. Leaks were identified on the flanges around the quench fitting quench oil circulation loop. When pre-sulfiding was conducted before the furnace feed in, the DMDS injection was overdosed, which increased the caustic tower loading. Boiler feed water from the deaerator contained too much oxygen at the beginning of the startup, which was caused by low stripping steam due to the steam control valve problem.

The startup propylene contained some water, which set back the propylene compressor startup. There was a delay on the ethylene compressor startup also, due to the startup ethylene quality. Methanol was injected at the locations where plugging was observed and the low point drains were opened periodically to purge it out of the system.

Startup naphtha inventoried in the quench water tower was too light. Some low boiling point components were carried over to the high pressure depropanizer. There was too much startup flux oil inventoried in the quench oil tower. The quench oil tower liquid level was high. The quench oil heating capacity was limited and the quench oil tower temperature was around 90°C.

After the furnace feed in, the quench oil tower bottom temperature was about 90°C at the beginning. Quench oil contained water and water flashing caused cavitation in the quench oil pumps. Only one quench oil pump was running to maintain a high liquid level in the quench oil tower. Gasoline product was high, due to the light naphtha feed. The quench water bottom gasoline chamber liquid level was high until more Pygas product was transferred to the OSBL.

The CGC fourth stage flow rate was higher than design. It was identified that there was flow passing through the anti-surge valve when it was in close position. The same was observed on the ethylene compressor anti-surge valves. When the compressors were running stable and the anti-surge valves were in 100 percent close position, the upstream and downstream block valves were closed and the anti-surge valves were taken out of service. Welding rods and piping debris were found in the valves. The valves were cleaned and put back into service. The propylene compressor fourth stage flow rate was higher than design. It was identified not an issue because during the startup, the ethylene vapor product was not exported to the OSBL users. There was about 50 t/h liquid ethylene product to storage which added to the propylene refrigeration consumption.

During the startup, it was difficult to achieve HP depropanizer overhead butadiene spec initially because the liquid feed to the tower was low. The operation team added liquid propylene to the tower to help establish the vapor liquid equilibrium, which accelerated the process to bring the tower overhead butadiene on spec. Leaks were identified at the demethanizer overhead man hole. The system pressure was lowered in the demethanizer prefractionation tower, the demethanizer tower, and expander inlet until the leaks were fixed. The expander shaft temperature was high at 8500-10000 rpm. The vendor was contacted to resolve the issue. Some filters in the cold section showed high differential pressure. Operators had to monitor them and changed out the filters frequently. Some rust was found on the filters.

Water hammering occurred at the propylene tower steam reboilers. Drains were opened to mitigate the water hammering. Condensate header pressure was built up with makeup nitrogen. Because the cracking feed contained some MTBE, methanol was generated in the cracking furnaces and carried over to the propylene tower. Operators had to draw more propane product from the tower to control the overhead propylene product methanol content.

ZPC has a very experienced operation team. Most startup issues were quickly resolved. After the startup, the plant was stabilized to check out all of the operation issues, and then the plant capacity was gradually increased to 100 percent within 4 weeks of startup.

## Conclusions

The ZPC Phase 1 ethylene plant integrated all of the advanced modern mega cracker design features. The plant was designed with high operation reliability and flexibility. Nitrogen was used to establish the circulation in the plant before feed in. This helped the plant to achieve quick chill down and minimize the flaring during the startup. The ZPC project was a fast-track mega cracker project and the startup was very speedy and successful.

Paper Number: 23C

April 20, 2021

# ZPC 1400KTA Ethylene Plant Startup

**Yong Wang, Rui Chen, Astrid Han, Joel Guillaume**
***Technip Energies***



# ZPC General Information

1. **Zhejiang Petroleum Chemical Co., Ltd. (ZPC) is China's first and largest private led joint venture in energy sector**
2. **Located at the Xiaoyushan island, Zhoushan City, Zhejiang, China**
3. **Several mega-scale grassroots integrated refinery and petrochemical complexes on site**
4. **One of the world largest petrochemical bases**
5. **Technip Energies provided proprietary ethylene technology and process design**
6. **Phase 1 has 1400 KTA ethylene capacity with mixed feeds from heavy liquids to ethane vapor**
7. **42 months from start of PDP design to plant startup**
8. **On spec. ethylene product within 23 hours after feed in**

2

## Overlook of ZPC Phase 1 Ethylene Unit

3

# 01 Ethylene Cracking Furnaces

4

# Cracking Furnaces

1. Technip Energies Ultra Selective Conversion (USC®) U-coil and W-coil furnaces
2. Fresh Feeds from refinery include HTO, naphtha, raffinate, NC5, mixed C4s and mixed C2s
3. U-coil – short residence time for liquid feeds to gain high yields
4. W-coil – vapor feeds with high loading capacity
5. Seven 224U furnaces and two 64W furnaces (largest U-coil and W-coil furnaces in operation)
6. Twin cell furnaces with common convection section
7. High thermal efficiency around 95%

5

# Cracking Furnaces

USC® U-coil

USC® W-coil

6

# Cracking Furnaces

1. F-1110/1120 heavy liquid 224U furnaces
2. F-1130 to F-1170 light liquid 224U furnaces
3. F-1180/1190 vapor feed 64W furnaces
4. Sixty-four ultra low NOx burners on each furnace
5. SCR bed installed in the convection section – first SCR installation on ethylene cracking furnace in China
6. Combustion air preheater using QW installed to reduce fuel gas consumption – typically one preheater per burner
7. TLX-D quench exchanger – steam boiler when liquid feed to furnace; BFW heater when vapor feed to furnace

8

## Cracking Furnaces

Combustion Air Preheater

TLX-D Transfer Line Exchanger

9

# 02 Ethylene Recovery Section

10

# Recovery Design Features

1. First Low Temperature Quench Design in China
2. Provision to Maximize H2 Production
3. Dual Pressure Two-Tower C3 Splitter System
4. Innovative C2R System Design to Achieve High Rundown Rate
5. First Installation of Feed/Effluent Exchanger in Dryer Regen System in China

TEN TECHNIP ENERGIES

11

# Low Temperature Quench Design

- Light Feeds – Naphtha, Raffinate, C4/C5s, Ethane Gas
- Fuel Oil Production in Furnace Effluent not Adequate
- Low Temperature Quench
  - 2-section QO Tower (Quench, Rectification)
  - Single loop heat recovery system (LPS level)
  - Fuel Oil Stripper
- No Continuous Flux Oil Import during Normal Operation
- More SHP Steam Production from Furnaces
- Lower Capital Cost
- Better Overall Plant Reliability

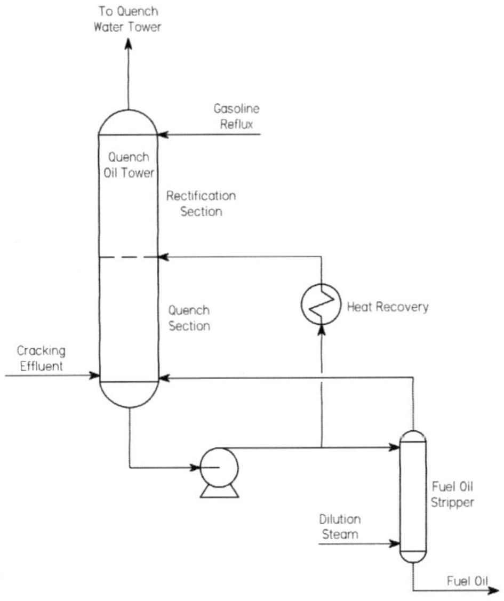


TEN TECHNIP ENERGIES

12

# Maximization Hydrogen Recovery

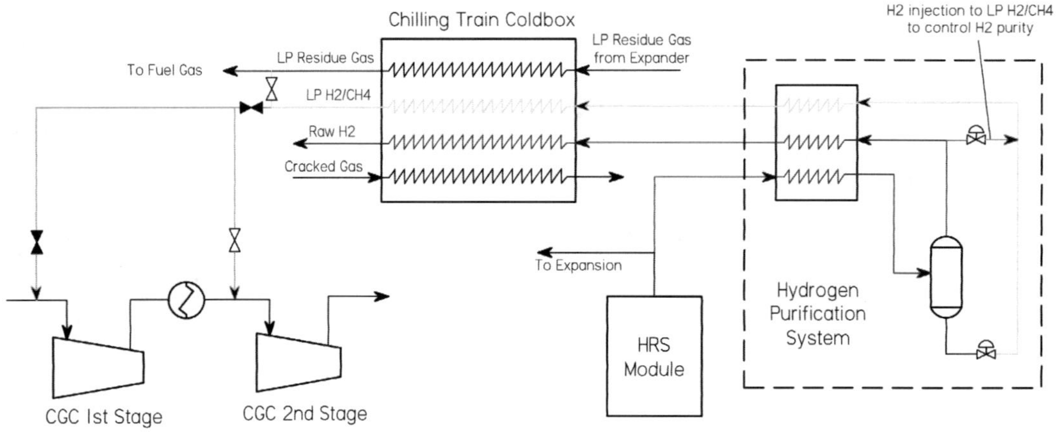


13

# Maximization of Hydrogen Recovery

- Hydrogen is produced for internal hydrogenation services
- Hydrogen can be exported as high value product
- Extra production of hydrogen is at the cost of additional energy consumption
- Option of maximizing hydrogen recovery was provided in the design for ZPC
- Hydrogen recovery > 85%

14

# Dual Pressure C3 Splitter System

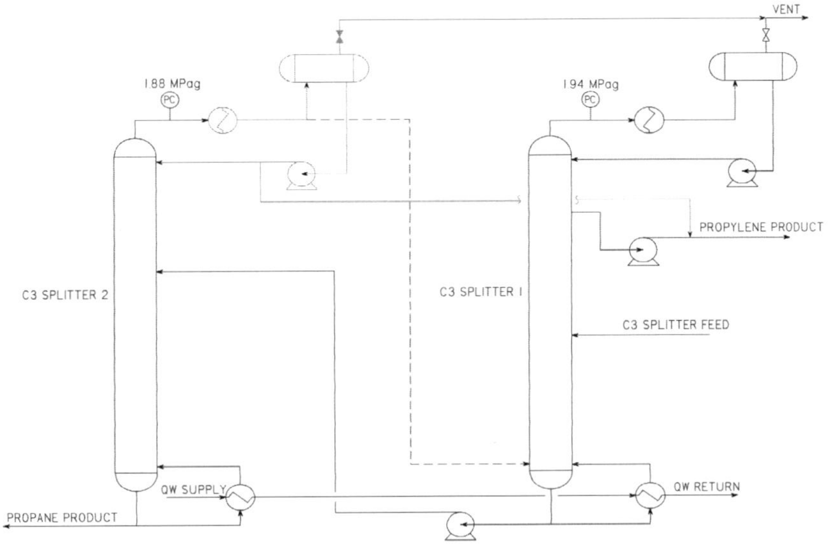


15

# Dual Pressure C3 Splitter System

*Light feeds result in the shortage of QW heat for C3 Splitter reboilers*

- Traditional C3 Splitter System:
  - Two towers in series
  - Pressure profile allow vapor from stripper overhead to rectifier bottom
  - QW heats the Stripper and Rectifier reboilers in sequence
- Modified C3 Splitter System:
  - Two towers are fed in series / overhead product draw in parallel
  - Independent pressure profile for each tower
  - Lower tower bottom temperature for splitter 2 compared with traditional stripper
  - Additional equipment

16

# Improvement of Ethylene Refrigerant System

- Six-stage ethylene refrigerant compressor
  - 1st – 3rd stages as user stages
  - 3rd and 4th stages as heat pump compressor integrated with C2 Splitter
  - 5th and 6th stages as product stages to produce LP and HP vapor product
- Ethylene product scenario of ZPC

| Phase | | Vapor (HP) | Vapor (LP) | Liquid | Cryo Liquid |
|---|---|---|---|---|---|
| Temperature | °C | 38 | 38 | -35 | -98 |
| Pressure | MPa(G) | 5.2 | 3.75 | 2.0 | 0.45 |
| Flow (Normal) | KTA | 300 | 1100 | - | - |
| Flow (Alternate 1) | KTA | 300 | 400 | 700 | - |
| Flow (Alternate 2) | KTA | - | - | - | 980 |

17

# Improvement of Ethylene Refrigerant System

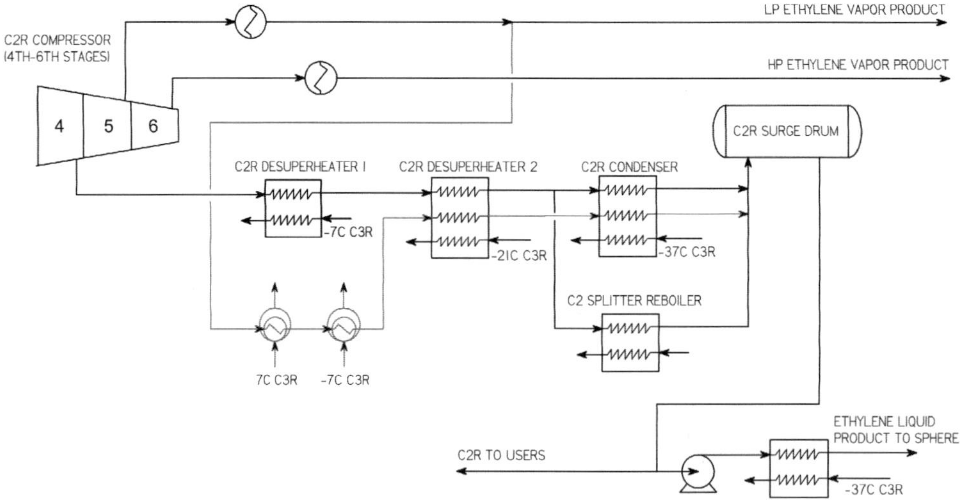


18

# Improvement of Ethylene Refrigerant System

*Increase ability of ethylene liquid rundown*

- Traditional C2R system:
  - Rundown goes through C2R desuperheaters and condenser
  - Rundown capacity up to 50%
- Improved C2R system:
  - Rundown goes through new S&T desuperheater and condenser
  - Rundown capacity up to 70%
  - Provides more operational resilience during the downstream user upset

19

# Feed/Effluent Exchanger in Dryer Regen System

- Feed/Effluent exchanger significantly reduces the HPS and CW consumption
- First time implemented in China

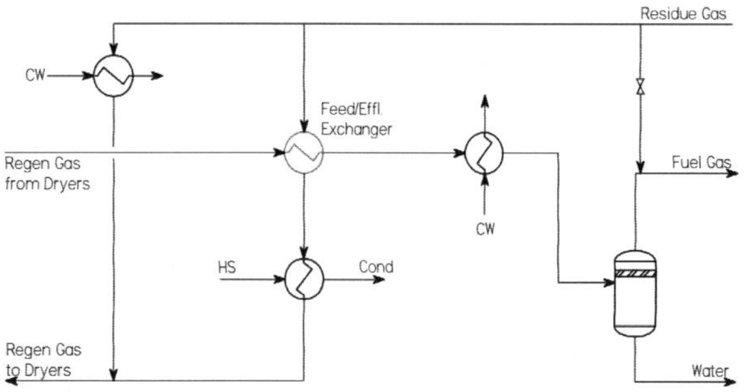


20

# 03 Ethylene Plant Startup

TECHNIP ENERGIES

21

# Pre-Startup

- Contract awarded June 2016, PDP finished Dec. 2016, Startup Dec. 2019
- Fully modularized furnaces and major towers shipped to site
- Furnace dryout and equipment test run
- QW pump vibration during precommissioning and inlet piping was modified

TECHNIP ENERGIES

22

## Pre-Startup (Furnace Modularization)

TEN TECHNIP ENERGIES

23

## Pre-Startup (Furnace Modularization)

TEN TECHNIP ENERGIES

24

## Pre-Startup (Furnace Modularization)

TECHNIP ENERGIES

25

## Startup

- **Dec. 17, 2019**
  - LPG brought to fuel gas system
  - Furnaces pilots lit to raise temperature
  - Flux oil introduced to quench oil tower
- **Dec. 18, 2019**
  - BFW introduced in F-1110
  - Nitrogen introduced to pressurize the QWT to 0.3 MPaG
- **Dec. 22, 2019**
  - Cracked Gas Compressor started up with nitrogen
  - More furnaces on HSSB and started exporting SHP steam to header
- **Dec. 24, 2019**
  - Propylene compressor started up
  - Started chilling of cold section with nitrogen circulation
  - Established total reflux in C3 splitter

TECHNIP ENERGIES

26

# Startup

- **Dec. 28, 2019**
  - Ethylene compressor started up
  - Cold box chilled to -120°C
  - Ethane rich gas introduced from F-1190 to purge nitrogen out of system
  - Established total reflux in C2 splitter
- **Dec. 29, 2019**
  - Started furnace feed in one-by-one
  - Butadiene on spec at HP Depropanizer overhead
- **Dec. 30, 2019**
  - Acetylene reator on spec.
  - Ethylene product on spec.
- **Jan 4, 2020**
  - Propylene product on spec.

TEN TECHNIP ENERGIES

27

# Startup Troubleshooting

- **Furnaces**
  - BFW quality poor, such as high O2 content
  - Leaks on steam system, such as phosphate injection flanges and steam drum man holes
  - Plugging at furnace attemperators
  - Leaks on the flanges around the quench fitting quench oil circulation loop
  - DMDS overdosing during furnace presulfiding
- **Quench**
  - Startup flux oil over inventoried in QOT
  - Startup napththa is too light
  - Quench oil heating limited – QOT bottome temperature 90°C before feed in
  - Quench oil contained water and caused cavitation in the QO pumps

TEN TECHNIP ENERGIES

28

# Startup Troubleshooting

- **Compressors**
  - Startup propylene and ethylene containing water, and methanol injected to treat plugging
  - C3R compressor 4th stage flow rate high – no ethylene vapor product export leading to high C3R consumption to condense ethylene
  - CGC and ethylene compressor anti-surge valves passing when completely closed
  - Expander shaft temperature high
- **Towers**
  - Leaks on Demethanizer overhead man hole
  - Difficult to achieve HP Depropanizer overhead butadiene spec with startup feeds – liquid propylene added
  - High differential pressure on cold section filters – frequent changeout and cleaning
  - Water hammering on C3 Splitter startup reboiler – blowdown from drains; condensate header pressure buildup with nitrogen injection
  - Methanol high in propylene product due to MTBE in cracking feeds – draw more propane from C3 Splitter bottom

29

# Conclusions

1. **1400KTA Ethylene Plant Integrated with Advanced Process Design**
2. **Fast Track Project with Maximized Modularization**
3. **Speedy and Successful Startup**
4. **Minimized Flaring during Startup with Nitrogen Circulation Established in the System**

30

# Thank you

For more information
Contact

rui.chen@technipenergies.com
yong.wang@technipenergies.com

## AICHE PAPER NUMBER: 23b

## INDORAMA VENTURES OLEFINS LLC

# CASE STUDY: CONVERTING A CRACKER IDLED FOR 15+ YEARS INTO A PROFITABLE PROCESS UNIT

*Paul Thomas Scott*
*Chief Technical Advisor – Process*
*KBR*

*Terry Gordy*
*Commissioning Manager*
*Indorama Ventures Olefins, LLC*

Prepared for Presentation at the 2021 Spring National Meeting
Dallas, TX, April 18-23, 2021

**INDORAMA VENTURES OLEFINS LLC**

# CASE STUDY: CONVERTING A CRACKER IDLED FOR 15+ YEARS INTO A PROFITABLE PROCESS UNIT

*Paul Thomas Scott*
*Chief Technical Advisor – Process*
*KBR*

*Terry Gordy*
*Commissioning Manager*
*Indorama Ventures Olefins, LLC*

This undertaking presented many challenges, and these will be summarized in this paper including the following:

1. Understanding the business strategy – IVOL goal was to prepare the operational and support staff for the start-up of the dormant unit.
2. Evaluating operational strategy – IVOL reached out to KBR for support on evaluating the start-up plan, including co-design options for plant improvements and training of operations personnel.
3. Delivering and implementing – KBR provided basic and detailed design for several parts of the plant. In addition, IVOL and KBR worked together evaluating and delivering the operational feasibility, pre-commissioning and start-up plans to put in place the necessary implementation steps to ensure a valuable and sustainable effort and practices for IVOL.
4. Results - The 440,000-tonne/year cracker entered commercial operation on Jan. 31, 2020. As of May 2020, the restarted plant already has achieved its design operating rate more than 100%, said Tony Barre, IVOL's site director.

Authors:
Alex Michinel – IVOL, Technical Director
Terry Gordy – IVOL , Commissioning Manager
Darryl Helton – IVOL, Operations Manager
Holt Lacy – IVOL, Production Specialist
Paul T. Scott – KBR, Chief Technical Advisor
Jim Walker – Retired, Independent Consultant
Umesh Shelke- Ingenero, Project Manager

## Introduction

The subject of this paper is the restart project for the former Oxychem Olefins unit in Lake Charles, Louisiana. The plant was purchased by, and is part of, IOD., Integrated Oxides and Derivatives, a wholly owned subsidiary of Indorama Ventures, LLC. The parent company Indorama Ventures is headquartered in Bangkok, Thailand, and includes operating facilities at 123 sites worldwide, and over 30,000 employees. IOD manages operations at a number of facilities in the USA including some recent acquisitions that were part of the former Huntsman Chemicals facility in Port Neches, Texas.

At the inception of this project, Indorama had one operating olefins unit in Nigeria, but did not have any olefins production in the USA. The primary driver for this project was to acquire some captive ethylene production to feed existing downstream processing units on the US Gulf Coast (vertical integration) and to take advantage of the SHALE gas revolution in the USA (lower cost ethane feedstock).

The decision to refurbish a mothballed unit as opposed to new grassroots construction was one of both schedule and cost, but primarily schedule. There were other competing ethylene projects already underway in the Lake Charles area and elsewhere on the US Gulf Coast region. IOD wanted a faster track to first product in order take maximum advantage of favorable economics.

In addition to the Olefins unit in Lake Charles, IOD is one of the largest producers of ethylene oxide and ethylene glycol in the USA and was also executing other grassroots construction projects worldwide at the same time as this project. This was a stress on company resources, but part of an overall plan for rapid growth of the company.

## Process Flow Scheme

The plant as originally conceived was a Butane-only feed cracker dating back to 1969. The process design was provided by C.F. Braun and Co. Seven cracking furnaces were provided by SELAS and the plant nameplate capacity was 500 MMlbs/year ethylene with substantial propylene production owing to the butane base feedstock.

The plant has been revamped a number of times with major revamps in 1985 and 1989. The 1985 revamp converted the feedstock from butane to a mix of ethane and propane and improved the energy efficiency of the unit; particularly in the furnace area where a common economizer and feed gas saturation were added. The 1989 revamp expanded the plant capacity further to a nominal capacity of 880 MMlb/year ethylene with some unique combinations of series and parallel processing units.

Indorama Ventures Olefins LLC (IVOL), now a division of IOD acquired the property in 2015 and the unit underwent a yet third major revamp with the objective of converting over to a 100% ethane feedstock base case while still retaining the same nameplate ethylene capacity.

C.F. Braun Co. was later purchased by Brown & Root, INC. which later merged with the M.W. Kellogg Co. in 1998 to form Kellogg Brown & Root (KBR). So this is a KBR legacy design for which KBR, with some modifications, is still marketing today for modern grassroot projects in Olefins manufacture. Needless to say, the concepts developed long ago by the C. F. Braun engineers were ahead of their time.

**Figure 1: Process Flow Scheme**

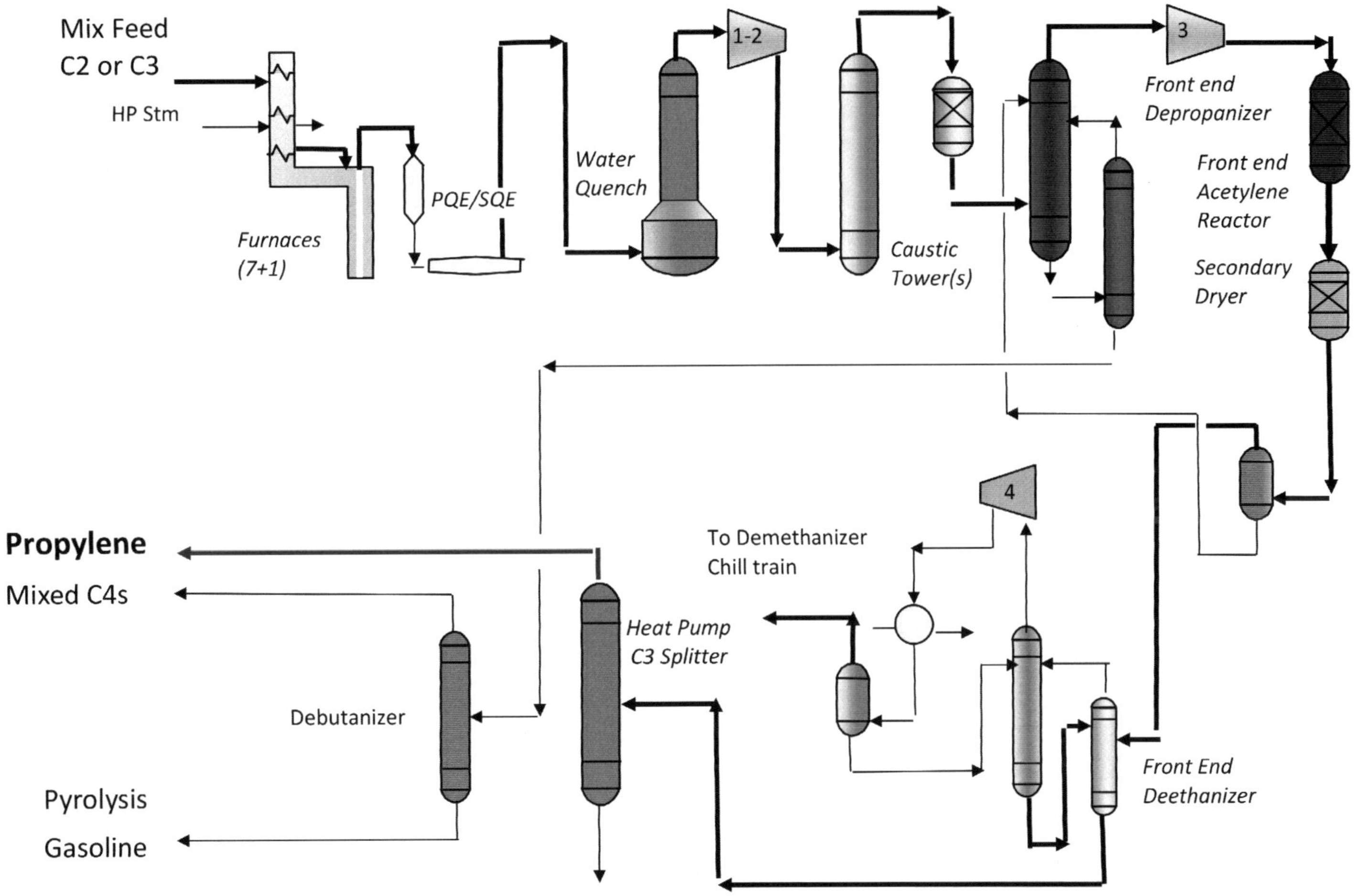

**Figure 1 cont.: Process Flow Scheme**

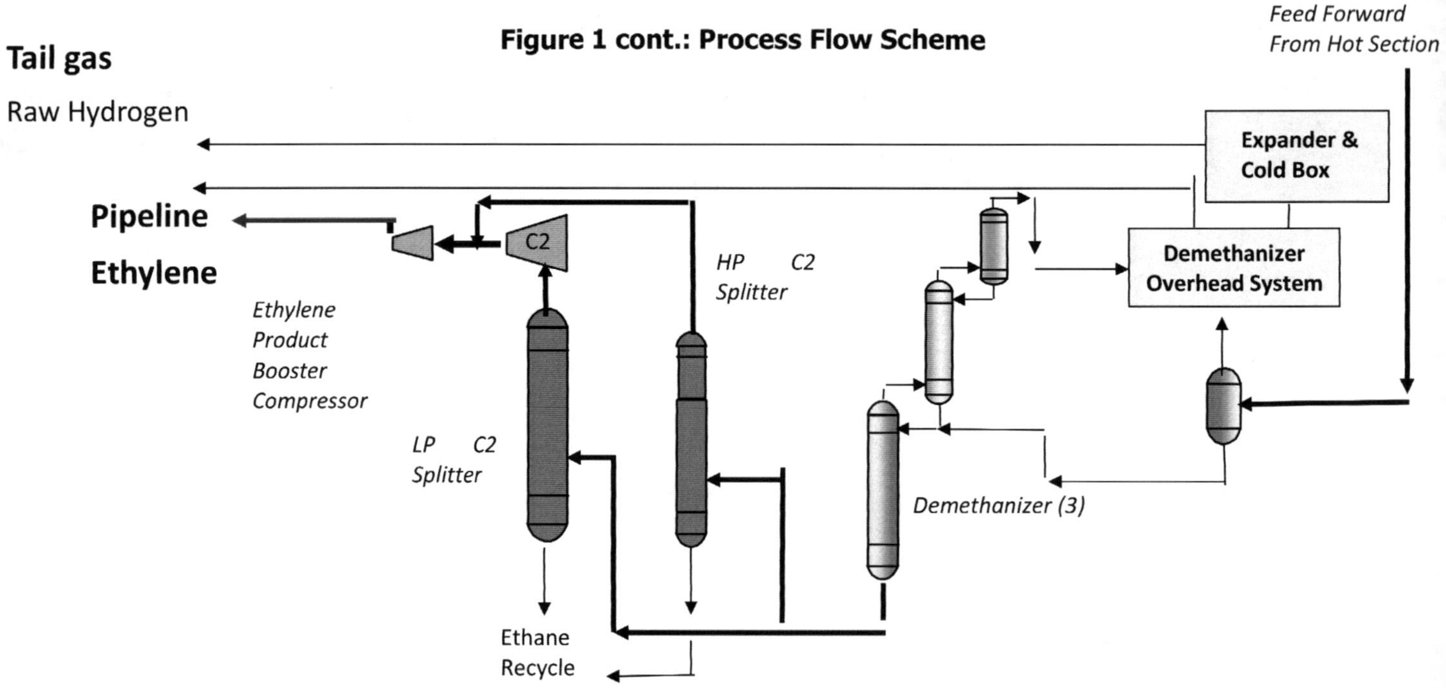

## Current Operation

The plant was considered commercialized on January 31, 2020, with the completion of a successful run of 30+ days of on specification ethylene production including 72 hours of continuous operation at 100% of design feedstock. The ability of the unit to operate on 100% ethane fresh feed has also been demonstrated. The energy efficiency of the unit is currently under evaluation.

## Project Scope

The project scope included the refurbishment and restart of the existing mothballed Oxychem Olefins unit which was decommissioned in 2001. Adding to the herculean effort required to restart a unit so long in mothballs, the project design basis was to convert the operation over to 100% ethane feed cracking to take maximum advantage of favorable economics. The unit was originally a 100% butane feed cracker so this was quite a move away from the original design basis. In addition, the feedstock source was changed from low pressure ethane vapor to dense phase ethane so Indorama wanted an efficient heat integration for that new feed stream at the different conditions.

The project included application for all new environmental permits, both air and water, and modifications to bring the plant up to current standards with regards to emissions. The previous air and water permits had been left to expire by the previous owner so there was no chance to be "grandfathered" under the previous regulations.

As you might imagine there was a primary construction contractor and a number of sub-contractors to handle the immense scope of the project and the refurbishment. A partial listing is given here with their general responsibilities.

KBR- Original flowsheet technology; initial feasibility study, H&M balances, specify process modifications for the all ethane case operation, perform basic / detailed engineering for select process areas and new equipment items, consultation services related to permit applications and startup activities
Ingenero- Detailed engineering contractor
Performance Contractors, Inc.- General Construction Contractor
Ethos Energy- Revamp of major compressors/ turbines
Apache- Insulation, scaffolding
MMR- Electrical, Instruments
Excel- Piping
Greenberry- Piping
PTS- Process Technical Services, commissioning, startup operators
SCR- Strategic Contract Resources, startup operators and specialists
SMWP- Specialty Manpower Works Projects, Pre-commissioning
Keystone - Railway siding, truck unloading

FTS- Furnace Tube Services, Cracking furnaces inspection and rebuild
CPL- Fired Boilers rebuild
Marley- New Cooling Tower
Lummus/ABB- New DCS control systems
Suez - RO unit (one new train added, refurbish existing train)
LINDE- A wet -air oxidation unit was added to treat the spent caustic stream for environmental compliance

IVOL owes debt of gratitude to all the contractors and subcontractors involved in the project, without their expertise and dedication, the project would not have been possible. Overall Construction Management was by IVOL.

**Project Execution Strategy:**

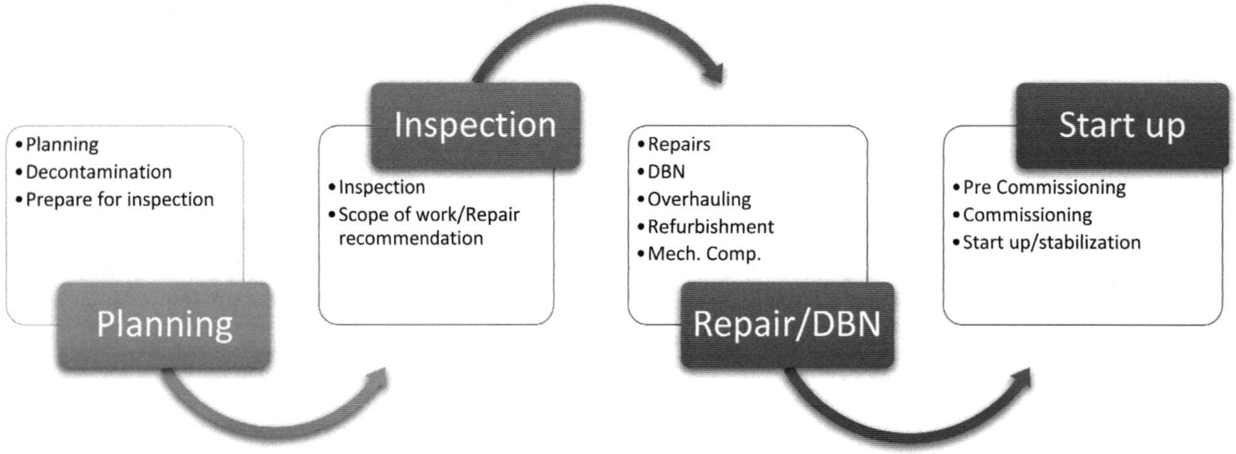


**Project Scope:**

- Repairs, Replace and DBN + Additions due to Regulatory Compliance

| Equipment | Scope | Repairs | Replace (after Inspection) | DBN (Rep+New) | Type of Repairs |
|---|---|---|---|---|---|
| Drums | 104 | 87 | 3 | 8+6 | Code, major & minor repairs, replacement on account of inspection or DBN |
| Exchanger | 135 | 97 | 12 | 19+7 | |
| Column | 18 | 18 | 0 | 0 | Code & Nozzle Repairs, Tray replacement; structure |
| Tanks | 25 | 21 | 3 | 1 | Demo, New tanks, Repairs |
| Piping | 3000 (Pkg.) | 2100 | - | (65,000 ID) | Demo, Repair, Replacement, new piping (repair 110,000 ID including small bore) |
| Cold Box | 2 | 2 | - | 0 | Core and piping change |
| Furnace (Convection Module, TLEs) | 8 | 1 | 7 | | Replacement of convection section with refractory |

| | | | | | |
|---|---|---|---|---|---|
| Cooling Tower | 7 cells | 0 | 7 | | Demo and erection of new CT |
| Pumps | 147 | 117 | 10 | 6+14 | Repair and replacement |
| Motors | 223 | 86 | 88 | 12+37 | Replace based on cost advantage to repair |
| Instruments | 900 | 500 | 300 | +100 | New, replacement of outdated |
| Control Valves& PSV | 1085 | 500 | 315 | 140+80 | |
| Electrical (Substation) | 10 | 10 | - | 0 | Panels retrofit, replacement |

## Project Schedule / Major Milestones

The major milestones achieved for the project were as follows:

**Project Schedule / Major Milestones**

| Key Activity | | Date |
|---|---|---|
| Acquisition | : | 23 Sep'15 |
| Team Mobilization | : | 20 Feb'16 |
| Plant Inspection Contract award | : | Mar'16 |
| Air Permit received | : | Aug'16 |
| Mech. Repair Contract award | : | Nov'16 |
| E&I Repair Contract award | : | Dec'16 |
| Mechanical Completion | : | Jan '18 (with exceptions of non-critical systems) |
| Pre-commissioning | : | June'18 (with exceptions of non-critical systems) |
| PSM / PSSR / Final Hazop | : | Nov'18 |
| Plant Startup/ First Product | : | Apr '19 |
| On Spec Full Production and Guarantee Test Run | : | Jan'20 |

## Major Challenges

What were the major challenges encountered, the unknowns that were uncovered which caused schedule delay, and the risks that were overcome?

- Decontamination of plant – Hydrocarbon free (took more than 2 months) before start of work.
- Documents – non-digitized, older revisions, no updates after last revamp, reverse engineering required in some cases.
- Had to start from scratch with the development of the operator training program and the standard operating procedures (SOPs).
- Regulatory compliance, asbestos handling – more than 60 trucks of asbestos disposed.
- PSM Compliance, This is a time and manpower intense effort for each system or sub-system
- Discovery Works- In many cases the work scope was driven by the inspections performed and could not be fully anticipated in advance.
    - There was a lot of corrosion detected in the furnace area in particular and the underground cooling water piping. The furnace convection sections were completely rebuilt, with nothing salvaged. The feed piping in the furnace area was also largely replaced. This was necessary in order to accommodate vaporization of the dense phase ethane feed at lower pressure.
    - All of the TLEs for the SELAS furnaces had to be replaced except one
- The cooling tower had been damaged in previous hurricane storms. It was determined that it was not economical to repair so it was replaced in kind.
- There were a number major modifications, including major challenges, related to the $NO_X$ mitigation strategy, the revamp of the feed system to accommodate dense phase ethane, and the revamp of the compressors.
- The drying of the recovery section was accomplished completely using portable air compressors and portable drying packs because the cracked gas compressor was not able to be used for that purpose
- New railcar and truck loading facilities had to be built or refurbished. A new truck unloading station for liquid propylene had to be designed and built.
- The completion of the HAZOP evaluation of the entire facility was a major undertaking, consuming a lot of time and resources. This also generated a number of modifications to be implemented which required additional engineering and modifications to the Distributed Control System (DCS).
- Staffing - There was difficulty in hiring qualified staff. There were two competing projects in the area on-going at the same time creating competition for experienced operators and engineers.

- Attrition- All of the first group of eleven operators quit to work elsewhere, most on the same day. There was a total of six commissioning managers, the first five having quit at some point due to the difficulty and stress of the job.

After feed introduction to the unit there were a number of operational difficulties encountered which caused some delays to making first product and to getting the unit to full production and steady-state operation.

- Poor vacuum in surface condensers,
- Vapor release at ethylene booster pumps,
- Line crack in C3 refrigeration compressor suction piping,
- Caustic contamination of the quench water system,
- Mechanical issues with the C-12, the methane compressor
- $CO_2$ in ethylene product due to regeneration of dryers with import natural gas
- The lack of a "forcing valve" for a new exchanger added in parallel to an existing exchanger service
- DCS grounding issues
- Loss of ethane feed from supplier

## Tools Used for Project Management

- Overall Schedule - Primavera P6 Schedule
- Micro-level schedules for critical areas using MS Project
- Progress measurement – using MS Excel Spread Sheets
- MS Excel Spread Sheets for Tracking Work Packages to System Turnover

## Phases of Project and Progress Tracking Methods

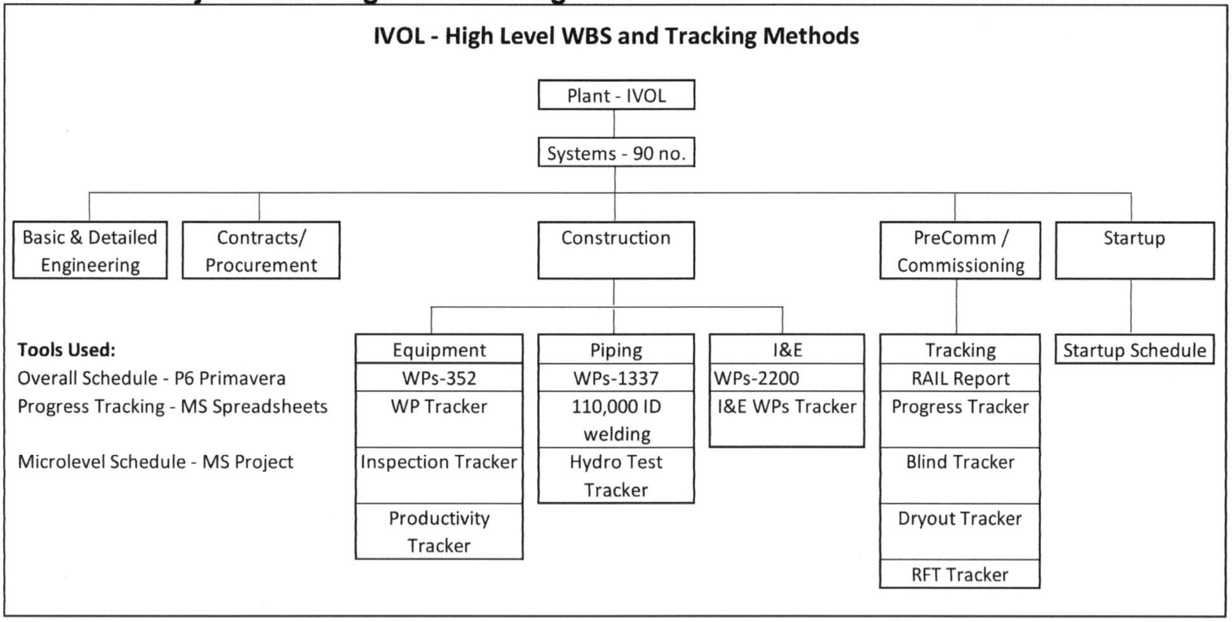


- **Planning Phase**
  - Work package tracker
- **Inspection Phase**
  - Inspection reports tracker
  - Work package tracker
- **Execution Phase**
  - Construction/Repairs
    - Daily report, weekly report, Productivity tracker
    - Work package tracker
    - Hydro test trackers
  - Packages Piping and Equipment
    - TPSL - Piping
    - TPSL - Equipment
  - Inspection
    - Inspection reports tracker

- **Precommissioning/Commissioning**
  - Rail Reports – PCI and MMR
  - SMWP – Progress tracker
  - Progress tracker – internal
  - Blind – tracker
  - Dry out tracker
  - RFT tracker
- **Start up**
  - Start up schedule and daily report

## Progress Curves

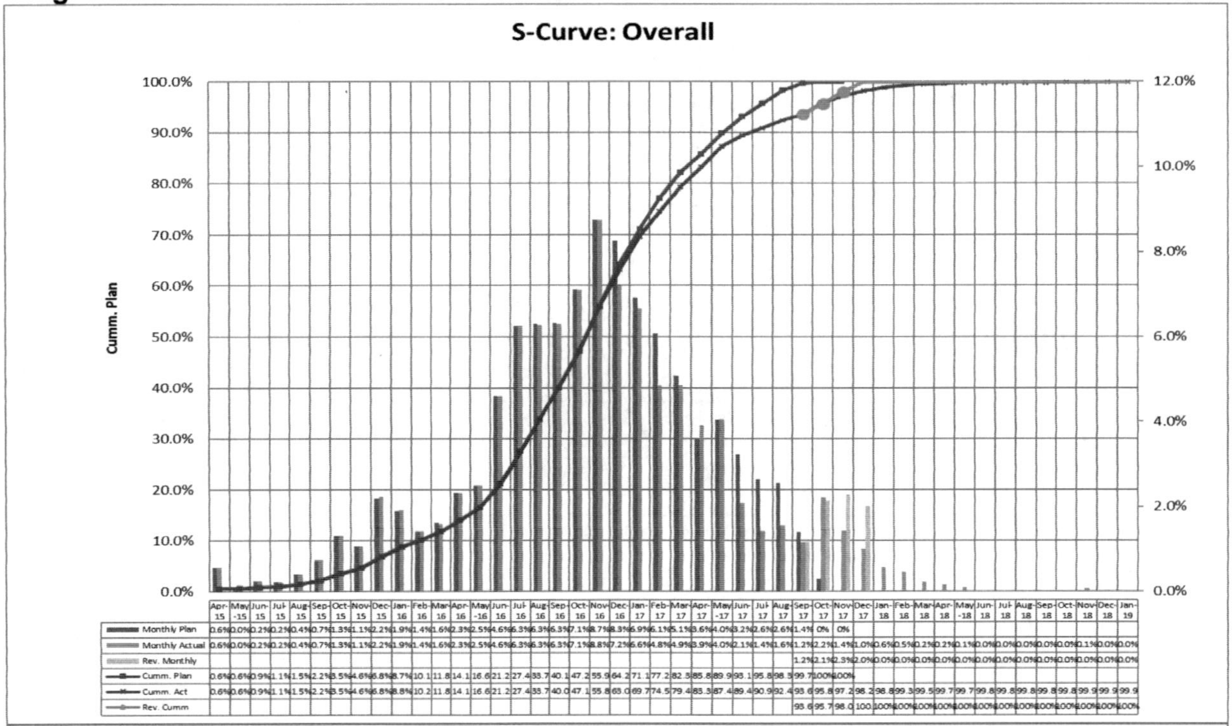


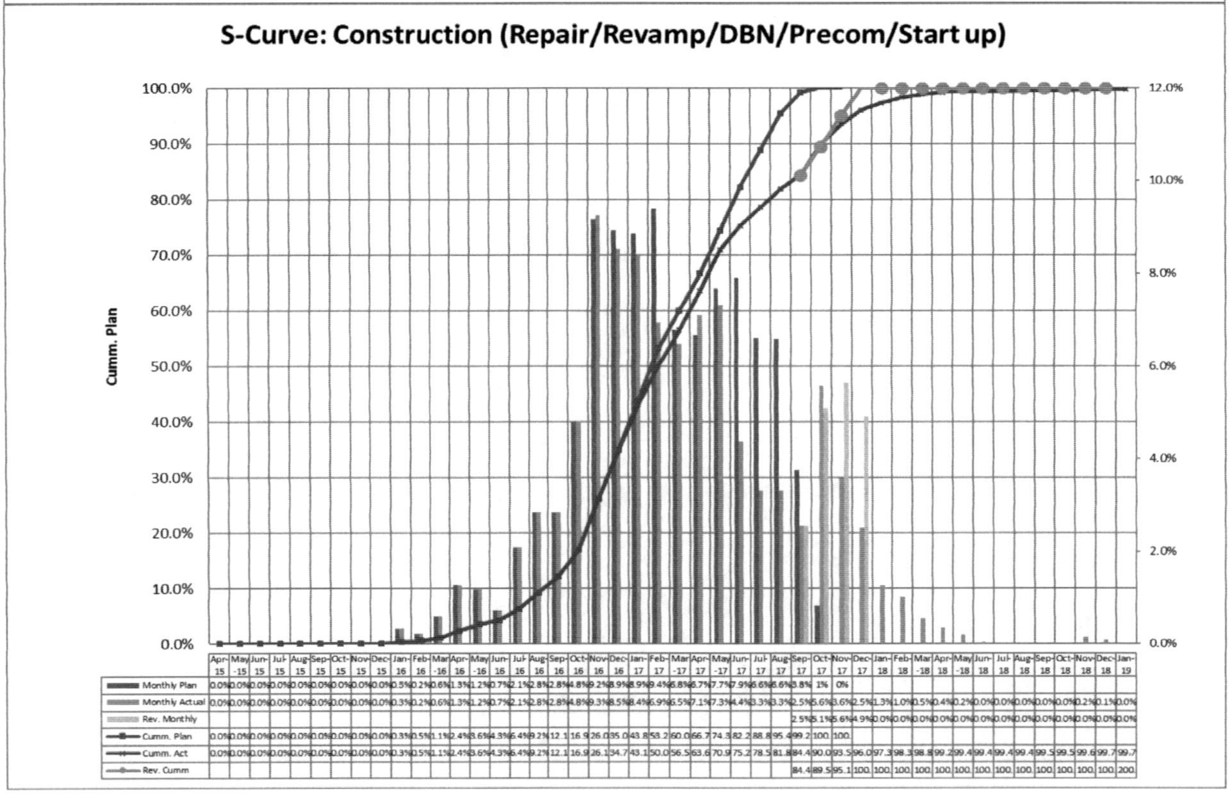

Furnaces during inspection

Furnaces during Execution

Furnaces – during startup

Miller
1-800-UR-RENTS

## Keys to Success

Many keys to the success of the project were identified.

Key No. 1 - Teamwork:
First and foremost, you have to assemble your team and develop the right attitude toward the work. It isn't operations against maintenance, or contractors against permanent hires, Indorama versus vendors, etc. You need to avoid this way of thinking. Everyone is on the same team. If there is a failure or a setback, it is not an individual failure, it is a failure of the whole team.

Key No. 2 - Training:
Start early putting together the operator training program.

Key No. 3 – Stay Focused on Project Objective:
You have to focus on the final or overall objective, don't get discouraged, expect setbacks and delays. Prepare yourself mentally for that inevitability.

Key No. 4 - HSE:
Keep HSE a priority at all times.

Key No. 5 - Recruitment:
IVOL was able to recruit a number of talented individuals from a facility and project in Mississippi which was being shut down at the same time, the so-called "Mississippi contingent". Supplementing the Mississippi contingent were a number of experienced contract operators. These formed the core group for the initial operations staffing.

Key No. 6 - Use of outside talent:
Key personnel were brought in from the Indorama facility in Nigeria, the engineering offices in Indonesia, the recently acquired Huntsman facilities in Texas, and there were a few former OXY employees and consultants (KBR, Ingenero) which were instrumental to the success.

Key No. 7 – Perseverance:
What does it take to succeed in a difficult project/ task?? In short, perseverance. When you encounter obstacles, rethink the situation, be creative in finding solutions, solicit expert help, double and redouble your efforts until you get there, never give up.

## Lessons Learned

If possible, complete more pre-inspection work prior to starting the project and defining the work scope and the cost estimates.

Start the work on the environmental and other regulatory compliance activities as soon as possible as those can cause large delay for the schedule, including a delay in the start of the major construction activities.

Hire the core operations staff as soon as possible in order to build the operator training program and the Standard Operating Procedures.

Complete a thorough HAZOP for the new facility and finalize all the drawings early in order to avoid rework and re-engineering effort.

Develop detailed plans for maximum integration (and maximum overlap) between the construction contractor, his work scope, and the personnel involved in the pre-commissioning efforts. Prioritize the order for systems' completion. Avoid rework, disassembly, and reinstallation. Only preform work one time. This is one of the keys to shortening the overall schedule for the project.

Expect some attrition (potentially a lot in this type of project) and plan for it

## Final Thoughts

What were the major accomplishments of project?

i. Did something never done before.
ii. Did it with no major injuries recorded; 6 Million man-hours without a Lost Time Injury.
iii. Started the project late after major competitors but finished before them, with being first into production.
iv. Changed peoples' lives permanently for the better, the employees and in the community;
    1. 1800 construction workers at the peak, 155 permanent full time employees and ~100 contract workers

**Cold Piping Event during New Plant Startup:**
**Do You Have All the Details You Need for Your PHA?**
**Paper 40b**

Daniel Euhus, John Pena and Hironori Miyaji
Shintech, Plaquemine, LA

**Abstract:**
As a part of the new 500 KTA ethylene plant constructed for Shintech in Plaquemine, LA, the operations team with support from the project team and engineering disciplines completed a comprehensive Process Hazard Analysis (PHA) of all parts of the plant including the hot side, cold side, utilities and co-product loading. A Hazards and Operability Study (HAZOP) process was utilized and vessel drawings, Piping and Instrumentation Diagrams (P&IDs) and other process information was utilized as much as possible. During startup, the Cold Board operator received a low temperature alarm for a section of carbon steel piping with the lowest temperature showing -80°C. The line was depressured at the same time that an in-field investigation was conducted to try and determine if the reading was real and if so, what the possible cause was. It was determined that liquid ethylene had built up in a line of pipe and leaked across two large block valves. The key detail is that the PHA had not identified this failure mechanism because it was equipment design specific at a level of detail that was not in the drawings available to the PHA team, and it could only occur during startup. Fortunately, there were no significant consequences for this event due to rapid recognition and response by operations and maintenance.

**Introduction:**
In February 2020, Shintech completed construction and commissioning on a 500 KTA ethane cracker in Plaquemine, LA and began operation. The plant was designed by Toyo as the "turn-key" Engineering, Procurement and Construction (EPC) contractor utilizing licensed technology from Lummus. The plant utilizes a front-end, low-pressure Demethanizer with a back-end acetylene converter.

This incident involved backflow of liquid from the feed line to the 168 tray ECL-431, the Ethylene Fractionator (or some might call it the C2 Splitter), backward toward the Green Oil Wash Column. Liquid flashed across two leaking isolation valves, causing a low temperature incident in upstream piping. The upstream piping is carbon steel which transitions to low temperature carbon steel after the isolation valves that leaked.

**PHA Completion and Risk Identification:**
In compliance with the Occupational Safety and Health Administration (OSHA) Process Safety Management (PSM) requirements, a Process Hazard Analysis was completed in 2017. The study used a combination of HAZOP for the entire process and Failure Mode and Effects Analysis (FMEA) for all trip systems. The HAZOP followed standard practices and items were risk ranked using a risk matrix developed by Shintech. The FMEA portion was done using the assumption that an individual component would fail and then determining whether the

outcome was safe or not. Likelihood of failure was not utilized for the FMEA portion. PHA software was utilized with facilitators trained by the software company leading the effort. The Initial PHA, completed in 2017, utilized issued for construction P&IDs. All changes after the Initial PHA were MOC'd (short for following the Management of Change required by OSHA PSM utilizing the Shintech Change Management work process). Appropriate Process Safety Information was available to the PHA team, including vessel detailed drawings. The PHA team was composed of engineers and operators experienced in process operations and the ethylene process specifically. In addition, an expert consultant in ethylene operations was brought in to provide additional perspective and experience.

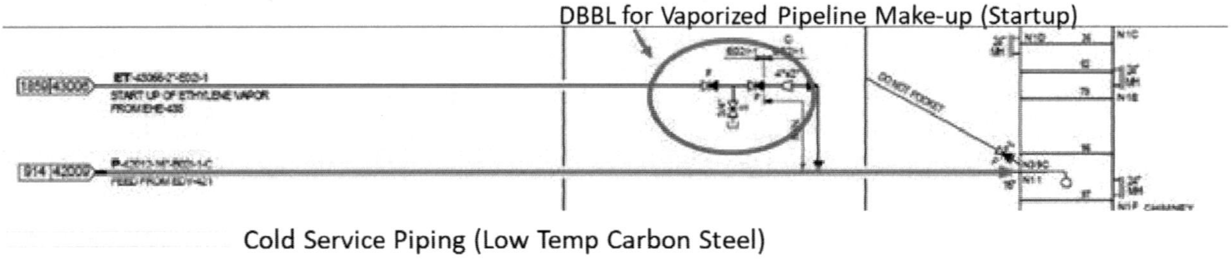


***Figure 1: P&ID Showing Feed Line into Ethylene Fractionator***

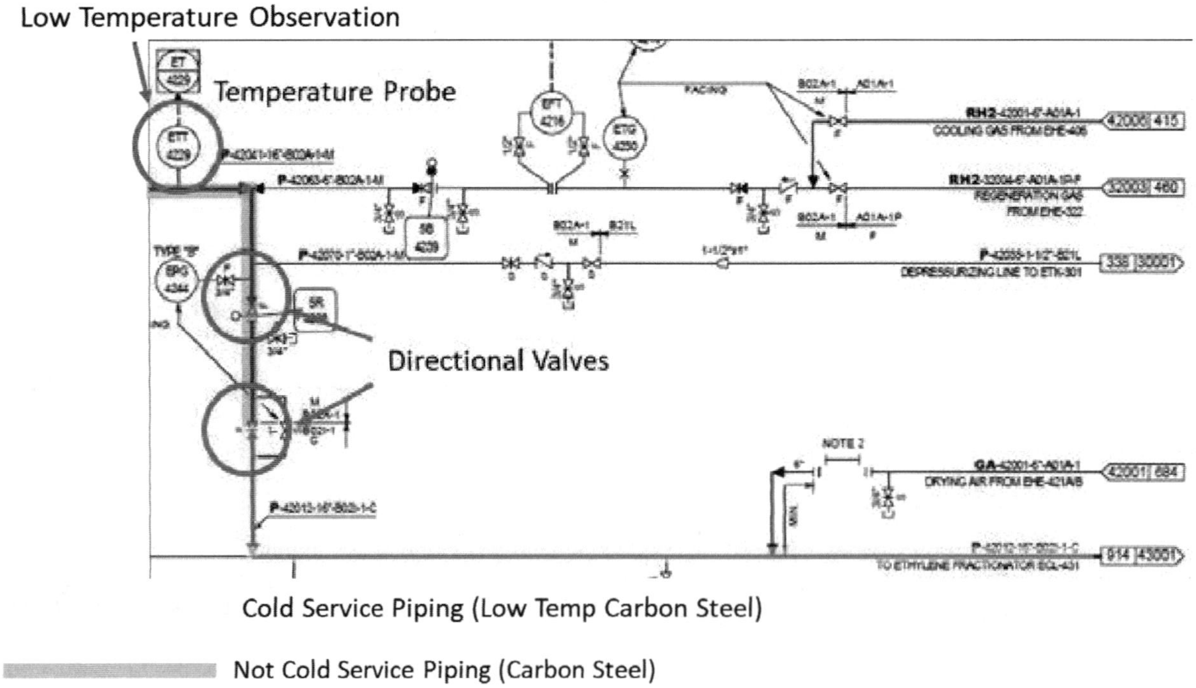


***Figure 2: P&ID Showing Upstream Section of Feed Line to Ethylene Fractionator***

As a part of the PHA, the plant was broken into areas with areas assigned to one of two PHA teams. The 400 area (product separation – where this incident occurred) was broken up into two sections for purposes of the PHA. Area 400A (Demethanizer, Deethanizer, Acetylene Converter and Fractionator) was completed by one team with no shared lines to other PHA areas. The line in question was a part of two different nodes in the PHA completed for Area 400A by the team.

The feed to ECL-431 is a vapor fed above tray 97. The PHA did not identify a risk of backflow on the feed line into the column. Design of the distributer was not part of the equipment package looked at by the PHA team. Also, startup operating mode (total reflux) was not considered separately in the PHA. Of these, the PHA should have considered the issue of total reflux, but without understanding of the valve design issues and the distributor design issues that will be explained below, the PHA would not have recognized the risk.

Figures 1 and 2 below show the relevant sections of the P&ID indicating the feed into ECL-431 (Figure 1) and the isolation valves that leaked by (Figure 2). Theses figures will be referenced with regard to the events that occurred in the rest of the paper.

**Startup Process Steps:**

Before going into the event, it is important to understand the process start up that occurs around ECL-431 while bringing the plant up. During a normal startup, The Propylene Refrigeration Compressor and propylene refrigeration system must be started first. The PRC is used for condensing the overhead of ECL-431. Once the PRC is in operation and the overhead condensers (two in parallel) are in service, ECL-431 can be inventoried with gaseous ethylene that is condensed in the overhead condensers until sufficient liquid level exists and the column is cooled enough to achieve total reflux operation on the column.

Once total reflux is achieved, the Ethylene Refrigeration Compressor and the ethylene refrigeration system can be inventoried from ECL-431. After inventory, start-up of the ERC allows the side reboiler on ECL-431 to be used. During Operation of ECL-431 on total reflux, the side reboiler serves as a primary condenser for the ERC system. This means that ECL-431 must be on total reflux before the ERC can be started.

After the ERC is started, ethylene from ECL-431 can be routed to several pumps and the cold train to help cool equipment to minimize flaring after feed in. Once this is complete, feed is introduced to the plant through the cracking heaters.

**Carbon Steel Cold Service Alarm:**

As shown in Figures 1 and 2, there is a transition from carbon steel to low temperature carbon steel across the "Directional Valves" circles in Figure 2. These valves are designed such that the only sealing surface is between the gate and the low pressure side (based on normal operating conditions, not startup conditions). The carbon steel section contains a temperature transmitter (circled in Figure 2) with a low temperature alarm at the low temperature limit for the carbons steel (-20°C). The morning of the event, the low temperature alarm sounded for this transmitter during startup activities. The DCS indicated the temperature was -80°C. At the time, no clear potential reason for such a low temperature could be readily identified (this section of the plant was at low pressure with nitrogen to maintain the lines oxygen free and dry for feed in). Significant effort went into identify why this alarm occurred, including swapped elements and transmitters, until ice was noticed on the inside of the thermowell. The directional isolation valves identified in Figure 2 were further tightened on the assumption that some undetermined hydrocarbon liquid must be flashing across the valves. The carbon steel portion of the line was slowly depressured to the flare and allowed to slowly warm back to ambient conditions (piping is insulated). The line never had more than a fraction of its rated pressure during the entire event.

**Operations Response:**
As noted above, the large diameter isolation valves indicated in Figure 2 were identified as the cause of the passing in the line. Operations was able to determine that liquid had backed up against the downstream valve (relative to normal operating flow), which flashed vapor by and then very low temperature gas had leaked across the second valve.

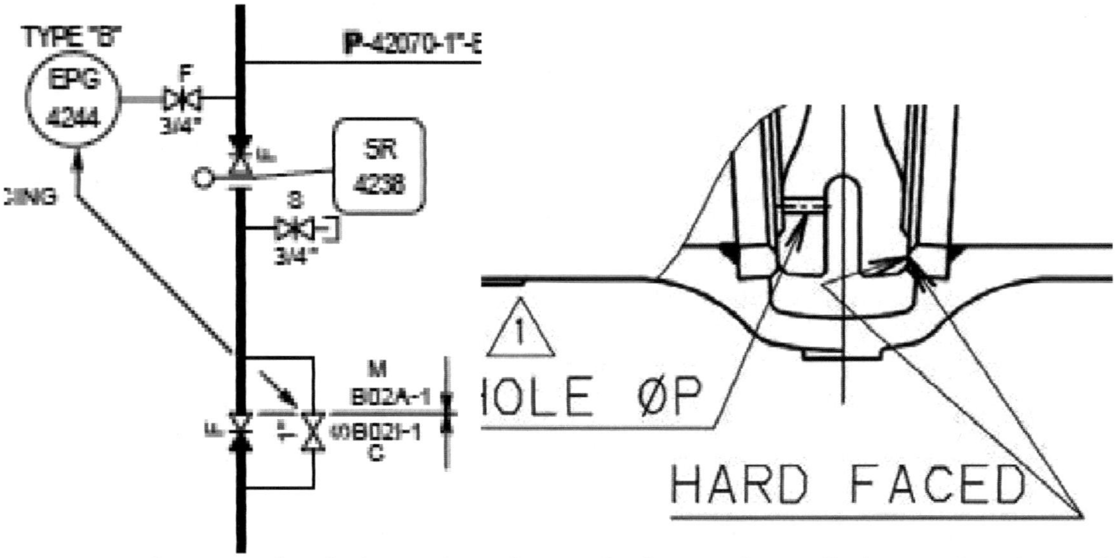


***Figure 3: Close Up of Isolation Valves that Leaked By and Detail of Face Showing Hole Drilled Out to Seat.***

GATE VALVE WITH BLEED HOLE (NOTE 2)

***Figure 4: Note 2 States that "Bleed Hole Shall Have Black Colored Side Which Is High Pressure Side in Valve Closed Condition". This means that the valve seat is vented by a drilled hole back to the high-pressure side of the valve. The only sealing surface is between the gate and the low-pressure side of the valve. And, these pressures are determined at normal operating conditions.***

The line in question has two block valves in series. However, they are the same style of gate valve and are not tight shutoff valves. In addition, the valves have holes drilled in them to vent between the gate and seat to the high-pressure side – effectively making them only have 1 sealing surface between the gate and the low-pressure side. A number of these valves in various services had to be reworked in a local valve shop after not providing satisfactory isolation in the propylene refrigeration loop. Operations was able to tighten the valves more and achieve enough of a seal across both valves to eliminate leakage into the carbon steel section. That section was allowed to depressure and warm up to atmospheric temperatures slowly (piping is insulated if normal operating condition is below 0°C). After conversations with our mechanical engineering team, it was determined that no further action was necessary.

**Unidentified Risk Findings:**

During and after the event both the operations team and the engineering and EPC support teams were rapidly trying to determine what could have led to liquid hydrocarbon being present against the isolation valve during the startup mode. As a part of a review of the construction of the fractionation tower, it was determined that the feed distributor was designed for vapor feed into the column. Rather than having holes along the bottom of the pipe, the distribution holes were drilled along the sides of the pipe as shown in Figure 5.

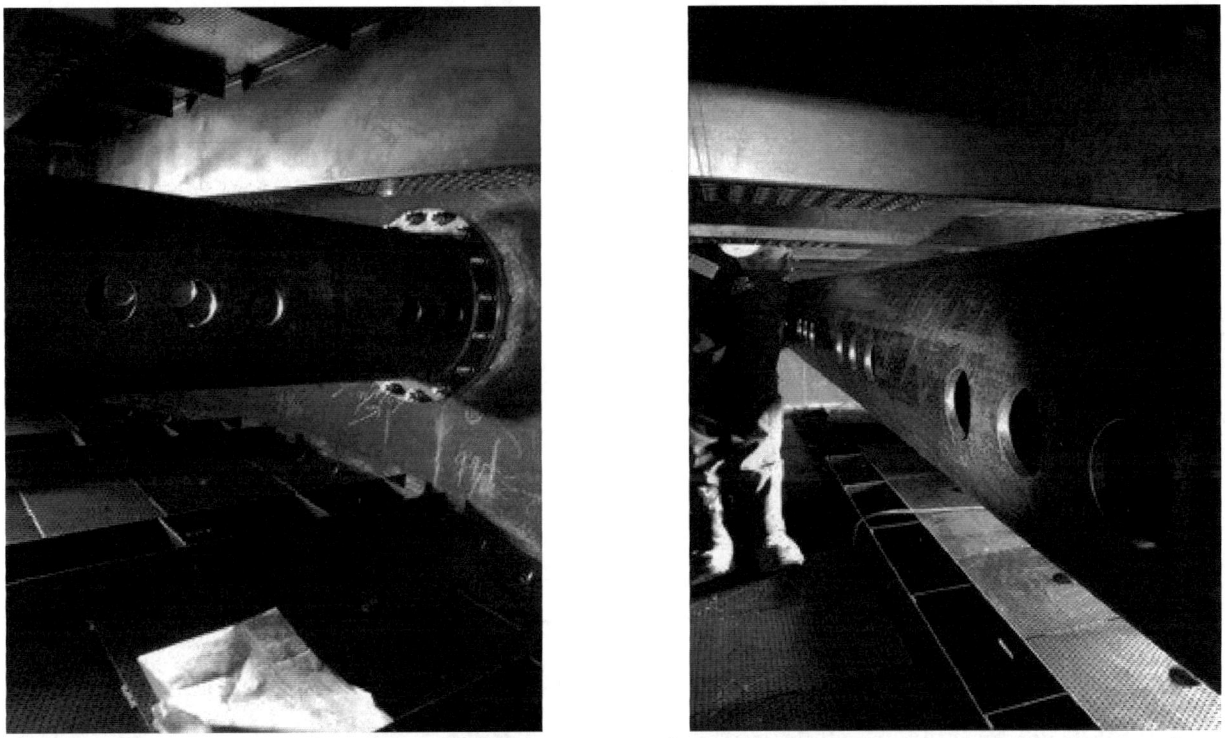

***Figure 5: Construction Pictures of the ECL-431 Feed Distributor Taken During Final Inspection***

During normal operation, the vapor flowing into the tower through the distributor will keep liquid from backflowing into the pipe. However, during startup operation with total reflux, liquid coming down from above will spill onto the pipe on the top and sides and then flow down into the holes on the sides of the distributor allowing liquid to flow backwards through the feed line to the tower, ultimately stopping at the isolation valves.

As noted, these valves have an inherent tendency to leak due to the design. This is especially true when the high-pressure side during normal operation is actually the low-pressure side during the conditions during startup before feed is introduced. Thus, the liquid flashed across the first valve, cooling and then leaked across the second valve to even lower pressure, likely cooling further. This led to the observed low temperature at the thermowell. And, -80°C was a logical temperature given that the normal boiling point of ethylene is -103.7°C and there was some pressure of nitrogen (approximately 15 to 20 psig) already present in the line. This flashing cooled the carbon steel well below its design temperature.

As noted previously, because the PHA did not recognize the risk of back flow or the potential shortcomings of the isolation valves, it did not identify the risk of having extremely cold vapor in this carbon steel. The good news was that other design considerations had led to the installation of the temperature measurement for the stream in the carbon steel section. This meant that an alarm was already in place and could be responded to in order to remove the risk.

**Operational Changes:**

Due to this event, operating procedures were updated to better identify this event and prevent it in the future. First, during startup, these two isolation valves are sealed as tightly as possible. With a new operating team, many of whom are working as operators for the first time, training to recognize when a valve has been "gotten down on" was required. Doing so will help to prevent leakage.

In addition, the bleeder between the double isolation valves can be evacuated to the flare by hose if any indication of back flow is noted. It should be noted that during two subsequent startups, this has not been required.

**Conclusions:**

Most importantly, the first response to an instrumentation alarm should be to assume that the alarm is real and protect the system from the alarm condition. This was not done, and fortunately did not lead to a larger issue. In the future, operators and maintenance personnel have been trained to assume the alarm is real and respond accordingly until it is determined that it is not.

Second, a good PHA may still miss some hazards that are inherent to the design and construction of certain pieces of equipment. This is particularly true if they involve components that may not be common in the experience of the PHA team. In this instance mechanical details of both the feed distributor and isolation valves had significant relevance and were not common in the experience of the PHA team (to our knowledge, none of the rest of the Shintech units have the type of isolations valves found in our unit where the gate seat is drilled out to the high-pressure side.

Finally, it is critical to consider non-standard scenarios like startup and shutdown during a PHA Particularly when liquified petroleum gases have a potential for reverse flow. This is something that will be corrected during the next PHA for the Shintech Ethylene Unit in 2022.

# IDENTIFICATION OF POTENTIAL STAGNANT AREAS IN AN ETHYLENE PLANT

**Debby Rossana, Ph.D.**
Nalco Water, Sugar Land, TX

**Greg Scherer**
Nalco Water, Sugar Land, TX

**Sheri Snider**
Nalco Water, Sugar Land, TX

Prepared for Presentation at the 2021 Spring National Meeting
Virtual Conference April 2021
Paper No. 60a

# Identification of Potential Stagnant Areas in an Ethylene Plant


## Abstract

Unexpected over pressurization, reduced rates and/or equipment damage can occur due to polymer build up in stagnant and low flow areas and dead legs. For this reason, identifying these stagnant/low flow areas before event occurs is critical. This paper discusses how to survey the system, to identify stagnant areas. This involves a detailed Process and Instrumentation Diagram (P&ID) and field review of the systems to identify the areas of concern. Several examples of what to look for are presented. Once these stagnant areas are identified, when possible they should be removed. When that is not possible, there are several ways to reduce the risk. These common practices will also be discussed.


## Introduction

Stagnant and low flow areas as well as dead legs in an ethylene plant can be places where polymer can build up. The plant could experience lost performance, reduced or no flow, equipment damage and in some cases rupture of equipment and chemical releases (1,2). This paper will offer some suggestions on how to identify these stagnant areas and provide a few mitigation steps.

It should be noted here that for the interest of time, this paper focuses only on distillation towers and associated equipment, but the learnings can be applied to any area in an ethylene plant where there are stagnant areas and reactive monomers.

## Definitions

Through out this paper, the terms stagnant areas, low flow areas and/or dead legs will be used. In many cases they are interchangeable and only one of these terms will be used to not be repetitive. However, these terms are each somewhat different and the brief definitions are shown below.

- Stagnant areas are those spaces that are exposed to vapors with little or no movement.
- Low Flow areas are those spaces exposed to vapors or liquids where there is some movement, but it is slow.
- Dead Legs are those areas where liquid or vapor can get trapped inside a pipe with no movement.

## The Audit

One of the first things that should be done during an audit of an ethylene plant is to look at the P&IDs. The goal of the P&ID review is to identify potential stagnant areas on paper prior to the field audit. For example, in **Figure 1**, a simplified reflux drum drawing is shown. The red circles high light those areas of possible concern. Of course, actual P&IDs at the plant are far more complicated and therefore more difficult to review.

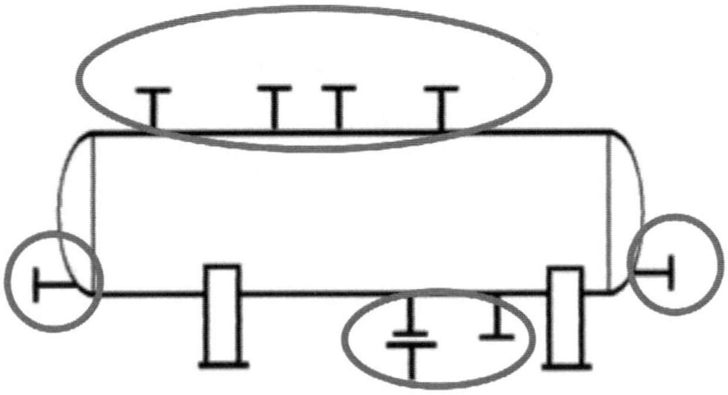

**Figure 1: Simplified drawing of a reflux drum**

When surveying multiple pieces of equipment, it is best to break them up into smaller pieces. The idea here is to be thorough and it is easier to be thorough when the survey is broken into smaller segments. Often these stagnant areas are hidden in plain sight so one must be meticulous in the survey.

Once the P&IDs have been reviewed, walking the unit is the only way to survey the equipment for stagnant areas and dead legs. Equipment can't be surveyed from an office or completely from the ground. The audit team will need to climb up to the condensers and reflux drum decks, etc.

While in the unit look for:

- Unused taps
- Instrument analyzer lines
- Rarely used lines such as start up or blowdown lines
- Level bridles
- Deep manways
- Low flow areas of exchangers
- Relief valve standpipes
- Areas where Teflon has been for gaskets and seals

- Recognize sources of oxygen, which can collect in low flow areas and increase fouling

During the inspection pay close attention to the various exchanger designs. Exchangers with large heads may have low flow areas where polymer can develop. And it is well known that horizontal reboilers in high fouling services have stagnant areas in the vapor spaces where monomer can condense and polymerize. **Figure 2** shows an overhead condenser with a deep channel head and polymer forming in that head.

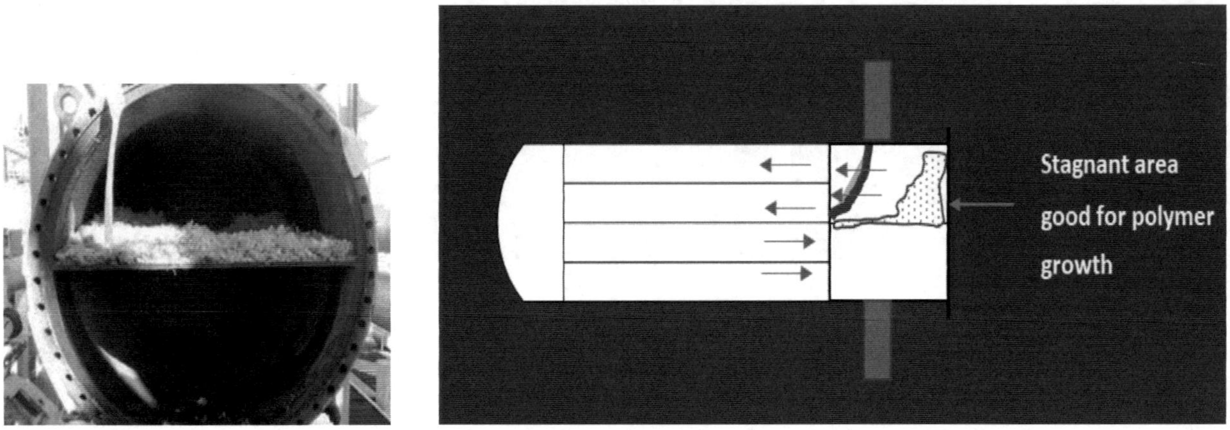


**Figure 2: Overhead condenser with a large channel head**

Another part of the survey should focus on the towers. Deep manways are often overlooked. **Figure 3** shows several examples of manways with polymer growing in stagnant areas. In many cases polymer in the manways is of little concern. However, there are situations where the polymer can either initiate further polymer formation or fall into the tower, potentially blocking downcomers or valves. If these areas don't cause any problems, then they do not need to be addressed but if those areas can create further problems then they need to be addressed.

**Figure 3: Polymer formation in manways**

Stagnant areas may exist in many areas and the goal of the survey is to find them all. Decanters, exchangers, reflux drums, relief valve standpipes, level gauges and those rarely used lines are all low flow or stagnant or dead legs where polymer can grow.

**Figure 4** shows some examples of low flow areas. The left-hand picture is a level bridle that has plugged and on the right is a relief valve stand pipe where the RV has been removed. In the middle is a drawing of a reboiler design where the reboiler line is valved off. When both valves are closed then the stagnant area is created. This material can form polymer while offline and cause issues. If the offline reboiler is out of service for an extended period, then the valves may leak-by causing liquid and vapors to invade the stagnant space. The monomers may then polymerize and foul the reboiler when it is placed in service. A drain line is needed to remove the liquid and replace the space with an inert gas such as nitrogen.

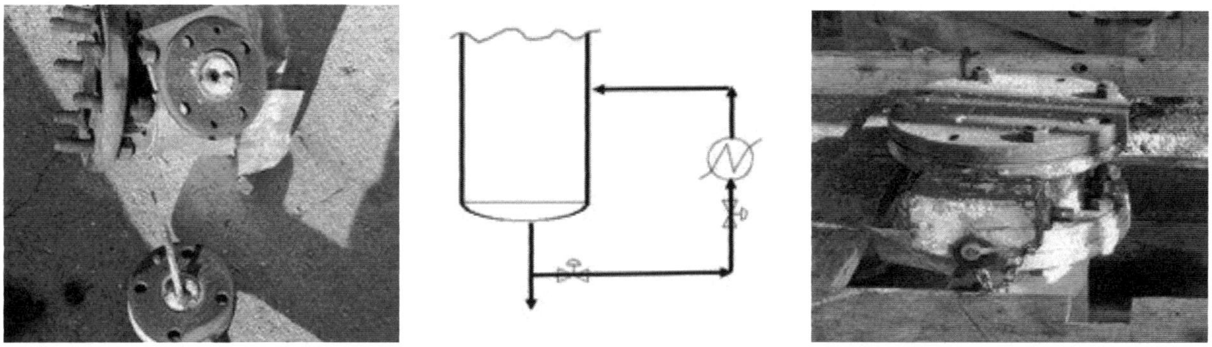

**Figure 4: Plugged level bridle (L), Reboiler with stagnant area (M), Fouled relief valve stand pipe ( R )**

## Review the fouling history

As part of the survey, equipment's fouling history should be reviewed. Here is an example of a plant who had historically long reboiler run lengths of about 2 years and then began to experience very short reboiler runs, eventually running only a few days before having to be taken out of service.

It was observed that the tube sheet was fouling but the tubes were clean, **Figure 5a**. There was nothing chemically different with this polymer than that from the past, but the polymer texture suggested that the polymer was growing in a stagnant area. The polymer appeared to be bumpy on one side and smooth on the other (3) **Figure 5b**.

This visual look to the polymer prompted operations to check the relief valve standpipes on the reboiler. The relief valve vent lines were found to be completely plugged. Polymer was falling from the relief valve line into the reboiler return line and sump of the tower and plugging the reboilers. Unfortunately, this polymer also plugged the drain line, **Figure 5c**, so the only way to remove the foulant was through the reboiler, **Figure 5d**. Even after cleaning the vent lines, the reboiler run lengths only marginally improved. The plant eventually took advantage of another outage in the plant to open the tower and clean the sump.

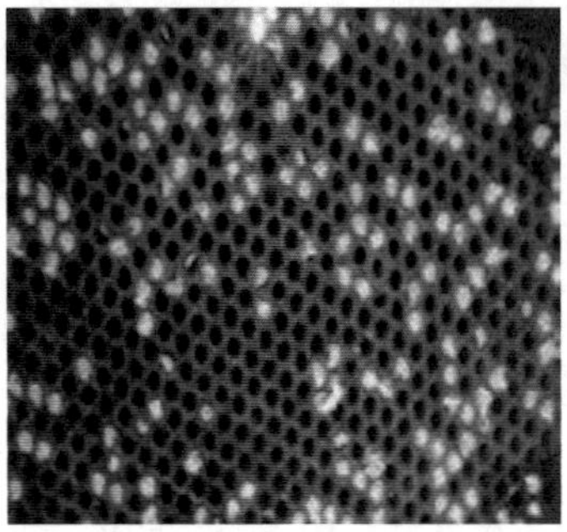

**Figure 5a: Plugged tube sheet** **Figure 5b: Polymer from tube sheet**

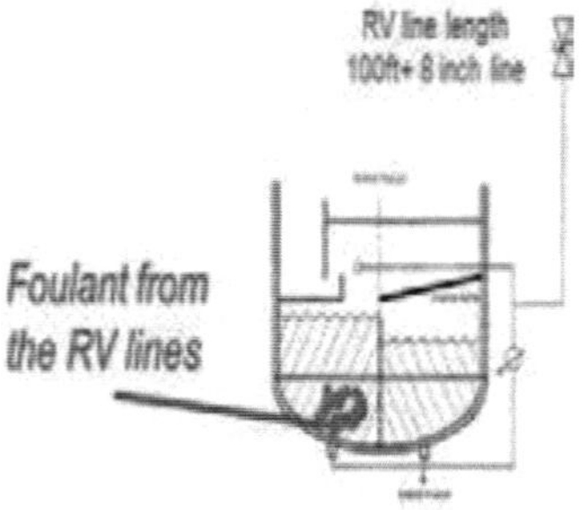


**Figure 5c: Plugged bottom line** **Figure 5d: Drawing of reboiler design**

Another contributing factor to the relief valve vent line fouling was the distance from reboiler to the relief valves. The relief valves for the fractionation section are located centrally on an elevated deck, **Figure 6**. This central location simplified maintenance activities but caused the piping from the reboilers to the relief valves to be more than 150 feet in length with up to 8 elbows and transitions from vertical to horizontal sections. The piping was between 8 and 10 inches in diameter.

**Figure 6: Central location for fractionation Relief Valves**

Another contributing factor to polymer formation in stagnant areas, is oxygen egress. Oxygen can collect in the same stagnant areas as the reactive monomers making fouling worse. In general, assume that a recycle stream from a refinery or polymer plant, a ship or a truck will have some oxygen and its risks should be accounted for, **Figure 7**. Maintenance activities such as reboiler cleaning or rust (ferric oxide) may contribute oxygen to the system.

**Figure 7: Means of oxygen ingress**

### Create a risk assessment

Once the survey has been completed, a risk assessment should be created, to identified the potential fouling risk associated with the stagnant areas, low flow areas and dead legs. The potential for occurrence should be determined as well as the potential for a problem. Once this is done, plans are made to control or eliminate the stagnant areas. **Figure 8** shows an example of a risk assessment.

| Do we have any of the following low flow or stagnant lines? | YES | NO | What can we do to mitigate this risk? |
|---|---|---|---|
| unused lines? | | | |
| dead legs? | | | |
| instrument or analyzer lines? | | | |
| start-up lines? | | | |
| blowdown lines? | | | |
| level bridles? | | | |
| deep manways? | | | |
| low flow areas in exchangers? | | | |
| PRV standpipes? | | | |
| **Do we have possible oxygen sources?** | **YES** | **NO** | **What can we do to mitigate this risk?** |
| import streams from off-site? | | | |
| recycle streams that could contain oxygenates? | | | |
| procedures that could allow oxygen/air in during equipment swaps? | | | |
| potential for CW leak? | | | |
| Teflon gaskets or seals in process equipment? | | | |
| **Do we have higher severity now than before?** | **YES** | **NO** | **What can we do to mitigate this risk?** |
| higher % monomer? | | | |
| higher temperatures? | | | |
| increased risk of oxygen? | | | |
| shorter runlengths due to fouling? | | | |
| more fouling than in the past? | | | |
| lighter colored fouling (white or off-white)? | | | |

**Figure 8: Example of a Risk Assessment**

## How does the industry address stagnant areas?

There are several diagnostic tools available to study stagnant/low flow areas and dead legs in a unit such as borescope, thermal scans, X-ray and neutron back scatter. When polymer is suspected in long runs of piping a borescope can be used, **Figure 9**. When there are several twists and turns the borescope may not be able to handle that many directional changes. A foam line pig has been used in some longer, multidirectional lines to ensure the line is clean.

Thermal scans can be used to determine if a line that should be warm, is cold, indicating a dead leg. Neutron back scatter is used to help determine if there is blockage and the extent of the blockage. X-ray, **Figure 10**, is also used to determine if a line is partially or totally blocked. In figure 10, the lighter areas are blocked with foulant and the darker areas indicate liquid flow.

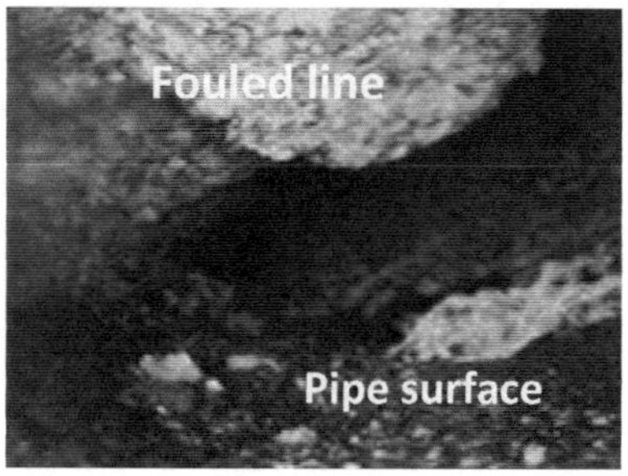


**Figure 9: Borescope inside a fouled line**

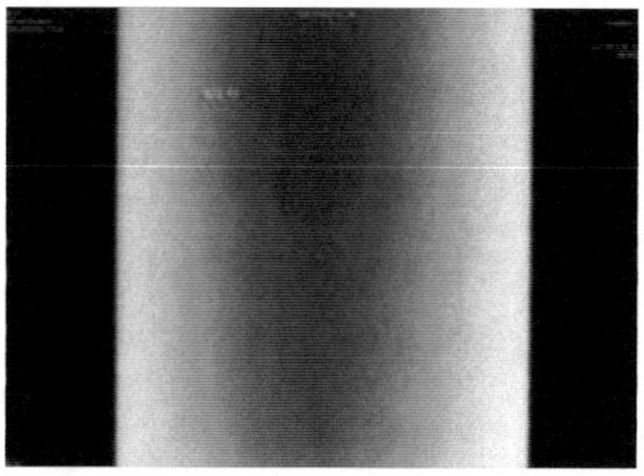

**Figure 10: Thermal scan of a partially blocked line**

Relief valve lines that are prone to plugging can be treated with a slip stream spray with or without inhibitor, natural gas, N2, etc. to help eliminate stagnant areas as shown in **Figure 11**. If a liquid is added, an aerosolizing quill as shown in **Figure 12** is recommended.

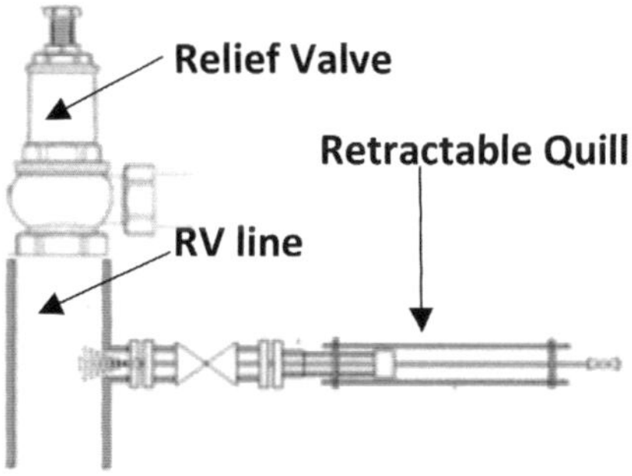


**Figure 11: Spray added to relief valve line**

**Figure 12: Aerosolizing spray quill**

A rupture disc, **Figure 13**, can be added upstream of a long run of piping in combination with a relief valve. This should be considered best practices for those relief valves that are positioned in long piping.

The rupture disc is positioned near the equipment and opens at a set pressure (4), which is slightly lower than the relief valve. Before the disc ruptures, the line between the equipment and relief valve is clear and free of reactive monomer and oxygen, essentially eliminating the stagnant area. After the disc ruptures, **Figure 14**, it should be assumed the line contains vapors that can polymerize.

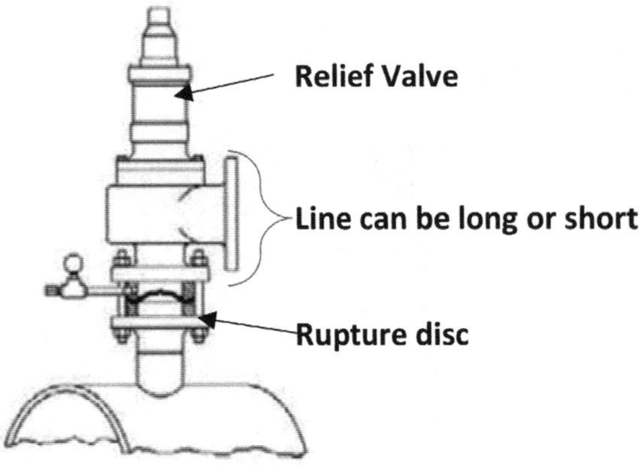


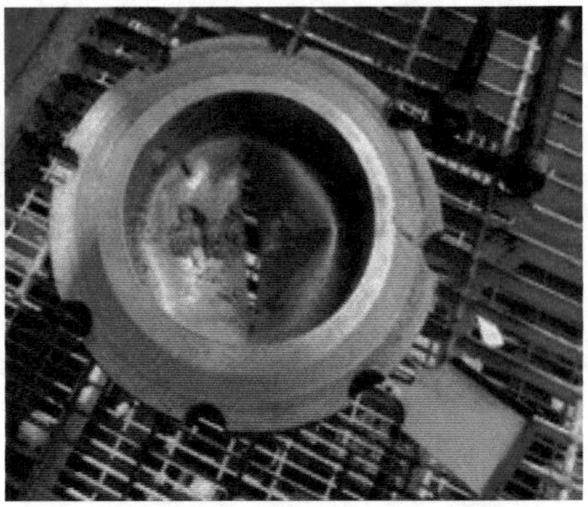

**Figure 13: Rupture disc on releif valave vent line**

**Figure 14: Ruptured rupture disc**

For those towers where polymer has accumulated in the manways, as shown in Figure 3 and is causing a problem or has the potential to cause a problem in the future, then at turnaround, the external manways can be outfitted with a spray nozzle, **Figure 15**. As with the spray to the relief valve, an aerosolizing quill is needed. In many cases, an aerosolizing quill will require a diluent stream to function properly. The bottom stream, reflux stream or some other inert stream may be utilized for the diluent flow needed for the aerosolizing quill. The idea is to keep the vapors that collect in those manways moving.

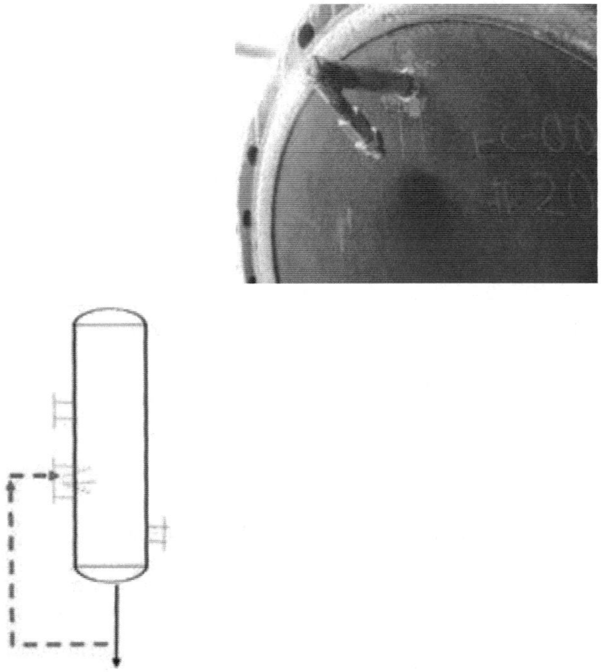

**Figure 15: An external manway outfitted with a spray nozzle in order to reduce stagnant areas in the manways**

Similar to the stagnant manway scenario, condensers with deep heads can create a stagnant area. In this case, the dollar plate can be outfitted with a spray nozzle at turnaround, **Figure 16**. If an inhibitor is being used for fouling control, the typical injection point is to the condenser feed. If a deep condenser head is creating a stagnant area, then the injection point may need to be moved to the condenser end plate or use both injection points (condenser feed and end plate).

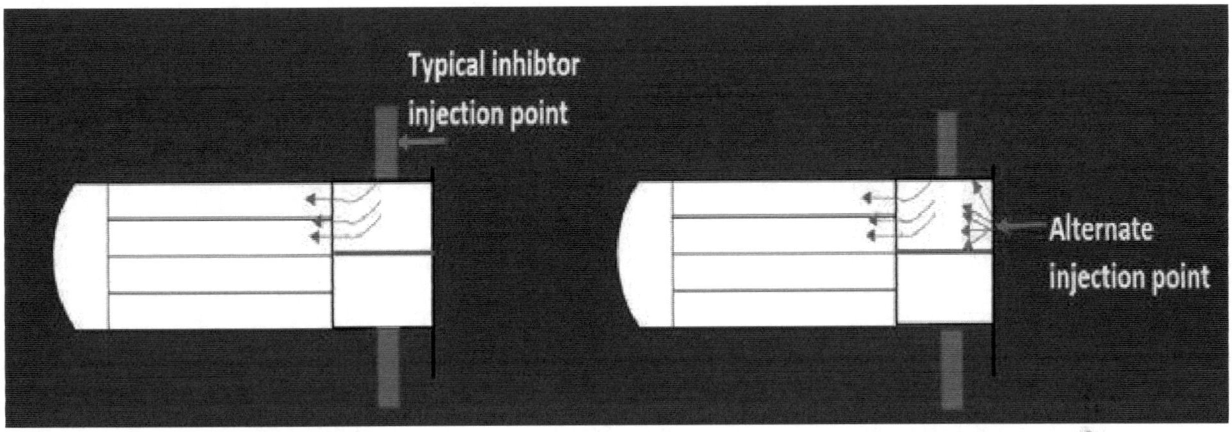


**Figure 16: Injection of a slip stream to reduce fouling in large condensers**

If the inspection reveals unused taps, then the taps should be removed, if possible. As with any equipment modifications though, care shoud be taken to ensure design ratings are maintained.

## Summary

Polymer can grow in stagnant areas causing equipment overpressure and/or damage. Surveying the plant, looking specifically for all the potential stagnant or low flow regions will turn up more areas than expected. Once identified, a risk assessment should be done to mitigate the impact that fouling may cause. Through the identification and mitigation of the effects these areas may cause, plant reliability can be improved, and process upsets reduced.

## References


1. AICHE paper 189g, Spring 2020, C. Cheng-Guajardo, et al
2. AICHE paper Session 2020, Spring 2008, S. Coleman
3. AICHE 35b, Spring 2019, Rossana, et al
4. "Use of Rupture Discs in Combination with Relief Valves", Fike white paper, R. Bours, 2013

# 618942 Putting on the Brakes: Potential Pitfalls of Unit Turndowns and How to Avoid Them

**Abstract:**

Whether dictated by adverse market conditions, plant constraints, equipment failure, inclement weather, or other considerations, often chemical manufacturers need to reduce production rates or even shut down the unit altogether. In an ethylene plant, unit turndown or shutdown can have significant impacts in nearly every area of the facility, from the quench to light ends fractionation. Many of these impacts can negatively affect operation due to increased potential for fouling, corrosion, emulsions, and foaming in distillation towers, heat exchangers, and other equipment and piping.

The purpose of this paper is to examine a selection of the most common, treacherous pitfalls that can occur during turndown and shutdown to determine what the impacts can be and how to mitigate them from a mechanical, operational, and chemical standpoint, thus providing for safer, more reliable operation and the protection of valuable assets.

# 60C

# Monitoring ppb Monomer Impurities and Catalyst Poisons with Process GC-MS

*John Wasson, MS*
*President*
*Wasson-ECE Instrumentation*

*Theresa Russell, PhD*
*Chemist, Process Analytics*
*Wasson-ECE Instrumentation*

Prepared for Presentation at the 2021 Spring National Meeting
Virtual, April 18-22, 2021

**Monitoring ppb Monomer Impurities and Catalyst Poisons with Process GC-MS**

*John Wasson, MS*
*President*
*Wasson-ECE Instrumentation*
*Theresa Russell, PhD*
*Chemist, Process Analytics*
*Wasson-ECE Instrumentation*

**Keywords:** gas chromatography, mass spectrometry, GC-MS, GC/MS, ethylene, propylene, monomer

## Abstract

Impurities in ethylene and propylene monomer streams are costly to both monomer producers and their customers. Consequently, real-time monitoring with process gas chromatographs is required for early detection and mitigation of contaminants during monomer production for and monomer feedstock evaluation. Traditional process gas chromatographs (PGCs) commonly use isothermal ovens, packed columns, and offer ppm detection levels of just a few contaminants per analyzer. Some critical analytes of interest, such as metal hydrides, are not addressable by typical PGCs and require measurement with laboratory instruments, which delays actionable results. A new process analytical approach is to employ an on-line instrument configured like lab instruments, namely using temperature-programmed capillary chromatography and mass spectrometry (GC-MS) to analyze monomer streams. Multiplexing the MS detector with pulsed discharge helium ionization and flame ionization detectors (GC-MS/PDHID/FID) enables quantitation of a broad spectrum of monomer impurities in a fraction of the time required to complete these analyses by laboratory instruments. Here we provide a technical overview of GC-MS, the GC-MS/PDHID/FID instrument, and share a data set obtained from the analysis of polymer-grade monomeric ethylene.

## Introduction

Competition in the ethylene market is fierce and demands for the quality of the ethylene are increasing. High-purity monomers are used to make polyethylene and polypropylene, the polymeric building blocks for two important plastics. Numerous potential ethylene impurities reduce the quality and value of ethylene by interfering with the subsequent production costs of polyethylene. Non-ethylene hydrocarbons could result non-ethylene polymers being produced. Contaminants such as oxygenates, sulfurs, and metal hydrides such as arsine and phosphine, have significant impacts on the metallocene catalyst required to produce polyethylene. For example, the catalyst is poisoned by even trace levels of common by-products such as hydrogen sulfide, acid gases and methanol. This poisoning reduces catalyst efficiency resulting in lower polymer yields per reaction. Catalyst bed poisoning also necessitates increased maintenance events and associated plant down-time, and, eventually, very significant catalyst replacement costs. For these reasons, polyethylene producers demand high-purity ethylene. Careful monitoring of ethylene process streams is required for early detection and remediation of contaminants to meet the product specifications.

The oxygenates, hydrocarbons, permanent gases, and sulfurs potentially found in monomer streams are very different compound classes and require multiple detector types for their complete analysis. Classical PGCs frequently house a single detector type which necessitates multiple instruments to address the complete analyte list. Laboratory GCs routinely use temperature-programmable convection ovens and capillary chromatography to achieve excellent analyte separation and resultant low sensitivities. In contrast, most process GCs are limited to isothermal air-bath ovens and using packed columns resulting in poorer detection limits for analytes of interest. It should be noted that the metal hydrides arsine and phosphine, which are potent catalyst poisons, are not addressable by typical PGCs. Thus, for the hydrides, or for low ppb quantitation of other process stream compounds, off-line analysis is required.

Off-line analysis poses several challenges for analysis of process streams, the riskiest of which is time. The additional time required to take a manual "grab" sample back to the lab increases the risk that changes in the process might affect product quality before action can be taken. The chemical composition of the sample is also at risk and might not reflect the true composition of the process stream. For example, light components can volatilize out of the sample and be lost in the head space of the sampling container. Reactive components can be lost on the surface of the sampling container. Precision of the PGC measurements is often greater than that of laboratory measurements because there are fewer changes to the sample and increased automation reduces measurement variation.

However, laboratory analysis does offer at least two benefits. Critical processes often have analytical redundancy to safeguard against loss of data in the event of instrument downtime and lab instruments often fulfill this role. Lab GCs have also been relied on to offer advanced methods and additional detector choices, such as mass spectrometers (MSDs), not included in typical PGCs.

MSDs are nearly universal and very sensitive detectors capable of quantifying analytes in the ppb range. Except for hydrogen and helium, an MSD can detect all analytes of interest. MSDs have a broad dynamic range ($10^6$) which allows them to be maximally useful in the plant where concentration ranges can vary considerably. The troubleshooting advantages of a mass spectrometer are unparalleled and allow for interrogation of "mystery" peaks, or alternatively, "the usual suspects" that show up from time to time in a process stream. Both types of anomalies are easy addressable by an MSD. Finally, as will be discussed in greater detail below, GC-MSD can offer triple verification of the identity an analyte. To the best of our knowledge, no other process gas chromatographs have offered GC-MS as an option.

The objective in the design of this instrument was to create a process gas chromatograph capable of lab-quality performance to address the full list of impurities of interest in ethylene and propylene streams. To achieve that goal, temperature-programmed capillary chromatography was combined with GC-MSD, GC-PDHID, and GC-FID. Complete characterization of the C1-C5+ hydrocarbons, permanent gases, common sulfurs and oxygenates, arsine and phosphine in an ethylene stream was achieved in 10 minutes.

## Basics of GC-MS

An in-depth review of GC-MS is beyond the scope of this paper and a basic understanding will be adequate to for the purpose of this discussion.

Gas chromatographs, using capillary columns carefully chosen for the chemical properties of their surface coating, spatially resolve analytes within a sample by virtue of the variable interaction of those analytes with the column coating. As the separated compounds elute from the column, they are directed to a detector for quantitation.

As the name suggests, the mass analyzer of a quadrupole mass spectrometer uses the mass of a molecule for identification and the abundance of that mass for quantitation. There are several important steps of this process outlined in Figure 1[2].

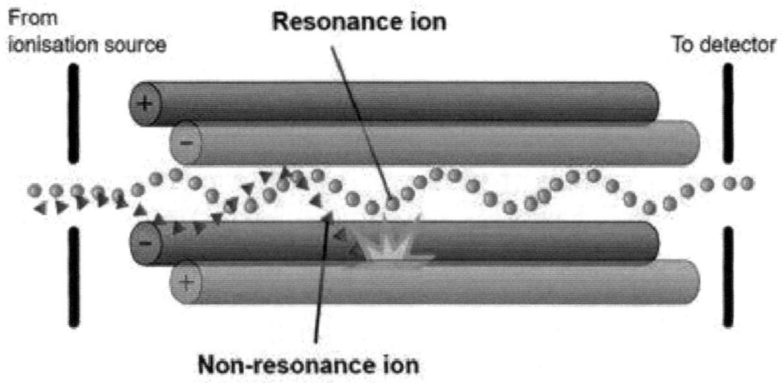


**Figure 1 - Quadrupole Mass Analyzer**

Compounds within the GC column eluates are delivered to the inlet of the MSD where they are exposed to electrons released by a heated filament. This interaction does two things. First, the electrons bombard the molecule and fragment it in a structure-dependent fashion. That is to say that the unique chemical structure of the molecule dictates the way the molecule will break apart under electron bombardment. This fragmentation pattern, as well as the relative abundance of the different fragments, is unique to the molecule. The impact of the electrons also strips an electron from the intact molecule and its fragments conferring a charge on them.

The charged intact molecules (called molecular or parent ions) and fragments (called fragment or daughter ions) are accelerated and guided through the instrument by a tunable, DC and RF electric fields in the quadrupole mass analyzer[2]. This field functions as a filter by allowing only selected ions, which resonate with the field, to make it to the detector while non-resonant ions are deflected away. It is the mass to charge ratio (m/z) that is selected but, because the charge is 1, the ion mass is readily determined. The selected ions then strike an electron multiplier which greatly amplifies the signal and sends it to the data system for quantification.

The observed m/z, the mass spectra, for methanol and diethyl sulfide are shown in Figure 2. These unique compounds have distinct spectra with characteristic peaks. Methanol has a parent m/z at 32 and a dominant fragment m/z at 31. Diethyl sulfide has a parent m/z at 90 and a dominant fragment m/z at 75. The fragmentation patterns as well as the ratios of the abundances of the fragments are constant and defining for these molecules. This allows for unambiguous identification of methanol and diethyl sulfide.

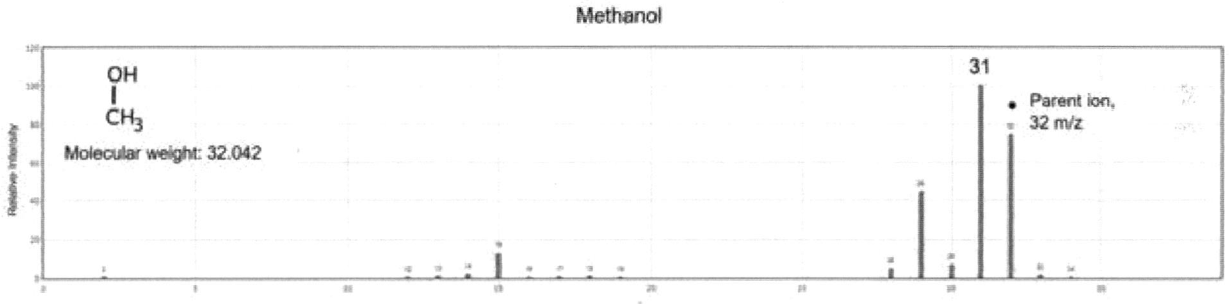


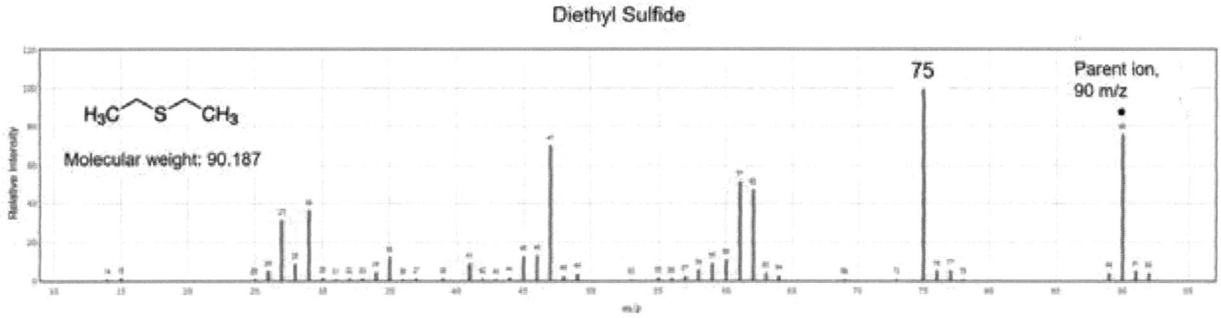


**Figure 2 - Example mass spectra**

Methanal (top) and diethyl sulfide (bottom) have unique mass spectra. Source: NIST Chemistry WebBook, SRD 69

The importance of this fact is highlighted in Figure 3. Methanol and diethyl sulfide often co-elute and are observed as a composite peak, seen at 8.84 minutes in this example (Fig. 3A). With an MSD, this composite peak can be separated by examining extracted ion chromatograms (EIC) which are individual chromatograms for a desired m/z (Fig. 3B, diethyl sulfide and 3C, methanol). As expected, the retention time of the individually extracted and resolved compounds matches that of the composite peak. EIC allow for quantitation of each individual contributor to the composite peak and clearly demonstrates one of the greatest advantages of the MSD.

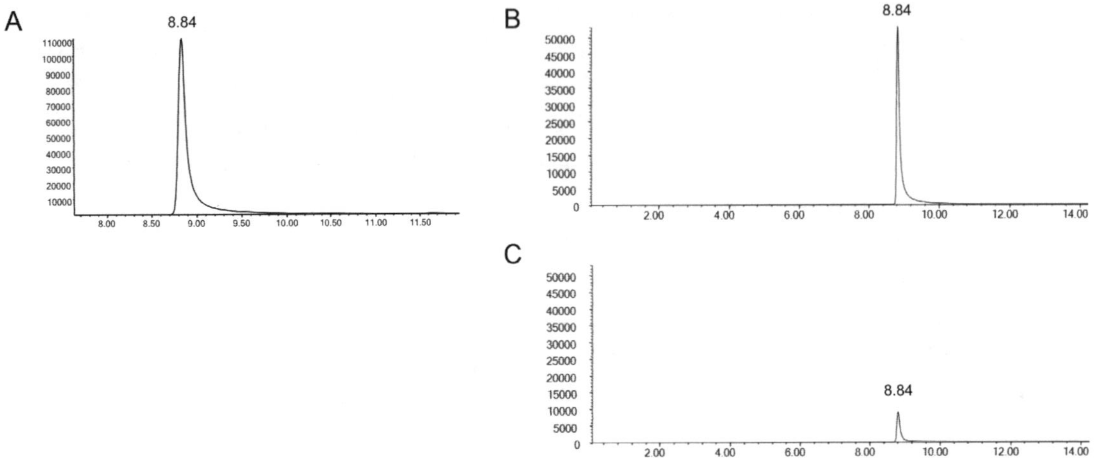


**Figure 3 – Extracted ion chromatograms resolve co-eluting compounds.**
A. Chromatogram showing co-elution of diethyl sulfide and methanol at 8.84 minutes.
B. EIC of diethyl sulfide. C. EIC of methanol.

MSDs also allow for the identification of unknown peaks that might appear in a chromatogram. The analyzer can be configured to identify the mass (m/z) of the peak which can then be found in a mass spec library. Alternatively, if the presence of a specific compound is suspected, the indicative masses (m/z) for this analyte can be quickly added to the analyte list to determine if the compound is present. Subsequent analysis of a prepared sample can be confirmatory and quantitative.

Lastly, GC-MSD can offer triple confirmation of an analyte by agreement of the expected retention time, correctness of the spectrum of the molecular and fragment ions, and the coincidence of those ions in the extracted ion chromatograms at the expected retention time.

## Case Study – Instrument Design

The process GC "Monomer Analyzer" used in this study was designed and built by Wasson-ECE Instrumentation, Fort Collins, CO (Figure 4). It was designed for one of the largest petrochemical companies in North America to speciate C1-C4 hydrocarbons and numerous common catalyst poisons ($H_2S$, COS, methyl mercaptan, ethyl mercaptan, methanol, arsine, and phosphine) in an ethylene matrix.

Temperature-programmed capillary chromatography was required to resolve and quantify such diverse analytes. Numerous capillary columns were housed in two patented, programmable, micro-convection ovens (MCOs) which wrap around the analytical columns to control temperatures precisely and rapidly. Packed columns were also used and housed in an isothermal oven. Electronic pressure and flow programming controlled the performance and resolving power of the analytical columns.

Column eluates were directed to three multiplexed detector trains (MSD, PDHID and FID). Combining the resolving power of capillary chromatography with the selective and discriminatory power of the MSD made GC-MS the best choice to address the oxygenates, sulfurs, and metal

A B

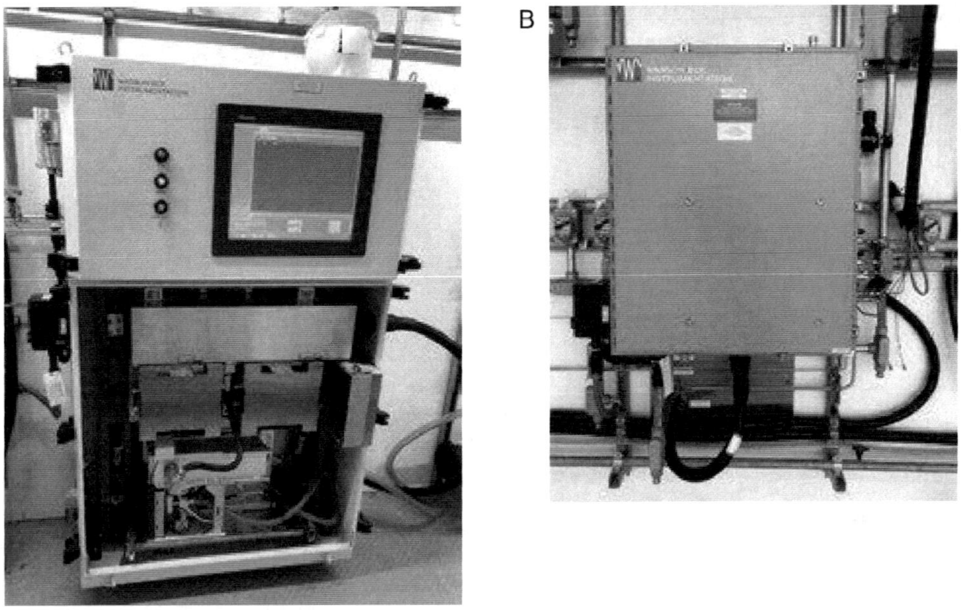

**Figure 4 – Process GC Monomer Analyzer and Sample System.**
A. Process gas chromatograph within the analyzer shelter. B. Sample conditioning system mounted outside of the shelter, directly opposite of the PGC. Samples are ported from the sample system to the analyzer through sample lines housed within a temperature-controlled heated bridge (black flexible tubing).

hydrides. An Agilent Technologies 5977B Mass Spectrometer was used. A Valco PDHID was selected for the permanent gas analysis to achieve the low ppb sensitivities requested[3]. The common combination of a methanizer and an FID causes tailing of the CO and $CO_2$ peaks which decreases the sensitivity for these analytes. PDHID does not suffer from this limitation. An FID is the best detector for hydrocarbon analysis and a Wasson-ECE FID was chosen for this application. The most common hydrocarbon impurities in monomer streams are hydrocarbons lighter than the monomer of interest. In the case of ethylene, methane and ethane are the most frequently observed contaminants. Heavier hydrocarbons were also potentially present and were characterized as well.

This instrument used a single chromatographic method to control sample acquisition and simultaneous analysis by the three detector trains. Unambiguous analyte detection and quantification was completed in under 10 minutes.

A custom sample conditioning system was designed to be mounted outside of the shelter and to accept eight sample streams: two ethylene streams and 6 calibration streams. Stream-selection valves delivered the chosen sample to sample-injection valves. Upon actuation, the injected sample was delivered, via a temperature-controlled bridge, to the analyzer inside the shelter (Figure 4).

Once inside the analyzer, column-selection valves directed the sample to the appropriate analytical column for chromatographic separation and elution to the appropriate detector for identification and quantification. This design, with the sample system outside of the shelter and the analyzer inside the shelter, allowed for the bulk of the hazardous sample to be kept away from potential ignition sources inside the shelter. Only very small aliquots of sample gas (0.1 -

1.0ml each) were delivered to the Class 1, Division 2 hazardous area rated analyzer. Where appropriate, inert tubing and hardware was used to prevent analyte loss within the instrument.

## Case Study – Results

A 50-meter capillary column in MCO2 resolved methanol, methyl mercaptan and ethyl mercaptan from a 10 ppm calibration blend (Fig. 5). The reference methods used were UOP 1015-17[4], UOP 1021-19[5], UOP 1022-18[6], UOP 1023-18[7], UOP 1024-18[8]. Calculated minimal detectable limits (MDLs) were methanol (25 ppb), methyl mercaptan (10 ppb) and ethyl mercaptan (15 ppb). Simultaneously, another capillary column in MCO2 resolved $H_2S$ and COS. The MDL for COS was 10 ppb while that of $H_2S$ was 40 ppb. Arsine, and phosphine (10 ppm each, Fig. 5) were also quantified with MDLs of 15 ppb and 30 ppb, respectively. Note that the retention times for many of these analytes are very similar and it would be difficult to resolve and quantitate them with another type of detector. However, the MSD's ability to generate an extracted ion chromatogram allowed for complete resolution and ppb level quantitation of these important compounds. The results of the process GC-MS system showed outstanding performance consistent with laboratory GC-MS systems.

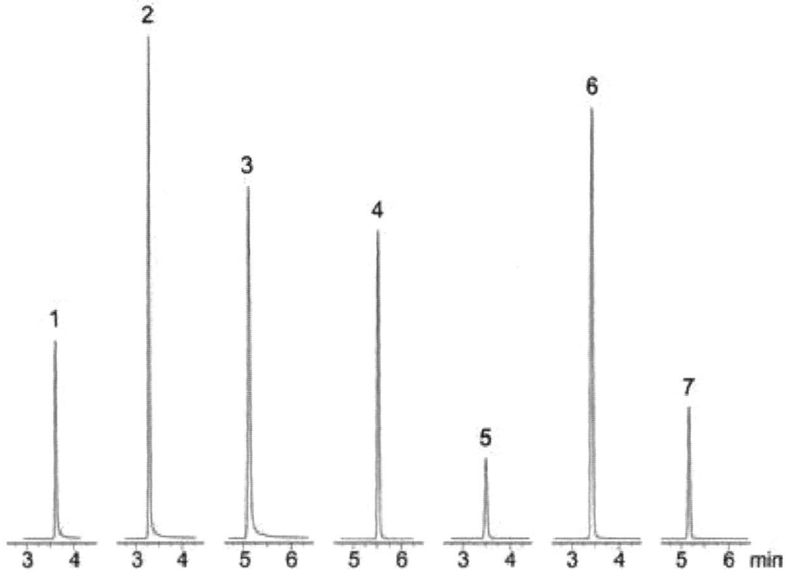


**Figure 5 – Sulfurs, oxygenates, and hydrides by MSD.** Extracted ion chromatogram results for 1) methanol, m/z 31; 2) methyl mercaptan, m/z 47; 3) ethyl mercaptan, m/z 62; 4) carbonyl sulfide, m/z 60; 5) hydrogen sulfide, m/z 34; 6) arsine, m/z 76; and 7) phosphine, m/z 34.

At the same time, the PDHID train quantified CO, $CO_2$, $H_2$, $O_2$/Ar and $N_2$ from a 10ppm calibration blend. The reference method used was ASTM D8098-17[9]. Two column sets in the isothermal oven, each with one packed and one capillary column, were used to separate the analytes (10ppm each, Fig. 6). The chromatogram shows two sections with a rising baseline at 3.7 minutes resulting from the switch between the column sets. The first column set resolved and quantified hydrogen, oxygen/argon, nitrogen, and carbon monoxide. The second column then set quantified carbon dioxide. The observed baseline shift is a consequence of switching column sets during the run. The MDLs for these analytes were hydrogen (100 ppb), oxygen/argon and nitrogen (500 ppb), carbon monoxide (20 ppb), and carbon dioxide (30 ppb). Methane was observed on each column set but analyzed by the FID (shown below).

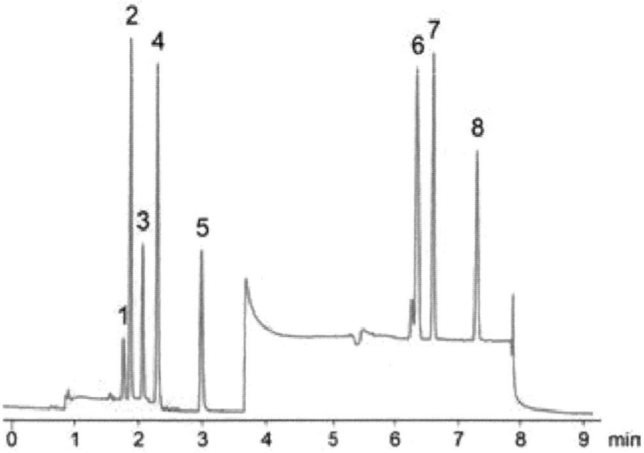


**Figure 6 - Permanent Gases by PDHID**
Chromatogram of PDHID demonstrating 1) hydrogen, 2) oxygen/argon composite, 3) nitrogen, 4) methane*, 5) carbon monoxide, 6) hydrogen/air composite*, 7) methane*, and 8) carbon dioxide. *Not quantified.

While the MSD and PDHID trains were analyzing their respective compounds, the FID speciated and quantified hydrocarbons. The reference method used was ASTM D6159-17[10]. Here, an ethylene stream was interrogated for hydrocarbons (200 ppm refinery gas mix) which

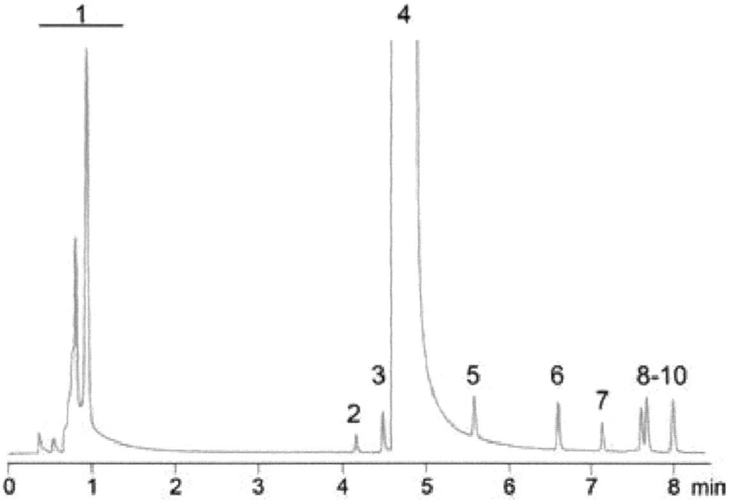


**Figure 7 - Hydrocarbons by FID**
Chromatogram showing results for 1) C5+ backflush, 2) methane, 3) ethane, 4) ethylene*, 5) propane, 6) propylene, 7) acetylene, 8) propadiene, 9) isobutane, and 10) n-butane. *Matrix

were resolved on a 50-meter column in MCO1 and speciated by a flame ionization detector (FID, Fig.7). C1-C4 hydrocarbons were speciated while heavier species were backflushed as C5+. The MDL was determined to be 2 ppm for these components.

The precision of the micro-convection oven temperature and electronic pressure controls were also examined in a separate engineering study (Table 1). Retention time reproducibility is a measure of how tightly controlled the temperature, pressure and backpressure in a system are. To examine the retention time reproducibility, data taken over two days from eight injections of a hydrocarbon mixture were obtained. The relative standard deviation (%RSD) of the results was calculated and found to be excellent for all eight compounds in the blend. Peak area reproducibility is measure of a systems' ability to repeatably deliver the same amount of a sample to the analytical column. It is a function of split inlet performance and mass flow control. To evaluate these parameters, 5% methane in argon was injected twelve times. Again, %RSD results obtained rival those obtained on laboratory GCs and are unparalleled in on-line PGCs.

**Table 1 – Micro-Convection Oven Performance**

A. Retention Times

| | Methane | Pentane | Hexane | Heptane | Octane | Nonane | Decane | Undecane |
|---|---|---|---|---|---|---|---|---|
| Day 1 | 2.744 | 3.619 | 4.687 | 6.645 | 9.581 | 13.197 | 17.099 | 21.035 |
| | 2.775 | 3.653 | 4.723 | 6.683 | 9.613 | 13.225 | 17.118 | 21.038 |
| | 2.746 | 3.625 | 4.696 | 6.66 | 9.603 | 13.223 | 17.12 | 21.047 |
| | 2.747 | 3.626 | 4.697 | 6.663 | 9.604 | 13.223 | 17.118 | 21.049 |
| | 2.747 | 3.626 | 4.698 | 6.666 | 9.606 | 13.224 | 17.121 | 21.049 |
| Day 2 | 2.747 | 3.625 | 4.694 | 6.659 | 9.598 | 13.213 | 17.108 | 21.036 |
| | 2.747 | 3.627 | 4.698 | 6.665 | 9.603 | 13.217 | 17.111 | 21.029 |
| | 2.745 | 3.624 | 4.694 | 6.662 | 9.601 | 13.215 | 17.106 | 21.033 |
| Retention Time | | | | | | | | |
| Avg. | 2.750 | 3.628 | 4.698 | 6.663 | 9.601 | 13.217 | 17.113 | 21.040 |
| StDev | 0.010 | 0.010 | 0.011 | 0.010 | 0.009 | 0.009 | 0.008 | 0.008 |
| **RSD%** | **0.373** | **0.285** | **0.225** | **0.157** | **0.096** | **0.070** | **0.046** | **0.037** |

Data from 8 injections of hydrocarbon blend over 2 days.

B. Peak Areas

| | Methane | |
|---|---|---|
| Peak Areas | 83767 | 83886 |
| | 84037 | 84020 |
| | 83813 | 83891 |
| | 83992 | 83966 |
| | 83727 | 83806 |
| | 83880 | 83859 |
| Peak Area | | |
| Avg. | 83887 | |
| StDev | 100.155 | |
| **RSD%** | **0.119** | |

Data from 12 injections of 5% methane in argon.

## Conclusion

The goal of this study was to design a single PGC capable of ppb detection of analytes of importance for ethylene and propylene producers and their customers. A novel approach using process GC-MS was developed to take full advantage of the sensitivity and specificity of MSDs. Tightly controlled and temperature-programmable ovens allowed for the use of fused silica capillary columns for the greatest resolving power. Parallel chromatography with the MSD, PDHID, and FID detector trains allowed for analysis of an extensive range of possible monomer contaminants at ppb levels not previously attainable with process GCs. Together, the data presented here clearly demonstrate that, when using the same technologies available to laboratory GCs, lab-quality performance can be achieved from a process instrument.

## References

1. Figure modified from the original, https://commons.wikimedia.org/wiki/File:Quadrupole_ion_trajectory.svg (accessed March 15, 2021).
2. Silverstein, R. M., Webster, F. X., & Kiemle, D. J. (2005). Spectrometric Identification of Organic Compounds (7th ed.). Wiley.
3. Wentworth, W.E., Vasnin, S.V., Stearns, S.D., and Meyer, C.J. (1992). "Pulsed Discharge Helium Ionization Detector," in Chromatographia, Vol. 34, No. 5-8 Sept./Oct. 1992.
4. UOP Standard 1015-17, 2017, Determination of Trace Oxygenates in Polymer Grade Ethylene & Propylene by Gas Chromatography Mass Spectrometry, Honeywell UOP, Des Plaines, IL, 2017, https://www.astm.org/Standards/UOP1015.htm
5. UOP Standard 1021-19, 2019, Determination of Trace Methyl Mercaptan, Ethyl Mercaptan, and Isopropyl Disulfide in Polymer Grade Ethylene and Propylene by Gas Chromatography Mass Spectrometry, Honeywell UOP, Des Plaines, IL, 2019, https://www.astm.org/Standards/UOP1021.htm
6. UOP Standard 1022-18, 2018, Determination of Trace Carbonyl Sulfide and Hydrogen Sulfide in Polymer Grade Ethylene and Propylene by Gas Chromatography Mass Spectrometry, Honeywell UOP, Des Plaines, IL, 2018, https://www.astm.org/Standards/UOP1022.htm
7. UOP Standard 1023-18, 2018, Determination of Trace Arsine and Phosphine in Polymer Grade Ethylene & Propylene by Gas Chromatography Mass Spectrometry, Honeywell UOP, Des Plaines, IL, 2018, https://www.astm.org/Standards/UOP1023.htm
8. UOP Standard 1024-18, 2018, Determination of Trace Permanent Gases in Polymer Grade Ethylene and Propylene by Gas Chromatography-Pulsed Discharge Helium Ionization Detector, Honeywell UOP, Des Plaines, IL, 2018, https://www.astm.org/Standards/UOP1024.htm
9. ASTM Standard D8098-17, 2017, Standard Test Method for Permanent Gases in C2 and C3 Hydrocarbon Products by Gas Chromatography and Pulse Discharge Helium Ionization Detector, ASTM International, West Conshohocken, PA, 2017, https://www.astm.org/Standards/D8098.htm
10. ASTM Standard D6159-17, Standard Test Method for Determination of Hydrocarbon Impurities in Ethylene by Gas Chromatography, ASTM International, West Conshohocken, PA, 2017, https://www.astm.org/Standards/D6159.htm

# 620136 Using Seeq to Empower SMEs in Ethylene Production

**Abstract:**

*The fourth industrial revolution dramatically increased the number IIoT sensors present in chemical manufacturing, corresponding to a wealth of time-series process data. Advanced analytics software applications like Seeq have become transformational in organizations, leveraging this wealth of data and turning information to insight. Seeq's low-to-no-code environment empowers process subject matter experts (SMEs) to apply advanced statistics and machine learning concepts to data from any of their connected historian or contextual data sources. Using Seeq has empowered SMEs in Ethylene production facilities to solve many previously unsolvable use cases. The use cases covered in this presentation will include: Analysis of Furnace De-cokes, Heat Exchanger Maintenance Prediction, Modeling Downstream Quality Parameters from Upstream in-process Data, and Advanced Asset Monitoring.*

# 619238 Applying AI/ML in Ethylene for Real Time Intelligence

**Abstract:**

## Abstract

**Title: Applying AI/ML in ethylene for real time intelligence**

**Author: Sameer Thorat**

Digital Transformation is driving major improvements in the process manufacturing industry. This is particularly seen in ethylene plants. The inherent complexities in the ethylene process makes application of data analytics a significantly challenging process but the insights that create the returns are many. With right domain expertise, in-depth process knowledge and appropriate data science techniques combined with first principal models, a customized data analytics solution can be applied in the ethylene process to meet individual plant needs and effectively drive improvements. Digitalization has made it possible to implement innovative solutions that efficiently provide real time intelligence for operations.

Digital transformation principally aims to enhance the traditional technical and manufacturing skills sets with new advanced techniques that together drive manufacturing process to new levels of excellence by identifying opportunities and eliminating inefficiencies in the process. Advance analytics provides the right information at right time to drive the right decisions without overwhelming operations with “data overload”. Driven by technology, Industry 4.0 is transforming the ethylene process and reshaping operation paradigms to create greater efficiency and synergies in operating strategy and solve challenging problems.

This paper discusses the various opportunities in the ethylene process and the enormous improvements possible from application of AI/ML solutions. Challenges in achieving the desired objectives of various stakeholders and approaches to ensure successful application of AI/ML models will be addressed. Demonstration of apt usage of technology to successfully drive benefits from the analytics based solutions will be provided.

The Industry 4.0 Digitalization initiative is driving industry specific advanced analytics to leverage both historic and real-time data to achieve improvements that are difficult to accomplish using traditional approaches. Substantial economic gains can be achieved when data analytics is coupled with the overall plant operating strategy. This paper will cover some of case studies to demonstrate on how manufacturers were able to improve their ethylene unit performance effectively by the “Digital Transformation” process.

# 617356 Improved Fouling and Corrosion Control in New Ethane Cracker Reduces Operating Expenses By $1.4 Million per Year.

**Abstract:**

Braskem Idesa, in collaboration with Dorf Ketal and Solenis, reports on steps taken to improve efficiency and reliability of Cracked Gas Compressor (CGC) and downstream units at the Coatzacoalcos, Mexico ethylene plant. This ethane cracker, commissioned in 2016, started with wash oil and wash water as additives to control CGC fouling. Compressor fouling issues led to a change in wash oil treatment that improved fouling control in the compressor, but created an emulsion in the CGC knock-out drums. This emulsion caused corrosion and fouling problems in the Dilution Steam Generator and Quench Water systems, increasing costs for energy, reboiler maintenance, filter replacements, and effluent treatment. This paper discusses actions taken to reduce operating costs by an estimated $500,000 per year without capital investment. Root cause analysis, and modifications in cleaning procedures, monitoring, and treatment chemistry are presented.

# 619471 2021 AIChE Spring Meeting & 17th GCPS Wednesday Keynote: Advancing the Circular Economy

**Abstract:**

The past several decades have illustrated the integral role plastics play in our society. From enabling modern conveniences to playing a role in the advancement of our sustainability goals, plastics are key.

Looking to the future, LyondellBasell is taking a leading role in advancing the circular economy. Ken Lane, LyondellBasell's executive vice president of Global Olefins & Polyolefins, will discuss the company's ambitions and comprehensive efforts to eliminate plastic waste in the environment.

# 618961 *Steam Blow Line Split – Cause, Corrections, and Recommendations*

**Abstract:**

*Standard industrial practice for cleaning steam circuits is the use of a low-pressure continuous steam blow, where a steam flow is controlled at conditions defined to create pipe wall forces above what is expected under normal operation. The steam control is commonly accomplished with the use of an orifice plate or a 'sacrificial' valve. During the 2017 commissioning of an ethylene production facility in Freeport, Texas a steam line failed and split open while performing a low-pressure continuous steam blow. This paper discusses the details of the failed steam blow, the cause of failure (line irregularities, acoustic induced vibration), and the leverageable corrections made to effectively and safely re-establish and complete the steam blow.*

**Ethylene Producers' Conference**

**Ethylene Plant Operations I**

**Paper Number: 82C**

# Energy Savings for Steam Cracking Furnaces by Combustion Air Preheating Technology


***Fanxu Meng***
*Technip Energies, Houston, Texas, USA*

***Yong Wang***
*Technip Energies, Houston, Texas, USA*

***Miguel Maldonado***
*Technip Energies, Houston, Texas, USA*

***Joel Guillaume***
*Technip Energies, Houston, Texas, USA*

***Mingwei Shi***
*Beijing Aerospace Petrochemical EC and EP Technol. Co. Ltd. (BAEEC), Beijing, China*

***Songlin Shao***
*Beijing Aerospace Petrochemical EC and EP Technol. Co. Ltd. (BAEEC), Beijing, China*

***Xiaofeng Li***
*Beijing Aerospace Petrochemical EC and EP Technol. Co. Ltd. (BAEEC), Beijing, China*

***Xiaofeng Song***
*Beijing Aerospace Petrochemical EC and EP Technol. Co. Ltd. (BAEEC), Beijing, China*

***Yue Liu***
*Beijing Aerospace Petrochemical EC and EP Technol. Co. Ltd. (BAEEC), Beijing, China*



Prepared for Presentation at the 2021 AIChE Virtual Spring Meeting & 17th Global Congress on Process Safety Houston, Texas, April 18 - 23, 2021

## Abstract

Cracking furnace energy savings is a major objective of petrochemical companies. Combustion air preheating technology for the ethylene cracking furnace burners allows for the recovery of some of the low-level waste heat rejected in the furnace effluent stream. Effective recovery of this low-level heat is challenging because its quality is not high enough to be otherwise effectively utilized and is usually removed by cooling with water and/or air.

The patented combustion air preheating technology discussed in this paper is a high-efficiency heat transfer technology. It is designed based on an understanding of the furnace operation, the plant heat balance, and experimental research. This technology utilizes low-level heat sources, such as low-pressure steam (150 ~ 200°C), quench water (60 ~ 90°C), and low-pressure condensate (100 ~ 200°C), to preheat the combustion air to the cracking furnaces, thus reducing total firing duty. As the fuel gas consumption is reduced, $CO_2$ emissions from the flue gas to environment are also reduced. No additional equipment (such as pressurized water pump, forced draft fan, motor, etc.) is required. The system does not require any automatic control. The air preheater of each burner can be easily switched on and off with the furnace on-line. The combustion air preheater needs only simple maintenance and its working life is more than 20 years. By 2020, this technology has been applied in more than 300 ethylene cracking furnaces and other negative-pressure heating furnaces in revamp and grassroot plants in China. It has proven its economic benefit, emission reduction, and flexible and reliable operation.

This paper discusses the fundamentals of the combustion air preheating technology. According to many years of practical experience, the energy savings of adding combustion air preheating system can be significant. A case study in Northeast China shows combustion air temperature is heated to 70°C from an atmospheric temperature of - 5°C by preheater, as 81°C quench water is used as the heat source. Energy savings, which are achieved by utilizing low-level waste heat, are better when combustion air is heated up to a higher temperature. The normal operation of cracking furnaces is not substantially affected after air preheaters are added. Case studies will demonstrate the cracking furnace thermal efficiency improvement, fuel reduction benefit and the need for a special design for each project because of the individual heat sources and site conditions.

## 1. Introduction

Ethylene cracking furnaces are the most significant contributors to the specific energy consumption in an ethylene plant, as they combust fuel gas in a firebox to provide high temperature for hydrocarbon cracking into ethylene and other valuable products. The fuel gas is fired through burners under the control of radiant coil outlet temperature (COT) to reach desired cracking severity. For 100% floor-fired cracking furnaces, the burners are mounted in the firebox floor and fire vertically upwards. The burners will all be capable of firing internally-produced residue gas in combination with supplemental/make-up fuel, as well as start-up and back-up fuels, with excess combustion air to avoid sub-stoichiometric operation.

Combustion air is typically pulled in across the burner damper by the negative pressure at the furnace floor, either with or without an induced draft (ID) fan maintaining a negative pressure at the radiant roof (arch). The air is controlled by adjusting the burner dampers or the draft set point. Operating at a higher draft will pull in additional combustion air. The flue gas, consisting of excess combustion air and stoichiometric combustion fuel gas products, exits from the furnace stack. Before exiting, heat in the flue gas is utilized in the firebox to provide the heat of cracking and residual heat is recovered in the furnace convection section to pre-heat hydrocarbon feedstock and to produce super-high-pressure (SHP) steam. Additional fired duty, with associated fuel gas consumption, is required to heat up excess combustion air in the firebox. If combustion air is preheated, fuel consumption is reduced when fuel is combusted with the same amount of excess air. Consequently, the total $CO_2$ produced by stoichiometric combustion of fuel gas is reduced.

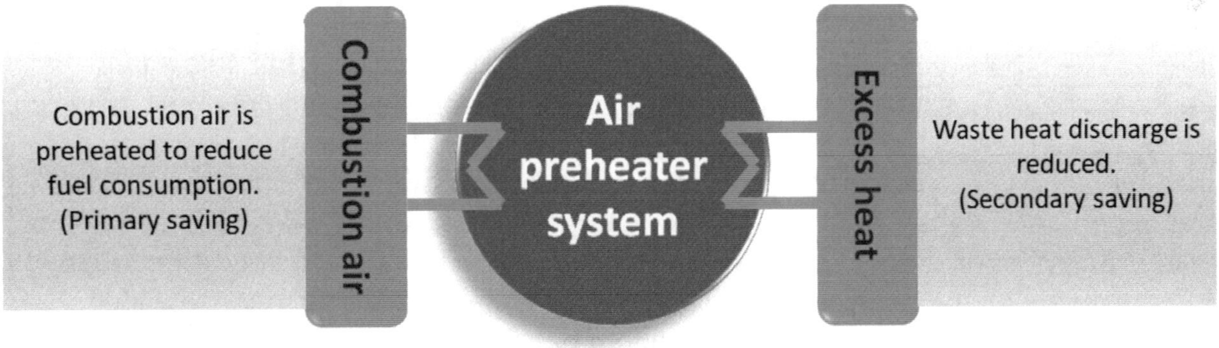


**Figure 1.** Air preheater system can save energy by reducing both fuel consumption and waste heat discharge.

The ethylene plant provides a great opportunity to integrate and optimize processes to recover energy effectively. For example, quench exchangers are used to recover high-level heat in cracked gas from furnace, producing SHP steam for utility. However, some of the low-level waste heat rejected in the furnace effluent stream is difficult to be effectively recovered. This part of energy, ultimately derived from fossil fuel, is wasted. To comply with operational procedures and environmental regulations, low-level waste heat is usually removed by cooling water and/or air, which adds operational and environmental burdens. Combustion air preheating technology allows the recovery of some of the low-level waste heat. Therefore, it can save energy by reducing both fuel consumption and waste heat discharge (**Figure 1**).

## 2. Air Preheater Fundamentals

Different types of low-level heat sources, such as quench water, low-pressure steam, and low-pressure condensate, have been proven useful for combustion air preheaters in actual ethylene plants. The applicability and energy-saving potential of these three types of low-level heat sources is determined by their temperature range (**Table 1**) and availability on an actual plant. **Figure 2** shows a schematic of air preheater using quench water in a gas cracker.

**Table 1.** Potential low-level heat sources for air preheater

| Heat Sources | Typical Temperature Range |
|---|---|
| **Low-pressure Steam** | 150 – 200°C |
| **Low-pressure Condensate** | 100 – 200°C |
| **Quench Water** | 60 – 90°C |

Hydrocarbon feedstock is converted into ethylene and other products in the cracking furnaces at high temperature, the heat for which is provided by fired fuel gas in the firebox. The cracked gas contains not only the valuable chemicals, but also a large amount of heat. This heat is recovered by a series of processes and units, one of which is the Quench Water Tower (QWT). The QWT uses quench water to cool the cracked gas upstream of the cracked gas compressor. Quench water is circulated and cooled and routed back to the QWT. The low-level waste heat in quench water can be used by a limited number of users in an ethylene plant, as its temperature is only about 60 to 90°C . Coolers are necessary to remove the excess heat before the quench water returned to the QWT. This provides an opportunity for addition of air preheater to utilize a portion of low-level waste heat in quench water. Combustion air is heated up through air preheaters and drafted into furnace burners, where it is mixed with the fuel gas and combusted. This sensible heat recovered through air preheaters combines with the heat release from fuel gas combustion.

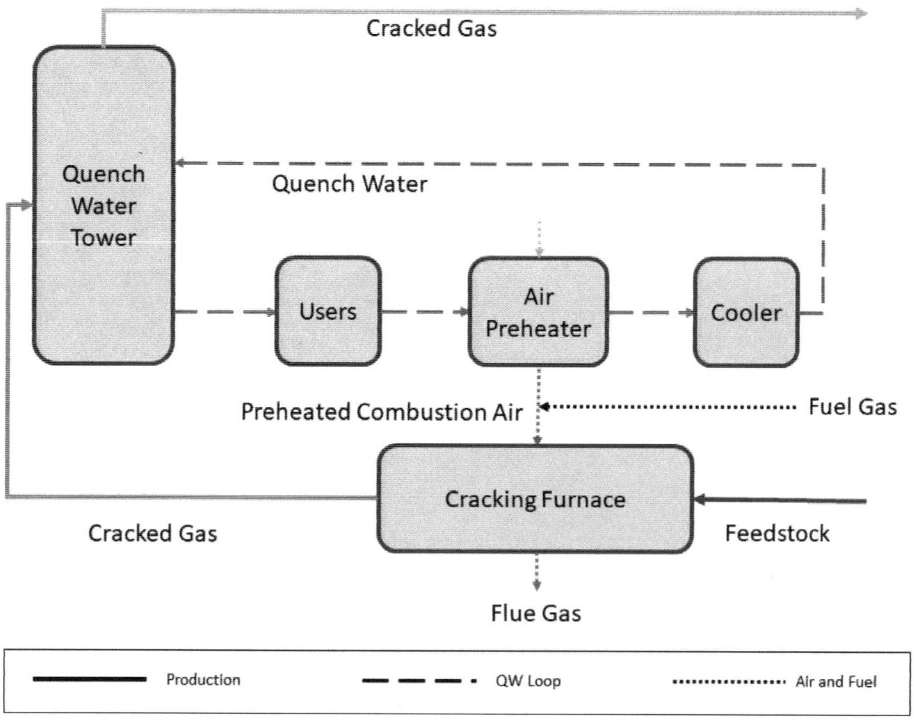


**Figure 2.** Schematic of air preheater using quench water (other major equipment and processes are not shown)

### 3. Case Study on Cracking Furnace

The furnace in the case study has the operating conditions shown in **Table 2**. The fuel savings and energy efficiency increase with combustion air preheaters implemented in an ethylene plant in Northeast China (see **Table 3**). This furnace has air preheaters to increase combustion air temperature to 70°C from average ambient temperature of -5°C. Accordingly, fuel gas consumption is reduced by 4.5%. The furnace operation maintains 10% excess air after adding air preheaters. Consequently, the flue gas flowrate decreases by 4.5%. The client claimed the payback time of the combustion air preheater was less than one year. In addition, $CO_2$ emissions in the flue gas should decrease proportionally. Based on experience, increase in NOx emissions is limited when combustion air is preheated no more than 100°C.

The same cracking severity under unchanged feedstock flowrate is essentially achieved, as COT is set to the same target. Process absorbed heat by coils in firebox and convection section remains substantially the same. Lower fuel gas consumption results in higher fired duty efficiency, as this efficiency is defined as the ratio of absorbed heat to fired heat. A 3.2% increase of fired duty efficiency is observed in this case study.

**Table 2.** Furnace operating conditions

| | Range | Unit |
|---|---|---|
| Dilution Steam to HC Ratio | 0.3 | |
| Coil Outlet Temperature | 835 | °C |
| Coil Outlet Pressure | 0.07 | bar-g |
| Hydrocarbon Mass Flow | $9 \times 10^4$ | kg/h |
| Residence Time | 0.2 | second |
| Excess Air | 10 | % |

**Table 3.** Comparison of utility and efficiency of furnace with and without air preheat (APH)

| | No Air Preheat (-5°C) | Air Preheat (70°C) | Change percentage with APH |
|---|---|---|---|
| Fuel Gas, ton/h | 10.81 | 10.32 | -4.5% |
| Flue Gas, ton/h | 217.3 | 208.4 | -4.5% |
| Fired Duty Efficiency[1], % | 94.5 | 97.5 | +3.2% |
| [1]Fired Duty Efficiency is defined as the ratio of absorbed heat to fired heat | | | |

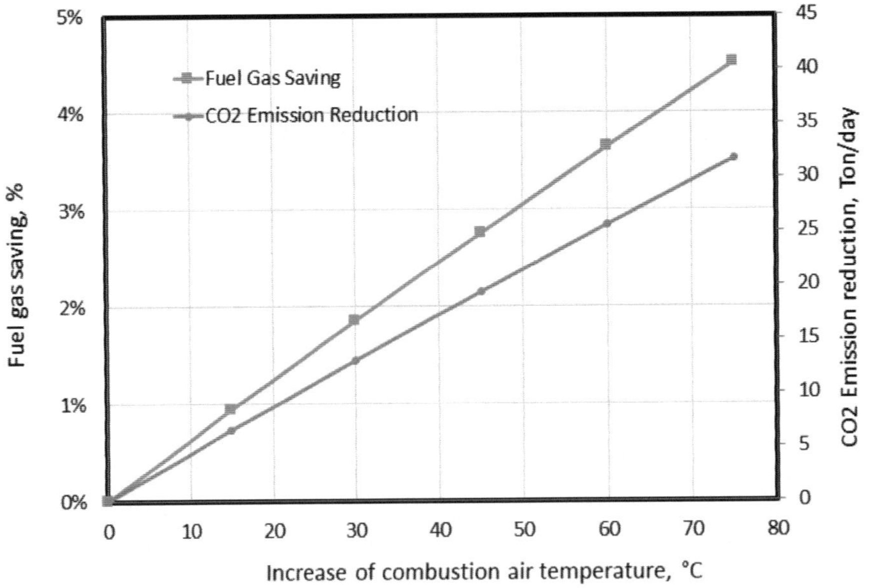


**Figure 3.** Increased combustion air temperature can save fuel gas and reduce $CO_2$ emissions.

A sensitivity study is conducted to demonstrate how an increase of combustion air temperature can enhance fuel gas savings and $CO_2$ emission reduction (**Figure 3**). However, the temperature of preheated combustion air is restricted by two factors:

1) Available low-level heating sources: Usage of low-level waste heat sources for air preheaters limits the highest available temperature to preheat combustion air.

2) Impacts to furnace operation: If combustion air is heated up to above 100°C, attention should be paid to the increase in NOx emissions.

A study is conducted as below, showing changes of key furnace operating parameters after adding combustion air preheaters.

**Figure 4** shows the change overview after combustion air preheaters are added. The changes of duty, flowrate, pressure drop, and efficiency are presented in percentages. The change of temperature is presented in absolute value. In summary, operating parameters of cracking furnaces do not change significantly. Fuel gas and fired duty decrease by 4.5%, as does the flue gas flow rate. The feedstock flowrate remains the same. As the COT and cracked gas effluent outlet temperature are similar, quench exchangers' duties remain basically unchanged. A lower flowrate of flue gas passes through the convection section, resulting in less heat recovered by the convection section. Each bank duty decreases, though the percentage is different from bank to bank. Hydrocarbon plus dilution steam (HC+DS) bank is at the bottom of the convection section. Even though the flowrate of flue gas decreases, the high log mean temperature difference still drives the efficient heat transfer, so this bank has a lower duty reduction (1.7%) than the overall convection duty reduction (4.1%). As less waste heat is available in flue gas due to lower flow rate and temperature entering the super high-pressure steam super heater bank (SHPSS), duties in these banks decrease and less boiler feed water (BFW) is required to make up the steam production. Accordingly, less SHP steam is generated by the furnace.

As the flue gas flows up the convection section, the available waste heat becomes less and less and duties of banks all decrease accordingly. The stack temperature decreases as well, which means higher overall furnace thermal efficiency is achieved. Attention needs to be paid to the lower stack temperature to prevent condensation from the flue gas. The overall pressure drop across the convection section decreases due to the lower flue gas flowrate. The ID fan has enough margin for additional pressure drop across added air preheaters. Figure 4 shows the comparison between furnaces with and without air preheaters. It also shows that a furnace revamped with air preheaters has a lower crossover temperature (XOT) than an existing design. For a new design, the XOT can be optimized to avoid the drop.

Low pressure steam is used as the heating source in one project in East China. Combustion air is heated up to 90°C from 5°C by using 200°C low pressure steam. Pressure drop of the combustion air side is 30 Pa at the max operating condition. Natural drafted burners are operated with 10 – 15% excess air.

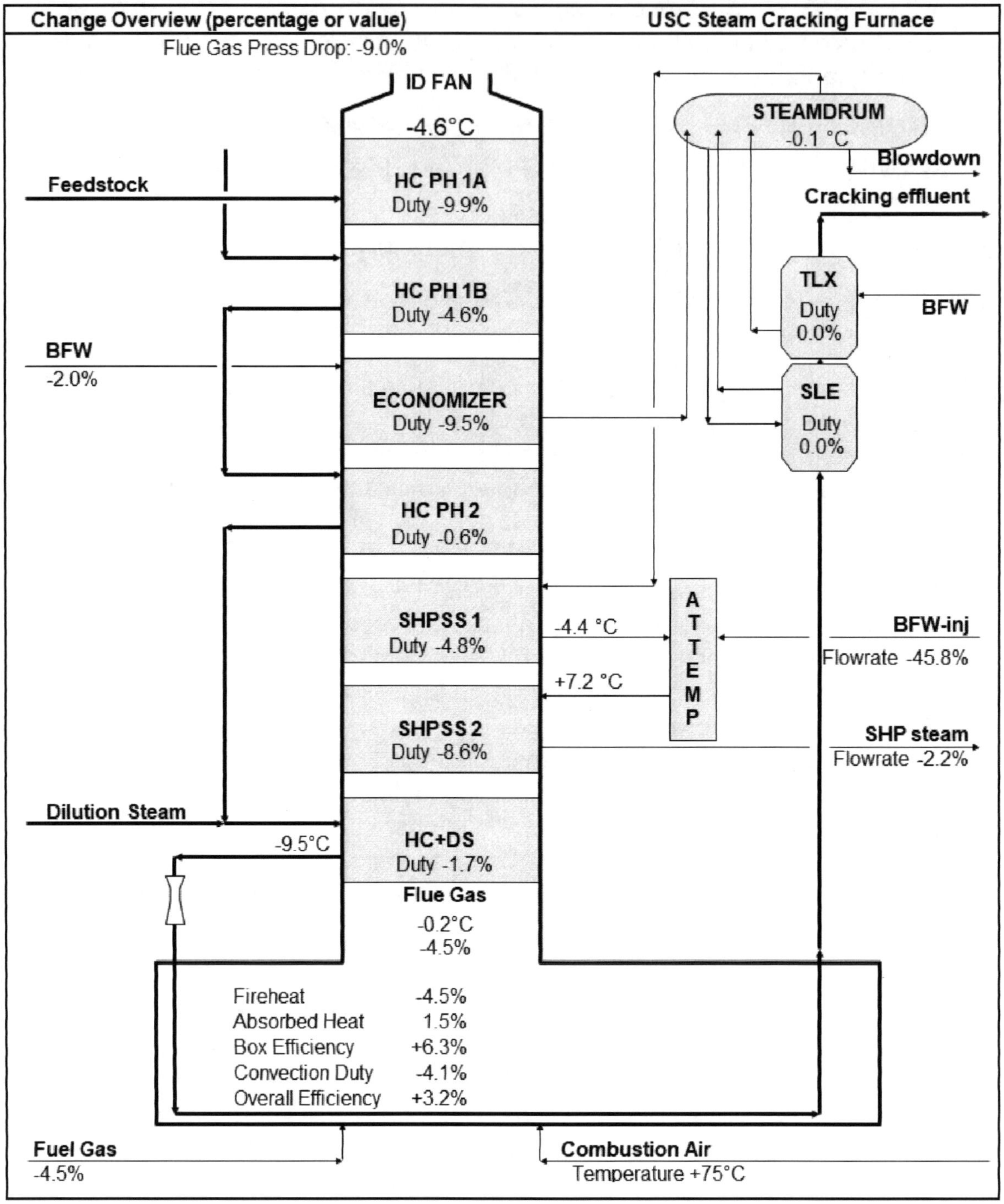


**Figure 4.** Changes (deltas) of key furnace operating parameters after adding air preheater (COT remains the same in the study)

## 4. Air Preheater Design

**Figure 5** shows the proprietary design of Beijing Aerospace Petrochemical EC and EP Technology Corporation Limited (BAEEC) combustion air preheaters. The preheaters (i.e., finned tube heat exchangers) are located at the burner combustion air inlet. The compact design allows for each of combustion air inlet to be equipped with an independent dedicated air preheater. The low-level waste heat medium, such as quench water, distributes inside the tubes from a header. Combustion air passes through the shell side and is heated up before entering burners and mixing up with fuel gas. Thermocouples are mounted on the preheater shell side to monitor the combustion air preheating temperature. Valves are installed on the heating medium piping, controlling in-and-out heating medium through the heat exchangers. A bypass damper is provided on the air side for the purposes of control, maintenance, and flexible operation. As the air preheaters are located next to the burners, they are easily accessible.

The proprietary design of air preheaters is unique for each application. Specific considerations about ethylene furnace must be included, such as the combustion air flow drafted by negative pressure, available space, effects to burner operation, safety, etc. The BAEEC combustion air preheater design minimizes the pressure drop across the heat exchanger. There is very little additional draft requirement from furnace existing ID fan. There is therefore no requirement for a forced draft fan to blow the air through the preheating system. Complicated air ducting is also not required. The implementation of the BAEEC combustion air preheater is quite simple. It can fit in both grassroots and revamp projects. Regarding the air preheater revamp system, fully understanding the plant operation and heat balance is essential, as the low-level waste heat sources need to be introduced from the existing process to the preheaters. This consideration ensures the high energy savings and low revamp investment.

The air preheater system consists of multiple furnaces and hundreds of air preheater devices. It is more like an integrated system rather than a single unit, as seen in **Figure 6**. This proprietary design has two remarkable major technical features:

1) No automatic flow control element is required even though the entire system consists of hundreds of air preheaters with the same heating source. The integrated heating medium distribution system includes the main header as well as the first, second and third level branches to evenly distribute the heating medium to individual air preheater without instrumentation control. The unique proprietary design ensures the balanced flowrate distribution and the temperature consistency of the preheated combustion air in all preheater outlets.

2) The system is resilient to the heating medium fluctuations. When the total flow rate of the heating medium changes, the entire system can automatically adjust the flow rate distribution and settle out on a new balance because of its unique design. This means

the temperature consistency of the preheated combustion air is guaranteed, no matter how the heating source fluctuates.

**Figure 5.** Air preheater design

**Figure 6**. Air preheater system

## 5. Air Preheater Features for Application

This patented air preheater technology has been applied to more than 300 ethylene cracking furnaces and other negative-pressure heating furnaces in revamp and grassroot plants, as well as other industrial furnaces, in China. It has proved its economic benefit, emission reduction, and flexible and reliable operation in applications. Major features are summarized below.

### *5.1 No additional utility consumption*

The technology adopts the available low-level waste heat resources in an ethylene plant as the heating medium. The additional heat recovery system does not add any utility consumption, such as water, steam, and electricity. The air preheat system, requires minimal operation attention. No additional equipment (other than the air preheater) is required (such as pressurized water pump, forced draft fan, motor, etc.).

### *5.2 Multiple benefits*

Air preheaters for ethylene cracking furnace can save energy by reducing fuel consumption (primary saving) and reducing waste heat discharge, which requires additional energy to cool down (secondary saving). As an example, quench water has to be cooled down with cooling water and/or air. If additional quench water duty is utilized in the combustion air preheaters, cooling water duty and/or electrical power will be reduced. This can specially benefit ethylene plants located in dry environments where water supplies may be scarce. Furthermore, as the fuel consumption is reduced, the $CO_2$ emissions from the furnace stack are reduced, which is beneficial to environmental protection efforts. Up to 2020, energy-saving revamps with this patented technology have saved about 300,000 tons of fuel per year (calculated by standard coal with heating value of 7,000 kcal/kg), reduced carbon emissions by about 500,000 tons, and created economic benefits of about 80 million dollars.

### *5.3 High efficiency of heat transfer in preheaters*

The design of the air preheater device is important to enable high-efficiency heat transfer, so that the low-level waste heat is fully utilized. The high-efficiency heat transfer technology is a patented design by BAEEC, which has broad industrial experiences with this application.

### *5.4 Good adaptability and resiliency to heat source*

The technology has good adaptability to use different heat sources, such as chemical plant low-pressure steam (150 ~ 200°C), quench water (60 ~ 90°C), low pressure condensate (100 ~ 200°C), sewage from boiler and other sources. The system design enables automatic balancing with the heating source fluctuations.

### *5.5 Flexible operation*

The device does not require an automatic control system. When an independent air preheater system supports multiple furnaces in parallel, the preheater bundles of each furnace can be switched on and off easily, providing flexibility. Notably, each air preheater in a furnace can be turned on and off independently from other preheaters in the same firebox. Although the combustion air temperature and fired efficiency are different for a furnace with or without a preheater, that furnace can be operated functionally no matter whether the preheater is on or off.

### *5.6 Long working life*

The air preheater revamp system does not require additional rotating equipment, allowing stable and simple operation and less maintenance. The whole system does not create any new waste discharge. It is proven to have a working life of more than 20 years and multiple benefits with a one-time investment.

## 6. Conclusion

Air preheaters for cracking furnaces display great features to save energy for an ethylene plant, without increasing utility consumption or complicated design. The operation impact is limited and predictable using SPYRO. This opens a door to revamp the ethylene furnace in an efficient and economical way, looking for energy savings and potentially environmental benefits. Integration of the heat balance of the cracking furnace and the balance of the plant is the key to designing an effective air preheater system.

TECHNIP ENERGIES

Paper Number: 82C

2021 AIChE Virtual Spring Meeting & 17th Global Congress on Process Safety Houston, Texas, April 18 - 23, 2021

# Energy Savings for Steam Cracking Furnaces by Combustion Air Preheating Technology

**Fanxu Meng[1], Yong Wang[1], Miguel Maldonado[1], Joel Guillaume[1]**
**(1)Technip Energies, Houston, TX**

**Mingwei Shi[2], Songlin Shao[2], Xiaofeng Li[2], Xiaofeng Song[2] and Yue Liu[2]**
**(2)Beijing Aerospace Petrochemical EC and EP Technology Corporation Limited (BAEEC), Beijing, China**

# Agenda

1. Introduction
2. Air Preheater Fundamentals
3. Case Study on Cracking Furnaces
4. Air Preheater Design
5. Air Preheater Features for Application
6. Conclusion

TECHNIP ENERGIES

# Furnaces are the major energy consumers in ethylene plants

- High temperature for hydrocarbon cracking
- Fuel gas is fired through burners
- Excess air to avoid sub-stoichiometric operation
- Heat in flue gas is recovered by convection section
- Preheat hydrocarbon feed and boiler feed water (BFW)
- Generate superhigh pressure steam (SHP)
- Heat in cracked gas is recovered by processes
- Quench exchangers
- Quench towers (oil and/or water)

TECHNIP ENERGIES

# Air Preheater Systems save energy

- Reduces fuel consumption
  - Consequently reduces flue gas and $CO_2$ production
- Reduces waste heat discharge
  - Low-level waste heat needs extra cooling otherwise

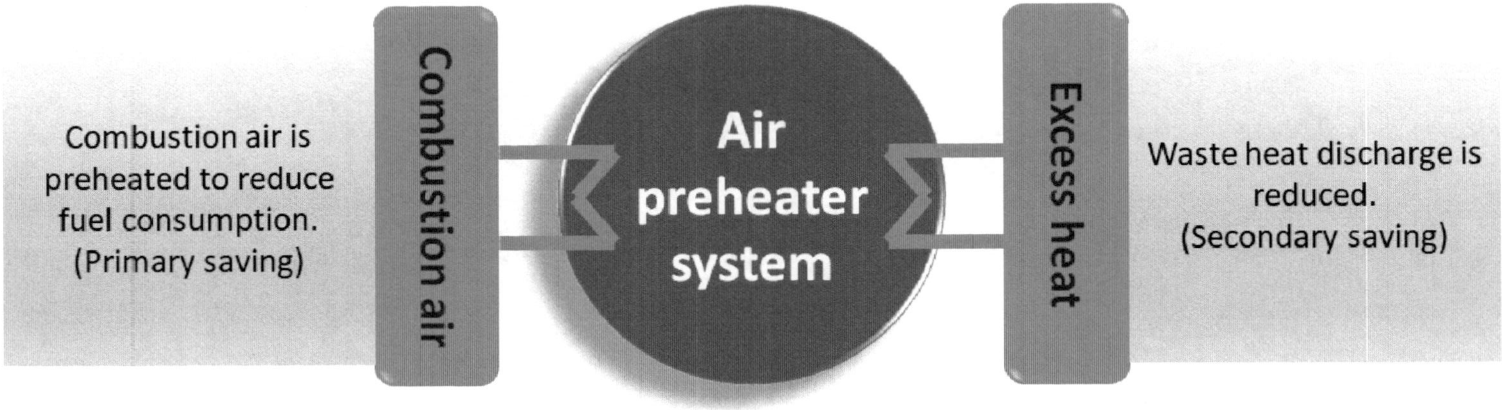


TECHNIP ENERGIES

Energy Savings for Steam Cracking Furnaces by Combustion Air Preheating Technology

# Air Preheaters use low-level heat sources

| Heat Sources | Typical Temperature Range |
|---|---|
| Low-pressure Steam | 150 – 200°C |
| Low-pressure Condensate | 100 – 200°C |
| Quench Water | 60 – 90°C |

TECHNIP ENERGIES

# Schematic: Quench water is used

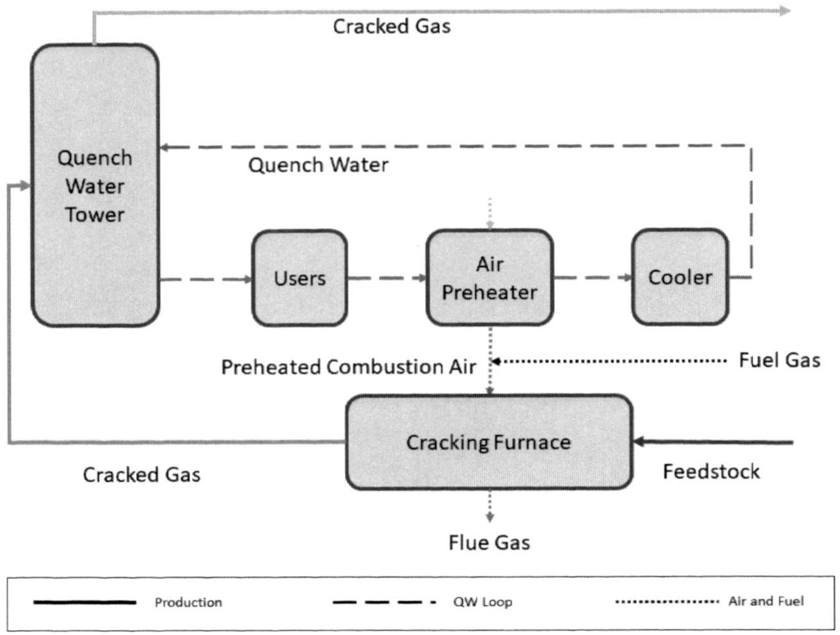


TECHNIP ENERGIES

# Benefits: Quench water is used

## Quench water

- Process uses quench water to cool cracked gas before it enters compressor
- Quench water is circulated and cooled before returning to quench tower
- A significant amount of air or water cooling duty is required to cool quench water before it is returned to the quench water tower

## Air preheater

- Cool quench water to reduce air / water cooling duty
- Recover low-level waste heat to reduce furnace fired duty
- No forced draft fans or extensive ducting required
- No additional furnace trip scenarios
- Conventional burners are used, making retrofits simple
- ID Fan has margin for additional pressure drop of added air preheaters

# Case Study: Look into benefits and effects

## Furnace conditions in the case study

- Based on an actual ethylene plant project
- Analyzed by SPYRO simulation

| | Range | Unit |
|---|---|---|
| Dilution Steam to HC Ratio | 0.3 | |
| Coil Outlet Temperature | 835 | °C |
| Coil Outlet Pressure | 0.7 | bar-g |
| Hydrocarbon Mass Flow | $9 \times 10^4$ | kg/h |
| Residence Time | 0.2 | second |
| Excess Air | 10 | % |

TECHNIP ENERGIES

# Air Preheaters save energy for furnaces

- Combustion air temperature is heated to 70°C from -5°C
- 10% excess air is maintained

| | No Air Preheat (-5°C) | Air Preheat (70°C) | Change percentage with AP |
|---|---|---|---|
| Fuel Gas, ton/h | 10.81 | 10.32 | -4.5% |
| Flue Gas, ton/h | 217.3 | 208.4 | -4.5% |
| Fired Duty Efficiency, % | 94.5 | 97.5 | +3.2% |

# Increased combustion air temperature enhances benefits

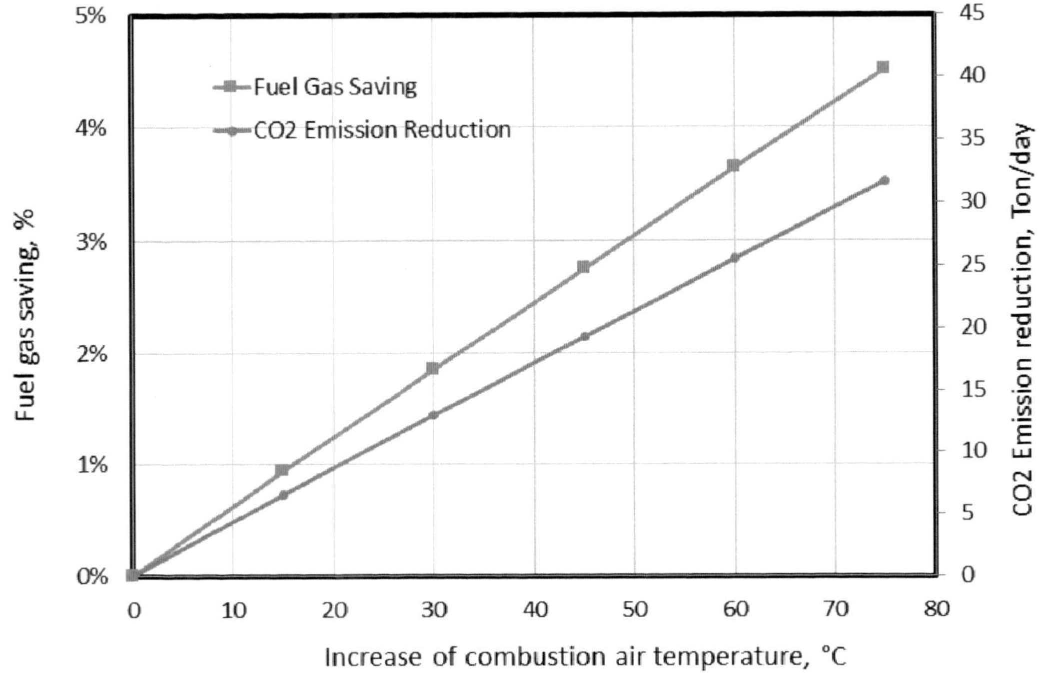


TECHNIP ENERGIES

# Combustion air temperature is restricted

## Available low-level heating sources

- Low-level waste heat sources are typically used

## Effects on furnace operation

- If combustion air is heated up to above 100°C, attention should be paid to the increase in NOx emissions.
- Based on practice, increase in NOx emissions is limited when combustion air is preheated less than 100°C.

## Change overview after revamp to add air preheaters

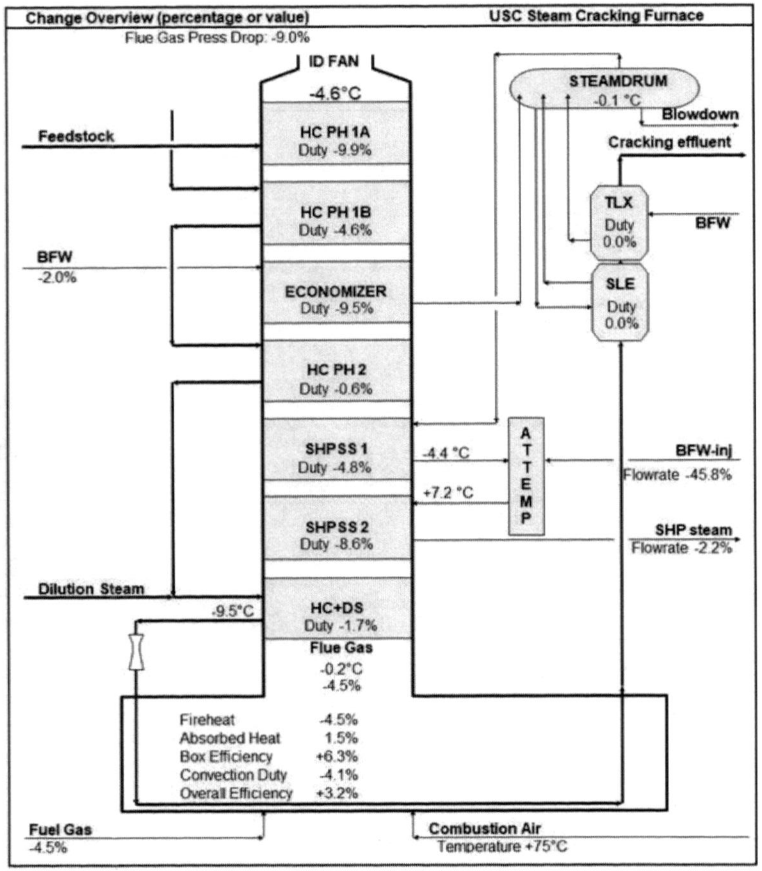


TECHNIP ENERGIES

Energy Savings for Steam Cracking Furnaces by Combustion Air Preheating Technology

# Air preheaters design

TECHNIP ENERGIES

# Air preheater system

**Balanced distribution**

- Hundreds of preheaters
- The main header and several levels of branches
- Without automatic flow control element
- Without instrumentation control

**Resilient system**

- Ensures the temperature consistency of the preheated combustion air
- No matter how the heating source fluctuates

# Air preheater features for application

- No additional utility consumption
- Multiple benefits
- High efficiency of heat transfer in preheaters
- Good adaptability to heat source
- Flexible operation
- Long working life

TECHNIP ENERGIES

# Conclusions

- Simple air preheaters using quench water for cracking furnaces display great features to save energy for ethylene plant
- No additional utilities consumption
- Simple design – no forced draft fans and complex ducting
- Limited effects on furnace operation, which is predictable by SPYRO
- Efficient and economic way to revamp ethylene furnaces
- Reduced $CO_2$ emission
- Reduced cooling water and/or power for air cooling
- Furnace and balance of plant design need to be integrated

T.EN TECHNIP ENERGIES

# Thank you!
# Questions?

**AIChE Paper Number: 98a**

# ANTI-EROSION COATING FOR IMPROVING TRANSFER LINE EXCHANGER DURABILITY

*Dr. Michael Olbricht*
*Engineer, Research and Development*
*SCHMIDTSCHE SCHACK | ARVOS GmbH*

*Christoph Rudolph*
*Sales Manager, Ethylene Business*
*SCHMIDTSCHE SCHACK | ARVOS GmbH*

Prepared for Presentation at the 2021 Spring National Meeting
April 18 - 23, 2021

## AIChE Paper Number: 98a

# ANTI-EROSION COATING FOR IMPROVING TRANSFER LINE EXCHANGER DURABILITY

*Dr. Michael Olbricht*
*Engineer, Research and Development*
*SCHMIDTSCHE SCHACK / ARVOS GmbH*

*Christoph Rudolph*
*Sales Manager, Ethylene Business*
*SCHMIDTSCHE SCHACK / ARVOS GmbH*

**Abstract**

Transfer Line Exchangers (TLEs) are affected by erosion, with most erosion damage occurring on the gas side. Damage can be caused by improper decoking or unfavorable operating conditions, but can also happen during normal operation under certain conditions. Erosion damage can lead to increased maintenance downtime, repair expenditure, operational upsets, and decreased run lengths. Special attention must be paid to weld repairs caused by erosion damage, which can be challenging and lead to failures if not carried out correctly.

To overcome this issue of erosion damages, the application of a protective layer is suitable at relevant parts of the TLE. The right choice of the material for the erosion resistant layer is crucial to ensure proper wear protection and long equipment lifetime. Therefore it is important to consider the particle properties, the interaction between particles and wear resistant material, and the microscopic wear mechanism. This is done by the fabrication of test pieces, their metallurgical assessment, and the experimental determination of the erosion properties of the individual protection coatings under replicable laboratory conditions.

This paper discusses the selection and special fabrication procedure to apply erosion resistant material onto TLE parts, enhancing durability and improving overall reliability. It will give an overview on the identification of the appropriate erosion resistant material, the manufacturing procedure to meet specific plant needs and other options in case damage occurs.

## 1 Introduction

The economic viability of an ethylene plant strongly depends on the reliability of the process equipment. During the production of ethylene, coke is formed as an undesired but unavoidable byproduct. The coke is mainly formed on the inner walls of the cracker coils. The formation of coke as well as its properties depends on multiple factors like feed stock, furnace type, and mode of operation. A hard layer is built-up on the inner wall of the cracker coil during operation of the ethylene furnace. The coke spalls off periodically during normal operation or to a higher extent during decoking or upset conditions. The hard coke particles are accelerated by the process gas stream and can lead to severe erosion wear in the Transfer Line Exchanger (TLE) downstream of the reactor coil, which can lead to uncontrolled operating conditions.

Operators got in contact with SCHMIDTSCHE SCHACK (SCS), the manufacturer of TLEs, due to erosion problems on the gas outlet side of the TLE and asked for a strategy to reduce material loss and to increase reliability of the TLE. Hereafter, the operators' experience with erosion on the TLE outlet side and the circumstances that led to the damages are reported. The methodology SCS used to mitigate this erosion problem and to apply an erosion resistant coating on the parts of the TLE that are prone to erosion is presented.

## 2 Background

### Operators' experience

Erosion wear by coke on process equipment is a known problem in the Ethylene industry. The erosion potential is associated to the operational mode and process conditions used in the furnace and TLE. Under regular operating conditions with decoking cycles in frequent intervals and without any upset conditions only little erosion is reported by operators. The severity of erosion is dependent on the utilized feedstock. The coke formed during the pyrolysis of ethane is known to be harder than the coke formed during cracking of naphtha (1). Due to the lower hardness of coke produced by the pyrolysis of naphtha, no significant erosion problems are reported for this feedstock if no undesired process conditions appear. In contrast to that, continuous but slight erosion on the gas inlet side of the TLE is observed during the utilization of ethane as feedstock. This kind of erosion is usually reduced by means of an erosion grid directly upstream of the TLE tube sheet and is limited to an acceptable level as per design by the choice of an appropriate flow velocity in the TLE for normal operation. The flow velocity and therefore the impact velocity of the coke particles on the TLE have a dominant effect of the severity of erosion.

High velocities and unusually high amounts of coke are favored by upset conditions and the choice of an improper operational mode. Operators reported significant erosion issues on the gas outlet chamber of the TLE for such events. The TLE with outlet chamber circled in red is pictured in figure 1 to indicate the position of the problems described hereafter.

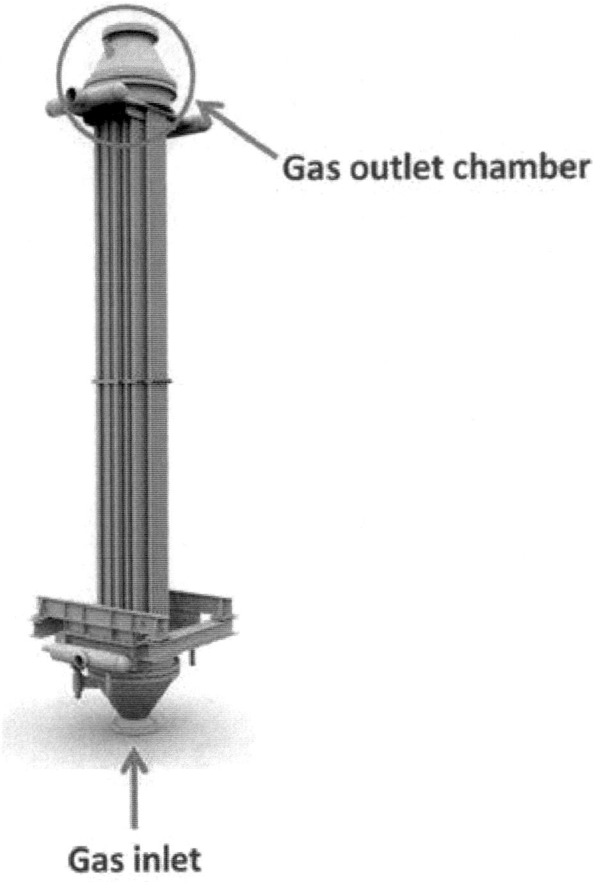


**Figure 1: Picture of TLE with highlighted outlet chamber**

An operator reported excessive fouling and plugging of TLE tubes during a prolonged run with ethane as feedstock. The total gas stream flowed through the remaining open tubes. The flow velocity is proportionally increased with the portion of the plugged tubes. The erosion potential increased disproportionately high with increasing flow velocity. As a consequence the gas outlet chamber was perforated and the furnace was forced to shut down. An example of how the global erosion potential is increased with the portion of plugged tubes is shown in figure 2. A properly designed TLE may thermally operate well with a reduced throughput of gas even if a significant number of tubes are blocked. However, in a case where the mass flow is increased again during a decoking procedure the erosion potential also increases drastically. Although the magnitude of disproportionality shown in figure 2 is based on SCS' experimental investigations, the erosion potential can be even higher in other cases due to the influence of the particular process conditions and hardness of coke formed in the cracking process. Local effects due to deflection, flow maldistribution and partial blocking of individual tubes may lead to an even higher local erosion potential.

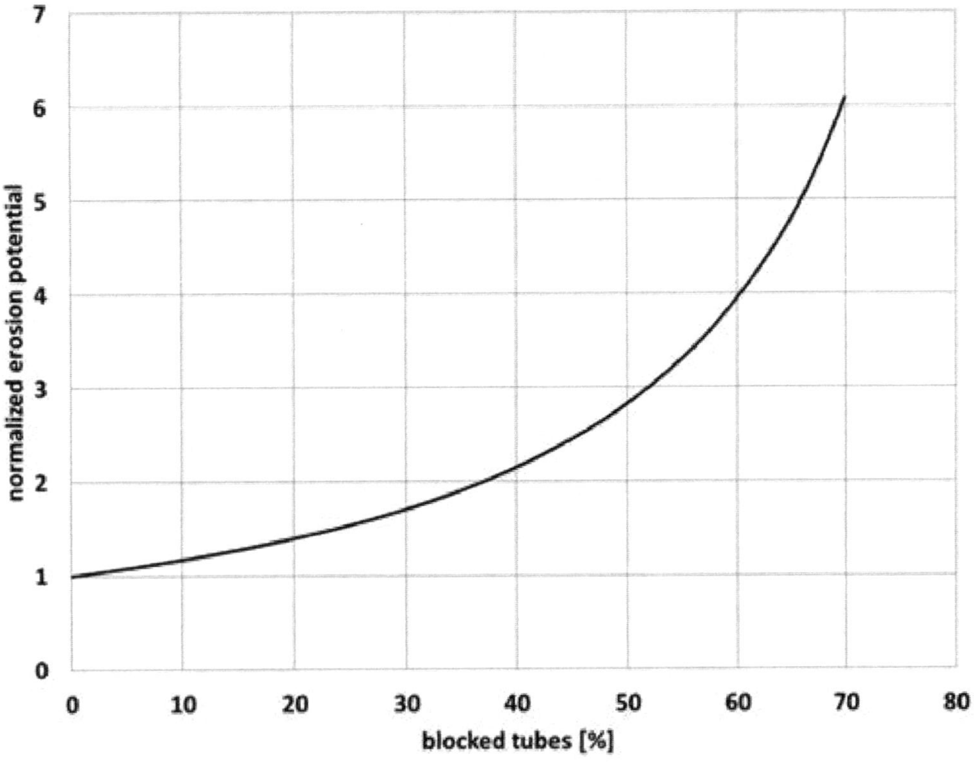


**Figure 2: Increase of erosion potential by blocking of tubes based on measurement with coke from a naphtha furnace**

A second erosion issue occurred in a furnace operated with naphtha. The operator reported an emergency shutdown of the furnace due to a plant steam system failure. A massive formation and spalling of coke from the radiant coils lead to plugging of a large number of TLE tubes. Again a perforation of the TLE outlet chamber by erosion was observed as a consequence of the increased flow velocity of the cracked gas through the tubes which remained open.

In both cases, the operators assumed the damage happened rapidly upon switchover to decoking mode. The typical appearance of damage in the outlet chamber caused by the combination of an increased flow velocity with a massive coke spalling is shown in figure 3. The location at which the outlet chamber cone was perforated is clearly shown. Although the hole appears to be comparably large, the operator mentioned that the damage arose in a short period of time. A precursor of this type of failure was also found by a customer during a regular inspection of a TLE. A thinning on the inner side of the wall of the outlet cone as shown in figure 4 was identified. The local, trough-shaped material removal and the elongated shape of the wear pattern in the main direction of gas flow are typical for erosion wear. This means that unfavorable operating conditions in terms of erosion can also lead to a damage of the TLE outlet chamber over a longer period of time. This is possible even if the operating conditions are less harsh than in the two cases reported before.

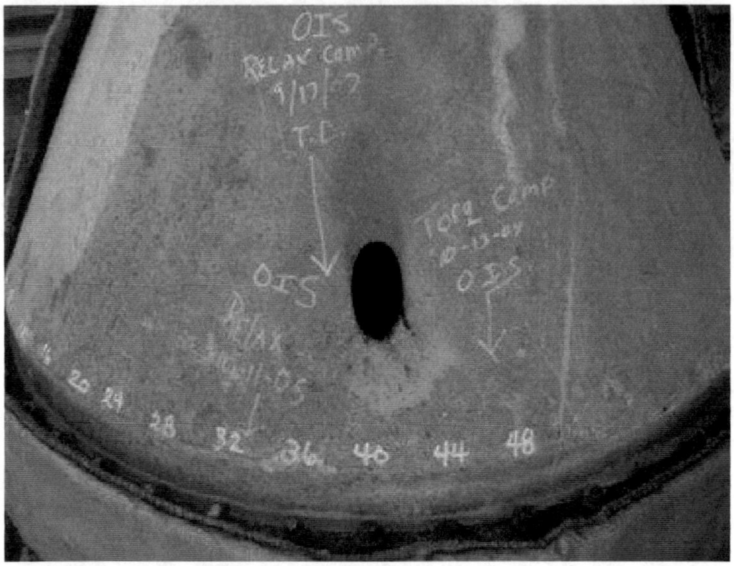


**Figure 3: Conical part of outlet chamber perforated by erosion**

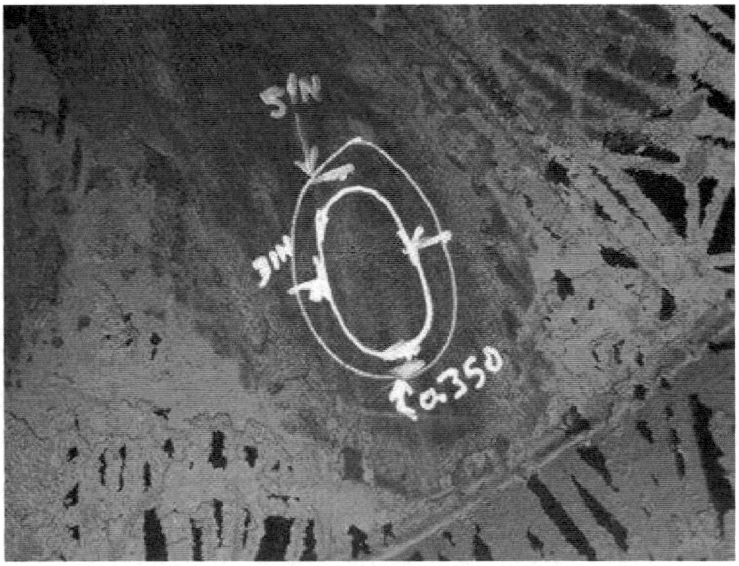


**Figure 4: Wear pattern on the inner side of the TLE outlet chamber**

## Additional information about erosion with respect to the TLE outlet side

To take an appropriate action to prevent the TLE from experiencing the damage described above, some further information is needed to get a complete overview about the influencing parameters that affect erosion wear. Beside the velocity and the hardness of the coke, the impact angle of the coke particles and the hardness of the target material that is shot by coke, are of major importance for the lifetime of component parts exposed to erosion. The erosion rate, which is defined as the fraction of the mass loss of the target material by the mass of particles shot on the target material, is used to quantify the erosion resistance of a specific material.

The impact angle under which the maximum material loss of most construction materials appears is in the range between 20° to 40° and depends on both the target material and the material of the particles as reported by Sheldon (2).

The impact of the hardness of the target material on the erosion rate was investigated by Wellinger and Uetz (3), (4). They found a complex relation for the interaction between target material and particles. As a simplified rule of thumb, however, it can be summarized that to achieve long-term erosion resistance, the target material must be harder than the particles. An increase in the hardness of the target material, but still below the hardness of the particle has only a minor, and in some cases even negative influence on the erosion resistance of the target material.

The temperature of the target material has also an effect on its ability to resist erosion wear. The material loss by erosion is usually increased at elevated temperatures as shown by Shayler and Yee (5). However, this effect is of minor relevance. Also the size of the particles that cause erosion has an effect on their erosion potential. Raask (6) showed that no erosion wear was observed for particles with a diameter smaller than 5µm. For particles with a size between 5µm and 40µm a linear increase of the erosion rate with the particle diameter was observed. The erosion rate was found to be independent from the particle size for diameters bigger than 40µm. This means that for a systematic experimental evaluation of the erosion properties of different materials in relation to the application in ethylene production, the correct choice of particle size is crucial. This must be uniformly larger than 40 µm for all tests in order to obtain comparable results.

## 3 Approach to solve the erosion problem

The approach to solve the erosion problem was to apply an erosion resistant coating on the parts susceptible to erosion. The aim was to develop a protection layer that can be applied, and bonds well, to commonly used TLE materials of construction. To achieve this goal, the methodology illustrated in figure 5 was used to identify the right coating material and parameters of the manufacturing process. After a pre-selection of materials, sample pieces were fabricated to investigate the impact of different process parameters on the properties of the manufactured protection layer. The investigation was done by a mechanical and metallurgical assessment of the sample pieces. To evaluate the erosion resistance of the individual coating, all sample pieces were tested in an erosion test rig. This overall step in the development of the erosion resistant coating was done iteratively as indicated in figure 5. For the best candidates that came out of this test procedure, an economic evaluation of the overall manufacturing process was done to identify the most commercially viable coating material.

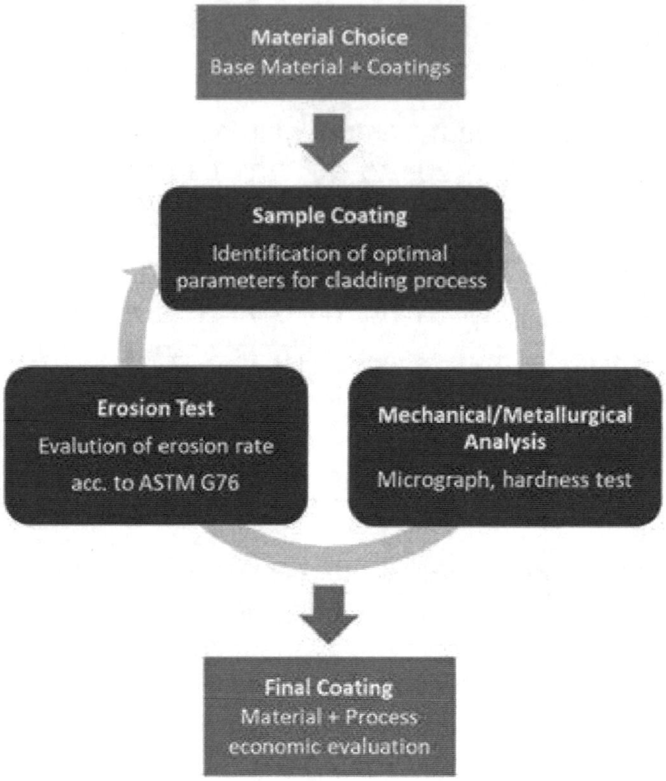


**Figure 5: Methodology to identify optimal coating material and manufacturing process parameters**

### 3.1 Selection of materials and manufacturing procedure

The connection between the erosion resistant layer and the base material has a huge impact on the lifetime of the layer. In comparison to other manufacturing processes, weld cladding provides a good adhesion of the layer due to metallic continuity. As a result of a pre-study, laser cladding was identified as the best technology to apply a weld cladded coating on parts of the TLE. Very thin layers of material can be applied on the base material by this technology. The benefit is the induction of less stress due to welding in the base material. The automated laser cladding process also provides a consistent quality of the manufactured layer.

From earlier field experience, the hardness required of the protective layer in order to provide long term protection against erosion by coke impact was known. Based on this information, different hard facing materials that fulfill the hardness requirement were pre-selected for the further development of the erosion resistant coating. As base material the most commonly used material of construction for TLEs was used. A transfer of the welding parameters between other materials often used in TLE manufacturing can be done without large effort due to similar properties in terms of weldability.

During the manufacturing of the samples, parameters such as laser power and feed speed as well as the laser focus were varied. A micrograph of a cross section of the produced weld beads was made and investigated with respect to the formation of pores and cracks, layer hardness, adhesion to the base material, and the microstructure in the heat affected zone. A typical test matrix for one hard facing filler material is shown as an example in figure 6. It becomes obvious that the weld penetration and the amount of material applied per welding bead is increased with laser power and decreased with feed speed. However, a low feed speed and a high laser power appears to be beneficial for manufacturing on the first view, these process parameters are not the best for other properties of the layer, e.g. hardness. Thus, a detailed analysis was made to find the best process parameters for every hard facing filler material.

With the optimal manufacturing process parameters, a sample piece was made for the further mechanical assessment. The behavior of the erosion resistant layer was tested under dynamic load. A cyclic three-point-flexural test was performed to simulate the load cycles during the lifetime of a TLE. The sample was elastically deformed almost up to the point where plastic deformation starts during each cycle. The layer was evaluated after the test in terms of spall off and growth of cracks.

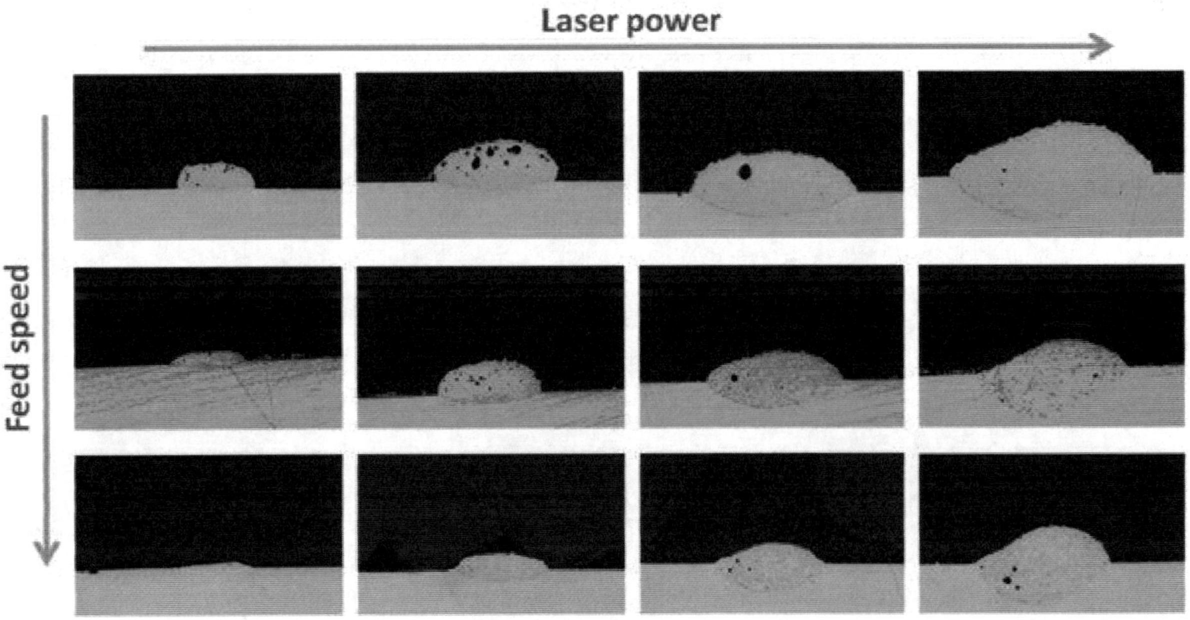


**Figure 6: Test matrix for sample manufacturing with varying feed speed and laser power**

### 3.2 Erosion test

To evaluate its wear resistance, each best parameter sample was tested under laboratory conditions by an erosion test procedure following ASTM G76 (7). The test rig is shown in figure 7. The main components of the test rig are a particle conveying system, a blasting nozzle and a specimen holder. The particles are transported with a high velocity gas stream from a reservoir to the blasting nozzle. The velocity of the particles is adjusted by the gas velocity. The particles are shot through the nozzle onto the target. The impact angle of the particles on target is adjusted by the orientation of the specimen holder relative to the nozzle outlet. To prevent the system from a dust explosion, the nozzle and specimen holder are placed in an inert box.

During the test, a defined amount of coke particles taken form an ethylene plant was shot onto the sample. The erosion rate was determined by measuring the weight difference before and after the test. Mainly particles with a diameter bigger than 100µm were used for the test to neglect the impact of the particle size on the erosion rate. The test was done with the same particle velocity for all samples. The particle velocity was in the range up to 150m/s. The velocity was measured by a high speed camera and a ruler. All test runs were performed at ambient temperature. These test conditions are chosen to fulfill ASTM G76 and limitations of the test rig on one hand, and to get as close as possible to the operation conditions in the TLE on the other hand. Every test was repeated three times with the same test parameters to obtain a high confidence in the result.

An example how the test pieces are damaged by coke under an impact angle of 90° is shown in figure 8. On the left side the uncoated base material is shown. A deep crater is eroded in the surface with a massive loss of material, such as what could appear in a short period of time at the outlet cone in case of upset conditions. On the right side the test piece with erosion resistant coating is shown. The test conditions are the same as that of the uncoated sample. In contrast to the uncoated sample, only minor microscopic scratches can be observed. The surface is still in good condition, and consequently, if this surface would be applied on a part of the TLE, this part would still be intact even after experiencing such a severe coke impact.

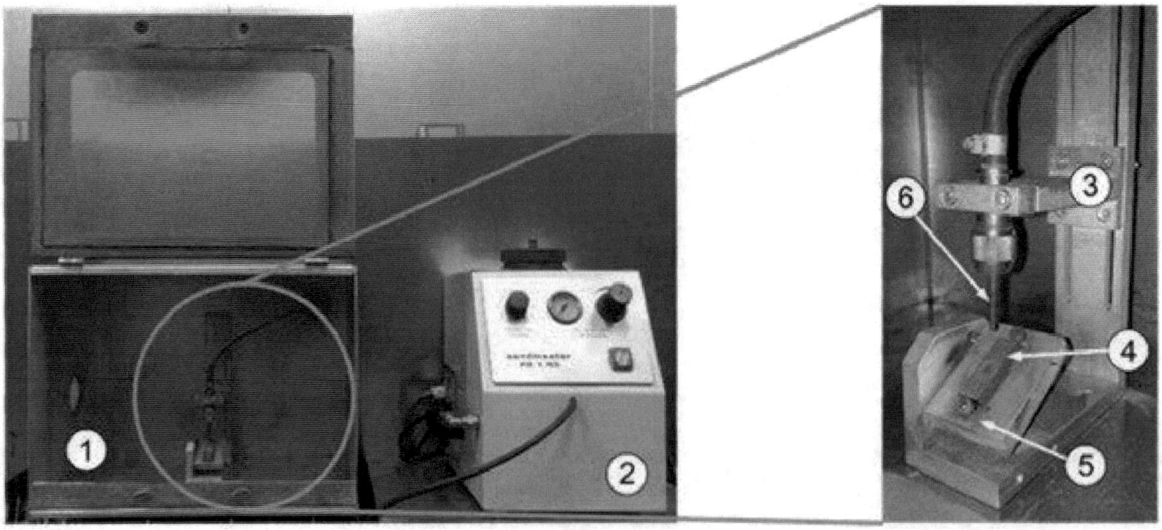


1: Inert box, 2: Pneumatic conveying system, 3: Nozzle holder, 4: Specimen, 5: Specimen holder, 6: Blasting nozzle

**Figure 7: Erosion test rig**

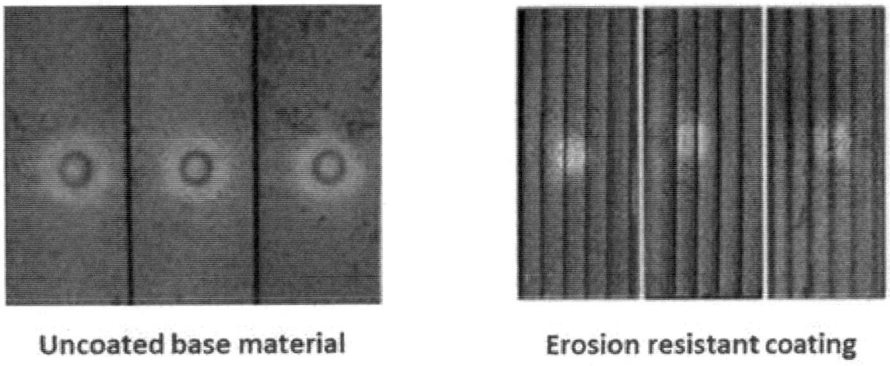


**Figure 8: Specimen after erosion test (impact angle 90°)**

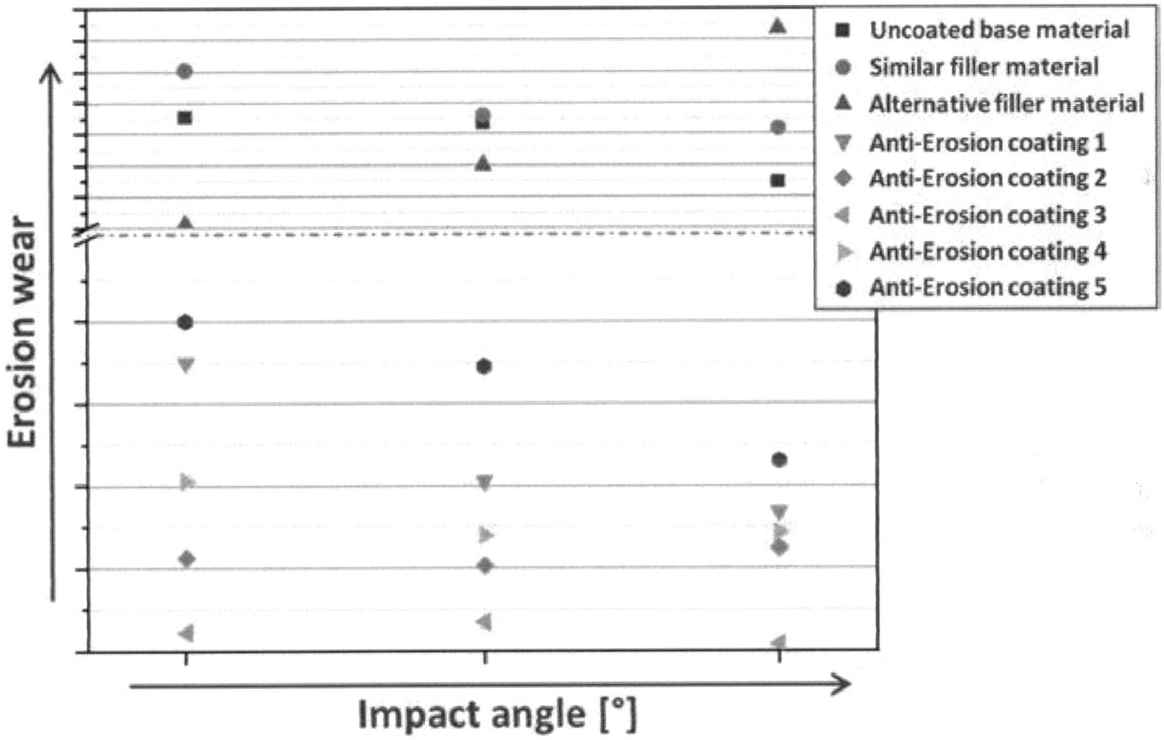


**Figure 9: Results of the erosion test**

The results of the erosion test of all hard facing materials are shown in figure 9. The material loss by erosion wear caused by coke particles is shown as a function of the impact angle. All tests were done with the same particle velocity. In addition to the erosion resistant materials, also the results obtained with the base material and with filler material similar to the base material is shown. The similar filler material was applied by manual metal arc welding. The results can be distinguished in two different ranges with respect to erosion wear. The higher erosion wear in the upper part of the diagram above the red line is attributed to the base material and the similar filler material. The results of the erosion test performed with erosion resistant hard facing coatings are shown in the lower part of the diagram below the red line. The mass loss by erosion is approximately one to two orders of magnitude smaller for all erosion resistant coatings in comparison to the base material and similar filler material. It can be summarized that no significant improvement of erosion resistance can be achieved by overlay welding with similar filler material. However, overlay welding with similar filler material is state of the art in repairing erosion damages, for example on the tube sheet of a TLE. A remarkable erosion resistance was observed for all Anti-Erosion coatings. Despite this general improvement of the erosion resistance, differences in the wear properties between the individual coating materials were identified. The best cladding material identified by the erosion test is "coating no. 3", which is indicated by the yellow triangles and shows the lowest erosion wear. The order in the ranking of the different coating materials was not affected by the choice of the impact angle. That means that the identified coating performed best under all conditions considered in the evaluation process.

It was concluded after economic assessment that the technically best performing material is, by coincidence, also the commercially most viable.

## 4 Manufacturing and operation of erosion resistant solutions

Depending on the operator's requirements, the erosion-resistant coating is applied to the components of the TLE that are severely affected by erosion. Two case studies are described below where excessive wear due to erosion led to unwanted plant shutdowns and required premature replacement of parts of the TLE on the gas outlet side. As a solution to the erosion problem, the service life of one gas outlet chamber and several inspection caps were increased by applying the erosion-resistant coating.

### 4.1 Retrofit of outlet chamber

An operator got in contact with SCS and requested help to solve the operator's erosion problems in the outlet chamber of a TLE. Due to significantly extended runs, there was an excessive buildup of coke in the coils and successive blocking of several TLE tubes. As a consequence, the erosion damage on the gas outlet chamber shown in Fig. 10, top right, resulted, in which the wall was perforated by the coke, similar to Fig. 3. Since the operator does not want any change in the operation mode and decoking cycles of the ethylene manufacturing process to eliminate or reduce the erosion problem, the operator asked SCS to solve the problem by providing an erosion-resistant gas outlet chamber. Since the primary focus of the new gas outlet chamber was erosion resistance, the geometric design of the gas outlet chamber was changed in addition to the application of the erosion-resistant coating. Due to the more favorable impact angle of 90° which leads to less erosion than smaller impact angles occurring in the conical chamber, the new gas outlet chamber was designed cylindrically with a flat lid, as shown in figure 10. The flat lid is shown for comparison before and after operation in figure 11.

The operator already has the new gas outlet chamber in operation. During a scheduled inspection of the TLE after 6 months of operation, the outlet chamber was inspected. As can be seen in figure 11 on the right, the erosion protection layer shows no signs of wear, despite the fact that the operating mode was not changed. Further inspections will be performed to document the long-term behavior of the protection layer.

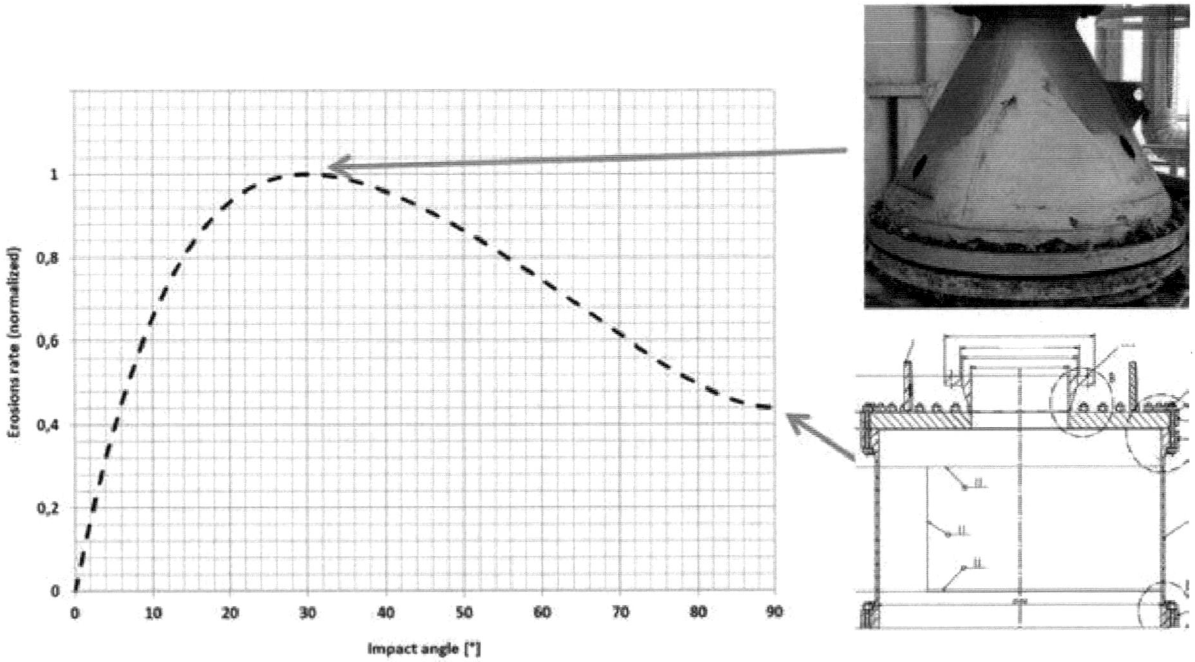


**Figure 10: Old and new design of gas outlet chamber**

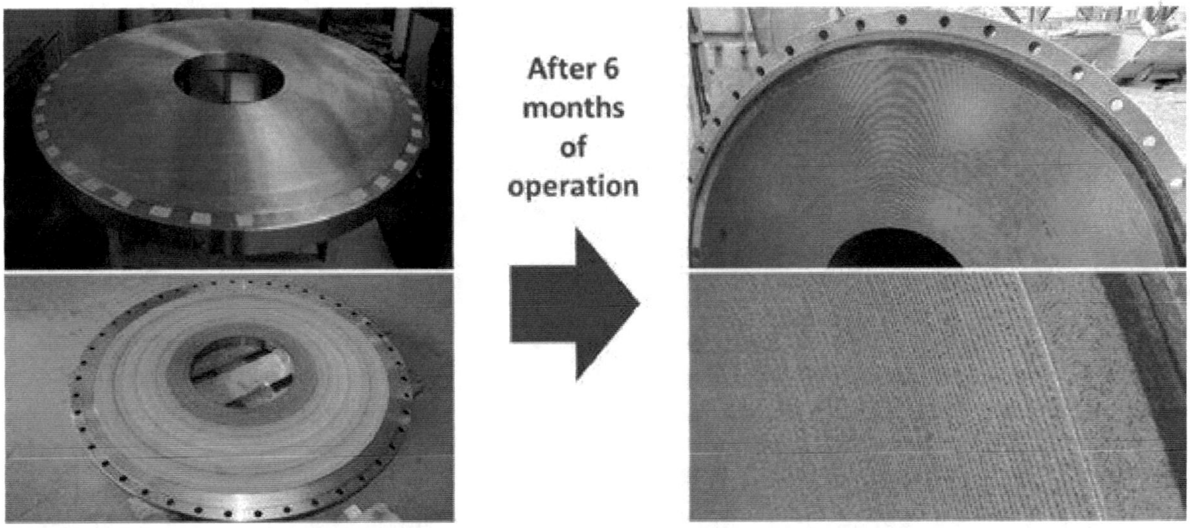


**Figure 11: Gas outlet chamber lid before and after operation**

## 4.2 Improved inspection nozzle cap for linear TLE

Another implementation of the erosion resistant coating was the application of a protection layer on spare part caps for the inspection nozzles of a linear TLE, as the caps used so far had to be replaced due to erosion wear. In the linear TLE, the heat exchanger tubes in which the cracked gas is cooled are arranged in a single row parallel to each other. Due to this design, the linear TLE does not have a circular gas outlet chamber but a gas outlet header in which all heat exchanger tubes end, as shown in figure 12. As can be seen, coaxial with each heat exchanger tube is a nozzle for inspection and cleaning purposes. Due to the position of the inspection nozzles at the outlet of the heat exchanger tubes, the caps of the nozzles are directly exposed to coke particles. The operator asked SCS to design the new caps in such a way that they can resist erosion for a longer time. For this purpose, the so-called erosion puck, which was already provided in the old design of the cap, was extended as far as the installation space and the thermal design of the surrounding parts would allow. In addition, an erosion-resistant layer was applied to the erosion pucks of the caps by means of laser cladding, which can be seen at the bottom right of figure 12.

**Figure 12: Coated pucks of linear TLE inspection nozzles**

## 5 Conclusion

Erosion on the outlet side of TLEs is particularly a problem when upset conditions appear. During such an event a massive amount of coke spalls off the inner cracker coil wall. By this coke, some TLE tubes are blocked and the flow velocity is increased in the remaining open tubes. With increasing flow velocity the erosion potential of the coke increases. As a consequence, the parts on the TLE facing the direction of flow are exposed to severe erosion wear. To address these problems SCS developed a solution to enhance the erosion resistance of these parts of the TLE by applying a hard facing layer via laser cladding. SCS identified the best parameters for the manufacturing process to apply a pre-selected variety of coatings and evaluated their erosion resistance under laboratory conditions. On customer request the gas outlet chamber of a TLE which is working under conditions of high erosion potential and inspection nozzles of a linear TLE were cladded with the coating material that was identified to be the best. Initial operation experiences show a good erosion resistance of the applied layer, optimizing TLE durability and overall lifetime economics.

## 6 Acknowledgements

SCS wants to thank the operators to be allowed to share their operation experience.

## 7 References

(1) Towfighi, J., Sadrameli, M., Niaei, A.: Coke Formation Mechanisms and Coke Inhibiting in Pyrolysis Furnaces, J. Chem. Eng. Jap., Vol. 35, 10, 923-937, 2002

(2) Sheldon, G.: Similarities and differences in the erosion behavior of materials, J. Basic Eng. Trans. ASME, 619-626, 1970

(3) Wellinger, K., Uetz, H.: Gleitverschleiß, Spülverschleiß und Strahlverschleiß unter der Wirkung von körnigen Stoffen, VDI-Forschungsheft 449, Forschung auf dem Gebiet des Ingenieurwesens, Ausg. B, Bd. 21, 1-40, 1955

(4) Uetz, H.: Abrasion und Erosion, 236-278, Carl Hanser Verlag München Wien 1986

(5) Shayler, P. J., Yee, K. H.: Erosion of AISI 303 steel by fine (about 5μm and about 50μm) ash particles, Wear 98, 127-140, 1984

(6) E. Raask: Erosion wear in coal utilization, 1. Edition, Hemisphere Publishing Corporation, 1988

(7) ASTM: G76 – Standard Test Method for Conducting Erosion Tests by Solid Particle Impingement Using Gas Jets; 2005

## AIChE Paper Number: 98b

# ENHANCING THE EFFICIENCY OF STEAM CRACKING FURNACES

***Dr. Dietlinde Jakobi***
*Head of Sales/R&D/Services*
*Schmidt + Clemens GmbH + Co. KG*

***Steffen Heyland***
*Steam Cracker Optimization Engineer*
*Schmidt + Clemens GmbH + Co. KG*

Prepared for Presentation at the 2021 Spring National Meeting
Virtual, April 18-23, 2021

# ENHANCING THE EFFICIENCY OF STEAM CRACKING FURNACES

***Dr. Dietlinde Jakobi***
*Head of Sales/R&D/Services*
*Schmidt + Clemens GmbH + Co. KG*
***Steffen Heyland***
*Steam Cracker Optimization Engineer*
*Schmidt + Clemens GmbH + Co. KG*

**Abstract:**

Energy efficiency increase is gaining more and more importance in a highly dynamic eco-economical environment. Steam crackers require considerable amounts of energy and thus, educated selection of equipment is essential. In a radiant coil it is key to reduce coke deposition and increase the heat transfer as well as prolong lifetime.

Iron and nickel based chromia-forming alloys had been installed in steam cracking furnaces for several decades, when alumina-forming alloys were introduced to the industry, featuring significantly lower coking rates and higher tube lifetime. Sulfur present in the feedstock strongly influences the coking rate and material deterioration. Due to a far better thermodynamic stability, alumina scales are more favorable as a means of protection in such severe conditions. In combination with an optimized coil replacement strategy, this technology increases plant efficiency by lower coke formation and longer tube lifetime.

Pushing the energy efficiency of steam cracking furnaces further means reducing the specific heat input per unit of ethylene. To achieve this, enhanced heat transfer tubes are becoming more and more common in the industry. Gas rotation inner profiles decelerate coke deposition, improve heat transfer and as such reduce the fuel consumption. In order to maximize the efficiency of the plant, it is recommendable to combine a 3D profile with an alumina forming alloy.

Performance studies and results from metallurgical investigations carried out on S+C proprietary alumina-forming alloys will be presented as well as a model calculation to estimate the efficiency impact of advanced steam cracking technologies. Furthermore, in 2016 S+C technologies were selected within the EU-funded IMPROOF project to engineer the most efficient steam cracker to date. Now that IMPROOF is completed, results from the technology evaluation stage will be shown.

## Introduction

Since radiant coils are an essential part of steam cracking furnaces, they have a huge impact on the efficiency, while operating under severe, alternating reducing and oxidizing conditions at temperatures even beyond 1100°C. Therefore, cast alloys installed in the radiant section need to provide not only high creep strength but also good resistance to corrosion attack (oxidation/carburization/nitridation), as well as excellent catalytic coking resistance. Coking is not only a limiting factor when it comes to cracking cycle lengths; it also increases the need for higher firing rates, thus emitting more $CO_2$. Hence, the coil components (centrifugally cast tubes, cast fittings) must be able to form a dense and continuous oxide layer on the inner surface, which should remain stable when exposed to steam cracking conditions, thus restricting the development of catalytically active sites where the unfavorable formation of coke filaments takes place.

Increasing the overall efficiency of a steam cracker means both optimizing reliability and production efficiency, apart from a sensible, mindful way of operation. At the baseline, it all starts with the appropriate selection of equipment, which addresses the requirements of the process as best as possible. The main requirements for a steam cracker coil can be assigned as high temperature corrosion resistance and mechanical strength to ensure best reliability.
From the process side, highest operation temperature and unrestricted heat transfer from the burners over the radiant tubes into the process gas is desirable.

Depending on coil design, feedstock and operating conditions, the main damage mechanisms for such components are a combination of carburization and creep [1]. Carbon can penetrate from the inner surface into the alloy; nitrogen can diffuse from the outer surface into the tube material. Both lead to the formation of brittle precipitates within the alloy's matrix (large carbides close to the inner surface and large carbonitrides near the outer tube surface). Increasing degree of carburization and/or nitridation, as well as formation of harmful phases during ageing, enhances the risk of material embrittlement with increasing process time. Exposure to operational transients, like emergency shutdowns, with such aged and embrittled alloys can easily lead to failure by brittle crack formation.

Over several decades, iron and nickel based chromia-forming alloys were the material of choice in steam cracking furnaces. These alloys are particularly susceptible to the above-mentioned failure mechanisms [1]. Therefore, research was focused on the development of alumina forming alloys. Alumina scales are thermodynamically much more stable under these conditions; thus, 18 years ago such alumina forming alloys were introduced to the petrochemical industry.

A dense and continuous oxide scale, which is stable up to very high metal temperatures, is able to protect coils against carburization and nitridation; this reduces ageing and prolongs tube life. Additionally, properly grown alumina scales do not contain catalytically active sites for coke formation (e.g., Fe- and Ni-particles). These important features reduce the amount of catalytic coking significantly, particularly in cases where considerable amounts of sulfur are present in the feed [2, 3, 4].

Such a stable and continuous alumina scale is formed on the inner surface of alloy Centralloy® HT E, which has already been installed in the radiant section of more than 400 steam cracking furnaces. Apart from an increased tube/coil lifetime, the alloy positively affects the day-to-day furnace performance: Depending on the operation strategy, the reduction of coke deposition increases the run-length, cracking chemistry and heat transfer.

However, coke deposition reduction is not the sole option to improve heat transfer, it can also be triggered actively. Thus, some years back S+C developed the internally profiled tube design "SCope" and validated that its benefits in plant applications agree with predictions from fluid dynamic simulations. Such a profile leads to a steadily rotating, very uniform, circumferential process gas flow pattern which improves the heat transfer due to boundary layer convection enhancement, enabling reduced tube metal temperatures. Moreover, it balances the heat distribution in the gas phase as well as in the tube shell, and finally, all effects contribute to a further reduction of the coking rate.

## Efficiency Gains Contribution of Alumina Forming Alloys

Feedback from the market on benefits of the alumina forming alloy has been previously reported with respect to improved production capability in terms of run length or throughput of plants [5].

In general, it can be observed that furnaces, which were partially upgraded with the alumina forming alloy in the outlet part only, provided a prolonged tube life: The material upgrade increased the coil life by up to 50%. The radiant tubes needed to be replaced later in time, when the colder inlet tubes (equipped with common chromia forming material) reached the end of their life. The upgrade of the entire coil metallurgy resulted in a life prolongation of up to 100%. However, these individual improvements depend on the process conditions as well as on the feedstock composition.

### *Influence of Materials Selection on the Limitation of Coil Lifetime*

In different furnace designs (e.g., 2-Pass & 5-Pass Coils) the inlet tubes were equipped with the standard chromium oxide forming alloy 25Cr/35Ni Micro and the outlet tubes (consisting either of 25Cr/35Ni Micro or 35Cr/45Ni Micro) were upgraded to the alumina forming material. Metallurgical investigations carried out after four years and five years in service, respectively, revealed that the inlet tubes in 25Cr/35Ni Micro failed due to heavy carburization (up to 2 wt.-% carbon), along with strong internal oxidation and finally creep damage – even though these tubes were exposed to much lower metal temperatures. The inner surface of the strongly carburized tubes exhibits pits due to the corrosion attack (see Figure 1a).

The hotter outlet tubes, equipped with an alumina former did not show any carburization or internal oxidation; the protective alumina scale on the inner surface was still present and intact after several years in operation and the alloy was still fit for service (see Figure 1b). Figure 2a shows the typical appearance of the inner surface of such carburized inlet tubes in common chromia forming alloys while Figure 2b displays the cross-section of the non-carburized alumina former exposed in the same steam cracking coil at higher temperatures.

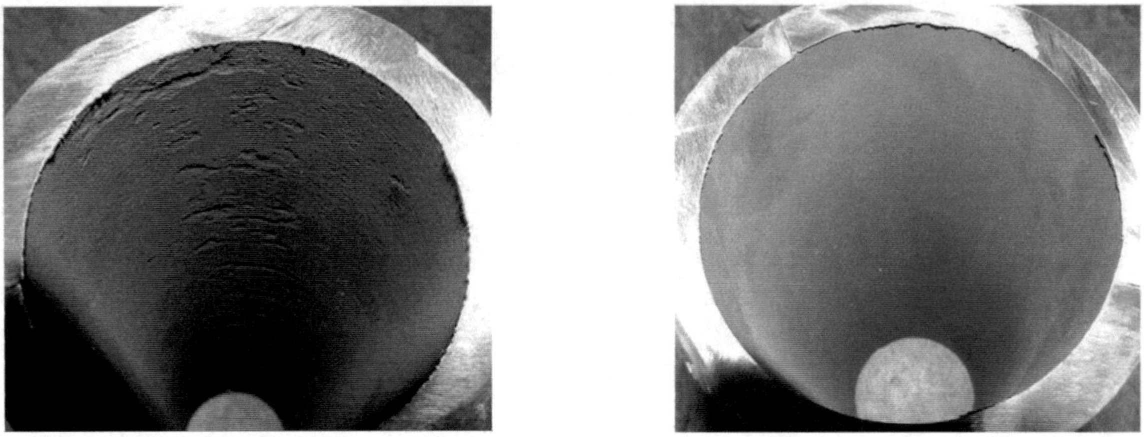

a) Inlet Tube: 25Cr/35Ni Micro alloy  b) Outlet Tube: NiCrAl alumina forming alloy

**Figure 1:** Appearance of the Inner Surface after Five Years in Service (2-Pass Coil)

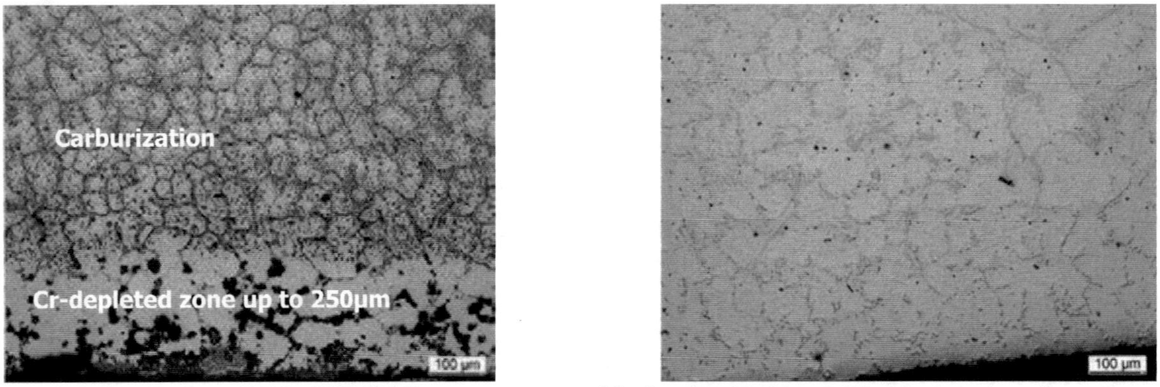


a) Inlet Tube: 25Cr/35Ni Micro  b) Outlet Tube: NiCrAl alumina forming alloy without carburization

**Figure 2:** Cross-Section of the Inner Tube Surface – Typical Appearance; Samples Removed After Four Years in Service (5-Pass Coil)

## ***Knowledge-based Coil Replacement Strategy for Increased Efficiency***

Very often a scheduled coil replacement is done without any prior tube inspection or metallurgical examination. The time in service might vary between four years and eight years, depending on operating conditions, furnace design and feedstock composition.

After the scheduled replacement of coils in alumina forming technology, plant operators and maintenance quite often decide to carry out metallurgical evaluations to determine lifetime consumption and material embrittlement, in order to draw conclusions about the weldability of the aged alloy.

In a particular case, the alumina forming alloy was removed from service after four years as this corresponds to the normal schedule for coil replacement. Several samples were taken from this furnace for metallurgical investigations and no corrosion attack (like carburization or strong nitridation) was detected. The very stable, protective oxide scale initially formed on the inner tube surface (and stabilized during the normal plant operation) was still present and successfully prevented carburization.

In order to determine the remaining coil lifetime, corresponding samples were creep-rupture tested. The material reflected that after four years exposure to service conditions in the steam cracking furnace, just 5% of the minimum creep life was consumed.

Many operators who replaced the coils during the scheduled furnace shutdown without prior destructive tube examination decided to carry out investigations on the aged material removed from service, especially in order to assess weldability. This provided valuable information to maintenance and repair personnel on how to handle the alumina forming alloy after several years in service. Neither carburization (inner surface) nor nitridation (outer surface) affect the materials properties negatively, so that the weldability of the aged alloy is comparable to new material.

As a conclusion, when switching from chromia forming alloy to alumina forming alloy, the replacement strategy should be adapted to make use of the increased coil lifetime and thus, reduce plant downtime for increased efficiency. At least as soon as the alumina forming alloy has exceeded the typical lifetime of the previously installed chromia forming alloy it is recommendable to carry out metallurgical investigations during each planned shutdown. Thus, the operator will gather knowledge about the possible extent of lifetime increase with the new technology in his specific furnace without taking any risks. Finally, the formerly scheduled replacements are turned into a dynamic, flexible replacement strategy based on solid knowledge provided by the metallurgical investigations. With increasing experience of the maintenance team, the conclusions can be transferred more and more to other furnaces in the plant, causing a reduction of maintenance activities and costs.

***Efficiency Benefits of Alumina Forming Alloys in Sulfur-Rich Conditions***

Steam cracker feedstocks can comprise a large spectrum of species. The range is from light, gaseous ethane up to heavy, liquid components like naphtha or even waxy distillates. A common characteristic of feedstocks entering radiant coils of steam crackers is the presence of sulfur, either as a deliberate constituent for gas cracking or its natural presence in the feedstock, such as naphtha. Basically, sulfur is needed to deactivate catalytically active particles on top of the tube inner surface: The day to day cyclic cracker operation at high temperatures causes the degradation of the protective oxide scales initially formed on the tube surfaces (radiant zone as well as convection box). Due to such material deterioration, small inclusions of iron and nickel particles from the base material are moved onto the surface oxide layer. When exposed to the hydrocarbon/steam atmosphere such catalytically active iron and nickel particles form CO, which might be a critical factor for the downstream section. The presence of a small amount of sulfur is known to poison the iron and nickel particles and effectively suppresses the CO formation. Furthermore, it is believed that the deactivation of catalytically active sites by sulfur even reduces coke formation reactions.

In 2018, D. Jakobi et al. [4] was the first to consider the influence of sulfur on both coking rates and material integrity. For 8 hours, a chromia former NiCrSi, 35/45 micro alloy surface was exposed to a realistic steam cracker atmosphere, which particularly reflects the thermal and chemical conditions of outlet tubes. Figure 3 shows the results of this experimental set-up with 250 ppmw addition of sulfur introduced as $H_2S$ (Figure 3a) and also without sulfur (Figure 3b).

It is obvious that the addition of this high amount of sulfur causes an extreme increase in the catalytic coking rate (coke filament growth), while the continuous (pyrolytic/process coke formation) coking rate was increased by a factor of 6.

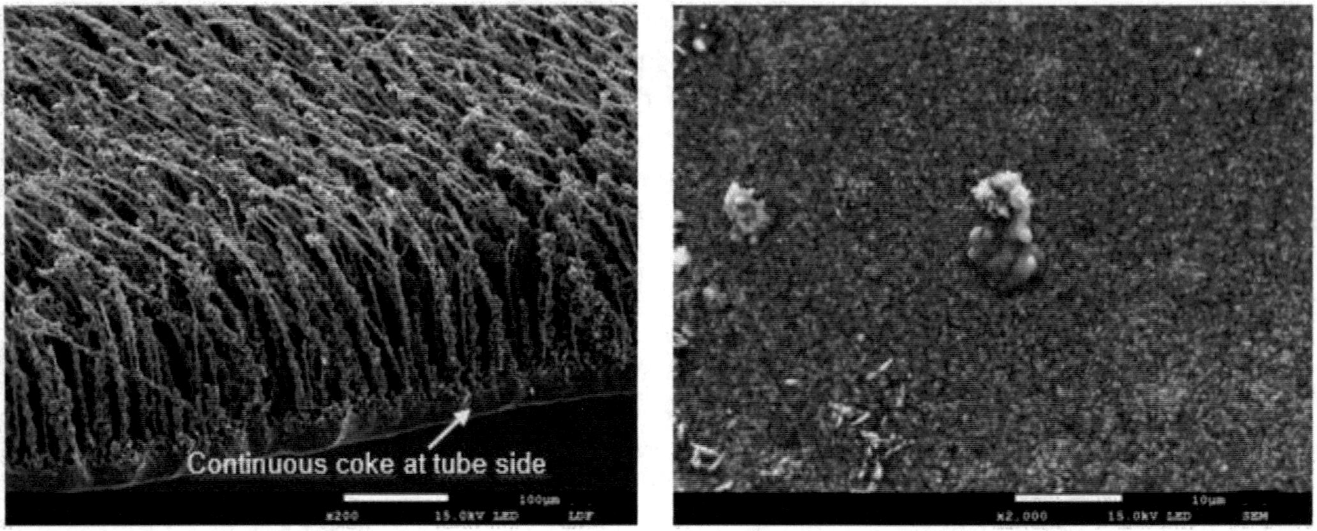


a) with $H_2S$ b) without $H_2S$

**Figure 3:** Coke growth on NiCrSi after 8h of cracking [4]

Although this testing procedure already reveals the influence of sulfur on coking, it is only valid for new material. In order to simulate the realistic, day-to-day operation including high temperature cyclic conditions and thus material ageing, a so called "stress test" was performed with two different materials, a NiCrSi (35/45) chromia former and a NiCrAl alumina former. This stress test includes a start-up in steam, three cracking cycles (14 hours) with 250 ppmw sulfur addition and subsequent decoking plus a final cracking cycle (8 hours), again with sulfur addition but without decoking, in order to analyze the developed coke structures on top of the surface.

From Figure 4 it can be concluded that even after exposure to strongly ageing conditions and in the presence of sulfur, the alumina layer remains protective and free of catalytically active particles that promote excessive coke formation. Solely a very thin layer (4µm) of continuous coke (pyrolytic coking) is present. The chromia former (35/45 micro alloy) however exhibits a voluminous and bizarre coke structure of 0.8mm thickness. Finally, it can be concluded, that the addition of sulfur massively increases the formation of coke on the surface of chromia forming alloys - particularly in an aged condition, while alumina forming alloys remain fairly unaffected.

a) NiCrAl b) NiCrSi

**Figure 4:** Coke growth on aged materials under sulfur-rich atmosphere [4]

Besides the investigation of coking behavior the study also revealed, that the presence of sulfur accelerates the deterioration of the chromia layers (formed on the NiCrSi alloys) by carbon- and sulfur-induced corrosion, strongly reducing tube lifetime expectation due to carburization and thus, embrittlement (loss of mechanical properties). The alumina layer of the NiCrAl alloy however withstands the sulfur-rich atmosphere and remains stable and protective. Both properties, reduced coke deposition and prolonged lifetime contribute significantly to an efficient plant operation, as will be shown later.

## 3D Inner Profile Tubes Address the Specific Needs of the Steam Cracking Process

3D inner profile tubes present a profiling shape modification of the inner tube surface that addresses typical shortcomings of conventional bare tubes. They aim to optimize the operation performance of steam-cracker furnaces.

In a steam cracker firebox, furnace tubes are mainly heated by radiation coming from two opposing directions. This leads to an inhomogeneous temperature distribution in the tube shell, as well as in the gas phase inside the tube (see Figure 6), i.e., the so called sunny and shady sides of tubes. Furthermore, heat transfer from the burners into the cracking gas is both limited by the maximum allowable tube skin temperature, as well as by the heat transfer from the inner tube side into the gas phase. The resulting effects trigger various issues that negatively influence the steam cracker performance: Sunny sides cause a local overheating of the tubes; therefore, the maximum design temperature is reached faster and consequently the length of the cracking cycle is reduced. Additionally, the local overheating results in temperature gradients from the sunny to shady side. These can induce mechanical stresses across the tube shell, which reduce the tube lifetime.

In bare tubes the constrained heat transfer through the tube shell into the gas phase in combination with high gas velocities produces a radial gradient in gas temperature with a minimum along the tube axis (Figure 5a). The consequence is an overheated gas phase boundary layer at the inner tube surface and a relatively cold temperature in the core flow of the gas phase. Hence, the temperature driven gas phase reactions result in overcracking in this boundary layer. This increases the coking rate on the inner tube wall, whereas chemical activity is low in the cold core flow. As a result, both the furnace cracking cycle length and the product yield of the cracking gas are reduced.

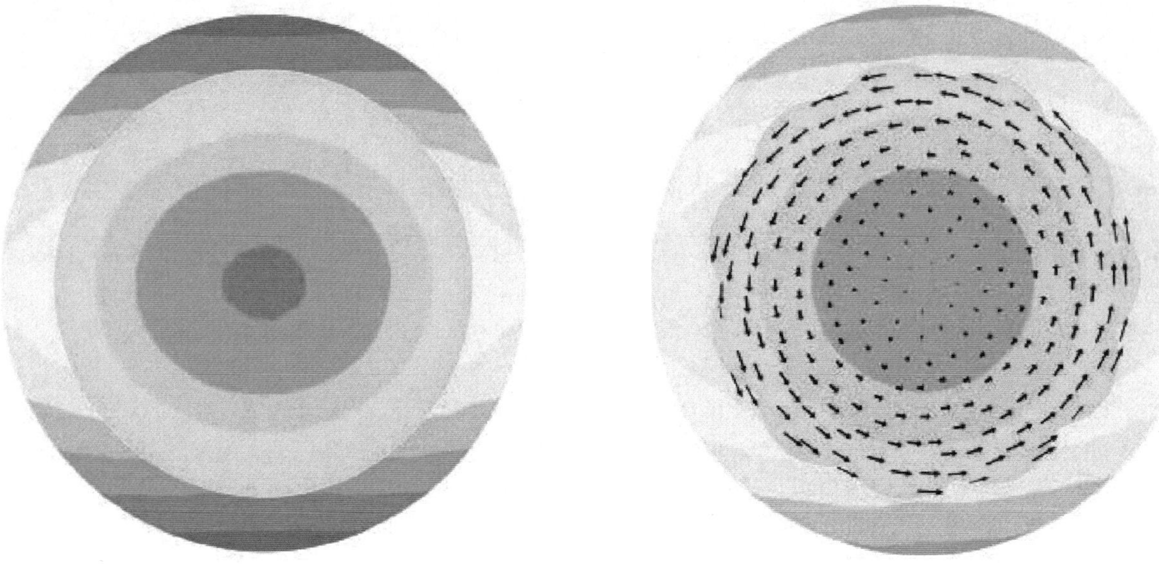

a) bare tube b) gas rotation 3D inner profile

**Figure 5:** Temperature Distribution and Gas Rotation after 2 m

All these physical factors should be taken into account when designing a profiled tube layout to modify the gas-flow characteristics in a beneficial way to overcome the inhomogeneous gas-temperature distribution. A gas rotation 3D inner profile induces a smooth gas phase rotation that increases the convective heat transfer at the inner tube side, reduces the maximum surface temperature of the tube and balances the temperature gradients in the tube shell as well as in the gas phase (Figure 5b & Figure 6).

The profile possesses fins with a specially designed shape, height and inclination to the tube axis. The careful adjustment of these profile parameters enables a quick buildup of a fully developed gas phase rotation and prevents elevated coking rates in the valleys of the profile, so that the process impact of the profile is maintained even at extended run length [6]. Since an increasing pressure drop along the coil is a considerable constraint of process performance, special attention needs to be paid to provide all mentioned beneficial properties at the lowest possible increase in pressure drop to minimize the negative influences.

As mentioned previously it is important that surface quality of the profile is equal to the conventional pull-bored tube surface of a bare tube. Therefore, and especially with NiCrAl material, complete protection by the oxide layer provides unrestricted anti-carburization and anti-coking properties.

**Figure 6:** Gas Rotation

The positive impact of a profiled tube with the described layout has already been proven in operation. The information and data obtained from field installations reflect results similar to the fluid dynamic simulations: lower tube metal temperature levels at start of run open up the production window and additionally a lower slope of temperature increase over time stretches the run length further.

The effect of reducing the over- and under-cracking in the gas phase was also investigated in field applications based on an elaborate and thorough on-site product yield analysis. It reflected an enhancement of olefin yield by more than 1% when operating with a naphtha feed. In this regard, such a profiled tube can be used either for increasing the run length or the throughput or the cracking severity and even a reduction in fuel consumption can be accomplished.

Special development work was performed by S+C in order to keep maintenance activities, especially related to the field replacement or repair works, unchanged: After beveling of the tubes by applying a special beveling machine, welding of the described profiled tubes can be done on site following exactly the same specifications as for bare tubes.

## Efficiency Increase Cases of Advanced Steam Cracking Radiant Coil Technologies

In the first sections of this paper, the beneficial properties of the advanced alumina former as well as the gas rotation 3D inner profile were presented and described in general. However, it is crucial to quantify the impact of the technologies on daily plant operation. For this reason, a calculation model was created, revealing the influence of these technologies on key performance parameters in the context of annual operation.

The main feature of the model is the dynamic adaption of process heat demand or fuel gas consumption respectively, taking into account the negative effect of insulating coke deposition and the positive effect of enhanced heat transfer technologies. The basic model input parameters are 8000 hours of operation (cracking + decoking) per year, 20 t/h propane feedstock, 33.95 % ethylene yield and a fuel gas requirement of 8.4 t/h at start of run conditions. Heat requirement during decoking is included with 10% of normal cracking operation.

Two typical industrial operation scenarios were analyzed in the model: "furnace limited" (FL) means, that the furnace is steadily operated up to the point at which coke deposition leads to approaching either maximum surface temperature or maximum pressure drop of the coil. On the other side, "scheduled decoking" (SD) means, that the cracking cycle length is predetermined and decoking is initiated at a preassigned point in time, probably well before reaching any design limitation.

| Parameter | NiCrSi | NiCrAl (FL) | NiCrAl (SD) |
|---|---|---|---|
| Cracking | 60 days | 86 days | 60 days |
| Decoking | 48 hours | 48 hours | 48 hours |
| Ethylene | 52.568 kta | 53.081 kta (+514 t) | 52.568 kta (± 0) |
| Fuel Gas | 67.821 kta | 68.433 kta (+611 t) | 67.130 kta (-692 t) |
| $CO_2$ | 171.800 kta | 173.348 kta (+1,548 t) | 170.047 kta (-1,752 t) |
| **$CO_2$/Ethylene** | **3.268 t/t** | **3.266 t/t (-0.002 t/t)** | **3.235 t/t (-0.033 t/t)** |

**Table 1:** Efficiency Model Results for NiCrSi versus NiCrAl

| Parameter | NiCrSi | NiCrAl+3D (FL) | NiCrAl+3D (SD) |
|---|---|---|---|
| Cracking | 60 days | 100 days | 60 days |
| Decoking | 48 hours | 48 hours | 48 hours |
| Ethylene | 52.568 kta | 53.255 kta (+687 t) | 52.568 kta (± 0) |
| Fuel Gas | 67.821 kta | 67.046 kta (-775 t) | 65.484 kta (-2,338 t) |
| $CO_2$ | 171.800 kta | 169.836 kta (-1,963 t) | 165.878 kta (-5922 t) |
| **$CO_2$/Ethylene** | **3.268 t/t** | **3.189 t/t (-0.079 t/t)** | **3.155 t/t (-0.113 t/t)** |

**Table 2:** Efficiency Model Results for NiCrSi versus NiCrAl+3D

The reference case in the model is a NiCrSi bare tube, which has a furnace limited cracking cycle length of 60 days under the given conditions. Comparison cases are NiCrAl technology (Table 1) as well as NiCrAl combined with 3D inner profile (Table 2), operated both as furnace limited and with a scheduled decoking. Based on field experience, the coking rate reduction of NiCrAl is set to 70% of NiCrSi (propane cracking), leading to a cycle length of 86 days while the additional cycle length extension of NiCrAl + 3D profile technology is assumed to be 100 days. This conservative assumption was made to keep the model results as universally applicable as possible. Taking into account customer-specific operation parameters or design limitations of a defined furnace design would very likely allow for a further increase in cycle length.

The model returns three key performance parameters that are supposed to be most relevant in the petrochemical industrial environment, such as ethylene production, fuel gas consumption and the related $CO_2$ emissions.

It is obvious that cracking cycle extension due to furnace limited operation with NiCrAl alloys considerably increases ethylene output while fuel consumption and thus, $CO_2$ emissions increase slightly because of the additional production days (and saved decokings). Nevertheless, specific efficiency in terms of $CO_2$/ton ethylene is even improved in this scenario (Table 1).

The combination of NiCrAl with a 3D profile technology (Table 2) further increases cracking cycle length and hence ethylene production up to additional 687 tons per year, while at the same time - due to the improved heat transfer - also absolute fuel consumption and $CO_2$ emissions fall below the reference case, indicating a massive gain in efficiency. The scheduled decoking operation magnifies the energy saving effects of advanced coil technologies due to a lower amount of detrimental coke deposition throughout the cracking cycle, so that finally up to 6000 tons of $CO_2$ can be avoided per annum.

The model does not account for efficiency benefits due to increased lifetime. These benefits can be achieved by installation of NiCrAl instead of NiCrSi and – depending on the operation scenario - prolonged cracking cycles (and thus a reduction of lifetime-consuming decokings) as well as lower average tube metal temperatures due to increased heat transfer when using 3D inner profiled tubes.

## Application of NiCrAl and 3D Profile Technology in the EU-funded IMPROOF Project

In 2016 IMPROOF was started, an EU-funded project that aims at improving the energy efficiency of steam cracking furnaces, while reducing emissions of greenhouse gases and NOx [7]. Within this project, the S+C proprietary NiCrAl alloy Centralloy® HT E as well as the 3D inner profile "SCope" were selected in order to fulfill the projects' efficiency targets. Before the industrial implementation of the considered technologies, they were tested and evaluated in lab and pilot plant scale at the Laboratory for Chemical Technology (LCT) of Ghent University.

### *Plug-Flow Reactor Tests with Chromia Former versus Alumina Former*

The performance evaluation of the NiCrAl was performed on coupons, featuring the original inner surface of an industrial tube. In a plug-flow reactor [8] the coupon is exposed to a typical furnace start-up procedure (formation of oxide layer), followed by high temperature stress conditions (TMT 1050 °C, ageing step) and a final cracking step in propane/steam atmosphere. For coupons made of chromia forming alloys HP+Nb MA (G 4852 Micro) and NiCrSi (ET 45 Micro) as well as the alumina former NiCrAl (Centralloy® HT E), the development of coke deposition during the final cracking step is continuously captured by an electrobalance.

Referring to the previous section, coke formation strongly depends on the catalytic activity of the formed oxide scales, but it is also influenced by the surface morphology. Figure 7 shows that at the beginning of the cracking cycle - when the bare inner tube surface is exposed to the hydrocarbon/steam atmosphere - the NiCrAl alloy features the lowest, so called "initial" coking rate, which corroborates the integrity of the alumina layer as it suppresses catalytic coke formation effectively, even in aged condition. Moreover, this observation can be transferred to the "asymptotic" coking rate, the continuous coking.

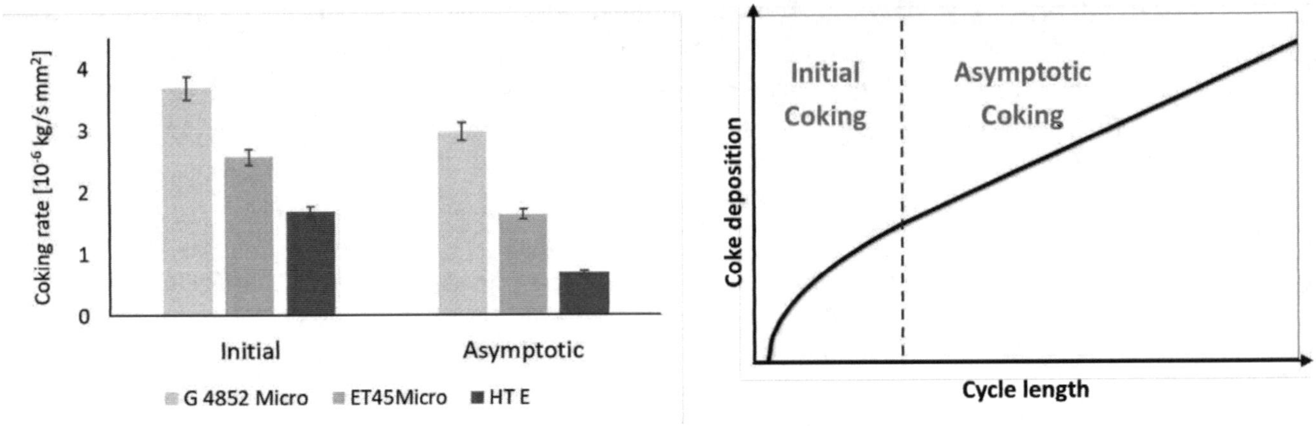


**Figure 7:** Initial and Asymptotic Coking Rate of Chromia Former and Alumina Former

The amount of coke deposited on the surface of an aged NiCrSi material is influenced by the effect of carbon induced corrosion, which was described in papers by Jakobi et al [3, 4]. This is an effect observed in aged chromia forming (NiCrSi) alloys due to the interaction of the chromia scale and the coke deposited on the tube surface: small worm like oxide particles grow on the sample surface together with the coke. During the subsequent decoking, the coke is burned off, but the oxides remain on the surface being stable phases under the decoking atmosphere. At the beginning of the next cracking phase, the coking rate is strongly increased due to the very large effective surface of the coral-like oxide deposits. Therefore, for chromia forming NiCrSi alloys, depending on the processing conditions, the initial as well as the asymptotic coking growth is increased both by catalytically active particles (present on the surface or on top of the coke layer) and the surface morphology (coral like oxide deposits). These two effects (formation of catalytically active sites as well as the surface area increase) are successfully suppressed by the protective alumina scale formed on the NiCrAl alloy.

***Pilot Plant Tests with Gas Rotation 3D Inner Profile***

The Laboratory for Chemical Technology (LCT) at the University of Gent operates a pilot plant test rig, which is a kind of downscaled steam cracker [9]. Thus, the pilot plant features all relevant steam cracker components and characteristics, such as heating up the steam/hydrocarbon feed in a convection section, cracking in a reactor coil which is suspended in a radiant firebox and finally a TLE cooling section with online product gas composition analyzer (Figure 8).

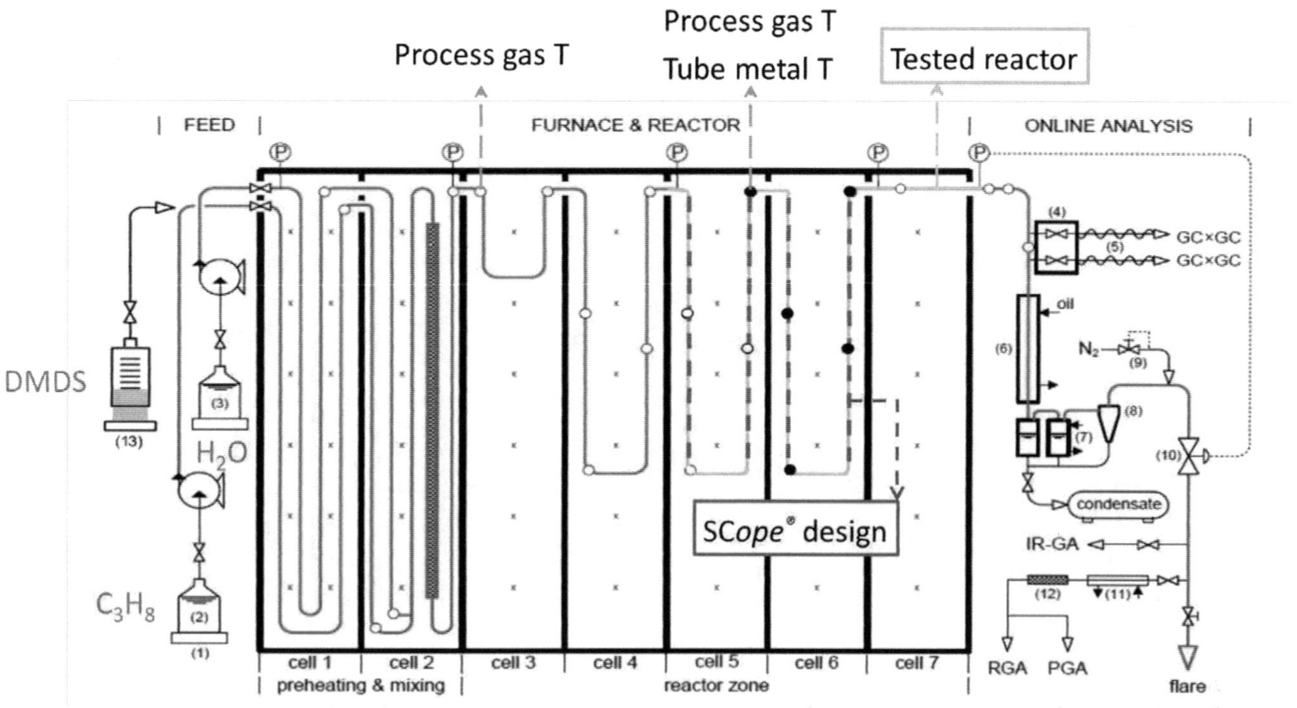


**Figure 8:** Pilot Plant Test Rig Design at LCT (University of Gent) [9]

As can be seen in Figure 9, a bare tube coil (NiCrAl) and the 3D profile tube coil (NiCrAl) were installed and tested. Just as in a real steam cracker, relevant operation modes like start-up, steam standby, cracking and decoking can be adjusted. The project target is to check the claim that 3D inner profiled tubes reduce the tube surface temperature by means of increased heat transfer. In combination with the homogenized gas phase temperature (due to better mixing) this effect is also supposed to decrease coke deposition.

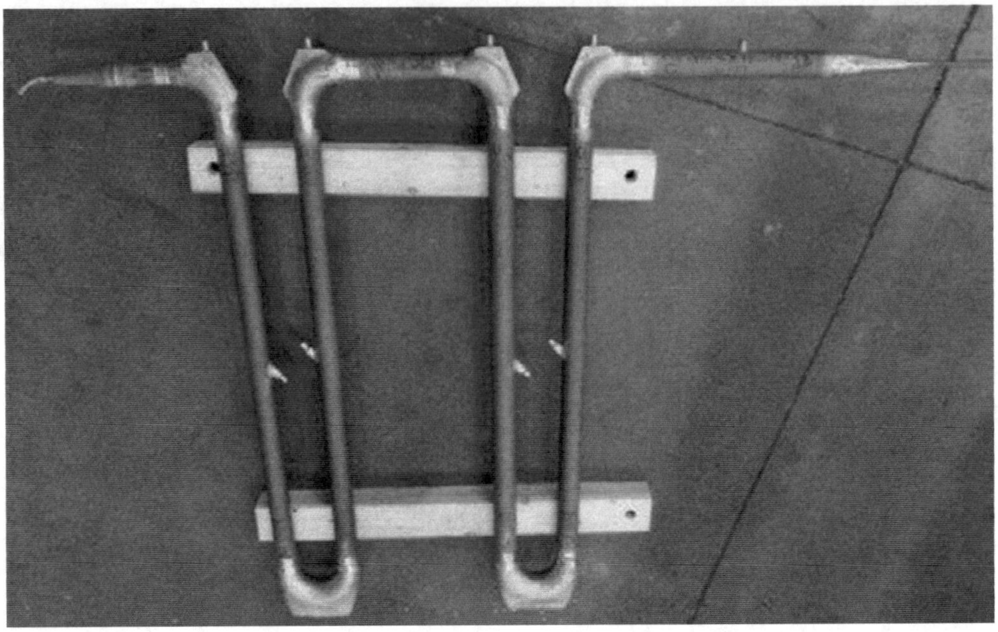

**Figure 9:** Gas Rotation 3D Inner Profile Coil before Installation in Pilot Plant Test Rig [9]

For tube surface temperature measurement the pilot plant features several highly accurate weld-on thermocouples along the coil and a pyrometer measurement, the latter representing the common standard in industry. Figure 10 shows the tube surface temperature measurement of the bare tube and the 3D profile tube during the third cracking cycle (6 hours duration) of the testing program. Despite some variances in the measurement the temperature difference between the bare tube coil and the 3D profiled tube is well above 30°C, which clearly exceeds the expectations from previous CFD simulations. Hence, heat transfer improvement is obvious and according to that, a fuel gas reduction was also found, namely -10%. Fuel gas reduction is expected to be over pronounced in this test rig, as – contrary to an industrial cracker - there is no coupling between the convection section and the radiant section, so heat recovery from the radiant section is not available.

The coke deposition during cracking is reconstructed by $CO/CO_2$ measurement in the subsequent decoking, meaning that a continuous coking rate recording in cracking mode is not possible. Therefore, the cracking cycle is split into two cycles with the first cycle running for 2h hours and the second cycle running for 6 hours. Each decoking is providing the corresponding coke deposition information so that finally a graph with initial coking (0 to 2 hours operation) and asymptotic coking (2 to 6 hours) can be generated. Figure 11 confirms, that the special 3D layout of the inner tube surface reduces the initial and asymptotic coking rate at least by a factor of 2 compared to a conventional, non-profiled tube.

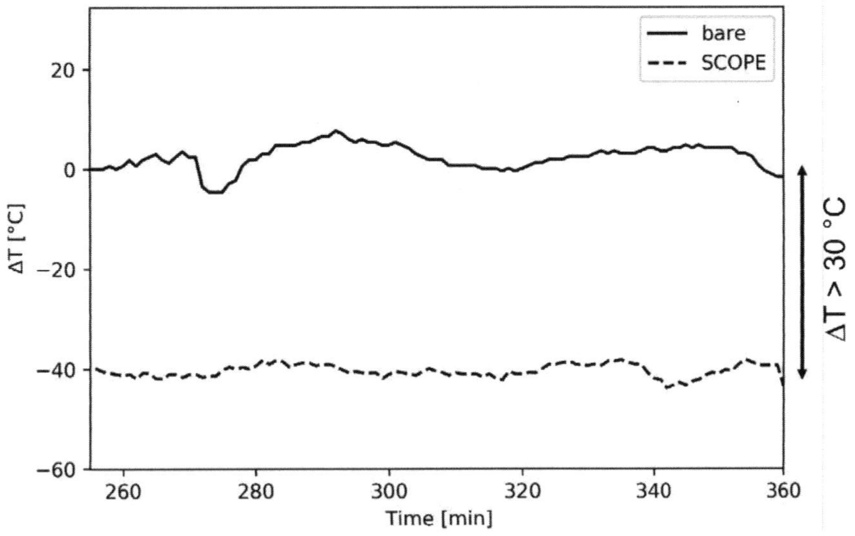


**Figure 10:** Tube Surface Temperatures of Bare Tube and Gas Rotation 3D Inner Profile Tube

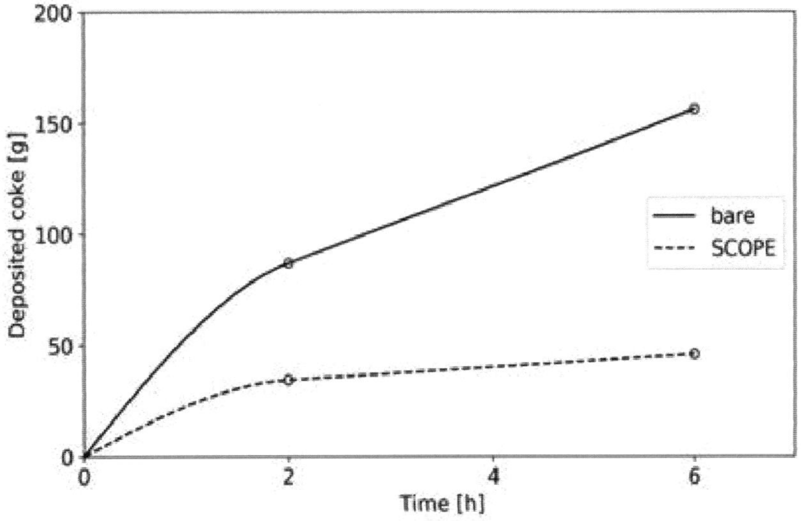


**Figure 11:** Initial and asymptotic Coking Rates of Bare Tube and Gas Rotation 3D Inner Profile Tube

## Conclusions

In a steam cracker coil both material selection and smart shaping of the inner tube surface can increase plant efficiency considerably. The key parameter is the optimization of heat transfer by means of coking rate reduction as well as convective heat transfer enhancement in the process gas phase. A further contributing parameter is the increase of coil lifetime and thus, plant availability.

Innovative NiCrAl alloys are superior to conventional NiCrSi alloys: especially in sulfur-rich steam-/hydrocarbon atmospheres typical for this industry, the presence of a dense and continuous alumina scale on the inner tube surface enables lower coking rates and higher coil lifetime. 3D inner profiles systematically increase the heat transfer and further reduce coking rates by avoiding overcracking reactions.

In practice - depending on the plant operation strategy - the technologies and ideally the combination of both, help to maximize ethylene output and/or reduce fuel gas consumption and $CO_2$ emissions, respectively. Whatever operation strategy is applied, a model calculation revealed that efficiency improvement in terms of specific emissions ($CO_2$ per ton of product) will always be accomplished.

A NiCrAl alloy and gas rotation 3D inner profile tube technology were selected within the IMPROOF project, which aims at engineering the most efficient steam cracker to date. In this project, both technologies proved their benefits and contributed to reaching the targets.

## References


1. D. Jakobi, R. Gommans (2003), "Typical failures in pyrolysis coils for ethylene cracking", *Materials and Corrosion 54*, pp. 881-887
2. D. Jakobi, C. van de Moesdijk, P. Karduck, A. von Richthofen (2009), "Tailor - Made Materials for High Temperature Applications: New Strategies for Radiant Coil Material Development", *NACE Corrosion 2009*, Paper No. 09155
3. D. Jakobi, P. Karduck, A. von Richthofen (2013), "The High-Temperature Corrosion Resistance of Spun-Cast Materials for Steam-Cracker Furnaces – A Comparative Study of Alumina- and Chromia-Forming Alloys", *NACE Corrosion 2013*, Paper No. 2287
4. D. Jakobi, P. Karduck (2018), "Behavior of High-Temperature Tube Materials in Sulfur-Containing Steam-Cracking Conditions", *NACE Corrosion 2018*, Paper No. 11191
5. D. Jakobi, J. Weigandt (2014), "Benefits from Educated Selection of Advances Materials for Radiant Coils – a 10 Year Field Performance Review of Alumina Forming Alloys", *AIChE Spring Meeting 2014*, Paper No. 152f
6. C.M. Schietekat et al. (2014), "Computational Fluid Dynamics-Based Design of Finned Steam Cracking Reactors", *AIChE Journal 60 (2)*, pp. 794-808
7. https://improof.cerfacs.fr/
8. Sarris et al. (2017), "Impact of initial surface roughness and aging on coke formation during Ethane Steam Cracking", *Industrial & Engineering Chemical Research,* pp. 5-6
9. S. Symoens, M.R. Djokic, J. Zhang, G. Bellos (2018), "Pas de deux" of high-temperature alloy and 3D reactor technology for steam cracking coils: impact on product yields and coke formation", *Chemreactor-23,* Conference

# Process Safety in Management by Walking Around - Strengthening Leadership in Process Safety

*Michelle Brown, Jose Ignacio Sanz*

*FMC Corporation, Philadelphia, United States, Michelle.Brown@fmc.com*
*FMC Corporation, Barcelona, Spain, jose.ignacio.sanz@fmc.com*

## Abstract

Management by Walking Around (MBWA) is a key safety leadership tool we have implemented in our manufacturing sites over the past few years. MBWA is an informal front line safety leadership technique which puts leaders on the shop floor to observe behaviors & conditions, listen to suggestions & concerns, engage in impactful discussions and look for & address signs of complacency.

In early 2019 – FMC utilized the MBWA technique to focus on life critical standard implementation (e.g. Line Breaking, Isolation, etc.) and occupational safety systems with great success.

In late 2019 – we began to wonder if we could incorporate process safety into the MBWA process. After reviewing our Loss of Primary Containment (LOPC) data – we determined a great place to start would be to have leaders looking for LOPC precursors – conditions and behavior which could lead to a LOPC event (e.g. vibration, unsupported pipe, etc.). Our Process Safety team created the education materials for this campaign in 6 modules:

- General awareness of process safety incidents
- LOPC from Pipes & Components
- LOPC from Major Equipment
- LOPC from Portable Equipment/Containers
- LOPC – Preventive Layers of Protection
- LOPC – Mitigative Layers of Protection

This paper explains the antecedents for this initiative, the process we followed and how it has helped to strengthen leadership of process safety across our organization.

## 1. Antecedents

Many process safety incidents and specifically loss of primary containment (LOPC) can be identified before the leak occurs. Analysis of past 3 years of PS events showed an important number of near misses and minor LOPC that resulted or could have resulted in releases of hazardous materials such as corrosive acids and bases or flammable materials. No significant consequences occurred due to mitigative layers of protection in place but there was a potential for risk to health, environment and equipment.

Whereas several management systems had been in place to prevent LOPC events such as PHA, asset integrity, process safety incidents and near miss reporting and investigation, workplace inspections, etc., a new additional focus on leak prevention was necessary to address and reduce the number of these events.

Leaks and LOPC are symptoms of equipment failures. Processes are designed to function in a prescribed manner and operate under defined set of conditions and with a defined maintenance plan. If we perform design, operation and maintenance well, we can expect equipment to have a predictable life. Issues with design, operation and maintenance can lead to premature failure and deficiencies can be from original conditions or due to changes. The problems are most often seen during the operation of the equipment, but prevention and solution can be related to any of the three functions.

What observations and actions before a leak occurs can prevent a leak and what actions employees could take to prevent LOPC and minimize risk? How can we identify potential Loss of Primary Containment (LOPC) events before a release occurs without implementing a "separate" process?

2. Management by Walking Around

As part of safety leadership training provided by a consulting company, FMC had implemented the program Management by Working Around (MBWA). This was mainly utilized as a safety leadership tool to show managers in the workplace and observe/identify safety concerns.

This program had successfully been implemented as part of EHS but never used specifically for process safety.

MBWA is an informal front-line leadership technique which puts leaders on the shop floor to:

- Observe behaviors & conditions
- Listen to suggestions & concerns
- Engage in impactful discussions
- Look for and address signs of complacency

The program had demonstrated that leaders appreciate very much the value of real-time observations and the ability to address the concerns and issues seen at the workplace rather than learning about an issue after the fact via an incident report and a root cause investigation. Leaders are busy and very frequently it is difficult to find the time to meet with workers at the workplace and address the issues and concerns. This aspect can also become frustrating for the leaders.

MBWA is just about preparation and planning to recognize what leaders can observe when they visit the workplace.

But how could MBWA work for process safety? Can this preparation and planning be made available for our leaders to conduct MBWA on PS?

3. The Program: PS in MBWA

The program started with analysis of data: where are the concerns for LOPC, what are the impacted equipment, precursors and causes.

The program will have then to identify areas of attention, develop training materials, educate and prepare our leaders. Finally get feedback and verify effectiveness. The process identified the following steps:

1. Analyze LOPC Data
2. Identify Observable Precursors
3. Develop PS MBWA Process
4. Educate Organization
5. Gather Feedback
6. Verify effectiveness

Data analysis showed that the majority of LOPC events was due to equipment damage versus human error or nonspecifically identified.

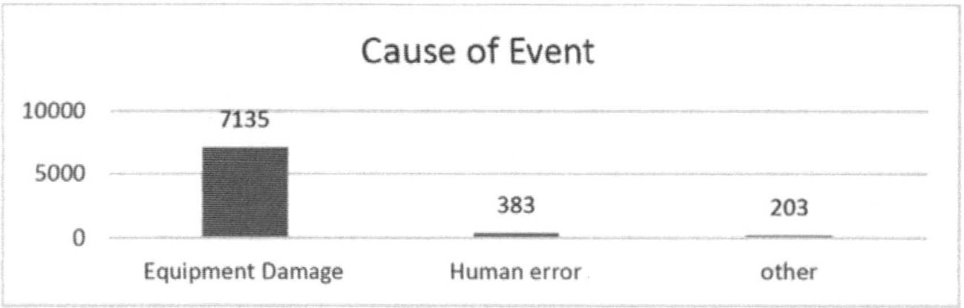


Figure 1: Data analysis of LOPC past 3 years. Cause

Additional analysis of LOPC events showed the distribution of equipment involved, like pipes and components, major equipment, mobile containers, hoses, etc. Examples of data analysis are shown in figures 1 and 2.

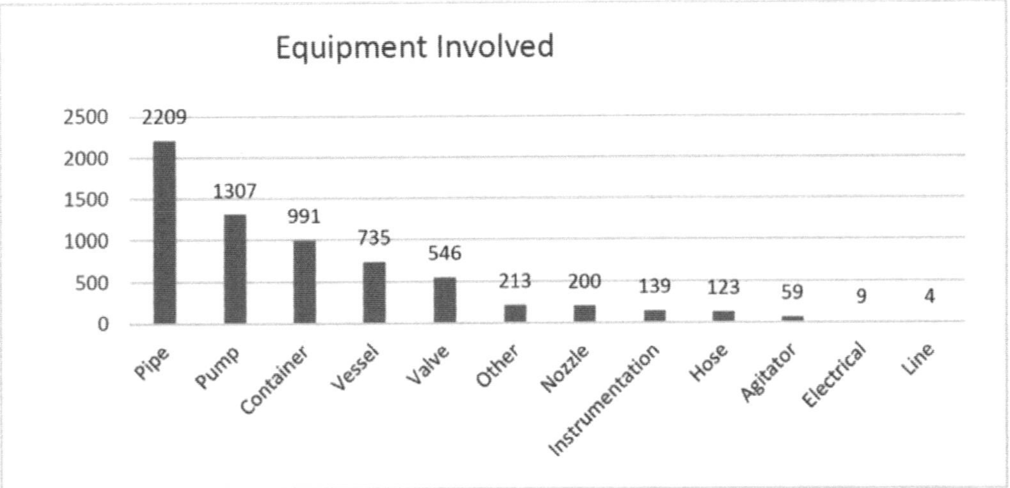

Figure 2: Data analysis. Equipment involved

But what are the precursors related to this equipment that could lead to a loss of primary containment event? Some of the examples found in the analysis of causes are listed below:

- Vibration
- Short bolting
- Poor support
- Corrosion
- Misalignment
- Missing caps/plugs
- Visible damage
- Incorrect material of construction
- Seal/gasket damage
- Inadequate or damaged preventative layers of protection
- Inadequate or damaged mitigative layers of protection

Could these precursors be identified in MBWA? Could leaders be prepared for visits to the workplace and identify these precursors before the LOPC happens?

## 4. Developing the Program

Program development required different steps to ensure preparation of leaders and conducting observations. Dividing into different modules will allow smooth implementation of the process and the required steps. These modules were delivered and executed with intervals of 2 months approximately. That allowed also to accommodate the same development sequence for each module as shown in the list below and in figure 3.

- Develop training materials
- Conduct virtual training to all leaders
- Site managers cascade information to their leadership team
- Conduct observations
- Take actions
- Evaluate effectiveness

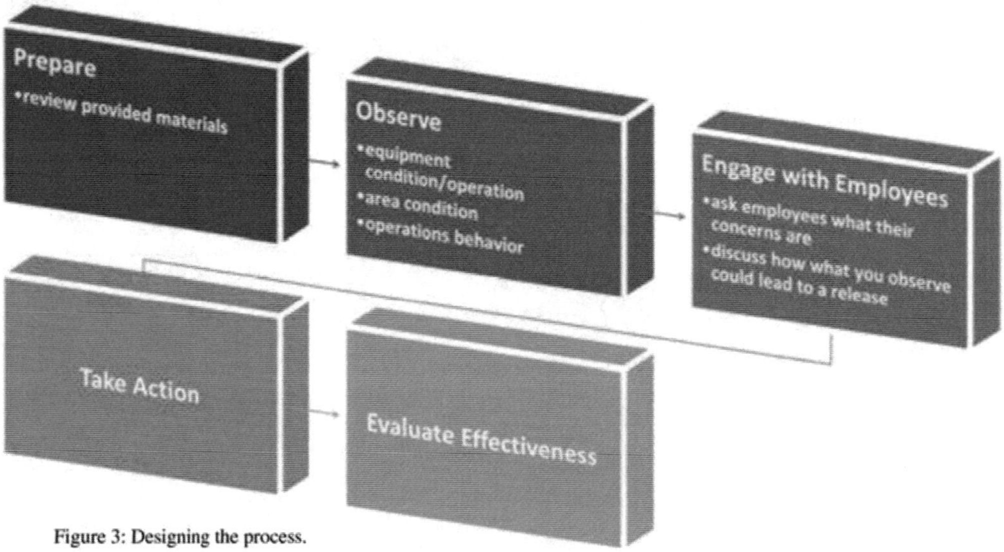


Figure 3: Designing the process.

Next step was to determine specifically the contents of each module for the planned observation. Table 1 shows how the modules were divided and the scope of each of them.

| Modules might include: | |
|---|---|
| 1 | Awareness of PS and PS incidents: concepts, importance of reporting all releases, overview of MBWA technique |
| 2 | LOPC from Pipes & Components: vibration, misalignment, valves with no cap/plug, bolting, hose failure |
| 3 | LOPC from Major Equipment: vibration, misalignment, coupling conditions, seal damage, bolting, bottom valves, flanges, powder handling equipment, pressure relief devices |
| 4 | LOPC from Portable Equipment/Containers: hoses & couplings, waste containers, portable containers as process equipment, containers and temperature, container transport, container storage, tank containers |
| 5 | Preventive Layers of Protection: overfill protection systems, SOP with operator response, signs/labels |
| 6 | Mitigative Layers of Protection: secondary containment, sprinklers, safety shower/eyewash placement and condition, signs/labels, hazardous area classification, LEL/gas detection & alarms |

Table 1: MBWA. LOPC modules

5. Educate the Organization

The training process should follow a logic and visual description of the potential problems associated to the subjects being observed but also the understanding of the hazards and the specific things to observe for each of the problems. This structured way of presenting the training materials (Problem-Hazard-What to observe/look for) as shown in figure 4 was always repeated for all the modules. This was projected as a more visual and consistent way to memorize the training and recognize the precursors of LOPC during the MBWA.

Figure 4: How training materials were structured for each of the modules

6. Training and Tools

Training materials were prepared by the Corporate Safety Team and presented remotely during regional sessions to all site leaders and others in the global organization.

Training materials used slides containing the structured mentioned before followed by set of examples with pictures showing the correct and incorrect situation.

Additional tools were developed such as a checklist to be available during the observations. They contained the observation check points consistent with the problem-hazard-observation structure of the materials provided in the training.

Some examples of the training materials have been captured in figures 4 to 7

Figures 4 and 5: Examples related to Hoses and couplings belonging to module on LOPC from Pipes and Components

Figures 6 and 7: Examples related to container transport belonging to module on Portable Containers.

## 7. Results

After implementation of the program, sites are reporting the most common issues found and actions taken. In some cases, observations led to implementation of corrective campaigns on specific concerns such as bolting, hose connections, dust processing equipment, missing end caps or transport of portable equipment.

A survey of the program was also conducted to collect feedback among participants. Quality of materials, frequency of observations, employee participation, level of knowledge of the concepts provided and main areas of concern detected during inspections were evaluated. Results showed very positive feedback on the program, contents and learnings. It showed also that frequent observations were conducted during the period and relatively high number of people participating in MBWA.

8. Next Steps

Site leadership teams and senior leaders have seen the value of recognizing hazards and precursors of LOPC during MBWA. Further analysis of data in the near future regarding actual LOPC events will show effectiveness of the program and the next things to focus on in a future program. In the meantime, FMC pursues another round of modules applied to operations where LOPC can occur. Continuation and improvement of the program in the near future will be based on data analysis.

9. Conclusions

- MBWA is a useful technique for leaders to learn and observe what is happening at the workplace, engage with employees and show leadership in EHS.
- Process safety can be applied in MBWA. There are plenty of opportunities to identify precursors and avoid the next LOPC and potential impact on health, safety and the environment.
- PS in MBWA is an excellent tool to strengthen knowledge and leadership in process safety.

# 619010 Vertical Thermosyphon Reboiler Operation Troubleshooting

**Abstract:**

Vertical thermosyphon reboiler stall can result in a loss of duty and/or a loss of unit capacity. The cryogenic demethanizer reboiler is susceptible to stall during transient or turndown operation, typically resulting from a relatively wide boiling range of the tower bottoms composition and a relatively small difference between the reboiler's inlet and outlet densities. This paper will discuss methods for correctly identifying a reboiler performance problem as stall, and will examine and share field techniques for un-stalling a reboiler.

# 619007 Furnace Modernization: An Economical Way to Upgrade Your Existing Furnace to Improve Energy Performance and Reduce CO2 Emissions

**Abstract:**

Shell is investing in a large multi-year furnace upgrade project whilst remaining on full production, resulting in a significant decrease of energy use and $CO_2$ emissions.

Shell investigated several options to maintain security of the production capacity of its 16 cracking furnaces after 50 years of operation, at its Moerdijk petrochemicals complex in the Netherlands. Following the outcome of a study, the selected options to choose from were a one-to-one replacement of 16 existing furnaces or upgrading 8 of those furnaces by doubling their capacity.

After detailed evaluation it became clear that the modernization of 8 furnaces was the most optimum path forward. Key drivers for selecting this option were:

- Improve energy performance and emissions
- Reduce environmental footprint (about 10% $CO_2$ reduction of Moerdijk's annual total), an important step to the Shell energy transition targets of being net carbon neutral in 2050 or earlier in step with society.
- Further process safety improvements
- Fast implementation within 5 years, without impacting production
- Lower investment cost compared to green field investment

The main process design challenge on this project is doubling the furnace capacity and improving on overall efficiency while re-using the existing firebox and burners, and maintaining existing foundations. The main project execution challenge is implementing these revamps outside of a plant turnaround to minimize the impact on the Shell Moerdijk lower olefins production, without any concessions to safety in the running plant.

This paper describes the furnace revamp design and the revamp execution strategy that minimizes field labour on site and maximises a modular approach.

Shell's technological know-how, project execution and operational experience combined with TechnipFMC's Process Technology and EPC capabilities has led to the current furnace design, enabling a 100% capacity increase within the same firebox.

99C

# HOW EFFICIENT IS MY CHARGE GAS COMPRESSION?

***Emerentino Quadro***
*Olefins Process Engineer*
*Braskem*
***Joana Arnaldo Ribeiro***
*Olefins Plant Engineer*
*Braskem*

Prepared for Presentation at the 2021 Spring National Meeting
Virtual, April 18-22, 2021

# HOW EFFICIENT IS MY CHARGE GAS COMPRESSION?

***Emerentino Quadro***
*Olefins Process Engineer*
*Braskem*
***Joana Arnaldo Ribeiro***
*Olefins Plant Engineer*
*Braskem*

**Abstract:** Energy efficiency is key for the Olefins plant competitiveness, additionally, the environmental aspects related to energy consumption are getting more importance. This way, monitoring and controlling the energy efficiency properly should be a primary skill of the process and operation engineer. Despite that, we have found a gap of understanding for the Charge Gas Compressor (CGC) energy efficiency in many teams. The CGC is intensive in energy consumption and it is present in all Olefins plants. Currently, many Olefins plants are facing the challenge of campaign increasing; excluding the legal aspects, by design, an Olefins plant campaign should be limited by the CGC performance, therefore, the CGC energy efficiency is a key issue for that. The present paper discusses several aspects related to the CGC energy efficiency and presents the main measuring to keep it on the right track.

## Introduction

Braskem is a Brazilian petrochemicals manufacturer, producing monomers, resins and chemicals specialties. In 2019, Braskem supplied 64% of the Brazilian resin demand.

Braskem's ethylene capacity is currently 5,000 kt/y, distributed across seven Olefins plants, located on five different sites, three of them based on naphtha steam cracking and two sites based on ethane steam cracking, including 200 kt/y from ethanol to produce the green polymer, from renewable resources. Four of these sites are in Brazil and one of them is in the port city of Coatzacoalcos (Mexico), built together with the Mexican group Idesa.

The steam cracker plant is intensive in energy consumption and the Charge Gas Compressor (CGC) energy consumption varies typically from 7 to 10% of the total Olefins plant energy requirement. Therefore, in order to improve the plant competitiveness, the energy efficiency optimization of this area of the plant is a key issue. Additionally, many crackers aim to minimize the environmental impact of the operation, in consonance with this thought, Braskem has announced the commitment to achieve zero carbon for 2050. The action plan to do that includes initiatives to use energy from renewable sources and improve the plant's energy efficiency.

However, the first step of the optimization process is to measure properly the energy efficiency. The Charge Gas Compressor is present in all Olefins plants and it is intensive in energy consumption. Despite that, measuring and analyzing the CGC efficiency is not a trivial task and, based on our experience, there is a gap of understanding about that in the process and operation teams.

This paper will focus on sharing our experience at Braskem (Camaçari Complex, localized in the northeast of Brazil) and studies related to this matter.

## Brief Process Description

The UO-1 Q1 BA has a typical naphtha steam cracker design.

The Charge gas Compressor is characterized by large volumes, high mass flowrates, low pressures in the initial sections and the need for multiple sections to meet the required discharge pressure. They typically have four to five sections across two or three compressor bodies of varying configurations and the compression ratio for each section varies from 1.5 to 2.5 in order to keep the section discharge temperature in control and minimize the polymerization rate.

The cracked gas (a mix of hydrocarbons with a number of carbons in the molecule between 1 to 7, hydrogen and water) leaves the quench water tower overhead at about 38 °C and is fed to the first section of compression and condensing by using an intercooler and knockout drum, which is the primary stage of separation. The gas resulted from the knockout drum is fed to the next section and the process is repeated at a higher pressure. The suction pressure for the first section is usually low (around 0.5 kgf/cm$^2$g) in order to improve the pyrolysis furnace operation by reducing the COP (Coil Outlet Pressure). The CGC suction pressure is controlled by varying the turbine driver rotation.

A charge gas washing with a caustic solution to remove acid components present in this gas ($H_2S$ and $CO_2$) is present in the charge gas compression area. It is also present in this system a charge gas drier by using adsorbents to reduce the water concentration to values lower than 1 ppm. Figure 1 presents a typical charge gas compression system.

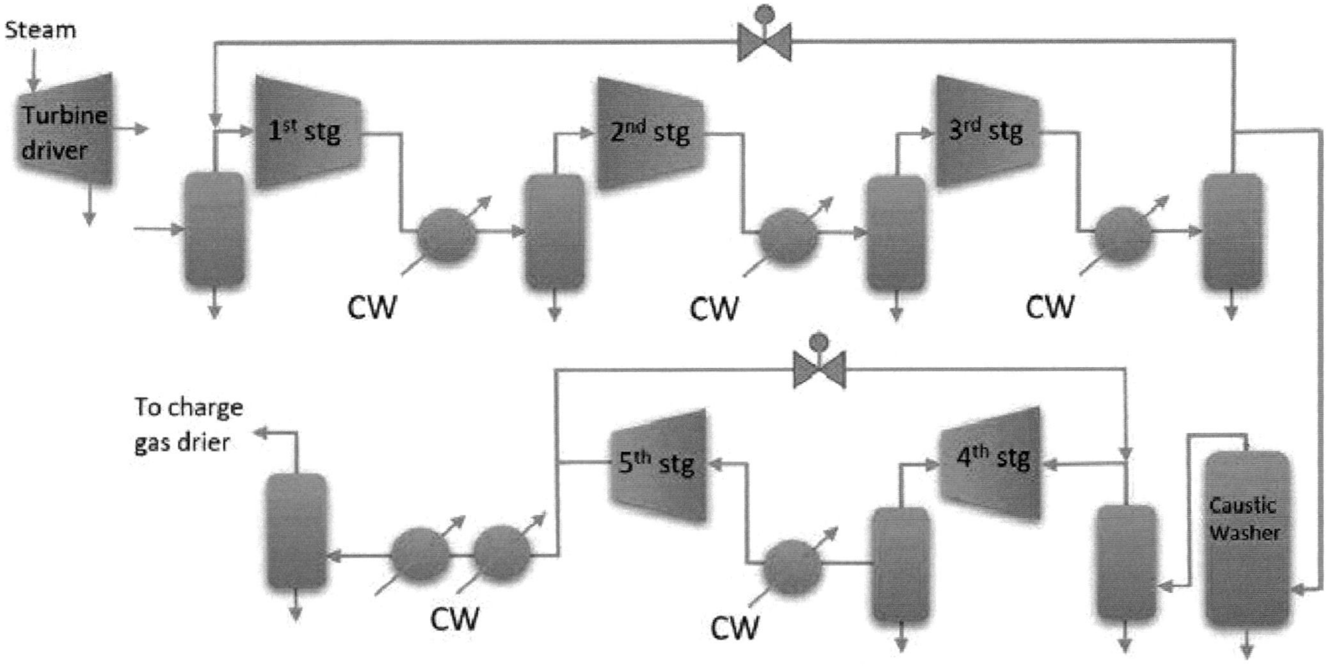


Figure 1. Croquis of a typical charge gas compression process scheme.

In fact, the true purpose of the CGC is to give energy to the charge gas, separate the light fraction and dry it in order to send it to the final cryogenic fractionation.

## Main Factors related to the CGC efficiency

The energy required by the CGC is affected by many variables, as follow:

- **Charge Gas Compressor (CGC) design**. Different machines require different recirculation to prevent the surge. The type of labyrinths could affect the energy consumption as well as the design of the intercoolers, pipelines and knockout drums. As a general design rule, around 8% of the total energy consumed by this system is due to the gas flowing through the intersection equipment (pipelines, intercoolers and knockout drum and filter);

- **Operation conditions**. Usually, low plant feed rates result in the necessity of CGC recirculation increasing in order to prevent the surge. High cold area backpressure affects the CGC efforts, resulting in energy consumption increasing, in the same way, the CGC suction pressure, typically determined by the pyrolysis furnaces requirement, impacts the CGC energy consumption;
- **Polytropic efficiency**. Oligomers deposits, internal recirculation through the labyrinths and charge gas composition affect the thermodynamic efficiency;
- **Lack of focus on recirculation optimization**. This part of the gap of energy efficiency is related to the operational actions in the day-by-day, particularly with the using of the existing facilities to control the recirculation in a minimum value to guarantee a proper distance of operation point from the surge line

From the operational team point of view, these are uncompressible factors, except the last one (focus on recirculation control). In fact, the operational team has no action on the machine design neither operational condition (plant feed rate, CGC backpressure etc.) and has limited action on the machine polytropic efficiency evolution along the campaign. However, it is important to understand how these factors affect the CGC energy consumption.

## How to measure the CGC efficiency?

A KPI frequently used to monitor the overall compression system efficiency is the Specific Energy Consumption (**$I_{SEC}$**), defined as the total energy consumed by the turbine driver divided by the flow rate of a stream representative of the plant capacity usage, for instance, the charge gas flow rate to cold area. This KPI is defined by the equation presented in Figure 2, where **$e_a$** and **$e_e$** represent the mass specific enthalpy of the steam in the admission and exhaust of the turbine driver. The advantage of this approach is the facility of calculation; it could be easily automatized in the DCS. On the other hand, once the general KPI cover all the factors cited above, it can not indicate where the energy has been drained.

$$I_{CGC} = \frac{F_{steam\ in} * (e_a - e_e)}{F_{charge\ gas\ to\ cold\ area}}$$

Figure 2. Specific Energy Consumption definition.

Another limitation of this approach is its restricted using to analyze the evolution of a certain compression system, this way it is not possible to use it to compare different machines. In fact, it makes no sense to compare the specific energy consumptions of different charge gas compressors because (i) different CGC designs will require different recirculation flow rate for the same percentage of capacity usage; (ii) eventually, different machines may be being compared at a different percentage of capacity usage; (iii) due to the restrictions in other parts of the plant, machines from different units may be operating under different operating conditions (for instance, suction pressure, backpressure from the cold area or overhead temperature of the quench water tower ) and, finally, (iv) CGCs may be being compared at different times in their campaigns, so that the drop in polytropic efficiency throughout the campaign may compromise the representativeness of the comparison. This way, it is possible to imagine that a certain machine with $I_{SEC}$ 0.90 GJ/t would be being operated in a more optimized way compared to another machine with $I_{SEC}$ 0.75 GJ/t.

Therefore, the $I_{SEC}$ increasing could be due to the operational condition changing (for instance, a period of low market demand for monomers or modification of CGC suction pressure for any reason), or unnecessary recirculation. In these cases, the $I_{SEC}$ evolution does not mean, in fact, a changing of the intrinsic energy efficiency of the system.

Another approach frequently used is to determine the polytropic efficiency (thermodynamic efficiency) of each section. Many times, it is not easy to monitor polytropic efficiency. A usual error is to use the design composition of the charge gas. A change of the value of the polytropic coefficient (Cp/Cv, dependent on the gas composition) from 1.20 to 1.22 results in an increase of polytropic efficiency calculated by 6 percentage points. Therefore is highly recommended to collect and analyze chromatographically the gas composition every time the polytropic efficiency is measured. Additionally, this calculation requires expunging the effect of water eventually injected in some charge gas compressors to reduce the discharge temperature, it could be performed by a thermodynamic simulation; usually, this water injection reduces the discharge temperature around 10°C. Additionally, many plants have not facilities to get these samples neither proper instruments to carry out these calculations. Besides that, the polytropic efficiency only reports the evolution of thermodynamic efficiency of each section in itself and does not take into account, for instance, a lack of efficiency of the turbine drivers or the effect of a flowing resistance increasing through the intercoolers.

In order to overcome these difficulties, this article proposes a methodology to monitor the compression system energy efficiency evolution.

## Polytropic Efficiency Evolution Monitoring (PEEM)

Is it being a CGC system with a SEC 0.70 GJ/t operated in a more efficient way compared to a previous scenario, when the $I_{SEC}$ was 0.80 GJ/t? Not necessarily.

In order to overcome this kind of issue, we developed the PEEM (Polytropic Efficiency Evolution Monitoring), this KPI is based on a non-phenomenological approach to estimate the $I_{SEC}$ of the system by using a lot of operational variables from the CGC and compares it with the $I_{SEC}$ calculated in according to the equation presented in Figure 2. The basic PEEM concept is presented in Figure 3.

$$PEEM = I_{CGC\,\boldsymbol{measured}} - I_{CGC\,\boldsymbol{predicted}}$$

Figure 3. PEEM definition.

This way, PEEM will provide a trend of thermodynamic efficiency evolution for the whole compressor. Initially, it will be close to zero and will increase along the campaign. PEEM do should be to calculated automatically by linking it with the operational database.

The PEEM model is developed for each system. In our case, this model was trained in a period of six months by using commercial statistical software, when the compressor was known to be clean.

The application of this methodology in a Braskem cracker resulted in a $I_{SEC}$ mean absolute error prediction close to 0.46%.

| Postprocessed results | Model fit | Predictions |
|---|---|---|
| Number of observations | 3463 | 866 |
| Max. negative error | -0,0348033 | -0,0280657 |
| Max. positive error | 0,029712 | 0,0274982 |
| Mean absolute error (MAE) | 0,00449016 | 0,00459973 |
| Root mean square error (RMSE) | 0,00592015 | 0,00601128 |
| Residual sum | 2,9914E-10 | 0,33798 |
| Standard deviation of residuals | 0,00592015 | 0,00599859 |
| Coefficient of determination ($R^2$) | 0,994869 | 0,994579 |
| Correlation | 0,997431 | 0,997298 |

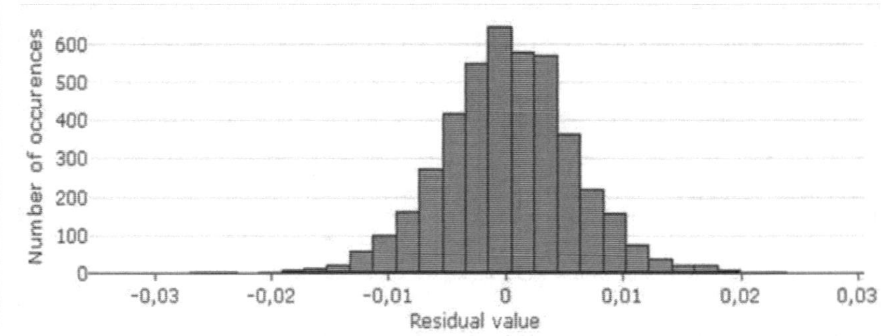


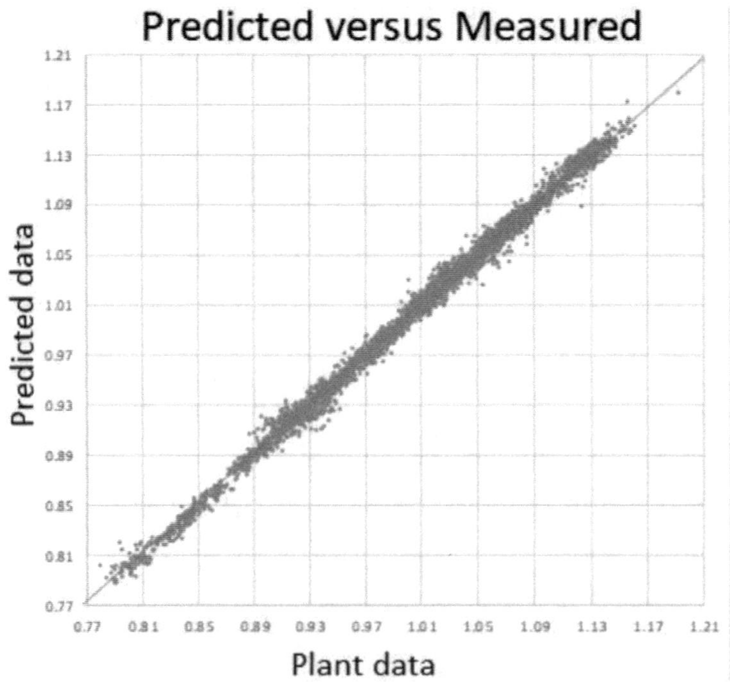


Figure 4. $I_{SEC}$ prediction deviation.

This methodology allows the identification of the most important operational variables for $I_{SEC}$. The statistical analysis shows the relative impact of each variable in the $I_{SEC}$ prediction, this way, it is possible to identify the most effective process adjustments to bring the energy consumption to the target value.

In order to validate the obtained model, 20% of the operational data was used to validate the model and only 80% of the available operational data was used to model training. Once PEEM is based on a non-phenomenological model, the PEEM is strictly valid only for the operational window in which it was trained.

The application of this methodology in one of the Olefins Plants of Braskem resulted in the graph presented in Figure 5. It shows the energetic efficiency of this CGC has not been modified significantly since the moment when the model was trained.

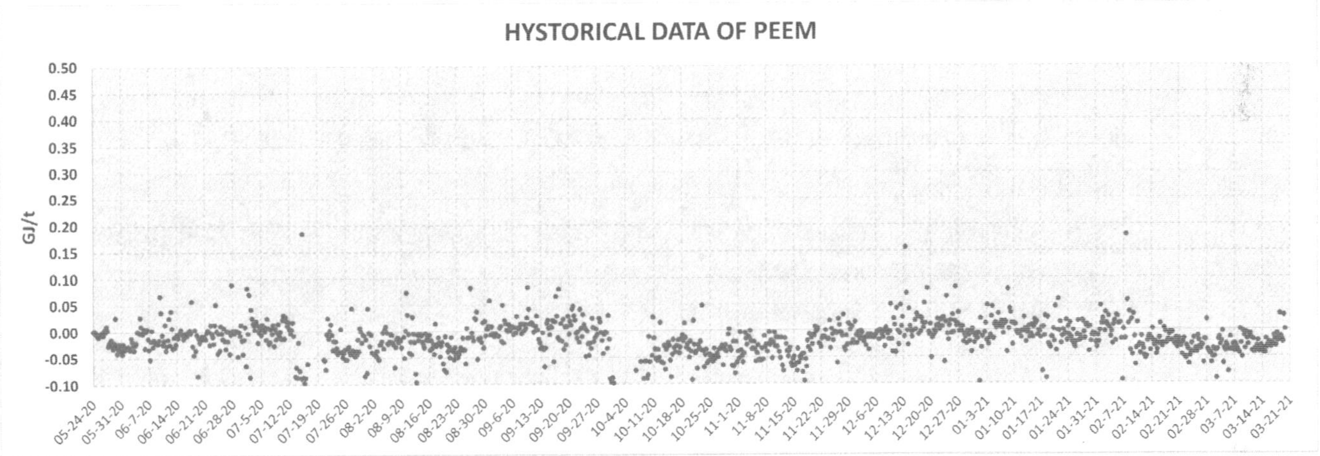


Figure 5. ISEC prediction deviation.

## How to keep thermodynamic efficiency on control

In addition to measuring the thermodynamic efficiency, we would like to control it. In order to do so, the following should be practiced:

- Keep the section discharge temperature below 90°C, by spraying continuously water in the suction or in the case. High temperatures result in high polymerization rates. Based on Arrhenius law, the fouling rate could double, case the temperature increases 8-10°C;

- Inject a wash oil in the suction or in the case of each stage. The desired characteristics for this oil are high aromaticity (preferentially higher than 90%), high ASTM D86 final boiling point (higher than 250°C) and low olefins content (bromine index lower than 1000). The availability in the Braskem Camaçari Complex of this stream with excellent chemical characteristics is due to the catalytic reforming installed in this site, however, this stream is supplied by some vendors;
- In order to minimize undesirable internal recirculation, the material of the compressor labyrinths should be resistant to chemical attack and erosion. Modern materials have been developed with this aim.

## Conclusions

- The approach presented allows for monitoring in a practical way the evolution of the CGC polytropic efficiency and verifying its trend over time with a relatively low error, regardless of the operational conditions.
- If a trend of significant reduction is observed, a more detailed analysis should be done and preventive actions are taken to prevent the further loss of efficiency. Options include:
  - Chemical injection to solubilize the oligomers inside the compressor and prevent its formation
  - Proper wash oil injection
  - Adjustment in the water injection to cooling down the discharge temperature etc.
- This initiative is particularly useful as many crackers are considering extending their campaign to 8 or 9 years.
- This methodology, in theory, could be adopted by any other CGC with minor adaptations, dependent upon the available instrumentation

## References

1. Snider, S., Ethylene Plant Cracker Gas Compressor Fouling, AIChE Spring Meeting, Houston, TX EUA, 2006.

2. Shaojun, L., Prediction of Cracking Gas Compressor Performance and its Application in Process Optimization. Chinese Journal oh Chemical Engineering.20 (6), 2012.

3. ASME Standart, Multistage Centrifugal Compressor Fouling Evaluation at High Power Setings. International Gas Turbine & Aeroengine Congress & Exhibition, Stockhlom, Sweden, 1998

4. De Haan, S., The Impact of Fouling and Corrosion in Ethylene Plant Efficiency. AIChE Spring Meeting, San Antonio, TX EUA, 2010.

# 619818 Quick Hit in Decarbonization – Start with the Flare

**Abstract:**

The pressure for sustainable, low carbon emissions operation is growing from the public, investors, and regulators. While decarbonization is a popular buzzword, practical programs for achievement are difficult to find. The logical starting point is to look at the flare and think about where the emissions originate. An effective flare management plan is a practical driver for reducing emissions. It can also improve the economics of a facility and enhance operational flexibility.

In this paper/presentation, we look at how chemical processing plants can improve flare management practices to maximize hydrocarbon recovery, minimize carbon emissions, and reduce operating costs associated with continuous flaring. Good flare management practices include (but are not limited to):

- Flaring assessments to determine current routine, non-routine, and emergency flaring operations
- Examination of flaring root causes and identification of reduction opportunities
- The development of a flare minimization plan with clear and achievable flaring reduction targets.
- Maintenance of the plan with proper change management practices to accommodate updates to the facility.
- The identification and monitoring of flaring key performance indicators (KPIs) to promote the right actions by facility stakeholders.

The paper/presentation will outline common flaring challenges that are typically seen in these facilities and discuss available mitigations that have both economic and operational rational for adoption. We will also provide a framework for a sustainable drive towards a low-emission operation with systematic and periodic assessment of an operator's flaring operations to identify opportunities and implement changes that reduce demand on the facility's flaring systems.

# AIChE Paper Number: 114b

# Electrification of Steam Cracking and Related Challenges

*Gunther Kracker*
*Group Lead Cost Estimation Petrochemical Plants*
*Product Management & Cost Estimation Petrochemical Plants*
*Linde Engineering*

*Mathieu Zellhuber*
*Process Design Engineer*
*Process Engineering Petrochemical Plants*
*Linde Engineering*

*Jens Becker*
*Teamlead Monomer & Utilities Technology*
*LyondellBasell*

Prepared for Presentation at the 2021 AIChE Spring National Meeting
EPC Technology Session: Environmental Session

Virtual presentation, April 21, 2021

## Abstract

With increasing societal and legislative pressure on our petrochemical industry, especially with regards to Green House Gas (GHG) emissions, there is a clear demand for viable technological solutions which effectively reduce GHG emissions from steam cracking. Far beyond merely storing and reutilizing $CO_2$, we need to directly reduce $CO_2$ emissions from steam cracking and correlated steam supply. A promising key technological option to do so is the electrification of equipment and utilization of renewable power from carbon neutral energy generation.

The paper and associated presentation provides a general overview on potential electrification measures specific to steam cracking technology and discusses the state of their developments. It deep-dives on selected electrification measures within the steam cracking process, such as the electrification of large turbo-compression machinery. Specific challenges and limits related to these concepts as well as fundamental pre-conditions for their technical application are assessed.

Some of the individual measures discussed are substantiated by practical examples and experiences from an existing European petrochemical complex, providing guidance for their technical feasibility in real world applications. This will include a review of reliability data based on long term experiences.

Finally, the paper elaborates on different combined overall plant electrification concepts over two distinct time horizons, i.e. mid-term solutions for partial electrification and long-term solutions for full electrification of steam cracking technology

# AIChE Paper Number: 618398

# Paper 114c - THE END OF THE STEAM AGE?

*Author & Presenter:* ***Nick Rogers***
*Technology Manager*
*Technip Energies, Milton Keynes, U.K.*

Prepared for Presentation at the 2021 Spring National Meeting
April 2021



**Abstract:**

The low energy, low emission cracker of the future is likely to see a major shift in the way that energy is developed and utilized within the ethylene plant. This presentation will look at what to expect in future cracker designs where low carbon electrical power is imported to drive the main compressors within the cracker.

The drive to reduce $CO_2$ intensity of crackers will lead to the re-design of cracking furnaces, so that the same fired duty goes into the cracking process but less fuel gas is fired for the purposes of making steam. This reduces overall fuel gas consumption which can lead to new opportunities for excess fuel gas, whilst the reduced furnace VHP steam production enables large steam turbine drivers to be replaced with electric motors.

Impacts on all major utility systems on a new liquids cracker design is illustrated as well as the implications for revamp projects.

Discussions on potential uses for any excess steam and fuel gas are shared.

Some high-level implications from switching turbines to motors will be presented based on recent vendor feedback.

## 1. Introduction

For many years, the role of steam has been integral to the design and efficient operation of ethylene plants around the world. The cracking furnaces have utilized the convection section in order to raise large quantities of HP or VHP steam that is then used to provide the driving force for steam turbines that power the main compression services. This system works fine, but the furnaces are fired from light hydrocarbons, which when burnt lead to large emissions of $CO_2$ from the furnace stacks.

If $CO_2$ emissions from our industry are to be reduced, whilst ethylene production is still forecast to increase, then a new approach will be needed in order to achieve these aims. Concepts for $CO_2$ reduction will need to be applied both to existing ethylene plants, in order that they remain able to operate, and for new ethylene plants such that they are sanctioned for investment and contribute to the $CO_2$ reduction targets set by either operating companies themselves or from national and international regulatory measures introduced to enforce the targets set.

In this paper, the key factor under consideration will be the electrification of the main compressors, based on the assumption that imported electricity will increasingly be generated from renewable or alternative technologies that do not emit $CO_2$ at the rate seen from the burning of fossil fuels.

This will be just part of the solution, along with existing and emerging technologies that will not be considered here, such as:

- Pre- or Post-combustion carbon capture and storage
- Electrification of furnaces
- Catalytic or novel cracking technologies

Electrification of the main compressors is not widespread, but is a relatively mature technology that is available now, so the impact on the overall plant utility systems is developed for two examples;

- revamp, where it is assumed that furnaces are not immediately replaced or modified
- new plant design, where new furnace designs can be tailored to minimize steam production, and lead to lower $CO_2$ emissions for the same capacity

The potential savings in $CO_2$ are readily calculated from the $CO_2$ Emission rate for generation of electricity of the technology utilized on site, in the particular region, or as published for each country. Typical values used in this analysis are:

| Technology | T $CO_2$e / GWh (Ave) |
|---|---|
| Coil | 888 |
| Oil | 733 |
| Nat Gas | 499 |
| Solar PV | 85 |
| Biomass | 45 |
| Nuclear | 29 |
| Hydroelectric or Wind | 26 |

Ref [2]

## 2. The Need for Change

The global ethylene capacity in 2015 was around 150 MT/a, producing approximately 200 Mt/a of $CO_2$ as direct emissions. Future predictions are that by 2030, the global ethylene production will be approaching 200 Mt/a, with a potential increased $CO_2$ emission rate of 275 Mt/a. In most crackers, 80-95 percent of all $CO_2$ emissions are from firing of fuel gas in the cracking furnaces, the residual is from boilers for steam generation and occasionally small amounts for process heating.

It is recognized in Europe, and beyond, that action to reduce greenhouse gas emissions is essential and legislation is in place that will incentivize companies to meet the targets that are being set and often reviewed and revised.

If we look at Europe as an example, the European Union (EU) has put in place legislation to decrease $CO_2$ emissions for all of Europe by 40 percent by the year 2030, when compared to those seen in 1990. Slightly above 20 percent has already been achieved; therefore another 20 percent is required in the next nine years. The EU Green Deal goes further, with the objective to be carbon neutral by 2050; therefore, the 40 percent reduction by 2030 may yet increase to 55-60 percent.

There are insufficient emission rights to cover for all emissions in the Emissions Trading Scheme (ETS). Crackers will get a part of their $CO_2$ allocations for free based on the cracker benchmark (the current ETS benchmark value is 0.681 T of $CO_2$/T of HVC for the $4^{th}$ period 2021-2030).

Data below in Figure 1 shows that most European crackers are above the $CO_2$ benchmark – for most plants a 20-30 percent reduction in $CO_2$ emissions will bring them below the benchmark Operators which make significant reductions in their emissions due to technology change, may therefore end up having more allowances than they are emitting and can sell these extra allowances (or buy fewer allowances). The price of $CO_2$ in the ETS is expected to rise, the current price (as of the end of 2020) is ~30 Euro/Tonne) [1]

**Fig 1 - $CO_2$ Emissions from EU Crackers 2016-7 (Data from CEFIC/EEPC)**

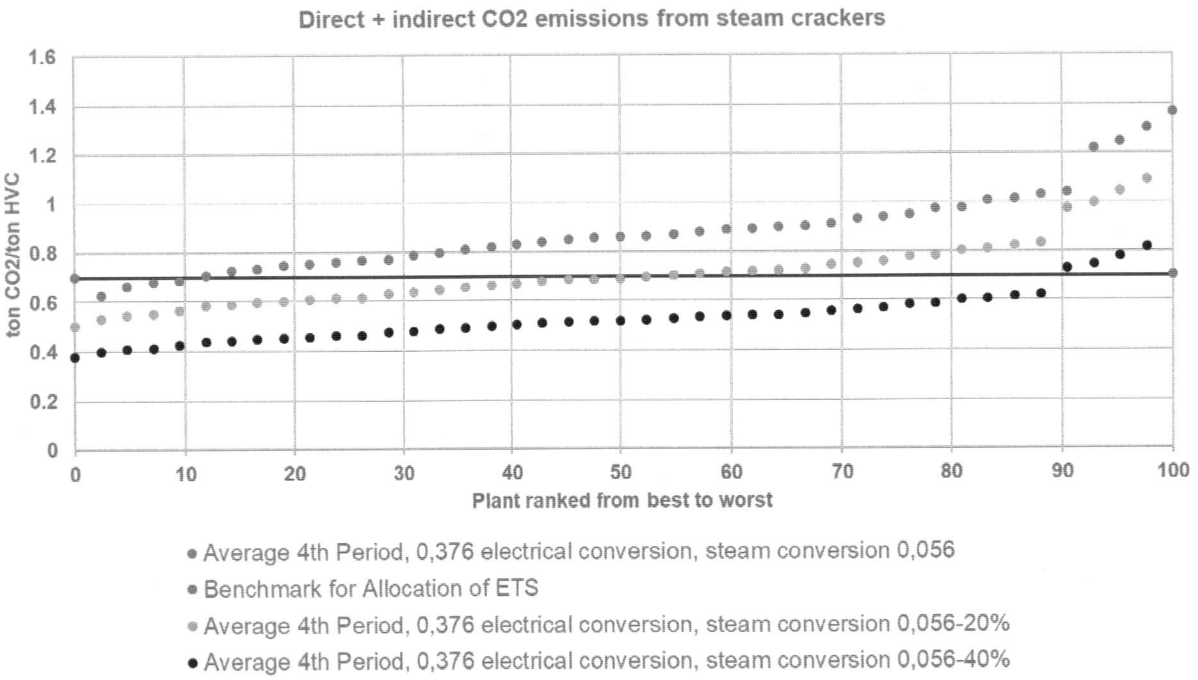


## 3. Impact of a low $CO_2$ design on a new Ethylene Plant

A 1500 kta Liquid Feed Cracker has been selected from which to study a conventional design of furnaces and utility systems against a potential low $CO_2$ design.

### A. Furnaces

The development of furnace designs to enable the same cracking, but with lower overall fired duty and lower VHP steam production will be presented at this conference by AIChE Paper Number: 617345, "LOW-EMISSION CRACKING FURNACE". This technology demonstrates that the VHP steam production can be reduced by up to 60 percent. For the purposes of the study on the impact on the balance of plant utilities, a slightly lower figure of 40 percent steam reduction is utilized.

### B. Compressor Drivers

For a new plant design there is no limit on the number of compressor turbine drivers that are converted to electric drive. The choice becomes dependent on the $CO_2$ footprint and generation capacity of imported electricity. The steam balance will then be adjusted to make maximum use of the steam that is still produced, taking into consideration the needs of the plant and the overall demand and flow of utilities across the whole petrochemical complex. Steam imports will not be required at VHP or HP level. Potentially an MP steam import will be required for start-up, though even this could be generated from the let-down of VHP steam once a number of furnaces are running in hot steam standby mode.

Typically, the main turbine drivers are as follows:

- Cracked Gas Compressor
- Propylene Refrigeration Compressor
- Ethylene Refrigeration Compressor / C2 Splitter Heat Pump
- C3 Splitter Heat Pump (if needed)

Some smaller turbines will be retained on steam, especially those serving critical pump services, such that a mix of electrical and steam drivers are available to ensure circulation in the event of an upset scenario such as power failure.

For the example given, the C3 Splitter Heat Pump is not present, and the Cracked Gas Compressor and the Propylene Refrigeration Compressor are chosen to be switched to electrical motor drivers. In order to keep all benefits clear, the revised steam balance assumes no imports or exports. Hence, the VHP utilized in the Base design is replaced with a smaller, cheaper, MP Boiler in the low $CO_2$ design.

| **Service** | **Base Design** | **Low $CO_2$ Design** | |
|---|---|---|---|
| | Turbine Power MW | Turbine Power MW | E-Motor MW |
| CGC | 90 | - | 90* |
| C3R | 44 | - | 44 |
| C2R/ C2HP | 24 | 24 | - |
| Power Generation | - | 62 | |

* CGC Power potentially split across two electric motors

The Fuel Gas balance shifts for the revised design, with much lower firing duty in the furnaces and the absence of VHP or HP boilers giving a much-increased FG excess that can be exported. Options for what to do with this excess hydrogen-rich gas are discussed later.

| | **Base** | **Low $CO_2$** |
|---|---|---|
| Furnace Firing Duty | 1281 MW | 995 MW |
| VHP Boiler Firing Duty | - | - |
| MP Boiler Firing Duty | - | 44 MW |
| Fuel Gas Available | 91 t/h | 91 t/h |
| Natural Gas Import [(a)] | ~ 2 t/h | ~ 2 t/h |
| Excess Fuel Gas | 5.9 t/h | 25 t/h |
| **t of $CO_2$ / t of HVC** [(b)] | **0.63** | **0.55** |
| **t of $CO_2$ / t of HVC** [(c)] | **0.62** | **0.48** |

(a) NG Import retained for furnace & flare pilots
(b) Based on ISBL firing + electric import at 376 t of $CO_2$/GWh (Standard EU figure)
(c) Based on ISBL firing + electric import at 26 t of $CO_2$/GWh (Wind/Hydroelectric)

Due to lower amount of surface condensers, the CW demand is reduced by 7200 $m^3/h$.

The most dramatic effect is seen in the Electrical Power Balance:

| Electrical Power | | Base | Low $CO_2$ |
|---|---|---|---|
| ISBL Process User Demand | MW | 8 | 8 |
| Compressor Driver Demand | MW | - | 90+44 = 134 |
| Generation | MW | - | 62 |
| Total Power Import | MW | 8 | 80 |

## 4. Impact of a low $CO_2$ design on an Ethylene Plant revamp

A 600 kta Naphtha Feed Cracker has been selected from which to study the impact on the utility systems with a potential low $CO_2$ design from partial compressor electrification.

### A. Furnaces

The furnaces are all existing, so it is unlikely that there will be wholesale changes or economic replacements purely for the purposes of $CO_2$ reduction. In this case the steam production will remain the same for the cases studied.

### B. Compressor Drivers

For a new plant design there is no limit on the number of compressor turbine drivers that are converted to electric drive, but when we look at an existing asset, then it is more feasible that limited driver changes will be made. The scope of changes will depend on the age and reliability of the drivers, the potential efficiency improvements and the $CO_2$ footprint and generation capacity of imported electricity. The steam balance can then be adjusted to make maximum use of the steam that is still produced from unmodified furnaces, taking into consideration the needs of the plant and the overall demand and flow of utilities across the whole petrochemical complex. Existing steam imports can be minimized or eliminated.

The main turbine drivers are as follows:

- Cracked Gas Compressor
- Propylene Refrigeration Compressor
- Ethylene Refrigeration Compressor

All other existing turbines will be retained on steam, so no change to the driver philosophy for those serving critical pump services.

For the example given, the Cracked Gas Compressor is initially chosen to be switched to an electrical motor driver since this is the largest driver (Options 1 and 2) below. Options 3 and 4 are provided for the case where the C3R machine is chosen, should this be more in keeping with the potential increase in low carbon electricity availability. In order to keep all benefits clear, the revised steam balances assume no steam exports. The C2R compressor is not switched to an electric motor driver because it is the smallest steam consumer of the main compressors and the compressor typically operates at a higher speed than the other compressors.

| **Service** | **Base Design** | **Low $CO_2$ Design** | | | |
|---|---|---|---|---|---|
| | | **Opt 1** | | **Opt 2** | |
| | Turbine Power MW | Turbine Power MW | E-Motor MW | Turbine Power MW | E-Motor MW |
| CGC | 35 | - | 35 | - | 35 |
| C3R | 23 | 23 | - | 23 | - |
| C2R | 5.3 | 5.3 | - | 5.3 | - |
| Power Generation | - | - | | 20 | |

| **Service** | **Base Design** | **Low $CO_2$ Design** | | | |
|---|---|---|---|---|---|
| | | **Opt 3** | | **Opt 4** | |
| | Turbine Power MW | Turbine Power MW | E-Motor MW | Turbine Power MW | E-Motor MW |
| CGC | 35 | 35 | - | 35 | - |
| C3R | 23 | - | 23 | - | 23 |
| C2R | 5.3 | 5.3 | - | 5.3 | - |
| Power Generation | - | - | | 10 | |

Option 1, with the just the CGC Driver being replaced by an electric motor, does not use steam efficiently and not surprisingly, the specific energy rises for this option. It does however, require the largest import of external electricity, so would give the greatest reduction in $CO_2$ emission, if the plant has access to sufficient low carbon electricity. The plant in this mode can reduce its MP steam import to zero.

If Option 1 is developed into Option 2 with power generation from the excess VHP steam, then the specific energy is improved from the Base Design as well as giving potential significant reduction in $CO_2$ emissions.

When the same exercise is repeated with the Propylene Refrigeration Compressor Driver being replaced with an electric motor, this gives similar results. Each variation gives a lower MP import demand and shows how the choice of which machine to electrify can be tailored to the available electric from low carbon sources.

Across all options the specific energy reductions are not significant. To achieve this then overall firing in the furnaces has to be reduced – which would require a program of furnace upgrades or replacements.

## 5. Fuel Gas Utilization

In the new cracker design presented above, the combination of enhanced furnace design and driver electrification gave rise to a large fuel gas excess for export. How this hydrogen rich gas is utilized is interesting as it has the potential to further improve site emission reductions.

The simplest option is to use it as a substitute for natural gas that is fired in other processes. When the cracker fuel gas is burnt, the $CO_2$ is reduced by around 35 percent per tonne of fuel consumed.

With cracker fuel gas being high in hydrogen, then this hydrogen can be extracted and sold as a premium product using PSA or membrane technologies. Similarly, for maximum hydrogen production the excess cracker fuel gas could be sent to a Steam Methane Reformer. SMR technology can now be combined CCS technology to produce blue hydrogen. With demand for hydrogen predicted to continue to rise for industrial and domestic usage, these are potential options outside the cracker scope.

The gas could also be a feed to methanol production, details of which will not be presented here.

The gas could also be utilized for power production. If the 25 t/h excess demonstrated earlier was fed to a combined cycle gas turbine (CCGT) unit, it could generate over 200 MW of power. This would generate electricity at around 315 T of $CO_2$ per GW/h of production. This cannot compare with low carbon alternatives, but would provide significant improvement in areas where power is derived from coal or natural gas.

## 6. Conclusions

Electric drivers will be a major consideration for new plants and existing cracker revamp opportunities.

Unlike some of the developments that are being considered, this is not new technology and can be applied today. Technip Energies has experience going back to the 1960s, where a number of plants utilized motor drivers due to the low cost of the electricity available local to the plant.

The principles highlighted in this paper are applicable to any number of compressor services. The limitation will be the quantity and reliability of the electrical supply.

A major review of ISBL and OSBL utilities is needed and particularly for revamps, this will be key for establishing the opportunities for steam exports. In other areas, the lower demand for CW gives opportunities for debottlenecking.

There is a potential to reduce $CO_2$ emissions by over 30 percent, depending on the renewable content of the imported power. For many older plants this could mean it will not be required to purchase expensive carbon credits to continue economic operation.

There is potential for further benefits from utilizing the excess cracker fuel gas for power generation, hydrogen production or chemical production.

## 7. References


1. "Price of polluting in EU rises as carbon price hits record high". Financial Times article, 11th December 2020. https://www.ft.com/content/11bd00ee-d3b5-4918-998e-9087fbcca3cd
2. "Comparison of Lifecycle Greenhouse Gas Emissions of Various Electricity Generation Sources", World Nuclear Association Report 2011

# 620258 Real-time analytics solution for augmented intelligence in ethylene

**Abstract:**

An Intelligent Real-Time Analytics Solution utilizes advanced analytics and AI/ML techniques to provide real-time insights. Customized to operations requirement, the solution with specific analytical features addresses essential needs including live gap analysis, improvement opportunities, instrument drifts, stability, performance benchmarks, predictive insights, and serves as a communication tool between engineers and operations. It also tracks performance and provide insights on a live basis for improving yield, efficiency and asset effectiveness.

A real-time solution, developed with a state-of-art technology and advanced analytics application for ethylene process industry results in several benefits.

- Benefits for Operators:
  - Provides more insightful information and prescriptive advice to make right decisions in a timely manner, avoid suboptimal operation and achieve significant profit improvement.
  - Provides a digitalized environment to address dynamic challenges and support safe and efficient operation
- Benefits for Engineers:
  - Data analytics provide a higher level of insights and understanding of process and equipment conditions
  - Helps engineers reduce data handling and model maintenance efforts and focus on plant improvement and troubleshooting

This paper discusses the various opportunities in the ethylene process and improvements possible from application of AI/ML solutions. The best methodology to address challenges in achieving the desired objectives and overall approach to ensure success.

SPRING21
+17TH GCPS
A Joint AIChE and CCPS Meeting

# Got Dust?
# Performing a Dust Hazard Analysis (DHA)

**Alfonso Ibarreta, Timothy Myers, Michael Stern, Sean O'Hern, Achim Wechsung**
**Exponent, Inc.**
**1075 Worcester Street, Natick, MA 01760**
**aibarreta@exponent.com, tmyers@exponent.com, mstern@exponent.com, sohern@exponent.com, awechsung@exponent.com**



Prepared for Presentation at
American Institute of Chemical Engineers
2021 Spring Meeting and 17th Global Congress on Process Safety
Virtual
April 18 - 22, 2021

# Got Dust?
# Performing a Dust Hazard Analysis (DHA)

**Alfonso Ibarreta, Timothy Myers, Michael Stern, Sean O'Hern, Achim Wechsung**
**Exponent, Inc.**
**1075 Worcester Street, Natick, MA 01760**
**aibarreta@exponent.com**

**Keywords:** DHA, Dust Hazard Analysis, Combustible Dust

## Abstract

Combustible dust hazards can be present at industrial facilities that handle or produce finely divided solid particulates (aka dust). National Fire Protection Association (NFPA) standards (including NFPA 652 *Fundamentals on Combustible Dust*) require facilities to perform dust hazard analyses (DHAs) on new and existing processes handling combustible particulate solids. The facility owner or operator is responsible for determining whether the materials being handled are combustible or explosible, and if so, characterizing their properties as required to support the hazard analysis.

The DHA is intended to systematically evaluate the fire, deflagration, and explosion hazards of a facility and the mitigation methods present in the facility. The DHA must be performed or led by a qualified person. Depending on the resources of the company, it may be performed by in-house staff, or may be led by one or more outside subject matter experts.

This presentation will outline best practices for performing DHAs, including how to best identify combustible dust hazards in facilities. The presentation will include discussion of the following topics:

- Identifying relevant combustible dust standards
- Identifying testing data required for material evaluation
- Selecting a hazard analysis methodology
- Identifying hazard mitigation strategies

# 1 Introduction

Combustible dust hazards can be present at industrial facilities that handle or produce finely divided solid particulates (e.g., powders or dust). Some industrial facilities intentionally handle powders as a raw material, intermediate, or final product. Other facilities generate dust as a byproduct or waste stream during operations such as cutting, grinding, or abrasive blasting.

National Fire Protection Association (NFPA) standards (including NFPA 652 *Fundamentals on Combustible Dust* [1]) require facilities to perform Dust Hazard Analyses (DHAs) on new and existing processes handling combustible particulate solids. Aside from the normal fire hazard posed by all combustibles, the unique hazards posed by a combustible dust are the following:

- Flash Fire: "A fire that spreads by means of a flame front rapidly through a diffuse fuel, such as dust, gas, or the vapors of an ignitable liquid, without the production of damaging pressure." [1]
- Explosion: "The bursting or rupture of an enclosure or container due to the development of internal pressure from a deflagration." [1]
    - Where 'deflagration' is defined as "Propagation of a combustion zone at a velocity that is less than the speed of sound in the unreacted medium." [1]

A combustible dust explosion hazard exists when all 5 elements of the explosion pentagon, shown in Figure 1, can be present at one time. If confinement is not present, a flash fire hazard could still be present. The goal of the DHA is to identify potential combustible dust hazards. The DHA team may also propose recommendations for how to prevent or mitigate hazards identified to have insufficient safeguards in place.

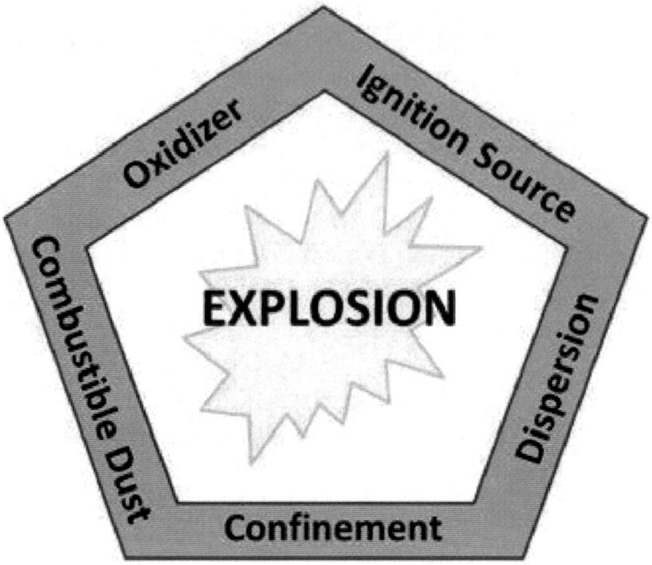


**Figure 1. Combustible dust explosion pentagon showing the five elements necessary for an explosion to occur.**

The NFPA standards that govern DHAs permit the owner/operator to follow multiple methodologies and strategies. The purpose of this paper is to summarize the DHA requirements and provide general guidance to owners and operators in how to prepare for and perform a DHA.

## 2 Combustible Dust Standards

The NFPA has developed a series of standards for the prevention and mitigation of dust explosions across a wide range of industries.[1] The NFPA has no enforcement authority, but these standards become regulations when adopted by an authority having jurisdiction (AHJ) such as when adopted by building codes, local, and state regulations. They are also typically considered good engineering practice, even when not required by regulations. NFPA combustible dust standards have been cited by OSHA when issuing citations per OSHA's General Duty Clause.

NFPA 652 *Fundamentals of Combustible Dust* provides a set of general requirements for the mitigation of fire and explosion hazards associated with combustible dusts and particulate solids

[1] The NFPA technical committees are currently working to consolidate these standards into a single document, NFPA 660.

across all industries and processes. The standard also directs the user to other commodity-specific NFPA combustible dust standards for additional requirements. These include:

- NFPA 61, *Standard for the Prevention of Fires and Dust Explosions in Agricultural and Food Processing Facilities*
- NFPA 120, *Standard for Fire Prevention and Control in Coal Mines*
- NFPA 484, *Standard for Combustible Metals*
- NFPA 654, *Standard for the Prevention of Fire and Dust Explosions from the Manufacturing, Processing, and Handling of Combustible Particulate Solids*
- NFPA 655, *Standard for Prevention of Sulfur Fires and Explosions*
- NFPA 664, *Standard for the Prevention of Fires and Explosions in Wood Processing and Woodworking Facilities*
- NFPA 850, *Recommended Practice for Fire Protection for Electric Generating Plants and High Voltage Direct Current Converter Stations*

These standards provide requirements to prevent the occurrence of ignitable dust clouds in the presence of ignition sources. Additionally, the standards contain requirements for mitigation methods that minimize the effect of dust explosions or fires if they do occur. A DHA will, in part, check for compliance with the applicable NFPA standards.

## 3 Determining Whether Your Dust is Combustible

Whether or not a Dust Hazard Analysis (DHA) needs to be performed in the first place depends on whether the dust produced or handled at the facility is combustible and/or explosible. According to NFPA 652, a particulate solid is defined as 'combustible' if it can propagate a flame or smoldering front when a 25cm-long train of accumulated material is subjected to a flame, and 'explosible' if it is "capable of propagating a deflagration when dispersed in air or the process-specific oxidizing media."[1] The facility owner or operator is responsible for determining whether the materials being handled are combustible or explosible, and if so, for characterizing their properties as required to support the hazard analysis.

While NFPA 652 permits the use of historical facility data or published data to determine whether a dust is combustible or explosible, it is important to note that variations in particle size and morphology, moisture content, composition, or other properties could affect the combustibility and explosibility of a material. Testing of a representative sample using the Combustibility Screening Test (Trough Test) and the Explosibility Screening Test (Go/No Test) is typically the preferred methodology [1]. The same material in different locations of the facility may exhibit different combustibility behavior. For example, plastic resin pellets used as a feedstock may not be combustible but small particulates of that resin accumulated on elevated surfaces or on vacuum pump filters may be combustible. Therefore, appropriate sample selection and/or sieving the sample to test the fine fraction of a sample are important for accurately assessing whether a combustible dust is present.

Once a dust is classified as combustible and/or explosible, additional tests may be required to analyze the potential hazard and determine the level of protection required. Some of the parameters that can be quantified include [2]:

- Deflagration Index ($K_{st}$)
- Maximum Pressure ($P_{max}$)
- Minimum Explosible Concentration (MEC)
- Limiting Oxygen Concentration (LOC)
- Minimum Hot Surface Ignition Temperature (MIT)
- Minimum Ignition Energy (MIE)
- Minimum Autoignition Temperature (MAIT)

Determination of the minimum ignition energy, minimum autoignition temperature, and hot surface ignition temperature are useful for determining the likelihood of ignition. Based on these results, safe operating conditions, such as maximum equipment temperatures, can be established. A material with a low minimum ignition energy may be susceptible to sparks and static discharges that necessitate certain precautions in handling.

The $P_{max}$ and $K_{st}$ values quantify the severity of a potential explosion and can be used to design explosion protection vents or chemical suppression systems. In some equipment or processes where practical, inert gas may be used to lower the oxygen content in the enclosure. Measurement of LOC value allows designers to determine the maximum permissible levels of oxygen in order to sufficiently inhibit deflagrations. Additional test methods are available for determining other properties of the dust.

It is typically not necessary to perform all tests on all dust samples, but rather specific properties can be measured as needed to perform the DHA or design protection systems.

# 4 Performing a Dust Hazard Analysis (DHA)

NFPA 652 requires that a DHA be performed at any facility handling combustible dust, and that the DHA be reviewed and updated every 5 years [1]. In addition, changes in equipment or operating conditions or materials should trigger a management of change review, which would address any hazards that are introduced by the changes [1]. Therefore, a hazard analysis is a necessary step in the implementation of any new or updated process that handles combustible dust.

## *4.1 DHA Methodology*

According to NFPA 652, "the DHA shall evaluate the fire, deflagration, and explosion hazards and provide recommendations to manage the hazards" [1]. However, NFPA 652 does not prescribe the evaluation methodology and allows the owner/operator to identify and utilize which methodology or methodologies to best meet their needs. A wide range of hazard analysis strategies are available, including:

- Hazard and Operability Study (HAZOP)
- Checklist

- What-If Analysis
- Failure Modes and Effects Analysis (FMEA)
- Event Tree Analysis
- Fault Tree Analysis
- Cause-Consequence Analysis
- Bow-Tie Analysis

These and other methodologies are described in detail in the Center for Chemical Process Safety books, *Guidelines for Combustible Dust Hazard Analysis* 1st edition [3] and *Guidelines for Hazard Evaluation Procedures*, 3rd edition, published by the American Institute of Chemical Engineers [4]. It should be noted that it is not the intent of the NFPA 652 standard to require owners and operators to follow Process Safety Management (PSM) as defined in OSHA regulation 29 CFR 1910.119 [5] in performing the DHA [1]. However, some of the elements of PSM are included in NFPA 652, and some dust handling processes may fall under PSM.

The DHA must be performed or led by a qualified person [1]. The qualified person should be familiar with conducting a DHA, as well as with combustible dust hazards. The DHA is normally performed by a team, which should include persons that are familiar with the process operations (e.g., engineers, operators, maintenance, safety).

### *4.2 Hazard Mitigation and Prevention Strategies*

As described above, the DHA evaluates the presence of hazards and the current safeguards designed to prevent or mitigate the hazard. Safeguards are required to ensure an appropriate level of risk to life safety but may also be warranted if the risk of operability disruptions or environmental releases exceeds the desired level. For example, an outdoor and remotely located dry material dust collector for non-metal combustible dust may not require explosion protection based on life safety considerations. Adding explosion protection, however, might significantly reduce downtime in the event of an internal deflagration within the collector.

Explosion mitigation and prevention systems function by eliminating one of the required conditions for a combustible dust explosion shown in Figure 1. The presence of combustible dust can sometimes be eliminated or reduced through substitution of the feedstock with a less combustible material or dilution of the combustible material with an inert material. The presence of an oxidizer can sometimes be eliminated or reduced through addition of an inert gas to partially displace the oxygen-containing air within equipment. Confinement, which is required for a deflagration to result in an explosion, can be reduced by adding explosion vents. Dispersion of the combustible dust can be reduced through the addition of dust collection systems and housekeeping procedures.

NFPA 652 provides multiple means to assess the compliance of the system being reviewed. Compliance can be assessed via a prescriptive DHA, a performance-based DHA [1], or a risk-based DHA. In a prescriptive DHA (such as a “checklist” analysis), the owner/operator ensures that all of the equipment, processes, and management systems meet or exceed each requirement listed in the appropriate chapters of NFPA 652 and any commodity-specific NFPA standards. A performance-based DHA evaluates meeting the performance requirements outlined in Chapter 6 of NFPA 652 and may not meet each outlined prescriptive requirement from other chapters. Furthermore, NFPA 652 permits the prescriptive requirements to be modified by a documented risk analysis, which allows for further flexibility in the meeting of the DHA objectives. Typically, a hybrid approach is used where most of the prescriptive requirements of the standard are followed, but where unique issues arise, a performance-based or risk-based analysis is performed.

Following the prescriptive requirements is one method to assess whether the risk of a combustible dust incident is at an acceptable level based on the properties of the dust and the safeguards present. In some cases, however, some or all the prescriptive requirements are unfeasible or undesirable for a given system. Alternatives to the prescriptive requirements may be permitted by the NFPA standards if they are an Equivalent Design, a Performance-Based Design, or based on a Risk Evaluation [6].

Performance-Based Designs must be evaluated by a qualified person based on design objectives including life safety objectives set forth in the NFPA standards and other criteria for operability acceptable to the owner/operator. The design objectives are then used to define performance criteria that ultimately determine the equipment and procedural thresholds and conditions that must be met in order to meet the design objectives. The development of these conditions and thresholds and their evaluation must be documented.

One example of an alternative design would be the use of administrative controls to eliminate or significantly reduce the probability of an employee being near a hazard at the time of a combustible dust incident. This alternative could meet life safety objectives if the probability of an employee being near a hazard in the event of a fire or explosion were sufficiently small and if the possible disruptions caused by a loss of equipment were acceptable to the owner/operator.

### *4.3 DHA Recommendations*

The results of the DHA review may involve recommendations to prevent or mitigate identified combustible dust hazards with insufficient safeguards currently in place. Recommendations should be actionable items that consider the feasibility and effectiveness of the proposed mitigation. Examples of recommendations often found in DHAs include:

- The facility needs to develop appropriate housekeeping procedures (including scheduling, regular inspections and unscheduled housekeeping procedures) to maintain the dust accumulations below 1/32 inch on horizontal surfaces, including elevated surfaces.
- The facility needs to develop a bonding and grounding auditing program to ensure that equipment remains properly bonded and grounded.
- The dust collector needs to be fitted with a properly-sized explosion vent that directs a potential deflagration to a safe outside location, in accordance with NFPA 68 (2018).

The owner / operator must document the results of the DHA and properly address the recommendations. Typically, there may be multiple methods to mitigate the risk of hazards identified by a DHA. Completing the DHA can be an iterative process as various solutions are evaluated until final responses to the DHA recommendations are prepared. Documentation of the response to the recommendations must include any necessary action items requiring change to the process materials, physical process, operating conditions, or facilities.

The owner / operator is responsible for maintaining appropriate management systems. This includes operating procedures and protocols, housekeeping procedures, hot work permits, inspection and maintenance programs, training and hazard awareness programs for personnel and contractors, emergency planning and response plans, and incident investigation systems. For each of these, a management of change process helps owners in considering safety implications before changes are made and a document retention program ensures that owners/operators keep proper records.

Lastly, the owner / operator is also responsible for reviewing and updating the DHA at least every five years or when material changes to the conditions occur.

# 5 Conclusions

NFPA 652 and other commodity-specific combustible dust standards require that a DHA be performed at facilities handling combustible dust, and that the DHA be reviewed and updated every 5 years. The first step in the process involves testing the particulate materials handled or produced at the facility, to determine if they are combustible and/or explosible.

The applicable NFPA standards allow the use of a variety of methods and strategies to ensure that combustible dust hazards are properly mitigated. Ultimately, it is the responsibility of the facility owner / operator to ensure the DHA is properly performed and the recommendations stemming from the DHA are addressed.

# 6 References.

[1] NFPA 652 *Fundamentals of Combustible Dust.* 2019 Edition.

[2] Michael C. Stern, Alfonso Ibarreta, Timothy J. Myers. *Assessment and Mitigation of Combustible Dust Hazards in the Plastics Industry.* 30th International Conference of the Polymer Processing Society (PPS-30), Cleveland, OH, June 8-12, (2014)

[3] Center for Chemical Process Safety (CCPS), "Guidelines for Combustible Dust Hazard Analysis." 1st Edition, Wiley & Sons, Inc., Hoboken, NJ (2017),

[4] Center for Chemical Process Safety (CCPS), "Guidelines for Hazard Evaluation Procedures", 3rd Edition, Wiley & Sons, Inc., Hoboken, NJ (2008),

[5] Code of Federal Regulations – Occupational Safety and Health Administration (OSHA): 29 CFR 1910.119 *Process safety management of highly hazardous chemicals.*

[6] Timothy Myers, Alfonso Ibarreta, Kevin Marr. "Prescriptive versus Performance-Based Mitigation of Combustible Dust Hazards" Proceedings of the American Institute of Chemical Engineers 2014 Spring Meeting, 10th Global Congress on Process Safety New Orleans, LA, March 30 – April 2 (2014).

SPRING21
+17TH GCPS
A Joint AIChE and CCPS Meeting

# HIRA in Mining Companies Is Necessary and a Big Challenge

**Presenter/Principal Author**
**Américo Diniz Carvalho Neto**
**RSE CONSULTING**
**Brazil**
**americodcn@rsem.com.br**

**Other Author(s)**
**Afonso Costa**
**RSE CONSULTING**
**Brazil**
**Afonso.costa@rsem.com.br**

Prepared for Presentation at
American Institute of Chemical Engineers
2021 Spring Meeting and 17th Global Congress on Process Safety
Virtual
April 18 - 22, 2021

# HIRA in Mining Companies Is Necessary and a Big Challenge

**Presenter/Principal Author**
**Américo Diniz Carvalho Neto**
**RSE CONSULTING**
**Brazil**
**americodcn@rsem.com.br**

**Other Author(s)**
**Afonso Costa**
**RSE CONSULTING**
**Brazil**
**Afonso.costa@rsem.com.br**

**Keywords:** HIRA Project,

## Abstract

The implementation of a risk-based process safety management system is a turning point for any company that intends to have a better understanding of the risks associated with their process. However, full implementation of all elements takes time, and in the meantime, catastrophic scenarios should be managed accordingly, without wasting time.

To fulfill both criteria, having long and short run implementation strategies, HIRA (Hazard Identification and Risk Analysis) process was developed for mining companies worldwide. Based on the Critical Control Management (CCM), HIRA process is focused on “quick wins” of process safety, management gaps that must be eliminated shortly, to keep high consequences scenarios from happening. Those scenarios are known as MUE (Material Unwanted Event).

HIRA for mining companies is based on the methodology developed by ICMM (International Council of Mining and Metals), which is comprised of 9 steps, 6 of them are only to identify scenarios and their causes, consequences and mainly the critical controls associated. The methodology intends not only to identify critical controls, but also to check their status (implemented or not implemented), identify items that can make them more reliable and to assign responsibility for the management of the control, throughout its lifetime, to guarantee they will perform their function when demanded.

To accomplish such tasks, HIRA process widely uses Bow Tie Analysis, a very intuitive and visual risk management tool, which allows even people who do not have a lot of experience in the subject to easily understand the graphic representation of the MUE, such as cause, consequence, top event and mainly the controls and their status. Bow tie sessions are followed by field verification of each critical control identified, to check coherence between records (data), integrity and reliability, of the controls.

As a result, both existing and possible controls (known as Future State) are identified, according to the MUE analyzed and, comparing information from field verification, a qualitative risk assessment is carried out for three different categories – Inherent, residual and forecast risk.

By the end of the HIRA process, many types of controls are identified, such as procedures, safety valves, inspection plans, guard rails, alarms, interlocks…and so on. They now will be managed differently, as they are classified as critical controls – items that are crucial do prevent the MUE form happening or mitigate its consequence.

# 1 Main section 1 heading (Introduction)

Many companies have improved their performance on safety-related issues using a wide range of rules, procedures, and standards; however, Material Unwanted Events, such as fatal accidents, rare catastrophic events and circumstances of over-exposure that are harmful to health, are still subject to occur. For this reason, the ICMM (International Council for Mining and Metals) developed an approach called Critical Controls Management (CCM), aiming at the continuous improvement of management over rare events, but potentially catastrophic, emphasizing critical controls. The methodology to improve the management control of these types of events, called Material Unwanted Events (MUEs) follow below:

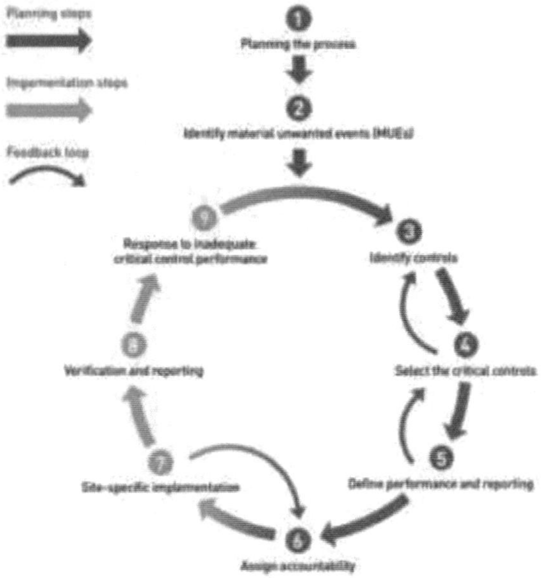


**ICMM Critical Controls Management Process (CCM)**

CCM's approach is based on some STEPS as follow:

**Step 1:** Develop a project plan that contemplates the organizational context, objectives, team responsibility and the individuals and business areas affected in the implementation.

**Key actions for STEP 1**

• Develop a plan that describes the scope of the project.

This includes:
– organizational context
– project objectives

– responsibilities
– business sections involved.

• Develop methods to:

– identify potential hazards and unwanted events
– assess risk
– review MUEs
– select critical controls
– assess objectives and performance of critical controls
– investigate critical control underperformance
– measure impact of the project
– identify ownership and accountability.

**Step 2:** Identify the MUEs to be managed and summarize the main information for each MUE. In order to identify, verify if they present material risk, evaluate the opportunities to eliminate it by improving the designs and describing the MUE completely (hazard, materialization mechanism and possible consequences).

**Key actions for STEP 2**

• Understand major hazards and identify potential MUEs.
• Apply selection criteria to MUEs with a focus on the consequences.
• Identify design opportunities to address the hazard, reducing the potential consequences and eliminating the MUE from the CCM process.
• Describe the identified MUE, including the relevant hazard, mechanism of release and nature of the consequences.

**Step 3:** Identify the controls, developing Bow Ties diagrams and evaluating them.

**Key actions:**

• Identify the controls.
• Prepare a bowtie diagram.
• Assess the adequacy of the bowtie and the controls.

**Step 4:** Select the critical controls for each identified MUE, summarizing its main information.

**Key actions For STEP 4**

• When identifying critical controls, apply the critical control definition and guidance in this section.
• Consider the performance requirements of the potential critical controls and how they could be verified.

• The final set of critical controls for an MUE should represent the critical few that, when managed using CCM, can effectively manage the MUE risk.

**Step 5:** Define the objectives and Critical Controls performance requirements, how they will be checked and the activities that will be affected by them.

**Key actions for STEP 5**

• Define objectives and performance requirements for each critical control.
• Identify current activities that affect the critical control's performance.
• Describe activities to verify performance and reporting requirements.
• Identify what would trigger immediate action to stop or change the operation and/or impose the performance of the critical control.

**Step 6:** Develop a list of those responsible for the MUE, for the Critical Control and for its verification, including the preparation of the final information report.

**Key actions for STEP 6**

• Assign owners for MUEs, critical controls and verification activities.
• Describe reporting plan for the health of critical controls.
• Assign owner for review of reports.

**Step 7:** Define a verification plan and reporting plan for each MUE, implementing it according to the specific needs.

**Key actions for STEP 7**

• Critical control information must be specific to a site or asset.
• Adjust the critical control definition, performance information and verification requirements as necessary to suit the local context.
• Site-specific planning for implementation may involve an iterative process.
• Site-specific planning should include establishing a foundation for CCM that includes leadership, communication and appropriate development of knowledge and understanding related to the critical controls.

**Step 8:** Implement verification activities and report the process, defining the status of each critical control.

**Key actions for STEP 8**

• Undertake verification activities for critical controls as described in MUE/critical control verification and reporting plan (developed in Step 5).
• Report a summary of verification activity results to the critical control owner.

• Report critical control verification status to the MUE owner.
• Reports should highlight priority information succinctly using traffic light system.
• Action initiated if critical control performance drops below the defined triggers (established in Step 5).

**Step 9:** Performance investigation if critical controls are performing below expectations.

**Key actions for STEP 9**

• Take action when critical control performance is inadequate (below the defined trigger threshold).
• Investigate the causes of unacceptable critical control performance.
• Information and data from the investigation should be used to continuously improve the CCM.

The companies risk map, in most cases, does not fully reflect the main risks to which the operation and the external population are exposed. This is due to the lack of a unified methodology to identify, quantify and measure risks. In view of this reality, one company can initiate the HIRA Project (Hazard Identification and Risk Assessment), to conduct a quick technical scan of risks in its operations to identify relevant and unmitigated situations.
HIRA was designed to quickly identify and verify the main risks of the company units operations. This is a global project to be implemented and will be at the center of the operational risk agenda in the coming months. The HIRA Project will focus on the very critical risks of severity in the company operations and its subsidiaries and its results, will be reviewed frequently by the Executive Committee for Operational Risks, the Executive Board and the Board of Directors. His results will guide the main actions to reduce the organization's risk profile.

## GOALS

The current study aims to identify and analyze operational safety risks with a very high-risk magnitude or very critical severity of a company site. In addition, the identification, definition of performance criteria and the establishment of criteria to guarantee the performance of associated critical controls are part of the scope of this work. From this, it will be possible to help the company to produce a complete and accurate inventory of the relevant unwanted events of the organization (MUEs) and, later, based on the identified MUEs, identify critical controls, associated performance requirements and means of verification.

For this purpose, this study will include:

1. A consolidated list of operational safety risks assessed with a very high risk magnitude or very critical severity (MUEs) for the unit;

2. Bow Ties based on the identified MUEs and a corresponding list of critical controls identified for each (following the suggested ICMM decision tree and based on the group discussion);

3. One-Pager for each critical risk control.

4. List of action plans to be developed based on the prioritization of risks.

5. Updated MUEs risk matrix of the unit.

## METHODOLOGY

The Bow Tie Analysis (BTA) can be used to identify and assess hazards with the potential to cause major accidents involving company site. This study can be conducted through presential or remote meetings through the TEAMS videoconferencing platform, by multidisciplinary teams composed of corporate company members and company site members and can be supported by external consultants.
Initially, happens a Workshop session, jointly between the consultants of RSE Consulting and company employees present at the meeting. The site MUEs were defined taking into account that the chosen MUEs would be among the most at risk in the facilities analyzed. In these sessions, MUEs were selected and evaluated from a previous list of more than 170 different MUEs, from different units of the company, including units in different countries which the HIRA project will be developed. Other new scenarios were also discussed and included as discussed with the local technical team. After determining the likely MUEs applicable to the site, these were then categorized as follow:

• Category 1: Scenarios in which, all the information presented during the Workshop, was sufficient for validation as a MUE. These are considered priority scenarios;

• Category 2: Scenarios in which, despite the information presented during the Workshop, these were not sufficient to validate the MUE in remote mode, thus requiring field visits and more information for validation. Also, in this category are included scenarios that can be considered as developments of scenarios classified as Category 1;

• Category 3: Scenarios in which, despite the information presented during the Workshop, these were not enough for validation as a scenario that meets the HIRA criteria, to be considered MUE. An on-site visit would be necessary to complement information and analysis. Thus, the team present at the MUE Workshop session with different specialists concluded in a list some (%)

MUEs classified as Category 1, that is, priority to the units of the complex. These meetings can be remotely, and the list was
subsequently validated by company leaders. It should be noted that the identification of MUEs needs to consider the historical events (Accidents and Near Miss) of the site, company and sector, in addition to predictable events, given the operations and activities at individual locations. Therefore, the identification of MUEs needs to include personnel with adequate experience and a review of relevant data.
Then, meetings were held remotely to analyze the MUE and generate project deliverables for the Complex. The analyzes were composed by members of the company corporate members, site regional members and external RSE consultants, where the consequences, threats and the respective existing controls for each MUE of the site were identified. During these meetings Bow Ties from other company units were analyzed with similar processes as a reference. Details about the Bow Tie technique are presented of this paper.

To identify the critical controls and controls of the MUEs, the proposal decision tree was followed:

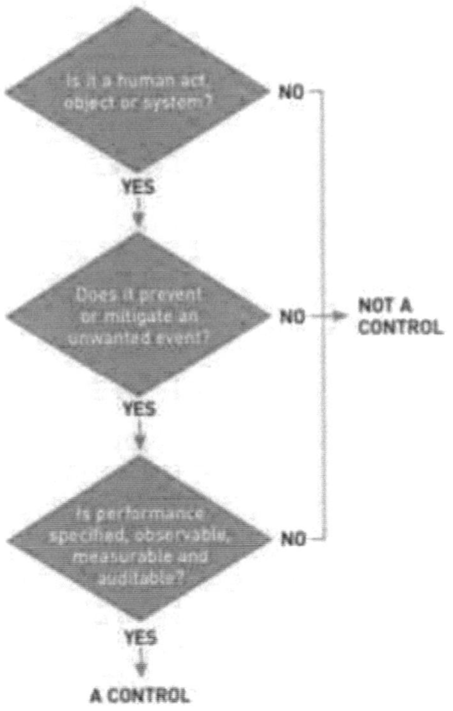


**Parameters for defining a control**

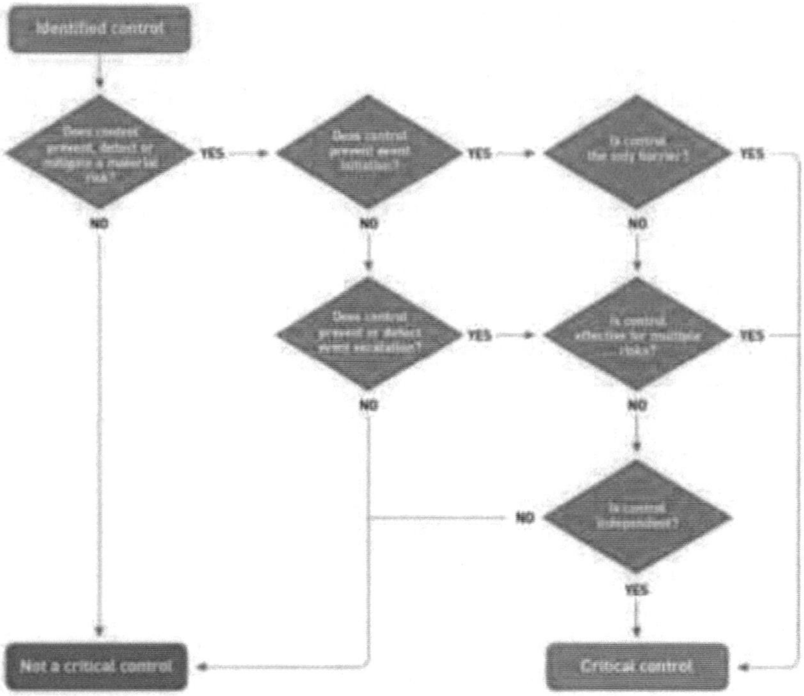


**Parameters for defining a Critical Control**

Once the critical controls were identified, a "One Pager" was generated for each one, that is, a quick briefing of just one page for each critical control (except "Future State" critical controls) containing the following information:

- Objectives of related critical control
- Performance requirements to meet objectives
- Activities that can enable or strengthen control
- Activities that can be verified to attest to the performance of a critical control
- Performance for critical control
- Trigger to stop the unit, revision or investigate, depending on the performance of the control
- Critical Controls attributes

The current levels of effectiveness (status) of performance of these critical controls were classified initially without field verification, based on the opinion of company specialists present at the meetings, to generate previous actions for the company, according to the methodology established for carrying out remote HIRA, during the COVID-19 pandemic. However, these classifications must be rectified / ratified in another phase of HIRA, when the field visit is done.
Remote risk classification meetings were also held, categorized as Inherent, Residual and Forecast, classifying them as follows:

- Inherent risk: premise = the existing critical controls to prevent each cause or mitigate each consequence are null (there are no controls);

• Residual Risk: premise = the existing critical controls to prevent each cause or mitigate each consequence are fully implemented or not implemented, depending on the classification of the status of each critical control determined through remote meetings by the Vale specialists team

• Forecast Risk: premise = critical preventive controls, mitigating controls, future controls (Future State) and all actions and recommendations captured during the meetings are fully implemented and functional, that is, they are compl ete and reliable.

It is important to note that risk assessments were carried out, remotely, without the field verification by the HIRA teams to validate, based on the One Pagers, the critical controls. Therefore, critical controls were evaluated in future scheduling, considered the next phase to be carried out at HIRA on site.

For the risk analysis of each MUE, the risk acceptability criteria of the company must be adopted. Below follows a brief explanation of Bowtie methodology selected in this Project.

**BOW TIE TECHNIQUE**

Bow Tie is a qualitative risk assessment method used to assess how high-risk scenarios develop. The essence of Bow Tie consists of plausible risk scenarios surrounding a certain hazard, and ways in which the organization prevents these scenarios from happening. The method takes its name from the shape of the diagram you create, which looks like a male bow tie.

**The Bow Tie methodology has several objectives:**

• Provide a framework for systematically analyzing a hazard.

• Helps to decide whether the current level of control is sufficient (or, for those who are familiar with the concept, whether the risks are as low as reasonably practical or ALARP).

• Helps to identify where and how to invest in resources that could have the greatest impact.

• Increases communication and risk awareness.

**General step by step construction of the Bow Tie diagram:**

Hazards Identification The purpose of hazard identification is to find, recognize and describe the risks that can help or prevent an organization from achieving its goals. Hazards can be operations / activities (operating rotating machines, driving a car),

I. substances (chemicals, hot fluids, etc.) or situations (a suspended load) that we deal with in the normal processes of our business. As long as these hazards are under control, they will not cause harm, but introduce the potential for harm.

II. Definition of the Top Event / MUE When control over a hazard is lost, it is usually possible to identify the exact moment when a normal situation changes to an abnormal situation, that moment is called Top Event. The Top Event is not yet a catastrophe, but the company is now exposed to the potential for harm from hazard.

III. Identification of Consequences The consequences are classified as the possible negative results directly linked to the top event, which can affect people, the environment, the facilities and the company's reputation.

IV. Threats Identification Factors that can lead directly to the top event are called threats. These are capable of causing the Top Event independently and are not associated with barrier failures.

V. Identification of Preventive Controls Risk management is about controlling risks. This is done by implementing barriers to prevent certain events from forming. A barrier (also called control) can be any measure taken that acts against some unwanted force or intention, in order to maintain a desired state. Preventive barriers, identified on the left side of the Bow Tie, can prevent the top event from happening or can eliminate the threat.

VI. Identification of Recovery Controls Controls or recovery barriers identified on the right side of the Bow Tie can prevent the top event from resulting in unwanted consequences or mitigate the consequences.

VII. Define Escalation Factors Once the controls have been identified, it is recommended, according to the Bow Tie method, to identify specific conditions or actions that reduce the effectiveness of the control. These are called escalation factors.

VIII. Identification of Escalation Factors Barriers Once the escalation factors have been identified, the last step is to look at what barriers we have to manage these factors. These barriers have the same concept as all the barriers previously discussed, but, now they do not prevent / mitigate a top event or consequence, they will prevent a barrier from failing. The BowTieXP software, version 9.2, developed by CGE Risk, was used to construct the Bow Tie diagrams. BowTieXP is the most widely used risk assessment software based on the Bow Tie methodology that allows you to easily create risk assessment diagrams.

## ANALYSIS RESULTS

Table attached shows the Material Unwanted Events (MUEs) that were hipotetical identified and jointly defined by RSE Consulting and the company team. As defined by the company, for an event to be considered as MUE it must have the potential to cause one or more fatalities or the potential to cause a major environmental impact that goes beyond the limits of the company.

In addition to these available scenarios (Category 1), the MUEs Workshop, conducted with a multidisciplinary team, generated potential scenarios for MUEs whose information presented was not sufficient to validate them in the remote mode, requiring additional information through field visits for validation (Category 2 and 3).

In total, 221 existing controls were defined to prevent and / or mitigate MUEs and 138 of these were classified as critical controls for these scenarios. These numbers consider the proposed future controls and the repetition of the same control between the scenarios.

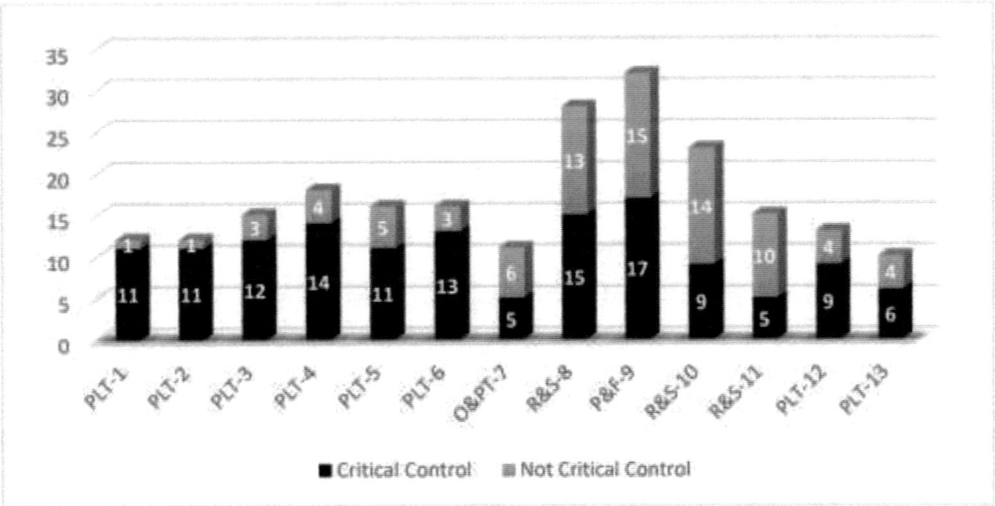


**Controls and Critical Controls**

According to the criteria adopted for the status of critical controls, the following were identified:

✓ 98 Critical controls classified as "Implemented"

✓ 09 Critical Controls classified as “Not Implemented”

✓ 31 Critical Controls classified as “Future State”

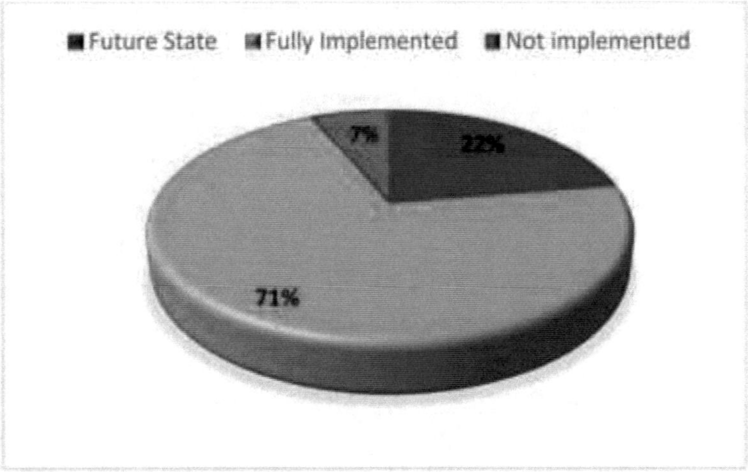


**Percentage distribution of the status of Critical Controls**

Regarding the type of critical controls, the following were identified:

✓ 104 Critical Controls classified as "Preventive";

✓ 34 Critical Controls classified as "Mitigatory".

This distribution demonstrates the greater effort of the professionals who participated in the assessments and identification of risks / controls in the Search for preventive solutions instead of mitigating unwanted scenarios, which satisfies the initial premise of HIRA.

Additionally, considering a larger sample of 19 sites, where 252 MUEs were analyzed, one can have a better understanding of the results.

• Critical controls - for the 252 MUEs evaluated in this sample, 4283 controls were identified, of which 2487 were considered critical controls. In this number are included the controls considered Future State, that is, the proposals identified by the group to reduce the risk of the assessed scenario. This means that 58% of the controls were considered crucial in preventing the occurrence or mitigating the consequences of the MUE, according to the evaluation team. This is always a point of attention when evaluating controls because it is not uncommon for members to tend to classify a large percentage of controls as critical. This is because companies prioritize resources to ensure the reliability of their critical controls, and the Operation team can use this criterion as a factor for prioritizing resources, going in the opposite direction to the methodology, which is a big mistake. It is up to the consultant to ensure that the flow of classification of critical controls is properly followed.

A rule used, which has served for the present project, is to consider as critical controls something between 45% to 55% of the controls. If the numbers indicate something greater than this, a ratification with the evaluation team is recommended.

For the same project, 1115 controls considered critical were classified as administrative controls, that is, they are basically controls that are highly dependent on human action to fulfill their required function. Examples of these controls are procedures and inspections. The latter is one of the main points of attention because it presents many opportunities for improvement as inspections are common not only in mining but in any company. However, many operational inspections carried out do not have the required reliability, that is, they are not able to identify deviations in integrity that can lead to the MUE.

These inspections are characterized by having a high frequency (sometimes daily) of execution and low level of details. On the other hand, there is the integrity inspection, which has a low frequency and a high level of details. The latter is usually performed by certified inspectors, has an associated procedure and inspection report issued at the end of each inspection. That is, depending on the type of inspection carried out, it is possible that this process is very effective in identifying structural damage or just a simple routine daily verification process, without the required level of detail.

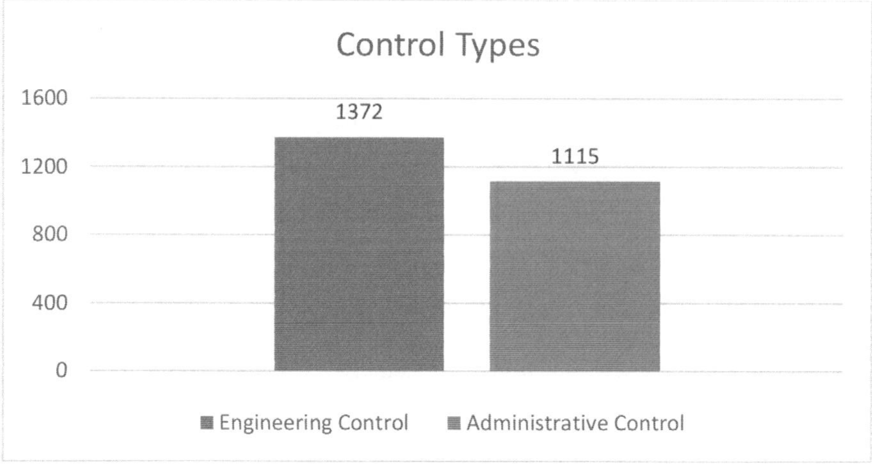


***Graph 2 – Types of Critical Controls***

It is highly recommended that this information about the items that can enable or strengthen critical control, in the case of operational inspection, are clearly described in the One Pager.

**Preventive or mitigating controls -** Considering the type of control as preventive or mitigating, again for the project considered, it was found that 1874 (75%) were preventive and 613 (25%) were mitigating. The mitigating controls are basically the emergency action plan and fire fighting system. Less frequently, dykes also appear as important controls for the MUE of loss of fuel containment.

It is important to say that, although they represent only 25% of the identified critical controls, mitigating controls are responsible for a higher percentage of recommendations. One of the most recurrent is the failure to consider the MUE in emergency pre-plans and drills.

**Conclusion:**

Throughout HIRA´s remote work of the company site, risk scenarios considered more critical were evaluated, with the potential for multiple fatalities or significant environmental impact or level 1 process safety event. Although the result of critical controls considered as implemented was in higher percentage, deviations were identified in the management of some pillars of Process Safety, such as Mechanical Integrity, and Risk Management. For a better understanding of the deviations listed, the list of actions is done.

A larger number of "automatic" or engineering critical controls type were identified in relation to those with human interface, such as operational procedures, inspections, maintenance plans, among others. For some of these controls, actions were proposed to ensure greater reliability on demand, activities such as systematic inspections, maintenance, calibration plan, looping control tests, fail-safe mode, etc.

About the critical controls identified with human actions, it is necessary the following requirements to the reduction of the failures: to reach deadlines of procedure's review, elaborate training plan/program with knowledge retention (evidence with records), recycling and periodic evaluation of the team performance, with actions for "under the established standard", as well as investigation records of near miss incidents.

About company site, it is important to highlight:

1. Treat the critical controls as critical items into the SAP or the specific management software.

2. Ensure proper inspection and maintenance for the automatic controls (safe interlock systems);

If the field verification of critical controls is performed in the company site made it possible to diagnose opportunities for improvement, strengthening and implementation of Process Safety Risk Management. Considering the concepts of hazard identification, risk analysis, change management, implementation of Process Safety Management System will be essential for the company site to manage its risks associated with the MUEs, to keep them at an acceptable level.

## 2 References:

[1] RSE Consulting HIRA methodology process, Americo Diniz Carvalho Neto, 2017.

[2] RSE Consulting Barriers Audit methodology, Americo Diniz Carvalho Neto, 2007.

[3] Health and safety critical control management Good practice guide, ICMM.

[4] CCPS. *Guidelines for Bowtie in Risk Management.* Center for Chemical Process Safety, American Institute of Chemical Engineers, New York, NY, 2018.

[5] CCPS. *Guidelines for Hazard Evaluation Procedures.* Center for Chemical Process Safety, American Institute of Chemical Engineers, New York, NY, 2008.

# 616590 Title: Silencing a Noisy Agitator Seal By Managing Stress

**Abstract:**

**Title:** Silencing a Noisy Agitator Seal by Managing Stress

**Abstract:** Did you know the sound that comes when you run your finger on the rim of a wine glass is referred to as "partial boundary layer lubrication"? Did you know this same phenomenon also occurs commonly on buffer and media tank agitators in bio-processing facilities? However, on agitator seals the sound can amplify to over 90 decibels, which exceeds OSHA's personal exposure limits.

Using an FEA model, John Crane found, that the noise from partial boundary layer lubrication is a result of vibration caused by the application of an even or consistent stress to the seal face, which is consistent with field trials since the seals rarely squealed at startup. Instead, the seals squeal after some run time when the right amount of stress is applied. This stress could come from the pressure of the vapor space in the tank or vessel, which also explains why the noise level and pitch can come and go depending on the pressure in the tank, agitator speed, as well as other scenarios.

This led to the discovery that the squealing can be reduced, or even eliminated, by managing the stress imposed on the agitator seal faces. Subsequently, John Crane developed a solution that results in an active seal face, meaning the degrees of flatness change with the stresses applied during operation. In other words, the rim of the wine glass changes as you run your finger around it preventing the noise condition from ever taking place.

**Take Home Message from Poster:** The noise from a media or buffer tank agitator seal can be reduced, making a quieter and safer workspace.

# 619172 Use of Data Analytics to Obtain Cost Reductions and Performance Improvements in a Regulatory Application


**Abstract:**

All operating nuclear power plants (NPPs) in the United States are required to monitor and assess the performance of plant structures, systems and components (SSCs) as specified in the Maintenance Rule (10CFR50.65). This rule requires that all licensees identify the occurrence of any SSC failure that results in a loss of safety function, often referred to as a Maintenance Rule Functional Failure (MRFF). All MRFFs are subsequently evaluated through the plant corrective action program (CAP) as required by 10CFR72.172. To ensure these requirements are met, all NPPs document events that occur which are evaluated to identify impact on the plant. A typical dual unit generating facility generates as many as 15,000 of these incident reports (IRs) per year which each must be evaluated by experienced plant staff. Of these IRs, only 0.1-0.2% typically are MRFFs. NPPs have found the monitoring of plant SSCs for compliance with the maintenance rule to be expensive and resource intensive.

An alternative approach which has the potential to reduce these costs is to automate the review of IRs through the use of data analytics and machine learning (DA/ML). Advances over the past several decades in computational capabilities and computer programming techniques / algorithms now permit application of DA/ML techniques to analyze the large historical data sets that contain information on the performance of SSCs over years of plant operation. Unfortunately, these techniques have not been widely adopted within the nuclear industry due to its highly regulated nature. A consequence of this structure is that any DA/ML models used in applications subject to regulatory oversight must provide a high degree of confidence in the predictions due to potential regulatory impacts associated with an incorrect determination.

The MRFF Analyzer is a DA/ML application which significantly improves the efficiency of the maintenance rule screening process. The model uses a combination of natural language processing (NLP), Bayesian statistics, and machine learning to identify the likelihood that the incident described by an IR represents a MRFF. The software was trained, tested, and validated on four years of historic incident reports (2015-2019) generated across Exelon Generation's fleet of operating NPPs. This data set consisted of approximately 650,000 IRs. Approximately 420,000 IRs from 2015 through the summer of 2018 were used to train the model, and 80,000 IRs from summer-winter 2018 were used to test the model. Model performance in the field was validated during 2019 with each IR generated by the sites analyzed by the software with review performed by on-site subject matter experts as specified in the plant program implementing procedures. Once the performance of the MRFF Analyzer software was verified as adequate to meet the requirements of the Maintenance Rule, the parallel use of review by human experts was discontinued.

This paper presents the development, training, and deployment of the MRFF Analyzer software across Exelon Generation's fleet of operating NPPs. The algorithms in use in the software are discussed in detail. The challenges dealing with highly skewed data are presented and the approaches used to address these challenges discussed. The results of a detailed sensitivity assessment are presented and used to estimate the expected statistical performance of the software. The overall performance of MRFF Analyzer since its deployment at Exelon Generation is discussed and estimates of the cost savings are provided.

# 617819 Search: A Symptom-Based Expert for Advanced Response to Chemical Hazards

**Abstract:**

Chemical exposure accidents pose a risk of enormous personal injury and property damage if initial response is not appropriate. In Korea, from 2014 to October 2019, 1,000 chemical accidents occurred, and 20 people died and 444 were injured. In addition, the number of chemical accidents shows a trend of increasing from 233 in 2014 to 277 in 2018. Existing chemical substance detection technology is focused on the development of response technology for leaks at known sites, and research on the detection at dynamically changing environment and knowledge service for hazardous chemicals considering symptoms is insufficient. Therefore, there is a need for research on a real-time hazardous chemical detection and diagnosis system and knowledge service that protects workers or exposed people from unexpected chemical accidents in the field, by detecting exposed chemicals quickly and helping to cope with dangerous situations early.

In this study, we develop an AI-based system for chemical identification from actively collected exposure symptoms, for real-time response to hazardous material incidents. In order to analyze the symptoms of chemical exposure accidents or contact with substances, information related to chemical substances (symptoms, toxicity, structural information, etc.) is collected and extracted from WISER, PubChem, MSDS, etc. to systemize the related information. The knowledge base of 1,000 core substances is built by designing ontology for symptom knowledge of accident-involving chemical substances. In case of WISER developed by NIH, chemical substances are estimated by simply matching symptom words manually inputted, but the proposed system actively collects the symptoms and predicts the candidate chemical substances (with probabilities) through the knowledge base inference considering the exposure situation. The knowledge base inference uses a knowledge graph as knowledge base that is easy to search, and uses a reasoning method such as the dynamic RDFS++ reasoner, which is easy to infer dynamic data provided by AllegroGraph, and OWL2RL materializer, which is easy to infer static data. In the version 1 of SEARCH, the dynamic RDFS++ reasoner is mainly used to proceed with inference.

Using Machine Learning (k-NN, Decision Tree, SVM, etc.) and Deep Learning, it also predicts toxicity from chemical substance information (structure, etc.) or symptoms from toxicity [1,2] for building expanded symptom KB. The accuracy of the chemical symptoms estimated from the structure and toxicity of the chemical substance is derived from the model with good performance among the previously used models, developing a system with an accuracy of 80% or higher that can estimate 43 symptoms for 1,000 chemicals. Through the combination of two network models that predict toxicity from the structure of chemical substances or predict symptoms from toxicity, it is possible to provide symptom knowledge to be used for symptom detection even for new chemical substances.

SEARCH provides a wide range of information on hazardous substances, including substance identification support, physicochemical characteristics, human health information, and containment and suppression advice. The prototype system is serviced with AI speakers as the main interface and demonstrated at the Gyeonggido Disaster and Safety Headquarters, and plans to combine the established knowledge base and chemical substance-symptom estimation model. The voice recognition interface using AI speakers enables effectively acquire information by saving time and effort rather than directly inputting and matching information manually.

# 617983 Application of Machine Learning Algorithm for the Analysis of OSHA Accident Database to Identify Trends in Performance Indicators of Process Safety Incidents

**Abstract:**

Many workplace injuries and fatalities occur in facilities dealing with hazardous chemicals due to unexpected leaks, exposure to dangerous process conditions and other loss of containment events. The U.S. Occupational Safety and Health Agency (OSHA)'s Process Safety Management (PSM) program has been in effect for almost three decades in the United States, requiring companies to establish comprehensive management programs to deal with such hazardous chemicals with the target to assure safety and health in the workplace. Analysis of the trend of process safety incidents in these industries can be a good performance indicator for the effectiveness of the PSM program.

For a long time, OSHA has been maintaining and publishing accident data collected from various industries since the 70's and currently has recorded information of more than 130,000 incidents involving fatalities and injuries arising from various causes. Analysis of such a large database to identify process safety incidents manually is extremely difficult and time consuming. Current study focuses on the development of a machine learning algorithm to screen the database and identify process safety incidents through use of the keywords and incident descriptions associated with each accident provided by OSHA.

In order to validate the machine learning process, the large OSHA database is filtered to analyze incidents from 1984 to current time and only chemical manufacturing and petroleum industries are considered through use of industry classification codes. More than 2,200 incidents are screened out in this manner and are analyzed to determine those that were process safety related incidents. The analysis was done both manually and through use of the developed machine learning algorithm to validate the capability of the latter. The keywords are classified into various categories to aid the algorithm in the learning process. This is the beginning step to the application of artificial intelligence to analyze big databases such as the OSHA database to determine common patterns behind injury and fatal incidents and provide industry with performance indicators that will provide an understanding of where the current laggings are and enable proactive measure to be taken to prevent future incidents.

# 617448 Game Based Learning – Making the Message Memorable

**Abstract:**

A perennial frustration for organizations working within high hazard industries is that despite going to great time, cost, and trouble to communicate important process safety messages across the entire workforce, incidents continue to happen. Making these messages "stick" is essential. This presentation will outline a novel approach to developing process safety knowledge and competence through Game Based Learning (GBL).

Based on research that suggests people remember 90% of what they 'do', GBL takes your message out of PowerPoint and broadcasts it interactively to the trainee. A GBL session typically lasts 30-60 minutes, is hands-on and, by using every-day analogies for technical subjects and grounding them in the real world, it is accessible to all personnel across an organization. Indeed, GBL for major accident awareness has been run for school children as part of Science, Technology, Engineering and Mathematics (STEM) teaching for years. GBL comes into its own in high hazard industries where, generally speaking, improvement is about small margins. This presentation will include some examples of games developed by Risktec, and cover a wide range of common process safety themes, to help prompt the audience's ability to replicate similar ideas in their own workplace.

SPRING21
+17TH GCPS
A Joint AIChE and CCPS Meeting

# Process Safety Cards – A Good Deal Safer

**David Hatch**
**Process Safety Integrity**
**Derby, UK**
**david.hatch@psintegrity.com**

Prepared for Presentation at
American Institute of Chemical Engineers
2021 Spring Meeting and 17th Global Congress on Process Safety
Virtual
April 18 - 22, 2021

# Process Safety Cards – A Good Deal Safer

**David Hatch**
**Process Safety Integrity**
**Derby, UK**
**david.hatch@psintegrity.com**

**Keywords:** Process Safety Cards, Bowties, Gamification, Microlearning, Workforce Engagement.

## Abstract

To improve the awareness of less technical personnel to potential Loss of Containment (LoC) scenarios and increase their vigilance to such events & their potential causes so they can be promptly addressed; memorable images are shown on familiar playing cards representing typical LoC events.

## 1 Introduction

Process Safety Cards have a simple goal – to improve the awareness of a broad range of personnel to potential major accident scenarios and increase their vigilance to such events & their potential causes so they can be promptly addressed. These memorable images represent typical LoC events to deliver knowledge literally into the hands of those who might contribute to and/or be affected by Process Safety Incidents.

With an ageing workforce and increased outsourcing of maintenance and operations activities, it is vital that time and knowledge poor stakeholders (who typically fall outside corporate process safety training strategies) are engaged and equipped to ensure they play their part in sustaining the integrity of assets.

## 2 Challenge

Process Safety Incidents, Major Accidents or Loss of Primary Containment Events continue to occur in the 21st Century despite better technology and bigger data. For example, on 28th January 2021, a liquid nitrogen release at the Foundation Food facility in Gainesville, US resulted in 6 fatalities and multiple injuries.

A review (by the author) of the 100+ CSB Investigations [1] indicates 200+ fatalities occurred of which approximately 40% were Contractors. This does **not** suggest or imply that the Contractors were at fault, however it does highlight that 3rd parties are often involved in major accidents.

Although Process Safety initiatives, typically referencing Process Safety Management frameworks including OSHA [2], CCPS [3] or the Energy Institute [4], are now commonplace, these may exclude external resources including outsourced Operations & Maintenance personnel who are therefore not as equipped as their internal counterparts to recognize and address potential Process Safety incidents.

We must also acknowledge that wisdom is being lost with the ageing workforce and that assets are getting older whilst those responsible for them are getting younger.

Are we possible focusing on the wrong people and/or protection measures?

How do we put knowledge into the right hands?

# 3 Workforce Engagement

In order to gain and sustain the attention and participation of the workforce, which may range from novices who need to be informed to veterans who need to be reminded, it is becoming more common to use Gamification methods. These employ game design elements e.g., point scoring, competition with others, rules of play to improve user engagement, learning and knowledge retention.

However, many games require considerable time to setup and play or often cannot be played in the workplace due to the cost and complexity of deploying electronic devices e.g., laptops or tablets to host the games safely (flammable atmosphere ignition potential).

Process Safety Cards offer an alternative approach which offers the benefits of gamification without the time and cost burdens and is based on Microlearning e.g., small learning units absorbed in small time units.

Players are encouraged and helped to learn and apply simple principles:

- REMEMBER Events,
- RECOGNISE Causes (Threats) of Events,
- RESPECT Controls (Barriers) which prevent or mitigate Events.
- REPORT Concerns about the presence or performance of the Controls/Barriers,

Thereafter there is an obligation on the duty-holder or owner/operator to:

- RESPOND to Feedback (concerns, questions, or suggestions).
- RECOVER Protection to its original state.

Using this simple & memorable 'R' model aims to embed these principles into the minds & memories of frontline personnel. It is critical that the RESPOND & RECOVER steps are performed otherwise there can be no faith that contributions are respected and that improvements are being made.

# 4 Deployment

Process Safety Cards represent 52 different LoC events that occur in the Process industries which are presented as familiar playing cards split into 4 separate suits.

- DEGRADATION Failure within design conditions or "under-strength"
- DISCHARGE Open route to atmosphere
- DAMAGE External impact
- DEVIATION Failure beyond design conditions or "over-stress"

These events are drawn from industry publications including the HSE [5] and CCPS [6] and whilst they are not claimed to be an exhaustive list, they can be attributed to familiar major accidents across a variety of industries.

Cards make it easier to communicate in all directions, top-bottom, side-to-side and inside-outside the organization by providing a common visual language. This gets and keeps the workforce onboard with the Process Safety or Asset Integrity mission.

Sample cards are shown in Figure 1 below:

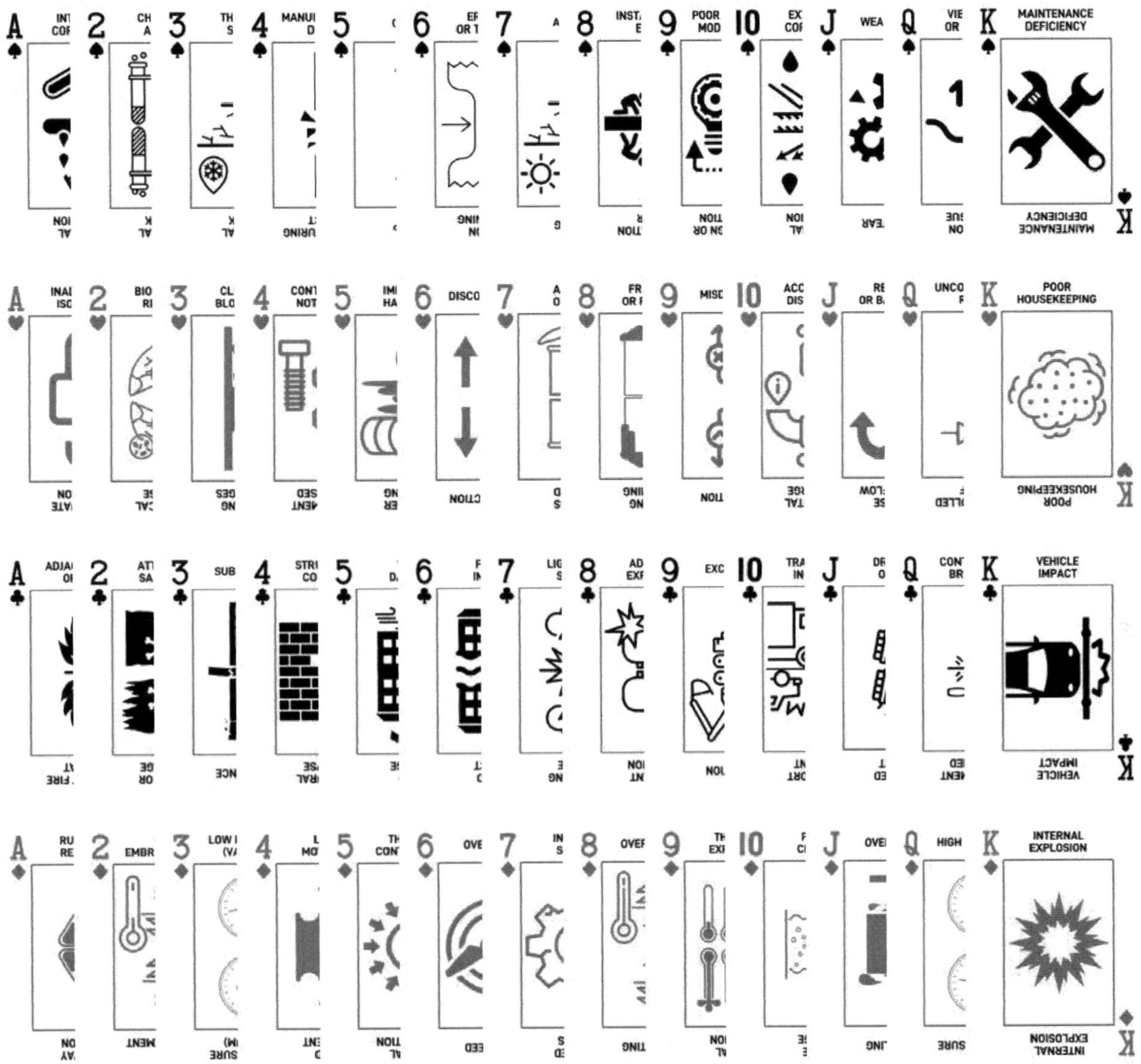


**Figure 1 Sample Playing Cards**

One of the core concepts of this approach is to "Know your Enemy" [7] where hazards are not the enemy (these are part of doing business) – it is Loss of Containment that is everyone's opponent.

There are several different opportunities to deploy the cards in operational arenas, because they are easy to carry & exchange:

- Simple passive learning (by osmosis) via familiar, informal card games.
- Toolbox talks, e.g., explain your experience of the card you've been dealt.
- 'Pop' or quick quiz e.g., name the event on the card (without title)
- Hazard spotting or surveys e.g., punch hole in cards and tag/tie them onto the plant at vulnerable locations

- Hazard reminder e.g., attach relevant cards to Permit to Work.
- Inductions e.g., visitors, suppliers or short-term contractors during outage, turnaround etc.

The cards are not limited to frontline personnel as they can also be used to engage engineering or management personnel.

- Safety moments e.g., discuss your role related to the card you've been dealt.
- PHA/HAZOP validation e.g., ensure all deviations (cards) have been discussed.
- Performance dashboards e.g., incidents categorized according to LoC type (or card)
- Hazard potential e.g., sum (add) the values of each card i.e., Ace = 11, King/Queen/Jack = 10 & other cards numbered
- Campaigns e.g., each week showcase specific events or cards.
- Relate cards to client or industry incidents e.g., CSB investigations [1].

Just as a pack of cards can create a variety of different games e.g., poker, blackjack, bridge etc. Process Safety Cards utilize the same playing cards elements (or 'pips') which offers an organic approach to gamification that can grow with peer or community feedback i.e., facilities, organizations or industries can develop their own games based on their technology and culture.

## 5 Scenario Summaries

Playing cards increase awareness of the potential LoC events, however in order to address these, an appropriate level of insight is necessary. Visualizing the LoC scenarios in bowtie format provides a clear, concise summary which is more easily understood and retained than textual or numeric data.

A one-page representation of each type of LoC event i.e., playing card, provides typical:

- Cause (Threat)
- Controls (Barrier)
- Effectiveness or Vulnerabilities (by barrier type)
- Degradation factors (by barrier type)

These are not intended to replace existing analyses and should not be considered sufficient to properly monitor & manage barriers, however they do provide memorable context with a particular focus on the Degradation Factors which create or enlarge the holes in the Swiss Cheese model and are controlled via asset and/or safety management systems e.g., competence, information, maintenance change etc.

A sample is provided in Figure 2 below:

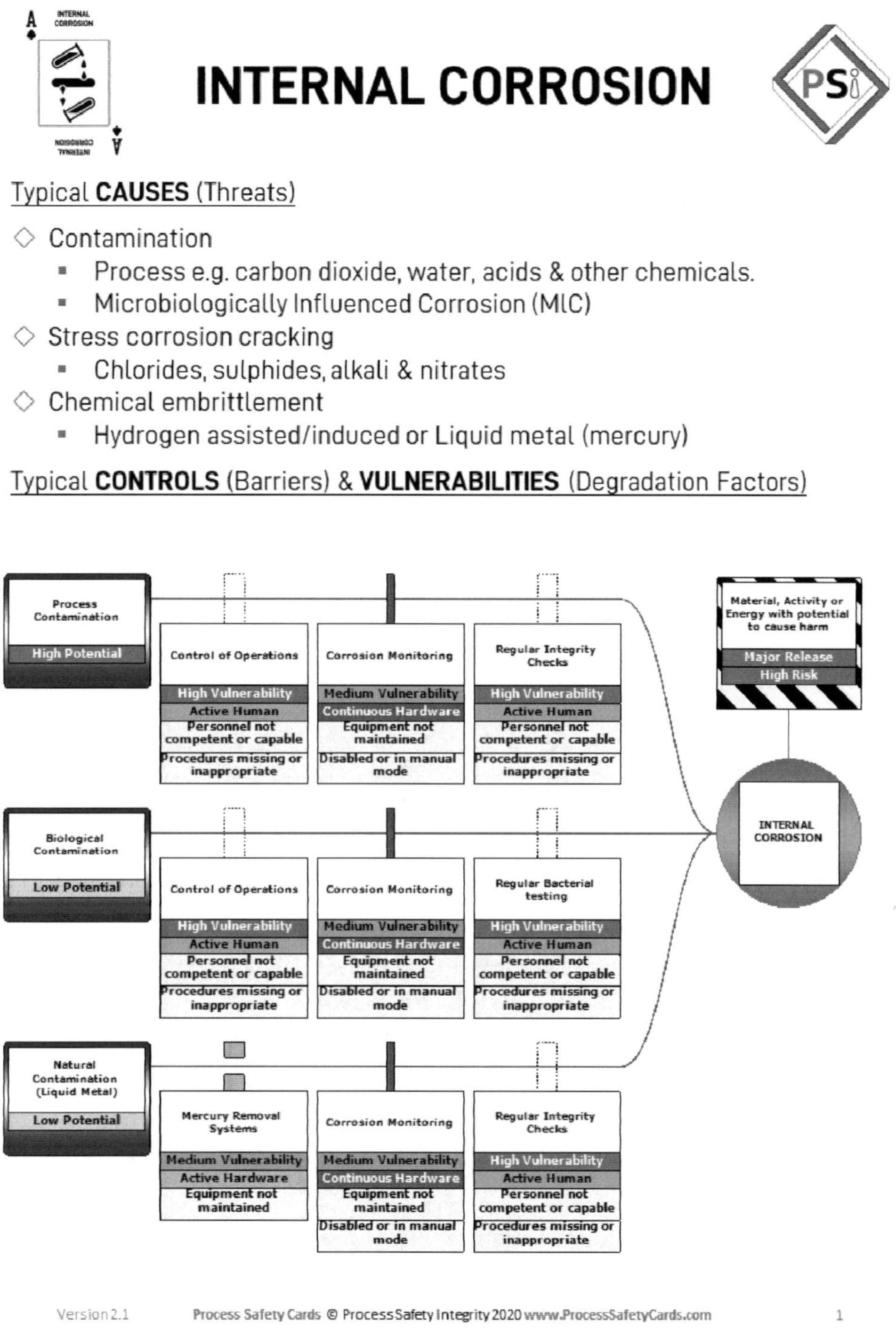


**Figure 2 Sample Scenario Summary**

Scenario summaries can be posted inside or outside populated or high traffic areas such as control rooms, canteens, workshops, or meeting rooms and rotated on a campaign basis e.g., a different LoC per week. They can also be published online e.g., intranet, SharePoint, or newsletter.

## 6 Online Collaboration

Although Process Safety Cards are designed to be physically handled, the current COVID-19 pandemic prevents or discourages such personal interaction, therefore digital solutions can be used to use or play the cards via collaborative platforms e.g., virtual whiteboards to connect remote or isolated team members.

## 7 Conclusions

Although based on gamification principles, the Process Safety Cards are not a game in themselves i.e., there are no rigid rules to follow. They simply provide a knowledge exchange toolkit to facilitate communications (in both directions) between management and frontline personnel to share learning using universal images and local languages.

These are not a new strategy (although they are a novel idea), they simply help deliver or reinforce established Process Safety (Asset Integrity) initiatives which are designed to 'Protect the Protection' i.e., look after the systems (hardware & humans) that look after you.

The cards increase awareness, the bowties enhance understanding, knowing your enemy improves vigilance which drives protection and asset integrity.

## 8 References

[1] U.S. Chemical Safety Board (CSB) Investigations (https://www.csb.gov/investigations/).

[2] U.S. Department of Labor, Occupational Safety & Health Administration (OSHA) Process Safety Management of Highly Hazardous Chemicals (29 CFR 1910.119) (https://www.osha.gov/process-safety-management).

[3] American Institute of Chemical Engineers (AIChE), Center for Chemical Process Safety (CCPS) Risk Based Process Safety (RBPS) Framework (https://www.aiche.org/ccps/resources/publications/summaries/summary-guidelines-risk-based-process-safety).

[4] Energy Institute High Level Framework for Process Safety Management (https://publishing.energyinst.org/topics/process-safety/leadership/high-level-framework-for-process-safety-management).

[5] UK Health & Safety Laboratory "Loss of Containment Incident Analysis" HSL/2003/07.

[6] American Institute of Chemical Engineers (AIChE), Center for Chemical Process Safety (CCPS) "Guidelines for Chemical Process Quantitative Risk Analysis".

[7] Sun Tze "The Art of War".

# 618386 Are Your Safety System Barriers Still Work for You?

**Abstract:**

Process Hazard Analysis is the tool to assess the systems in terms of the process safety scenarios and to determine the ALARP level of the risks that may arise. All the barriers in the Process Hazard Analysis is assumed to be working in the normal operations. However, how you manage to sustain the ALARP level? Do you know any deficiencies and by-pass during the operation? How do they effect the ALARP level? Are you still in the safe side?

All the barriers like Emergency Shut Down Systems, Pressure Safety Valves, Fire and Gas Detection Systems work for your system and you need to follow up in terms of any deficiencies or by-passes made on these. This process starts with reliability of these systems and continues to follow up during the operation. Emergency Safety Systems by-pass are undesired activities but during the operation there would be a need to do this. Here the important thing is how you manage the system with these by-pass and how you follow up these by-pass procedures. Industry and automation systems are improving and there are lots of ways to simultaneously cover all these changes and do the risk analysis to operate the systems safely during these by-pass times. So many companies would be using different ways of controls, in our refinery we have improved a system to follow up all the emergency safety systems with a hierarchy of control and approval automatically including the “Plan B” risk assessments for safe operation of this time periods.The important point here is the type of by-pass (conditional or non-conditional), duration of by-pass (approval hierarchy), what else do you have to operate safely, what other parameters should be followed up by the operator during the by-pass and does he know it?

These automated systems should be covered and controlled during the manual operation or by-pass time intervals in order to eliminate the human errors and not to have any kind of process related incidents. We do need to remember important Process Safety Incidents related with the by-pass of emergency shut down systems like Chernobyl. This solution and automated system would be a barrier not to have such incidents.

SPRING21 +17TH GCPS
A Joint AIChE and CCPS Meeting

# Stress–Complexity–Communication on Oil Platform & Mental Map Gaps, situations of Routine and Emergencies of JBH, the operator

**Salvador Avila Filho**
**UFBA**
**avilasalva@gmail.com**

**Luiz Fernando Lopes de Souza**
**COPENOR**
**lflsouza@globo.com**

**Lucas Menezes Pereira**
**UFBA**
**lucasmpereira13@hotmail.com**

**Rita of Cassia Spínola Avila**
**UFBA**
**ritaavila2@gmail.com**



Prepared for Presentation at
American Institute of Chemical Engineers
2021 Spring Meeting and 17th Global Congress on Process Safety
Virtual
April 18 - 22, 2021

# Stress–Complexity–Communication on Oil Platform & Mental Map Gaps, situations of Routine and Emergencies of JBH, the operator

**Salvador Avila**
**Ufba**
**avilasalva@gmail.com**

**Luiz Fernando Lopes de Souza**
**COPENOR**
**lflsouza@globo.com**

**Lucas Menezes Pereira**
**Ufba**
**lucasmpereira13@hotmail.com**

**Rita of Cassia Spínola Avila**
**Ufba**
**ritaavila2@gmail.com**

**Keywords:** Stress, Communication, Hazard Energy, Human Factors, Accident

## Abstract

Stress, complexity, and communication are human capillary elements that allow a passage of the flow of energy from diverse human factors, in all phases of the factor network, in all dimensions. These characteristics influence the appearance of technical, managerial, and human error, including the appearance of archetypes and possible habits. The phases of factors that affect human performance (PSF) start in Social Culture, indicating trends in omission and communication difficulties due to different linguistic aspects, passing through possible current social factors, such as the generation and gender conflict in addition to the effects of the multicultural aspects in the work environment. These social phenomena can be global or local in nature and can affect managers and workers differently. The first possible barriers to be structured are in the organizational and safety culture in addition to the prioritization between safety and production within this industry or risky activity. When approaching individual and group aspects, communication will be cited in practically all phases, as well as stress, but it is only from failure and lack of control that stress, as a state of work, can generate high consequences, even anticipating the accident. The communication is cited insistently in accidents involving oil platforms.

In this work we will study scenarios of accidents that have already occurred and probable causes in the elements that should block the hazard energy from the design of the oil installation. Complexity and communication will participate, or failures, in the construction of the accident. At some point, we will focus on the effect of behavioral gaps without dividing factor dimensions into layers. Considering the systemic failure format and its connectivity, we will try to identify, citing Johnny Big Head (JBH) in constructed events, how the danger energy flow occurs to the point of causing the accident or disaster as occurred in Pipe Alpha. In the first, it caused many human losses and the second caused extremely high environmental impacts, both considered disasters. We will rescue the definitions of concepts already worked on, the impact of the stressful environment on the operational decision and on the body, the quality of communication through the organizational pyramid and the complexity of the task that can be calculated to define the resultant of sociotechnical reliability. We will then rebuild the oil accidents already known in addition to the local accident. Boarding for platforms requires neutralization of the field of relations between operators and managers, and between contractor and contractors. The relationship of empathy will be discussed as a positive factor in the establishment of barriers, preventing the three capillary elements from developing their energy with a chain reaction effect. Also in terms of technological redundancy are the good Human-Machine Interface (HMI) design for emergencies and high-level communications, as well as physical visualization of primary elements for the reclassification of the event between operational lack of control, incident, accident, and disaster. The Methodology, therefore, involves scenarios built with phases of the accident, considering the events of the various human-organizational factors in the network. This causal diagram event will show how stress, complexity and communication can have a potentiating effect in the accident. We will try to investigate if the barriers installed are adequate in intensity, redundancy, and place. We find the places of hazard energy concentration and points of failure that these elements have capillarity. These scenarios are explained using the caricature of the JBH operator and the signaling of cognitive and social functions in addition to the characteristics of the process and the task trying to indicate the influence on communication interface factors with stress and the difficulties of address unknown dangers and cultural biases. This discussion can be quantified through the hazard energy units released using diagrams and equations to assist in the quantitative analysis of the impacts of these three capillary elements, stress, complexity, and communication.

# 1 Introduction

Stress, complexity, and communication (ESCOCOM) are human capillary elements that can allow the passage of the flow of energy originated and enhanced by human factors, in all phases and dimensions of the network indicated in Figure 1. ESCOCOM gaps influence in the appearance of failures and human errors in execution and in the decision. The emergence of archetypes and possible bad habits caused by cultural biases and affect the informal rules of the operator JBH in its workplace indicated in Figure 3.

The phases in the network of factors that affect human performance (PSF) start from the Social Culture Analysis, indicating trends for omission and communication difficulties due to linguistic aspects, going through possible current social factors, such as generational conflict and gender conflict, in addition to the effects of multicultural aspects in the work environment. These social phenomena can be global or local and can affect managers and workers in different ways.

The human elements indicated as barriers to the reduction-contention of hazard energy in Figure 1, are worked during the design phase and reviewed to maintain safety and organizational standards with the objective of maintaining cognitive functions and interfaces in stable operation, bringing effectiveness to the company's results.

The barriers structured in the project to meet the routine and emergency in risk activities (Figure 1) depend on the differentiated organizational moments (Figure 2) that activate the respective cognitive functions satisfactorily or with gaps (Figure 4). These organizational moments and the relationships between human factors (Figure 1) are influenced by organizational and security culture and by the respective decision procedures involving the conflict between safety and production. When addressing individual and group aspects, communication and stress are cited in all phases, from routine to emergency, in the risk industries and mainly in oil exploration and production. The level of stress causes high consequences from the moment it influences the decisions in the phase where the operation is uncontrolled, the mode of equipment failure, and the relationship with the accident (Figure 1). The breaking of the communication link is an insistent element as a cause of accidents involving the chemical process industries and activities on oil Rigs.

# 2 Literature Review

### *2.1. Swiss Cheese model, from qualitative to quantitative*

For [1], multidisciplinary is required to discuss the accident. According to [2], a set of barriers is necessary to prevent risks from causing harm to patients. Each barrier has holes (just like in a Swiss cheese). When, by chance, all the holes are aligned, the danger materialize in the accident format.

### *2.2. Human Performance Factors in The Operation and Human Elements in the Project*

Barriers to avoid failure should be based primarily on human factors rather than engineering aspects only. According to [3], these layers of Human Factors can be: Cultural (Regional and Organizational), Social Phenomena, Company, Management, People and Group, Operational Control, Failure Control, Accident Control and Disaster Control.
Safety barriers should be considered from the design of oil exploration units to avoid cognitive pitfalls. If social relations in the workplace are not considered at any time, there will be consequences that are likely to be identified in the event [4]. Barriers are interventions in the processes along the chain of actions for production, seeking new conditions to interrupt the flow of energy from danger. At each layer of human factors, the energy of danger is accumulated or drained defining the possibility of deviations, failures and accidents in the workplaces. The environment directly influences worker behavior. Since the boarding to oil platform, for example, the operator's memory may be confused or dispersed making it difficult to carry out work activities. According to Ávila in [4], it is necessary to humanize the platform areas that are critical spaces of isolation, favor communication and social contact of the worker.
Instructions, measurements, automation level, operations control, and remote conferences indicate that oil production tasks are critical operation. This operation is performed or controlled by few people who must have appropriate characteristics for decision: clear mental map, optimal level communication and clear alternatives with established risk dominance level. Risk perception varies depending on the cultural scheme in which, JBH is involved. The petroleum exploration group can generate human errors due to communication or language problems [4].
According to [5], are examples of accidents on oil rigs: fall and disruption of equipment materials, leaks, fires, and electric shock. Social coexistence within platforms and regions of production must be based on an explicit sense of justice and common sense in routine activities for the development of egalitarian leaders.
In activities on the platform or in the production field, the physical exertion can be high due to storming and cognitive effort is typically low leading to good routine control. When this routine is well set on arrival, the comfort is greater. If the manager has a good sense of justice in the team management, will occur harmonies bringing back the strays, to the next task. The tasks are performed more naturally and without haste "exaggerated".

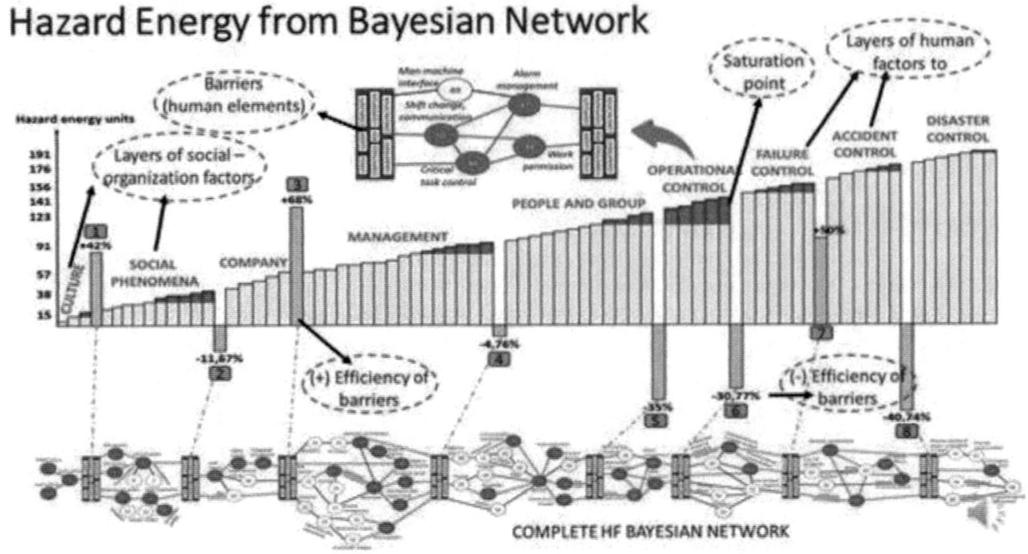


**Figure 1. Quantification of the Network of Human Factors and Elements. AHFE 2020**

### *2.3. The JBH operator in the Accident caused by human factors and failure of the barriers*

The movements of Operator JBH (Johnny Big Head) in the Organization influence and are influenced by his cognitive, intuitive, and affective functions and can cause failures of human capillary elements such as complexity, communication, and stress control. The experience of this operator is not being questioned but discussed how he is moving in the moments experienced in the workforce: leisure (free thinking), physical and organizational survival, application of concepts and skills, the moment of decision, almost impulse, impulse in the actions, competence formation, organizational integration, and social inclusion in work activities.
The functions performed by JBH for life and for work are integrated into these organizational movements and must follow specific functions that vary according to the technology design, organizational and security culture, regional aspects, leadership, and the external environment to their activity.
The key point here is to discuss JBH activated psychological-psychic-motor functions to meet the respective cognitive-behavioral gaps that can occur at each organizational moment. On the other hand, there is the analysis of JBH, in each position in the "spring" model, that discuss organizational experience, and maturity.
The study of the birth of failure and maintenance of correct work in the workplace begins by understanding which cultural biases may be modifying or guide the behavior of JBH. Complexity cannot be forgotten during risk analysis of activities without which, technologies do not marry mind maps. JBH's activities in oil production are related to different tasks dependent on maintenance responses that is usually outsourced, communicating in a 14-day shift in platform.
Knowing the risks and complexities, JBH has efficiency in its function and in its workplace that depend on specific parts of cognition and the organizational environment. In this case, memory, quality of information, correct skills in demand, and adequate perception, regardless of stress level, are characteristics that deserve tests to adjust this mental map.
At JBH operations, the behavior gaps and cognitive actions promote situations that cause human error may occur. We know that the level of stress in the operation of the oil industry may be related

to overwork or cognitive effort, so part of the indicators that indicate a tendency to human error can be measured by indirect factors such as the percentage of shift double work.
This discussion of JBH in the workplace is equivalent to the region of deviations where visibility is exceptionally low. From that moment on, with the appearance of human error, visibility increases with failures, incidents, and accidents on the oil rig.
The key point here is the functioning of JBH in the working system and the birth of human error following the script: cultural biases; risk and complexity; efficiency in the workplace during the performance of the task; measurements of cognitive functions; analysis of human performance factors; failure analysis of safeguards or human elements; classification of derived human error; occurrence of technical failure after human error; identification of process losses; accidents - direct cost and negative image; and the possibility of a disaster occurring.

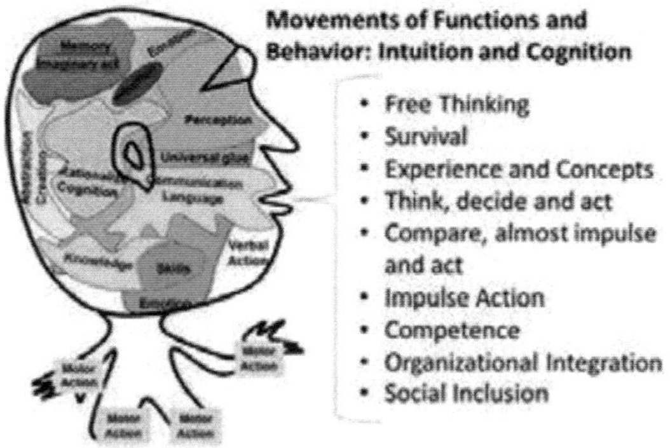


**Figure 2. Qualitative Assessment Functions and Organizational moments of JBH in Oil Rig**

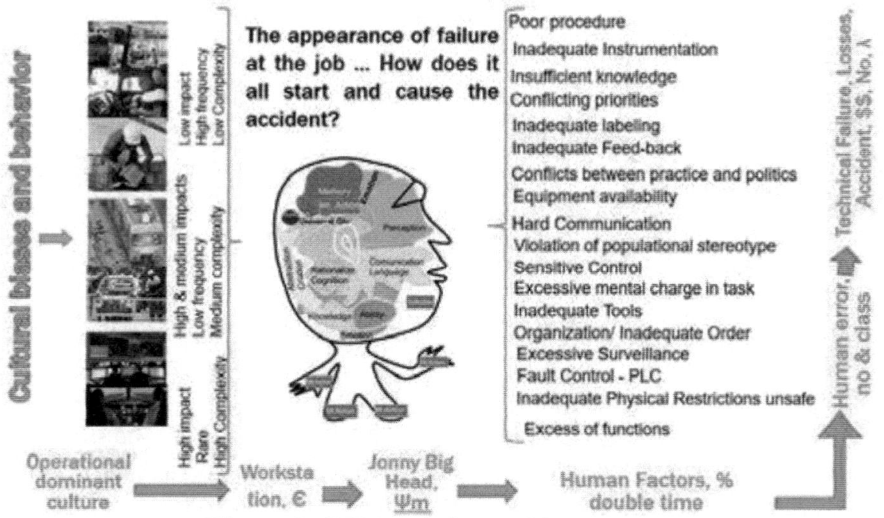


**Figure 3. JBH and the accident process**

### *2.4. Cognitive Quality of JBH*

In the book "Human Factor and Reliability Analysis to Prevent Losses in Industrial Processes: An Operational Culture Perspective" [6], Ávila suggests a product named as Cognitive Quality Analysis in two stages: in the first stage statically, the primary functions of learning and decision

are related to organizational, technological, and external biases that follow expected characteristics as indicated in Figure 4. In the second stage where social and individual emerging events are analyzed that alter the type of human error and priorities in the decision, is the case of cognitive gaps, without apparent explanation.
The JBH Routine is investigated for 2 to 4 months in a period of high process losses through bad habits, human errors and behavior deviations that cause the loss of process and hence it is suggested what is the causal link of these social-human and organizational events.
The key point is the cognitive gap caused by social, human, and organizational causes causing human error and loss. These gaps in cognitive function must be investigated, to strengthen with psychological safeguards, the quality of the task performed.

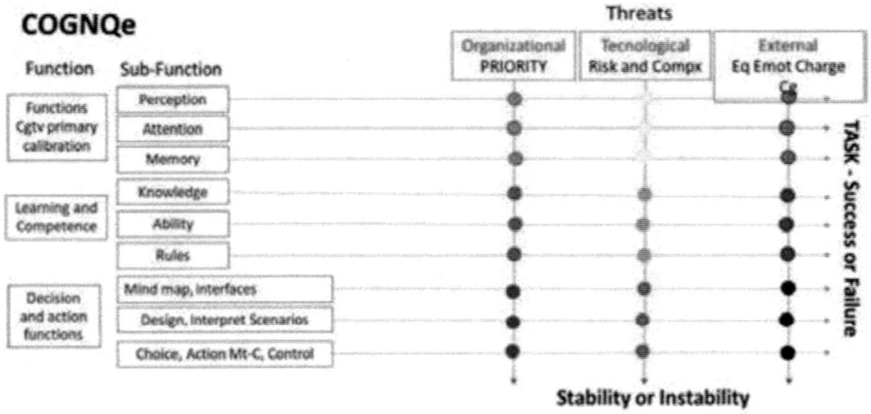


**Figure 4. Cognitive Quality and Primary Functions. Book Human Factor Elsevier [13].**

### *2.5. In ship collision*

Rasmussen, in [1], presents the analysis of shipwreck accident in ship using technical and human factors. Later, Ávila, based on the ship collision article [7], elaborates a scenario and includes human and technical factors in the model of networks and flow of danger energy and presents in the AHFE in 2020 [3]. This theme was the reason for critical analysis of simulation promoted by IBAMA (Brazilian Institute of the Environment and Renewable Natural Resources) discussing ship-to-pier shock in Brazil, State of Bahia, with subsequent oil spill.
In this simulated work Ávila presents the following deficiencies: (a) did not discuss the impact of the leaked product and possible events of fire and explosion and damage; (b) did not discuss weather events and difficulties in containing leaks; (c) the emergency leader had no attitude or knowledge to size and organize necessary external resources, such as isolation buoys; (d) the damaged physical part of the ship was not part of the discussion hindering the actions to stop the leak; (e) the emergency team, contingency organization and local operations did not use redundant oral communication; (f) there were errors in the identification of context, construction of scenario and communication (time, scenario, size of danger and suggestion of actions); (g) there was delay in communication making it difficult to anticipate actions and vision for sequencing the actions; and finally, (h) failures in the support functions for the emergency command.

### *2.6. In Social Relations on Platform*

Ávila, in [4] and [8], discusses social relations in oil rig from the discourse of maintenance teams during preparation course for shipment. The observations indicate that offshore work can has a higher level of stress and can cause the loss of emotional balance affecting routine work and also contingency actions in case of emergencies at sea, caused by environmental events, logistics accidents or accidents inherent in this activity. We draw attention to the following aspects discussed: (1) Platform manager's sense of justice to avoid isolation of groups separated by region of origin or function; and (2) Before returning from the platform, after 13 days shipped, or as soon as arriving on the platform, in the first day, care must be taken to keep the cognitive machine prepared for the activity, with adequate perception and level of attention, so as not to get lost on the horizon, or to know how to gather documents to understand the scenario.

# 3 Hypotheses

The Chemical Industry has spent a lot of time understanding and correcting human errors, but cannot access information about the events and environments that cause cognitive gaps. The questions to be asked are: How do you build cognitive failure including cultural aspects, primary functions of cognition? What are the factors that impact on emotional balance?

In this work, complexity is considered a human element or barrier to the technological and organizational project and depends on the systems of process control and security, the requirements and steps for the task and the social relations established in the organization. This complexity may require specific skills for the analysis and writing of robust communication procedures and tools so as not to run the risk that messages or language symbols will not be perceived.

The state of stress installed in the routine of the work can be influenced by characteristics of the culture (global and regional) and the management style in the organization. On the other hand, a greater complexity in the process or task, should be treated by managers and engineers, to avoid raising the level of basic stress in the work environment. Stress control and analysis of the environment and stressor vectors is a strategic subject for management and supervision. In the organization one can think of anti-stress rituals in the work routine, but, is this possible in any organization or technology? Metallurgy for example works with high temperatures and uses the color of the product for decisions, is it possible to improve thermal stress to avoid metal losses in the slag?

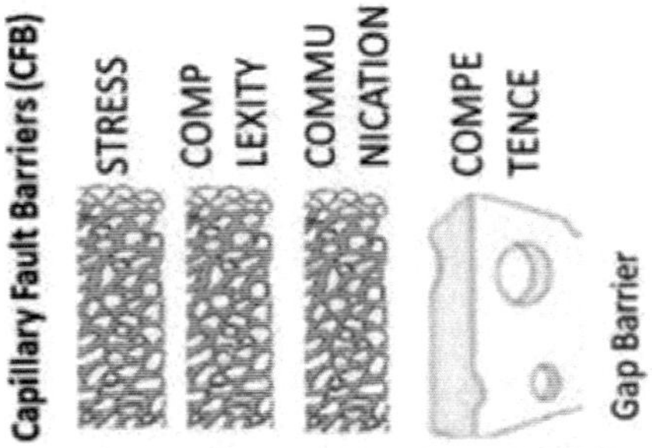


**Barrier Failure Detection: Visible GAP vs Invisivible CFB**

**Figure 5. Barrier Failure Detection**

The failures in human design barriers or elements can be easily detectable when incorrect delegation of operators who do not have sufficient knowledge or ability to perform activities causes human error. Thus, the gap of this barrier lies in the incorrect delegation, or in the lack of training, or even in the lack of skill. This gap can be resolved more quickly.
Communication tools are rooted in technology, task performing, and management. These tools drive decisions and is a dominant characteristic in determining the resulting behavior of the worker. Therefore, it is important to know how to measure C4t (complexity, cooperation, commitment, and communication) [9] in the routine of the operation and for emergency. The capillarity of the stress, communication and complexity indicate the difficulty of achieving the non-recurrence of failures in the timeline due to the diffusion potential of deviations resulting from the low quality of these human elements or project safeguards. This happens mainly on oil rigs, when managers eventually do not consider good social coexistence as a priority goal.

# 4 Methodology

The Methodology discusses the scenarios of the accident of the oil industry with the phases of the accident, considering the events of the various human-organizational factors in the network. This causal diagram event will show how stress, complexity, and communication can have a potentiating effect of the accident. We will try to investigate whether the installed barriers are adequate in intensity, redundancy, and location. It finds the locations of concentration of danger energy and points of failure that these elements have capillarity. These scenarios are explained from the JBH role and the signaling of cognitive and social functions, in addition to the characteristics of the process and the task trying to indicate the influence on communication interface factors with stress and difficulties in addressing unknown hazards and cultural prejudices.
The Methodology (Figure 6) is developed in the following steps: (1) Present the Scenarios where a given task is taking place, exploring a new well, or producing oil on a platform, with consecutive events in each situation; (2) Discuss the human elements of project as stress control, communication and complexity in each event that occurred in the accident; (3) Identify organizational skills and possible failure in ESCOCOM barriers; (4) To prepare a survey for validation of the ESCOCOM method and analysis; (5) Prepare matrix related type of event, cognitive function, moment of organization, failures and possible actions.
The time is truly short, and we will fulfill the preparation of the poll and design the analysis matrix, but we will not be able to validate the matrix with the application of this questionnaire to the personnel of maintenance and operation of oil rig.

| | | | | |
|---|---|---|---|---|
| 5.1.1 Task Scenarios, Accident Factors Events: Piper Alpha | 5.1.2. Flag and Discuss ESCOCOM as failures in barriers. | 5.1.3. Discuss Organizational Moment and flaws in Cognitive Functions vs ESCOCOM. | 5.3 Elaborate Survey for Validation of the ESCOCOM Method and Analysis. | 6.1. Prepare Matrix for ESCOCOM Analysis. |

**Figure 6. Methodology for ESCOCOM Analysis Matrix**

# 5 Cases Study

In this work, we will study scenario of accident that have already occurred and probable causes in the human elements that would block the hazard energy from the design of the oil installation. Complexity and communication will participate in the failures, in the construction of the accident.

At some point, we will focus on the effect of behavioral gaps and cognitive, without dividing factor dimensions into layers. Considering the systemic failure format and its connectivity, we will try to identify, citing the worker JBH in constructed events, how the hazard energy flows to the point of causing the accident or disaster as occurred in the Piper Alpha Rig. This accident caused many human losses, considered disaster, and where consecutive human errors occurred [4].
Ávila asked in [4]: why did so many mistakes occur in this accident? Some elements were suggested as possible causes and that deserved attention and investigation such as: omission, inadequate management, failures in decision-making under stress, lack of knowledge of procedures and planning of the task including human and social factors, confused mental map and absence of a united team.
The oil rig is a community, like a city, with formal and informal rules. The platform manager is the mayor of the city that, if adopting biases and preferences before the inhabitants, will bring to your industry-city, situations of credit or discredit that led to resilience in face of the crisis or the crisis without resilience. It is important to analyze physical and psychological stressors that are inherent to the regional environment and culture, in addition to the environment and organizational culture. Stressors are intensified when considering the physical environment and the risks involved with oil exploration in the pre-salt layer, for example. This situation brings the possibility of incidents and accidents because of human errors in performing the task, which may be intrinsic errors of man at work, errors in group action, errors in the planning of the task, or even errors in its execution [4]. The tasks depend heavily on the environmental climate and the organizational climate and can become complex when the search for power is strong between groups and the effects of isolation are not perceived by the leaders.
The written communication considered cultural differences, with different language codes. Reliable oral communication depends on the person's confidence in the spoken word, this for the effective realization of the practice. It is important to avoid the vices of digital communication as copy paste together with the tendency to omission in operation. Management communication should be clear and detailed or fast and imperative, it depends on the moments. Communication with contractors includes feedback on perception and listening and work requirements. Visual communication facilitating quick understanding of the situation. The training media seeks the harmony of the team for good coexistence in the work environment. Leaders should be structured with knowledge and skills in human and operational reliability including aspects of social relationship and multiculturalism in the task. Instructions, measurements, automation level, operations control, and remote conferences indicate that this is a critical operation. This operation is performed or controlled by few people who must have appropriate characteristics for decision: clear mental map, communication at optimal level, clear alternatives with established level of dominance. Risk perception varies according to the cultural scheme in which the operator is involved [4].
Competent management and clear and unified communication are important factors for the proper functioning of a platform. The company is expected to provide emergency equipment, and maintain a team prepared with competence and agility for evacuation in case of accidents [10]. On the other hand, it is expected that this team absorbs the exercises and has skill in the use of equipment, in stress control, thus avoiding unsafe behavior at the time of decision.

### *5.1. Piper Alpha Accident*

5.1.1. Scenario with Events.

The accident in July 1988 at the Piper Alpha offshore oil rig [11], located in Scotland in the North Sea, caused the deaths of 167 workers, as well as material losses. This accident was caused by accumulation of errors and failures in management that were impregnated in the organization, structure, procedures, and culture. Inspection and maintenance failures were also identified because they adopted procedures that were inadequate to the activities performed on the platform. In 1998, [12] described the technical and operational failures that triggered the accident on the Piper Alpha platform and the lessons learned after the disaster that led the offshore industry to invest in design and operation-related safety. With the implementation of these preventive measures indicated a 90% reduction in risk in disaster involving this activity but did not include in this percentage human or communication factors that are relevant to avoid the accident. Each of these basic events was influenced by a series of decisions and actions. Some of these decisions were incorrect, others came from incomplete analyses, acceptable at the time they were made, but catastrophic in conjunction with other events. At least some of these conjunctions could have been anticipated, decisions and actions, are labeled according to the stage in which they occurred: design, construction, operation and much more specifically, maintenance. The events occurred:

*E1 - Disturbance of the process; E2 - Failure of both condensate injection pumps; E3 - Failure of the blind flange assembly at the PSV; E4 - Undetected release of condensed vapors; E5 - First ignition; E6,7- Failure of gas detectors, fire protection (deluge) and emergency shutdown systems; E8 - Failure of the fire walls; E10, 11 - Tube breakage in module B and large oil leakage; EI2 - Fireball was spread; E14 - The fire spread to the storage of fuel; E15 - Failure of diesel power fire pumps; E16 - Rupture of the risers, first of the Tartan, then of MCP-01 and Claymore; E17 - Jet fire under Piper Alpha; E18, 19 - Immediate loss of electricity; failure of emergency light; E20 - Loss of the control room; E21 - Public address system failure; E23 - Decision to hire and promote the individual; E24, E25 - spread of fire and smoke across the platform; E26 (E34, E36, E44) - Victims on board; escape and rescue survivors; E32 - Ineffectiveness of the Tharos in fighting fire.*

5.1.2. Discuss Failures and the ESCOCOM Barriers

**Table 1. ESCOCOM Matrix**

<table>
<tr><th>Events</th><th>Complexity</th><th>Stress Control</th><th>Communication</th><th>Competence</th></tr>
<tr><td>E1 System overload and operator confusion</td><td colspan="2">Complexity. Physical interdependence and network management of the platform; signal lost; lack redundancy.</td><td colspan="2">Competence. Promotion of personnel to temporary critical position.</td></tr>
<tr><td>E2 Failure of both pumps</td><td colspan="4">Competence. Improper maintenance of pumps A and B; Maintenance fails to report night shift about pump A, it was out and PSV.</td></tr>
<tr><td>E3 - Blind flange failure</td><td colspan="4">Competence. Error in the fitting of the blind flange; without mounting inspection.</td></tr>
<tr><td>E4 – Condensate stems</td><td colspan="4">Complexity. Alert system, poor design of monitoring panels; Failure to read and interpret control signals.</td></tr>
<tr><td>E5 - First ignition</td><td colspan="4">Competence. Potential ignition source detection error.</td></tr>
<tr><td>E6, E7 - Gas &security detector</td><td colspan="2">Complexity. Operator failure to verify gas source</td><td colspan="2">Competence. Operator failure to verify gas source</td></tr>
<tr><td>E15 - Fire pump failures</td><td colspan="2">Complexity. Decision to shut down the automatic system to protect divers</td><td colspan="2">Stress Control. Turn off auto system to protect divers</td></tr>
<tr><td>E17 - Jet fire under Piper</td><td colspan="2">Control Stress. Underestimation of fire severity and positive event.</td><td>Communication - deficiency platform & ship</td><td>Competence. Decision to continue production</td></tr>
<tr><td>E22 – Radio telecom</td><td colspan="4">Communication. Lack of redundancies in communication system</td></tr>
<tr><td>E23 - Hiring of staff</td><td colspan="2">Control Stress. No organizational redundancy chain of command</td><td colspan="2">Competence. Insufficient training for this type of emergency</td></tr>
</table>

| | | | | |
|---|---|---|---|---|
| **E26** - Victims on board | Complexity. Nonalternative official authority when IOM is disabled | Stress Control. Escape emergency, counterinformation about survivability | Communication. IOM failure to give evacuation orders | Competence. Insufficient or inadequate evacuation training |
| **E32** - Ineffectiveness of the "Tharos" | Control Stress. Firefighting failure in "Tharos" equipment. | | Competence. Delay in master Tharos' decision to take command of oscillation in time. | |

5.1.3. Cognitive Functions and the Organizational Moments and Consequent Events

In the project of the activity, it is necessary to hire technologies and people who can maintain production in the desired capacity and quality. This means having capable process, controlled production cost and team available to perform adequate tasks on equipment and products.
The primary functions of cognition (perception, attention, and memory) are the basis for cognitive processing and action. These functions must be absorbed organizationally to meet specific steps such as learning and analyzing the scenarios for the decision.
The moments in the organization prepare and adapt the worker for the activity and the accomplishment of the tasks within the business objectives and under the influence of social culture. Thus, the lack of leisure; physical and organizational survival as biases for decision; inadequate learning for the application of concepts and skills in performing the task; the decision that suffers biases and preferences; the actions in urgency and emergency situation bringing automatism and trust in the simulated; also, organizational integration and social inclusion to ensure efficiency in work activities.
The elaboration of requirements for the hiring and development of operators operating on oil rigs is related to the static part of cognitive quality. The elaboration of emergency situations or with emerging social and technological hazards unknown becomes a dynamic situation.
The quality of primary functions, decision and learning ensures success in organizational moments that require competence, commitment, communication, emotional balance, team involvement, cooperation with attitude, for this the JBH operator must believe in work, leadership and the sense of justice installed.
The indicators present in the workplace are a reflection of cognitive quality and organizational moment. Thus, to manage human factors we need to measure and make adjustments in the efficiency of the workstation, from the quality of cognitive functions, the negative factors for human performance with the respective measurement, the failure of safeguards or human elements, the classification of the derived human error and the knowledge of the mode of failure that causes the losses in the process.

### *5.2. Poll for validation*

The objective of the survey, with simple questions, applied to a group of workers in the areas of maintenance and operation in oil rig, is to validate the statements described in the course of the text about the importance of communication, complexity and stress, which are probable causes of accidents. The ideal operator represented by JBH with its cognitive and social functions, experiences and perceptions has characteristics that will help you identify the concentration sites of danger energy and points of failure in the workplace to indicate possible interventions. The questions listed below begin in a direct way, and end by addressing issues that are part of the social context on the platform.

1. Do you believe that human and organizational factors should be considered in the procedures for successful evacuation on an oil rig?
2. Major accidents have occurred on oil rigs, such as Piper Alpha. In the reports it was considered that the layout of the plant was an obstacle to evacuation and activation of alarms, do you agree that the design of the plant is important?
3. Do you agree that human factors should be considered from planning and building platforms to avoid future accidents or incidents?
4. It is common knowledge that financial issues often speed up procedures and decision-making in the operation. Is the actual reason for the failure in the procedure is recorded in the shift book?
5. Natural events (storms, presence of rays) are out of man's control, but it is a real possibility. Is the stress caused by the unknown discussed in the preparation of platform operation teams? External or internal events.
6. Clear communication activates the cognitive functions of the operator. Is the cultural question (as regionalisms) an obstacle to be considered to homogenize routine procedures?
7. Social rituals begin even before the professional reaches the platform. They are rituals of embarkation and disembarkation, procedures of group routines (social), activities of different complexities (professional) that affect the performance of operator. Do you agree?
8. Do you think the word "humanize" platform can lessens the responsibility involve teams? It is an isolated place of work in conjunction with social activities, do you agree that it is necessary to humanize the platform? what do you think about necessity of awareness of the danger that this work environment poses?
9. Does the communication redundancy for safety procedures help keep the worker's cognitive alert?
10. In your opinion, what is the importance of Organizational Insertion to prevent failure of the oil rig? In your opinion, what is the importance of Social Inclusion in the group's activity?
11. Which of the following human elements related to platform design are more difficult to control to avoid failure or accident. Mark from the smallest to the largest and why. Reduce the complexity, control the stress, acquire, and apply the competence, communicate at the right time the appropriate information to avoid the accident.
12. In your opinion how to measure the stress level of an activity? How to stay in control? How to combat stress caused by an emergency and how to avoid chronic stress?
13. How to measure the platform manager's sense of justice? What impact on the team if there is no sense of justice?
14. Do stories of occurrences in the organization help to form beliefs? positive or negative? Or both?
15. Which of the following factors cause the greatest cognitive flaws or gaps: lack of organizational commitment, external environment, social emerging, technological risk, or complexity?
16. What is the influence of emotional balance on stress? What is the influence of the mind map on the decision? How to measure the quality of this information flow and the result of the task?
17. Why do human error and accident continue to occur despite current oil production teams being aware of the dangers involved and the procedures on the oil production task agenda?
18. In your opinion, do the problems of complexity in the task-process-social relationship, the difficulties of stress control and the challenge of communication incorporated into a fair culture affect the possibility of an accident? Does the probability of human error increase a lot when the factors are present in the routine of the operation? What about the Emergency to avoid the fire, for example?
19. Do you believe that the accident is caused by multiple dimensions? For example, Social, Human, Technological, Organizational? Or is it more correct to say that the accident is only technological (false instruments)?
20. The difficulty of avoiding failures in safeguarding communication, stress and complexity is due to what cause? If there is good competence in the emergency team, can we avoid incidents on the oil platform?

### *5.3. Discussion*

These questions do not cover all the problems in the Oil Platform as we presented considering the piper alpha accident, the discussion about cognitive functions and respective failures, the different movements in the organization, and the social relations in this important activity. We understand the necessity to develop the cognitive process to achieve better results in the task performance.
We understand the importance to analyze this multidimensional problem to guarantee that we will not repeat accidents as Piper Alpha and Horizon. In this way we began to discuss psychological, behavioral, and organizational challenges in Risk Situations. We need to confirm the opinion of specialists in practice and concepts in Oil production to prioritize the principal emergent latent failure including social aspects.
Even without validating with an opinion of the practitioners and specialists in the organizational and cognitive ways to avoid the accident in the critical task, we propose a matrix to guide future investigations to increase the cognitive quality, in organizational and technological environment.

## 6 Conclusion

The flow of Danger Energy originated and enhanced by human factors can have its passage allowed by the elements Stress, Complexity and Communication. Thus, the accident needs to be contained through the proper sequential positioning of a set of multidisciplinary barriers. The barriers suggested by the Network of Factors and Human Elements model are: Cultural (Regional and Organizational), Social Phenomena, Company, Management, People and Group, Operational Control, Failure Control, Accident Control and Disaster Control.
The chemical industry has not been able to seek information about the events and environments that cause cognitive gaps and that also affect emotional balance, although much time has invested to understand and correct human errors.
To validate the concept of systemic failure and the JBH character, it was explained how the danger energy flow occurs to the point of causing the disaster as occurred in Piper Alpha, in which there were consecutive errors for failures in management, inspection, maintenance, structure, procedures and culture.

### *6.1. ESCOCOM Matrix*

Here we present the ESCOCOM Matrix as a guideline for the elaboration of interventions in the workplace on the oil platform.

**Table 2. ESCOCOM Matrix**

| | Stress Control | Compe-tence | Communi-cation | Comple-xity |
|---|---|---|---|---|
| Organizational Issues, External Environ | | | | |
| Efficiency at the Workstation | | | | |
| Moment Organizational JBH learning, integration & inclusion | | | | |
| Cognitive Functions: Attention, perception, and memory | | | | |
| Learning Functions | | | | |
| Decision and Action Functions | | | | |
| Risk & Complexity in Oil Platform Logistic and social rituals | | | | |
| Human Errors, Failures, Incidents and Accidents | | | | |

## References

[1] J. Rasmussen. "Risk Management in a Dynamic Society: A Modelling Problem". Safety Science, vol. 27, n. 2/3, pp. 183-213, 1997, Elsevier.

[2] J.T. Reason. "Human error". Cambridge University Press, 1990.

[3] S. F. Ávila, J. S. Ávila, L. M. Pereira and E. R. F. Lima. "Dynamic View of Human Elements Design and Projection of Human Factors in Critical Task Operation Using Bayesian Networks". 2020 International Conference on Applied Human Factors and Ergonomics. July 16-20th, 2020.

[4] S. F. Ávila. "Risks on the platform: Socio-human guidelines prevent accidents and alter standards in oil production." Revista Proteção, 260th ed.

[5] S. F. Ávila. "Risk Management and Prevention Program to Avoid Loss of Process (LP3) Based on Human Reliability, Practical Application and Measurement". In: 2011 AIChE Spring National Meeting, 2011.

[6] S. F. Ávila, I. C. S. Cerqueira and C. N. Santino. "Human Factor and Reliability Analysis to Prevent Losses in Industrial Processes: An Operational Culture Perspective". Elsevier, 2021.

[7] S. F. Ávila, L. M. Pereira, R. Ávila, C. P. Pena, P. M. F. M. Arezes, E. R. F. Lima. "Reviewing Tools to Prevent Accidents by Investigation of Human Factor Dynamic Networks". 2020 International Conference on Applied Human Factors and Ergonomics. July 16-20th, 2020.

[8] S. A. Filho. "Human errors assessment, social environment at oil platforms". 4th Latin America Conference in Safety Process, 2012, Rio de Janeiro.

[9] I. C. S. Cerqueira, E.Drigo, S.A.Filho and M.Gagliano. GCPS "C4t: Safe Behavior Performance Tool", 2019.

[10] N. Norazahar, et al. "Human and organizational factors assessment of the evacuation operation of BP Deepwater Horizon accident". Safety science, v. 70, p. 41-49, 2014.

[11] E. Paté-Cornell. "Learning from the Piper Alpha Accident: A Postmortem Analysis of Technical and Organizational Factors". Risk Analysis, vol. 13, n. 2, 1993.

[12] D. D. Drysdale, R. Sylvester-Evans. "The explosion and fire on the Piper Alpha platform, 6 July 1988. A case study". Philosophical Transactions of the Royal Society of London. Series A: Mathematical, Physical and Engineering Sciences, vol. 356, n. 1748, pp. 2929-2951, 1998.

[13] S.F. Avila, I. S. Cerqueira, C. Santino. Human Factor and Reliability Analysis to prevent losses in industrial processes: An Operational Culture Perspective! Elsevier. Been launched in May, 2021.

Effective response to a power loss event is critical to the safety and mechanical integrity of a manufacturing plant. Power failures can lead to accidents, injuries, loss of primary containment (LOPC), government recordable events, mechanical failure and loss of production. Due to the unpredictability of a power event, there is no single solution or response plan that can mitigate all of the potential issues that a power event creates.

A power event can cause alarm showers/floods which prevent the operator from seeing and reacting to high consequence alarms that could prevent a serious incident if responded to in a timely manner. Based on this, we performed investigation into the last several power failure events at our sites with a focus on the distributed control system (DCS) and its impact on response to power events. This presentation covers the findings from this investigation which include when power events occur, alarm counts/metrics during the events, the severity of each different power event, and findings during root cause investigations (RCI).

We then discuss the different tools that have been implemented into the DCS specifically to aid in operator response to a power event. Some of the tools that are covered in this presentation are DCS graphics "overview" screens for power failure events which provide one-click capability to layer 3/4 process graphics, alarm hiding/suppression concepts, bulk alarm acknowledgement/unlatching techniques, alarm silencing, and automated actions in logic/programming to build more resiliency against an event.

# Development of Mobile App for Behavioral/Near Miss/PSE Reporting: Executive Summary

Ecolab, a global petro/chemical manufacturer wanted to improve their safety data reporting procedures, and create a consistent process for collecting/analyzing data.
The company also had the goal of incorporating more digital technology into their work practices and process production areas.

After a needs analysis, leadership recommended:

1. Development of a consistent Safety reporting procedure for all global plant sites to follow
2. Incorporate digital technology by use of a mobile app with shared database entry
3. Standardized analysis of safety data to quickly address trending safety events

## Development of a consistent Safety reporting procedure

Global plant sites weren't required to use a consistent tool or process for safety data reporting. The project team analyzed numerous safety data reporting practices at sites, to look for consistencies and best practices. These were used to develop a consistent reporting procedure.
The procedure included types of events for reporting, as well as roles and responsibilities of employees at the plant sites. Required data would include Behavioral observations, Process Safety Events (PSE) and Near Miss Events. The overall goal of the procedure was to eliminate unsafe conditions and behaviors (leading indicators) before they escalated to more severe events (lagging indicators).

## Incorporate digital technology by use of a mobile app with shared database entry

With the company goal of digital transformation, the project team looked for a solution that would eliminate paper-based practices at the sites. A mobile app solution was recommended, and pre-made options were explored. Nothing was found to meet the needs of the plant sites, so an internally developed app was recommended.
The project team used Microsoft Power Apps as a low-cost solution to develop an app in-house. The app included data entry fields for plant site, type of event, location event occurred, and many other important data points. The app was translated into several languages.
Safety events could be reported either using a mobile device, or on computer terminals located throughout the plant site. All reported safety events were downloaded from the app into a shared database for easy data analysis, and easy access by many different plant sites.

## Standardized analysis of safety data to quickly address trending safety events

Part of the reporting procedure included roles and responsibilities for analyzing safety data reports. Safety data was analyzed to identify trends. Specific recommendations/actions were made to eliminate the unsafe conditions/behaviors found in the safety data trends.
PowerBI was used to quickly identify safety data trends during the analysis process. Trends were communicated through the PowerBI dashboard by using graphs and other easily interpreted visual elements.

## Project Pilot

The pilot consisted of a 90 day "Safety Challenge" where a specific group of plant sites reported as many near miss and PSE observations as they could identify. Trends in the data were tracked through system generated reports, and displayed using a PowerBI dashboard. Plant site safety leaders immediately saw the benefits of a standardized safety reporting procedure, and ease of use with the mobile app.

The pilot has expanded to include additional plant sites in various countries. This has created a new set of challenges for deploying the Safety reporting procedure, expanding use of the app, translations, and identifying best practices for mobile devices. The pilot is set to complete in early 2021

SPRING21
+17TH GCPS
A Joint AIChE and CCPS Meeting

# Processed Reactive Metal Particulate Combustibility

**Robert Zalosh**
**Firexplo LLC**
**Wellesley, MA 02481**
**bzalosh@firexplo.com**



Prepared for Presentation at
American Institute of Chemical Engineers
2020 Spring Meeting and 17th Global Congress on Process Safety
Virtual
April 18 - 22, 2021

# Processed Reactive Metal Particulate Combustibility

**Robert Zalosh**
**Firexplo LLC**
**Wellesley, MA 02481**
**bzalosh@firexplo.com**

**Keywords:** reactive metals, metal swarf, dust explosibility tests.

## Abstract

Besides the usual particulate combustibility sensitivity to composition and particle size, processed metal particulate combustibility also depends on processing, handling, and additive complications which can alter the particle morphology and surface layer characteristics. These other complications render it difficult to generalize reactive waste particulate combustibility properties without testing to provide accurate property data. Examples of these considerations are given with regard to explosibility data for processed/waste particulate of different size and shape and moisture/additive levels for aluminum, manganese, titanium, and zirconium. As an example of the enormous variations in explosibility for these materials and particles, the data for normalized rate-of-pressure-rise, $K_{st}$, range from 2 bar-m/s to 658 bar-m/s.

## 1 Introduction

Most dust explosion incidents involve powders being processed or dust generated by particulate processing and handling. The distribution of equipment involved in dust explosions presented in the 2005 CCPS Guidelines [1] indicates that the overwhelming majority of incidents occur in dust collectors, grinders, dryers, blenders, and other processing and dust collection equipment. This is particularly true of metal dust explosions. Taveau et al. [2] cite statistics on German dust explosions showing that 65% of the metal dust explosions occurred in dust extraction and grinding systems. Danzi and Marmo [3] cite similar metal dust explosion statistics indicating 43% of these explosions occurred in dust collection systems, 18% in grinding/polishing equipment, 5% in milling equipment, and another 5% in mixers.

Processing of metal particulates often changes the metal oxide content as well as the particle size distribution and morphology. Many processes involve alloy particulates or generate alloyed dusts. Marmo et al [4] collected and tested fourteen metal alloy waste dust samples from different industrial facilities, and determined that their explosibility categorization could be correlated in terms of the metal cation weight fraction and the $d_{10}$ particle size. Since most of the dusts were either iron or ferrous alloys, it is not clear whether these two parameters would be sufficient to categorize more reactive metals and alloys.

Danzi and Marmo [3] obtained 43 metal dust samples from a variety of processing equipment and dust collectors in order to determine if and how the particular processing operation influences the dust explosibility. They note that most of the samples were mixtures of alloys, finishing materials such as abrasive media, and other contaminants, and were partially oxidized as a result of the processing or finishing operation. They use their data to discuss the processing operations and equipment that are more or less likely to produce severe dust explosion hazards.

The intent of this paper is to focus on specific reactive metal processed powders and dusts. Applicable combustibility data are reported from a variety of sources and references, with the data being related to the particulate process and handling conditions, including the influence of moisture and other process materials. In a few cases, the data are also related to reported dust explosion incidents.

## 2 Processed Aluminum Particulate Combustibility

### *2.1 Coarse Aluminum Floor Dust*

Aluminum cutting and trimming operations at an aircraft part manufacturing plant produced floor dust that was periodically swept up and cleaned with a portable vacuum. A floor sweep sample and a portable vacuum sample were obtained for testing. Table 1 shows the dust size distribution 10th percentile, $D_{10}$, values and the $P_{max}$ and $K_{st}$ data obtained from ASTM E1226 20-liter sphere tests. The values for the vacuum sample are relatively low, with the larger particle size floor sweepings sample being lower, such that the sample is considered marginally explosible.

Table 1. Aluminum Floor Dust Sample Test Data

| Parameter | Portable Vacuum Sample | Floor Sweep Sample |
|---|---|---|
| $D_{10}$ (μm) | 430 | ≈ 500 |
| $P_{max}$ (bar-g) | 5.3 | 0.9 |
| $K_{st}$ (bar-m/s) | 38 | 2 |

Clouthier et al. [5] have shown that the marginally explosible course aluminum sample ($D_{10}$ = 63 μm) with low $P_{max}$ and $K_{st}$ values (2.2 bar-g and 4 bar-m/s, respectively) obtained in 20-liter vessel testing with a standard 10 kJ igniter could not be ignited at all when tested in a 1-m$^3$ vessel. Therefore, the large particulate floor sweep sample is probably also noncombustible. Clouthier et al. determined that their moderate particle size aluminum sample ($D_{10}$ = 13 μm) actually produced larger $P_{max}$ and $K_{st}$ values when tested in the 1-m$^3$ vessel than in the 20-liter vessel. Therefore, the portable vacuum sample data in Table 1 may underestimate the explosion severity, and repeat testing in the larger vessel is warranted for a more definitive and precise determination.

### *2.2 Mixed Aluminum Fines from Grinding and Sanding Booths*

Aluminum and carbon fiber composite parts at the aircraft part plant are ground in one booth, and sanded in another booth. Dust samples were from the filters in those booths were collected and tested for both explosion severity and Minimum Ignition Energy (MIE) per ASTM E2019. Data for these samples, which appeared to be primarily aluminum, are shown in Table 2.

Table 2. Aluminum Grinding & Sanding Dust Sample Test Data

| Parameter | Grind Booth Sample | Sanding Booth Sample |
|---|---|---|
| $D_{10}$ (μm) | 30 | 45 |
| $D_{50}$ (μm) | 65 | 65 |
| $P_{max}$ (bar-g) | 9.8 | 10.0 |
| $K_{st}$ (bar-m/s) | 230 | 259 |
| MIE (mJ) | 20 | 19 |

Both Table 2 samples had $P_{max}$ values of about 10 bar-g, and $K_{st}$ values over 200 bar-m/s. These values are significantly higher than the literature values cited by Taveau [2] for freshly produced aluminum powders with median particle sizes greater than about 20 μm, but they are comparable to his cited values obtained with samples from the dust collector of an automobile factory that experienced a devastating aluminum dust explosion involving widespread propagation. The Table 2 MIE values of about 20 mJ are roughly twice the value cited from the automobile factory dust collector.

The relatively high $P_{max}$ and $K_{st}$ values for the Table 2 samples may be due in part to the morphology of the dust particulate. The samples appear visually (Figures 1 and 2) to have edges and angularities not present in freshly produced powders. Particle surface area effects are discussed in the next processed aluminum dust subsection of this paper, and shape effects are discussed in the sections on titanium and zirconium particulate.

**Figure 1 Aluminum grinding booth dust sample.**

**Figure 2 Aluminum sanding booth dust sample.**

### *2.3 Fine Aluminum Filter Cake Flakes*

An aluminum processing plant uses wet milling followed by a filter press to obtain the desired particle properties for further processing of the aluminum flakes. After experiencing a series of small fires involving the filter cake, the plant collected flake samples for explosibility testing. The dry flake characteristics and explosibility data are shown in Table 3.

Table 3. Fine Aluminum Flake Test Data

| Parameter | Sample 1 | Sample 2 | Sample 3 | Sample 4 |
|---|---|---|---|---|
| $D_{10}$ (μm) | 1.7 | 2.6 | Not Measured | 1.3 |
| $D_{50}$ (μm) | 6.2 | 8.7 | Not Measured | 4.4 |
| Specific Surface Area ($m^2/g$) | 1.8 | 1.1 | Not Measured | 2.7 |
| $P_{max}$ (bar-g) | 10.5 | 9.3 | 10.2 | 8.7 |
| $K_{st}$ (bar-m/s) | 658 | 546 | 471 | 631 |
| MIE (mJ) | < 3 | 5-10 | 5-10 | < 3 |

The high $K_{st}$ values shown in Table 3 are roughly consistent with the Bartknecht data plot of $K_{st}$ versus specific surfaces area shown in Figure 1.23 of Eckhoff's book [6]. Bartknecht obtained a $K_{st}$ value of about 1000 bar-m/s at an aluminum flake specific surface area of about 3 $m^2/g$, and much higher values at larger specific surface areas. The extremely low MIE values shown in Table 3 explain the susceptibility of these flakes to ignition by relatively weak electrostatic discharges as is thought to have occurred in the incidents at the plant from which the samples were obtained.

## 3 Processed Manganese Particulate Combustibility

### *3.1 Manganese Powder/Dust Explosibility Data*

Unalloyed manganese powder/dust explosibility data are not as prevalent as data for most other reactive metals. Table 4 shows most of the published and previously unpublished data known to this author. The wide variation in values for $K_{st}$, i.e., from 22 bar-m/s to 157 bar-m/s, can primarily be attributed to differing particle size distributions. The variation in Limiting Oxygen Concentration (LOC) may be partially due to uncontrolled surface manganese oxide layer thicknesses on the samples, since manganese ignitability is quite sensitive to oxide layer presence as discussed below.

Table 4 Manganese Dust Explosibility Data

| Parameter | Sample 1 | Sample 2 | Sample 3 | Sample 4 | Sample 5 |
|---|---|---|---|---|---|
| Data Source | Eckhoff [6] | Eckhoff [6] | Senecal [7] | Private | Private |
| $D_{50}$ (µm) | 16 | 33 | < 74 | 212 | < 74 |
| % < 74 µm | 82% | 70% | 100% | 16% | 95% |
| $P_{max}$ (bar-g) | 6.3 | 6.6 | 6.7 | 3.7 | 4.5 |
| $K_{st}$ (bar-m/s) | 157 | 69 | 150 | 22 | 130 |
| LOC (v% $O_2$) | No Data | No Data | 14% | 12.5% | 9.5% |

### *3.2 Manganese Oxide Layer Effects*

Manganese particulate explosibility and ignitability depend greatly on the presence and depth of a surface oxide layer, such that fine manganese powder can be easily ignited if it does not have surface oxidation. In fact, the CAMEO Chemicals Manganese entry states that manganese dust has been known to be pyrophoric [8]. This is an important aspect of manganese powder manufacturing facilities, which rely on nitrogen inerting system to prevent ignitions while the powder is being processed. Senecal [7] described an incident in which the nitrogen system was inadvertently shut-off while a fine powder was being milled. He describes how air entered the mill feed where an explosible powder cloud existed along with several possible ignition sources, such that a deflagration was initiated in that area and propagated to the downstream bucket elevator.

There have been several unpublished accounts of fire and explosion incidents in and around inerted manganese milling and transport equipment. Some accounts indicate that the milled manganese powder virtually auto-ignites when the manganese first encounters ambient air. Inerting systems in these facilities are now carefully controlled so that there is a low level of oxygen to allow a passivating oxide layer to form on the manganese. NFPA 69-2019 requires the oxygen concentration in facilities with continuous oxygen monitoring to be at least 2 v% $O_2$ below the LOC, so that concentrations should be less than 7.5% $O_2$, based on the lowest LOC value in Table 4.

### *3.3 Manganese Contaminant Effects*

Besides rapidly adsorbing oxygen, manganese and manganese oxide readily adsorb many other chemicals. This allows impurities to occur in some manganese supplies, and potentially in manganese processing. Manganese powder Safety Data Sheets list acids and bases among the incompatible materials to be avoided.

One impurity, polyvinyl alcohol, has been attributed as a likely cause for a fire in a manganese particulate warehouse. Carbon residue from the PVA was found with the partially burned manganese after the fire. The fire was extinguished by applying a combination of sodium chloride, METLX™, and sand, and the supplier was contacted to request impurity removal from future supply batches.

# 4 Processed Titanium Particulate Combustibility

## *4.1 Processed Titanium Dust Combustibility Test Data*

As described in the CSB report on the AL Solutions explosion incident investigation [9], several titanium dust samples were collected from drums of as-received scrap/swarf metal and from in-process material intended for conversion to pressed compacts. SEM micrographs of the samples showed a mix of large cuttings and crescent shaped swarf along with smaller particles. Some of these titanium samples were combined in order to have sufficient material for a complete set of ASTM E1226 explosion severity tests and ASTM E1491 dust cloud auto-ignition temperature tests. Results for this coarse particulate composite sample, together with results from a different sample with much finer particulate from a vacuum collector at another facility, are shown in Table 5. The $K_{st}$ data demonstrate the wide range associated with dry titanium dust, depending on particle size.

Table 5. Processed Titanium Particulate Explosibility Data

| Parameter | Drum Sample | Vacuum Sample |
|---|---|---|
| Median Particle Size, µm | > 425 | < 45 |
| Percent < 74 µm | 0 | 94.6% |
| $P_{max}$ (bar-g) | 6.3 | 8.2 |
| Kst (bar-m/s) | 75 | 164 |
| MIE (mJ) | NA | 51 |
| Dust Cloud AIT (°C) | 450 | NA |

In order to visualize the flame produced by the burning of a suspended cloud of fine processed titanium dust, a demonstration was conducted in a transparent 1.2-liter cylinder. The beginning of the observed flame propagation image, the first video frame after ignition, is shown in Figure 3. The intensely bright flame image is a manifestation of the 3720 °C titanium adiabatic flame temperature [10].

Figure 3 Fine titanium dust flame in 1.2-liter cylinder.

One of the AL Solutions coarse particulate drum samples was used to determine the effect of particulate moisture content. In this case, 20-liter sphere Go/No Go explosibility tests with two 5 kJ igniters were conducted at a dust concentration of 1000 g/m$^3$, which was the concentration that produced the largest $P_{max}$ value in the ASTM E1226 full set of tests. Figure 4 is a plot of maximum pressure versus moisture content from these tests. There is a steep drop in pressure between 17% moisture and 22% moisture, with pressures remaining in the range 1.6 bar-g to 2.5 bar-g for moisture levels between 22% and 33%. Although these high moisture contents produced pressures sufficiently high for the samples to be designated as combustible per ASTM E1226 criteria, the Clouthier et al [5] comparisons of 20-liter sphere data and 1-m$^3$ vessel data for a weakly combustible metal suggest that the 20-liter sphere data are overdriven by the 10-kJ ignition source, and the higher moisture content samples would probably produce No Go results in the larger test vessel.

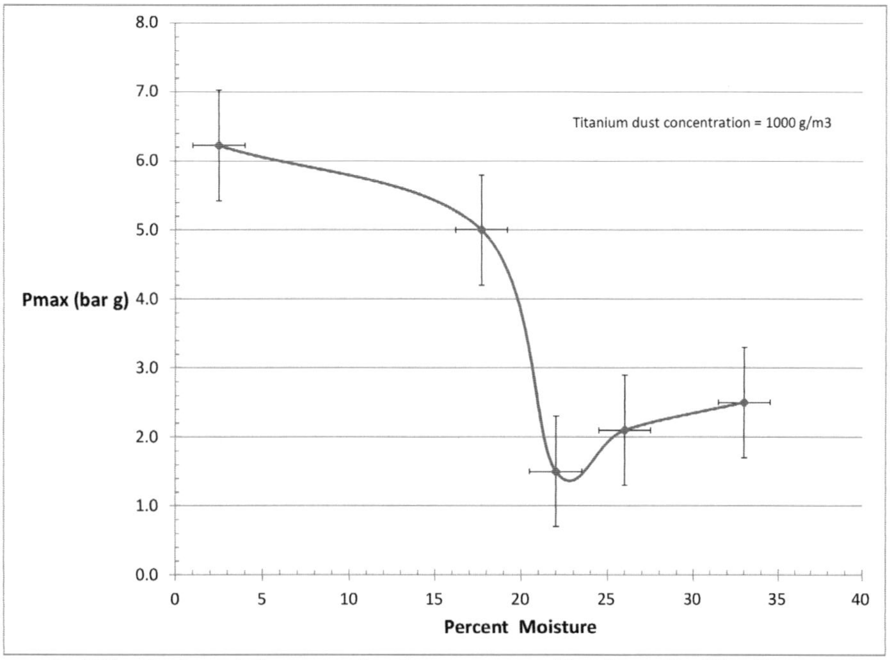


Figure 4. Effect of moisture content on coarse titanium Go/No Go pressures.

### *4.2 Processed Titanium Dust Hydrogen Formation*

Since the AL Solutions titanium particulate samples had a high moisture content, there was some concern about a possible titanium-water reaction. Lu et al [11] conducted a comprehensive study of effects of titanium exposure to water and to oxygen at different temperatures and concluded that there was negligible oxidation of titanium by water during short-term exposures at temperatures less than 300 K. They do note that there is slightly increased water-induced oxidation (with concomitant formation of titanium hydroxide and hydrogen) upon longer exposures.

Since there was a long delay between titanium sample collection and explosibility testing, a multi-gas detector with a hydrogen sensor was used to check the hydrogen concentration in half-filled sample jars about six months after the samples were acquired. The readings in two jars were 1% and 2% of the hydrogen LFL, i.e., sufficiently low for evolved hydrogen not to affect the dust combustibility test data described above. Most of that limited hydrogen generation was probably from air oxidation since the measured oxygen concentrations in the sample jars at that time were 20.7% $O_2$ and 20.6% $O_2$.

## 5 Processed/Waste Zirconium Particulate Explosibility

### *5.1 Zirconium Swarf Particulate Explosibility Data*

Zirconium swarf particulate explosibility is very different than the explosibility of fine zirconium powder. Fine zirconium powders with median particle sizes of about 10 μm and 20 μm have been reported to ignite spontaneously when dispersed as a dust cloud in the standard dust explosibility test apparatus and the Godbert-Greenwald furnace [12]. Slightly larger dispersed zirconium dust particulate (diameter less than 42 μm) required a spark igniter for deflagrations [13]. Zircaloy chips produced spontaneous sparks but not sustained flame propagation when dispersed in a 1.2-liter Hartmann test vessel, with these spontaneous sparks igniting fine zircaloy dust (smaller than 74 μm) when dispersed along with the chips. Larger zirconium swarf particulate described below does not generate either sparks or spontaneous deflagrations when dispersed as a cloud.

Table 6 shows particle size, moisture, and explosibility test data for three zirconium waste samples. One sample was crushed pellet particulate from the CSB AL Solutions investigation [9] and the other two samples were wet and dry zirconium turnings/swarf from another facility. The sieve-based particle size data is somewhat misleading since many of the swarf particulates had the characteristic sickle shape with widths in the range 70 μm to 90 μm, and lengths of 360 μm to 1300 μm. None of these samples exhibited any sparking or self-ignitions when subjected to compressed air dispersal. The ASTM E1226 $P_{max}$ and $K_{st}$ values for all three samples are relatively low, with the exception of the $P_{max}$ for the wet swarf. Clearly, a moisture level of about 23% did not reduce this zirconium swarf explosibility compared to the dry swarf.

Table 7. Zirconium Waste Particulate Explosibility Data

| Parameter | Crushed Pellet | Dry Swarf | Wet Swarf |
|---|---|---|---|
| $D_{50}$ (μm) | 570 | 280 | 280 |
| $D_{10}$ (μm) | 250 | 160** | 160** |
| Moisture (%) | 5.2% | 0.5% | 22.6% |
| $P_{max}$ (bar-g) | 4.3 | 4.1* | 7.5* |
| $K_{st}$ (bar-m/s) | 12 | 7* | 9* |
| Layer MIE (mJ) | NA | 1,000 – 10,000 | < 3 |

* These values were obtained with just one test series using both one 5 kJ and two 5 kJ igniters.
** Layer MIE tests were conducted with finer swarf particulate ($D_{10}$ < 75 μm).

The last row of data in Table 6 shows the layer Minimum Ignition Energy data obtained with the dry and wet swarf samples. The tests are conducted with a charged capacitor discharging to a grounded copper plate on which swarf layers of different thicknesses are deposited. The spark lowest spark energy to produce a visible flame is recorded as the Layer MIE. The 3 mm thick wet swarf sample produced a much larger and more intense flame, as seen in Figure 5, than the dry swarf, and at a much lower spark energy. The Table 6 data, showing the wet swarf to be more easily ignitable and producing a higher deflagration pressure than dry swarf, may not be generally applicable, since there are some reported and some anecdotal contradictory accounts of moisture effects as discussed below.

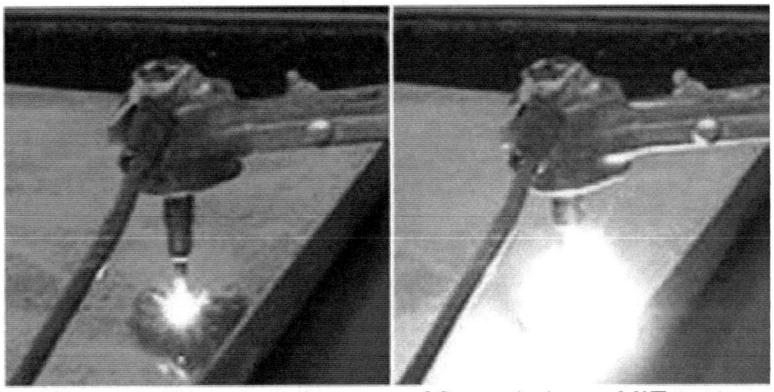

Figure 5. Wet zirconium swarf flame in layer MIE test.

### *5.2 Zirconium Swarf Moisture and Additive Effects*

Although the zirconium-water reaction is generally reported to be negligible at ambient temperatures, the high moisture contents and sample testing delay of the AL Solutions motivated a check on hydrogen concentrations developed in the half-filled sample containers. Multi-gas sensor readings for two zirconium swarf samples showed a hydrogen concentration of 2% of the LFL in one container, and a reading of zero in the other container. The oxygen concentration was 19.8% $O_2$ in the container with the 2% LFL hydrogen concentration, indicating that the hydrogen was generated by air oxidation, rather than any low-level zirconium-water reaction.

Matsuda and Yashima [12] reported zirconium dust cloud ignition temperature data as a function of zirconium powder moisture content. Their data showed the cloud temperature increasing from 360 °C to 400 °C as moisture content increased from 5% to 15%. They also showed room temperature ignition for a completely dry fine dust cloud, as mentioned above. Their reported dust layer ignition temperature for dry powder was about 272 °C for a 5-mm deep layer, and about 238 °C for a 15-mm deep layer. Anecdotal accounts of spark ignition dust layer tests conducted in other laboratories indicate that dry zirconium swarf can at least sometimes be readily ignited and produce rapid flame spread along the layer. Zirconium fines Safety Data Sheet information cited in the CSB AL Solutions report "recommends keeping zirconium fines either extremely dry (less than 5 percent water) or extremely wet (more than 25 percent water)."

One complicating factor influencing zirconium swarf combustibility is the frequent presence of additives such as cutting and grinding fluids, cleaning fluids, corrosion/oxidation inhibitors, and abrasive media. The fluids can produce a coating on the surface of the zirconium and other metal swarf and may partially alter the metal oxide layer underneath. An X-ray Photoelectron Spectroscopy (XPS) elemental analysis of the surface composition of the swarf used for the tests listed in Table 6 indicated high concentrations of sodium and carbon, with the former likely originating from a grinding fluid and the latter likely from cutting oil or from resin in grinding media. Many of the fluid additives are water soluble (fluids used in the zirconium processing for the swarf used in the Table 6 tests had pH values in the range 9 to 12) with alkaline components that can cause the swarf moisture to be a dilute alkaline solution. It is not clear to what extent these surface additives and dilute solutions may have influenced the results in Table 6, and the results for other zirconium swarf combustibility tests.

Some metal processing additives can have mitigating effects on swarf combustibility. There have been unpublished tests showing that the addition of a cutting oil can make it more difficult for a zirconium swarf layer to be ignited by sparks. Thus, there are varying effects of different additives, with some additives enhancing swarf combustibility and others ameliorating it. Therefore, laboratory combustibility testing that accounts for the possible presence of moisture and process additives should be an essential aspect of reactive metal wastes characterizations and waste handling determinations.

# 6 Conclusions

Besides the usual effects of particle size, composition, and moisture, processed reactive metal particulate explosibility test data are heavily influenced by particle morphology, oxide layers, and the possible presence of processing additives. Test data for processed aluminum particulates show $K_{st}$ values in the 230 bar-m/s to 258 bar-m/s for grinding and sanding booth dust, and values between 471 and 658 bar-m/s for fine flakes. Test data for processed manganese particulate show $K_{st}$ values in the range 22 to 157 bar-m/s depending on particle size, and Limiting Oxygen Concentrations in the range 9.5 % $O_2$ to 14 % $O_2$, with the corresponding need for small oxygen concentrations to maintain the passivating manganese oxide layer. Fine and coarse processed dry titanium waste exhibited $K_{st}$ values of 164 bar-m/s and 75 bar=m/s, respectively, with wet titanium being marginally combustible at moisture concentrations greater than 20%.

# 7 References


1. CCPS, Guidelines for Safe Handling of Powders and Bulk Solids, Center for Chemical Process Safety, AIChE, 2005.
2. Taveau, J., Hochgreb, S., Lemkowitz, S., and Roekaerts, D., Explosion hazards of aluminum finishing operations, Journal of Loss Prevention in the Process Industries 51 (2018) 84–93.
3. Danzi,, E. and Marmo, L., Dust explosion risk in metal workings, Journal of Loss Prevention in the Process Industries 61 (2019) 195–205.
4. Marmo, L., Riccio, D., and Danzi, E., Explosibility of metallic waste dusts, Process Safety and Environmental Protection 107 (2017) 69–80.
5. Clouthier, M., et al, Iron and aluminum powder explosibility in 20-L and 1-m3 chambers, Journal of Loss Prevention in the Process Industries 62 (2019) 103927.
6. Eckhoff, R., Dust Explosions in the process industries, 3rd Editioin, Gulf, 2003.
7. Senecal, J., Manganese mill dust explosion, J. Loss Prev. Process Ind., v 4, (1991).
8. https://cameochemicals.noaa.gov/report?key=CH16141
9. CSB, AL Solutions, Inc., New Cumberland, WV Metal Dust Explosion and Fire, U.S. Chemical Safety Board Case Study Final Investigation Report, July 2014.
10. NFPA 84, Standard for Combustible Metals, Table A.1.1.3(a), National Fire Protection Association, 2019.
11. Lu, W., Bernasek, L, and Schwartz, YJ., Oxidation of a polycrystalline titanium surface by oxygen and water, Surface Science 458 (2000) 80–90.

12. Matsuda, T. and Yashima, M., Ignition Characteristics of Zirconium Dust, Japan National Institute of Industrial Safety Report NIIS-RR-99 (2000).
13. Leuschke and Wuttig, X, Entzundung schnell bewegter Zircaloyteilchen durch mechanische Beanspruchung (Ignition of fast-moving zircaloy particles by mechanical stress), Staub - Reinhalt. Luft 43 (1983).

# 617938 Using Technology to Improve Safety

**Abstract:**

As the world grows and changes, our business processes evolve, and we must change to meet our new needs. Recent technologies like smart phones, tablets, computers, smart televisions, smart homes, and applications that can manage every aspect of daily life are now widely available in any employee's personal life. These devices and programs reduce the time needed to do routine tasks, providing new opportunities to focus on things the individual may feel are more important or fun. In some cases, these processes learn from previous transactions and tasks and require less of the individual's input as it continues to operate.

However, in a person's professional life, it is common to have a significant portion of one's day composed of filling out paperwork and performing hands-on tasks. Although work instructions, procedures, and policies for everyday tasks are now available electronically, a field employee needs access to a computer to view, access and print documents. This limits the productivity and effectiveness of employees since it forces them to access all information at a single access point or having them carry around paperwork that can easily be damaged by work conditions and the environment. One limitation is unexpected equipment failure. In such cases, it is rare that employee would have all technical information readily available to help troubleshoot in the field immediately, except for minimal visual HazCom labels and warnings. This decreases overall process efficiency by inadvertently creating restraints – specifically, having to go back and collect all required paperwork- on the employee which can be overcome with everyday technology.

Understanding how so much information is instantly available outside of work is important to replicate that level of access and ease in the workplace. At Rain Carbon Inc., our mission is to be "a progressive industrial company committed to collaborative science-driven innovations that result in safe, reliable and efficient operations to ensure outstanding results for our customers, partners, employees and investors." With this as our guide, a new system was implemented to target several areas: safety, reliability, information availability and efficiency.

At our Candiac advanced materials site in Quebec, Canada, integrative software was implemented in 2017 as a pathway to achieving the company's mission. This software provides instant access to process information through phones, tablets and computers, including work instructions, policies, procedures, ability to complete round check sheets and work permits, action assignment and tracking, picture and video capability, report unsafe conditions and incidents, equipment malfunctions, and the management of workforce skills. The capabilities provided by the software have proven to be most effective when used on mobile devices as it allows for information to be accessed directly from any location within the plant helping us improve in all the below points:

- Enhancing site management by providing a live view of notices, work requests and unsafe conditions to immediately communicate and respond to the situation at hand;
- Increasing employee involvement in safety due to higher visibility and lower response times to address unsafe acts/conditions;
- Reducing redundancies of having to go back and forth between paper and electronic work; and
- Decreasing unplanned downtime by providing all relevant technical information instantaneously enabling employees to address deficiencies without delay.

Rain Carbon has pushed forward with innovation and technology by implementing QR codes for all equipment linked to the integrative software solution. This adds a significant amount of efficiency to work processes on the shop floor. When unplanned work is to be performed, the employee can take any mobile device with the software installed on it and scan the QR code on the respective equipment. This will take the employee to a screen with all necessary information regarding the respective equipment, including the work permit and list of materials needed to perform the work in question. A process that would, in most cases, take a matter of hours is reduced to a few minutes and a couple of taps.

Although reliability and efficiency have increased since the system was implemented at the Candiac facility, the most significant improvement has been observed in safety. Prior to implementing this software, safety walks were completed in an electronic or printed version of a Word document, safety incident notifications would be accomplished through phone calls and emails, and near misses would be recorded in a separate system. The centralization of reporting and the enhanced information availability highly encouraged employees to report all safety items with the backing of a more engaged management group, thus enhancing employee safety. Reporting increased with the introduction of the program by 772% on the first year and by 2,150% on the second year. Similarly, during that timeframe, near misses decreased by 40% and first aid cases by 33% compared with the current year.

Results:

From 2017 to 2019, there was a clear increase in reporting after the implementation of the software. Through management support and commitment, we can see a clear decline from 2019 to 2020 YTD of overall identified items, increasing the quality of work and safety across the site.

Implemented in 2017, the program has significantly improved the safety performance of Rain Carbon's Candiac site by engaging management and ensuring all safety items are properly communicated and addressed with as little downtime as possible.

# 617740 Facility Siting Revalidation - a New Tool

**Abstract:**

Facility siting studies require periodic review at specified intervals during the life of a facility. As stated in 29 CFR Part 1910.119 (e) (6), revalidation is required every five years based on the previous full study completion date. While revalidation is a regulatory requirement, little is often known about the process of undertaking the initial study or review and, in some cases, this process can become challenging.

A new and simple revalidation tool is available that helps operators determine if their facility siting study needs a complete redo, a retrofit, an update, or a simple revalidation. This paper introduces and explores this simple revalidation tool, providing examples of how it can be used in a plant environment.

# 617756 Quantifying the Conditional Dependencies between Layers of Protections in Event Tree Analysis

**Abstract:**

An approach to include conditional probabilities when calculating probabilities of disaster in quantitative risk analysis is presented.[1] Since not all layers of protection are independent of each other, using conditional probabilities leads to more accurate calculations of probabilities of disaster in a chemical process. Based on event tree analysis, probabilities of disaster are calculated assuming that events are statistically independent (independence assumption), and separately considering that events affect each other (conditional probabilities), in three cases with increasing complexity and number of layers of protection. First, a simple alarm – operator case is studied, then two cases involving a chemical reactor and operator with increasing number of layers of protection are analyzed. For all cases, we considered multiple scenarios to show that results do not depend on the input probabilities of particular events. For the latter two cases, the input probabilities of each event were randomly sampled with Monte Carlo simulations. Probabilities of disaster calculated with the assumption of statistical independence were compared to probabilities of disaster calculated with conditional probabilities. In most scenarios of the three cases, findings show that the probabilities of disaster calculated using the independence assumption are lower than those with conditional probabilities, which can lead to undercounting the layers of protection that are needed. Process engineers need to be aware that assuming that events are statistically independent can lead to the possibility of undercounting of layers of protection needed for a chemical process.

Reference: Mott TC, Kivistik PM, Panorska AK, Cantu DC. Comparing conditional probabilities and statistical independence in layers of protection analysis. Proc Safety Prog. 2020;e12215. https://doi.org/10.1002/prs.12215

# 617799 Comparison of Methodologies for Dropped Objects Analysis in Shallow Water Drilling Rigs in the Gulf of Mexico.

**Abstract:**

The reform of the energy sector undertaken in 2013 by the Government of Mexico to empower foreign investment in the hydrocarbon sector led to the arrival of private companies with new projects and technology to the Gulf of Mexico. The projects managed in this region move continuously from the exploration stage to the production stage. During this stage transition, Offshore Projects require development of safety evaluations for the design of the platforms that were not common in the Mexican market; among these evaluations there is "Dropped Objects Analysis".

The accidental dropping of objects is considered among the main risks on offshore production platforms since its occurrence can affect personnel (injuries and / or deaths) and damage the facilities (causing spills and / or fires). Thus, as part of engineering design development, there is a need to assess the potential risk associated with impact from oscillating objects and loads that could fall during routine and non-routine lifting activities.

The analysis focuses on the possible damage to the area of operation of a given platform, structures, process equipment and subsea pipelines, evaluating the frequency of dropping combined with the type of damage that would result from the impact. However, for Dropped Object Analysis there are different methods that involve complex calculations, including solving integral equations, which for practical purposes can result in a large time investment.

This work addresses the use of a numerical method that allows a complex calculation routine to be carried out towards a solution with simple arithmetic, thus facilitating the evaluation of the frequency and impact of dropped objects on the platform and reducing the solving time.

As an application case, the Dropped Objects Analysis is presented in two offshore platforms for the exploitation and transportation of hydrocarbons installed in southeast Mexico by a private operator. During the study, the frequency and consequence of dropped objects were evaluated to issue protection recommendations in the design of both platforms and were also compared with the use of traditional calculation methods.

# 617993 CFD-Based Ventilation Analysis - Basis for Conducting Consequence and/or Risk-Based Assessments

**Abstract:**

The combination of powerful capabilities in the development of Computer-Aided Design (CAD) drawings; i.e., 3D models; and the performance of formidable computer processors confirm the execution of robust Computational Fluid Dynamics - CFD-based simulations in manageable short-duration periods of time; i.e., minutes. Using CFD tools allow considering the fluid-geometry interaction, key factor for identifying specific features of the site section being analyzed. This information is not possible to be foreseen when using screening tools based on experimental data correlated and/or phenomenological methods.

The present paper illustrates a case study that provides examples of how a CFD-based ventilation analysis can provide insight information of areas that require special attention when a consequence or risk-based assessment is to be performed. The identification of regions with air recirculation, stagnant flow areas with low wind speeds, congestion volumes, etc. verify leak locations and orientations to be evaluated when performing advanced dispersion, fire, and explosion analysis.

# SYSTEMATIC APPROACH TO PROCESS SAFETY MANAGEMENT ON ROAD LPG TRANSPORTATION

**Presenter/Principal Author**
**Américo Diniz Carvalho Neto**
**RSE CONSULTING**
**Brazil**
**americodcn@rsem.com.br**

**Other Author**
**Adriano Nonato**
**RSE CONSULTING**
**Brazil**
**ext.adriano.nonato@rsem.com.br**

Prepared for Presentation at
American Institute of Chemical Engineers
2021 Spring Meeting and 17th Global Congress on Process Safety
Virtual
April 18 - 22, 2021

# SYSTEMATIC APPROACH TO PROCESS SAFETY MANAGEMENT ON ROAD LPG TRANSPORTATION

**Presenter/Principal Author**
**Américo Diniz Carvalho Neto**
**RSE CONSULTING**
**Brazil**
**americodcn@rsem.com.br**

**Other Author(s)**
**Adriano Nonato**
**RSE CONSULTING**

**Keywords:** LPG transportation Riks; PSM in LPG distribution, LPG Transport

## Abstract

Process Safety Management is systematically adopted in sectors of Brazilian industry, such as oil and gas and petrochemicals, with clear operating safety gains, based on performance indicators. In common, they all use fixed structures and are siting centered to carry out their operations. Also important in the energy chain in Brazil, bulk LPG transportation supplies industrial chain, and business and residences as well. With all this importance requirements for robust process safety management are still missing in LPG distribution sector.

This work aims to implement a safety management system within a company currently responsible for the largest share of the bulk LPG market, considering the particularities of this type of operation, totally dependent on road transport for LPG distribution. To achieve this the first step was identifying the main activities that represent the bulk system's operation, involving bobtail trucks, as follows: supply through road routes; unloading, which is the supply to the customer and the operation of the bobtail truck itself. The strategy of implementing the safety management system considered sequential application of the following steps: 1) Reassessment of the bobtail truck project, aligning it with the latest technological solutions; 2) Conducting a hazard identification study for supply and unloading activities, proposing a customized hazard identification model for supply; 3) barriers auditing to evaluate unloading components reliability; 4) Auditing operating activities based on CCPS RBPS in order to identify existing gaps. The fifth and final step is the definition of the Process Safety Management approach aligned with road transportation best practices worldwide and in compliance with legal and corporate requirements.

The studies already developed demonstrate that a careful analysis of the main normative reference identified gaps in the bobtail project, with emphasis on passive safety systems, absent in the original project. In the supply risk assessment stage (routes assessment), a specific model for route risk analysis was developed using a customized Excel spreadsheet for risk analysis through predefined menus. This model reduced the assessment time required and the level of subjectivity, allowing the company to develop hundreds of analyzes with the use of video conferencing tools, essential in this case, due to its nationwide scope.

This is a work in progress and the next steps, when completed, will make the company achieve a degree of excellence in Process Safety Management through monitoring for the implementation of all RBPS elements that have been identified as absent or deficient. Many other industrial sectors that carry out transport of dangerous products can benefit from this methodology for implementing RBPS, through the use of the described models, adapting them to the specific modal.

## Introduction

It is recognized that chemical transportation risk management is more than just a service; but rather is a wide-ranging system of activities that involves numerous stakeholders with varying backgrounds and interests. To effectively manage transportation risk the entire supply chain, spanning the delivery and handling of raw materials through the offloading and unpacking of finished products, needs to be included and evaluated. In addition to safety and business continuity, the evolving threat of security concerns are now included as part of a comprehensive transportation risk management program. Given the complexity and global reach of many transportation networks, as well as the tradeoffs between safety and security, the management of transportation risk can challenge even the best business plans and programs.

Considering all these aspects and risks involved in LPG transportation, RSE Consulting developed a PSM project with one of the biggest gas distribution companies in Brazil, o order to develop a diagnostic, and create a PSM system to be implemented in the company.

# 1 Bulk LPG transportation risk assessment

This stage aimed to evaluate a pilot project of adequacy of LPG transport vehicle. The use of common risk analysis methodologies and verification of compliance with international benchmarks were applied.

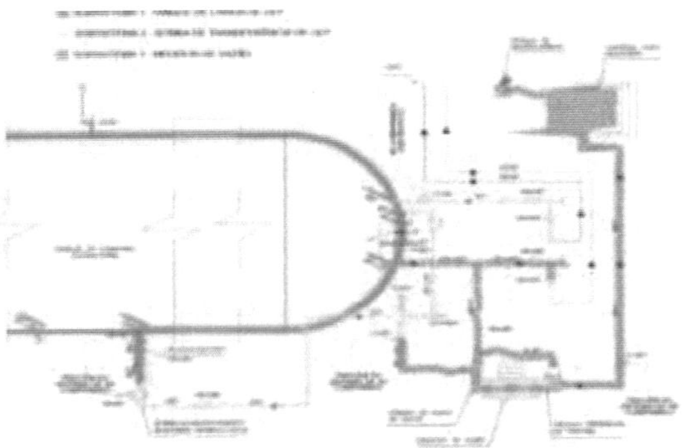

A HAZOP of the system was carried out to identify operational weaknesses

**A revised project was obtained considering improvements in management and safety elements, some highlighted below:**

Include the identified critical components into the inspection and maintenance plan (Bobtail). Ensure that inspection and maintenance plans have effective tests and tasks for identified and potential damage mechanisms.

Conduct training with operators on the new operating demands (Bobtail), reinforcing the concept of critical components, defining the points that should be evaluated during inspection routines, in order to ensure that the trailers transporting LPG meet the appropriate conservation conditions in relation to the Preventive Barrier of Periodic Inspection of Trailers.

Ensure that truck fleet will have a supply pump interlock system when the minimum level of 5 % of the volume of the LPG tank is reached, to avoid cavitation of the pump.

Every operational unit across the country should comply with managerial based actions concerning standardized execution of maintenance and inspection of critical components.

Include as managerial based actions the adoption of indicators for assets of integrity and reliability as stated on API 754

Define a maximum and minimum speed range to be adopted during the hose rewinding activity on the reel in order to ensure operation compatible with operator safety.

Ensure that a relief valve is installed on the swivel joint. This valve has the function of relieving pressure due to LPG expansion that can occur between two shutoff valves. Ensure that its construction and pressure class meet NFPA 58 requirements

Evaluate the feasibility of deploying a portable remote device for activate emergency activation system.

Modify emergency activation system so that the entire system is shut down or otherwise enforced to fail safe state (pump, pneumatic valves and engine and truck electrical system).

Reevaluate the removal of LPG liquid supply line check valve, which, as designed, will be replaced by a pneumatic-driven shutoff valve.

## 2. EVALUATION OF TRANSPORTATION ROUTE

The company under assessment supplies LPG for the vast majority of brazilian states

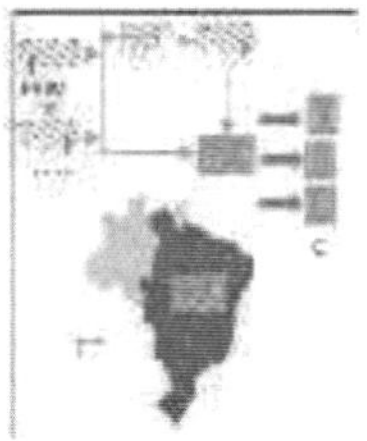

Process safety management on this company is still on its infancy ant still being circumscribed to the staff on Brazilian southwest headquarter. The staff engaged on conducting the evaluation had to overcome the lack of any previous knowledgeabout Risk Analysis of the operating team further hampered by mobility limitations imposed by the pandemic.

A method was developed for route risk analysis as recommended by the CCPS.

All relevant attributes that impacted the risk of the routes were identified and assigned weights for them.

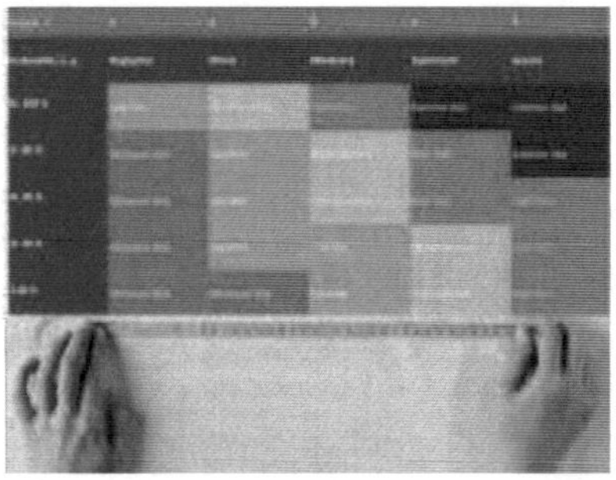

The weights have been calibrated to ensure consistency with the normative risk matrix.

An Excel spreadsheet was developed to assess risk, deploying the afore mentioned risk attributes, allowing teams to engage with maximum objectivity and time savings. Another advantage of the spreadsheet was its smooth learning curve, enabling the use for different teams of the company and possibility of future improvements to evaluate management elements

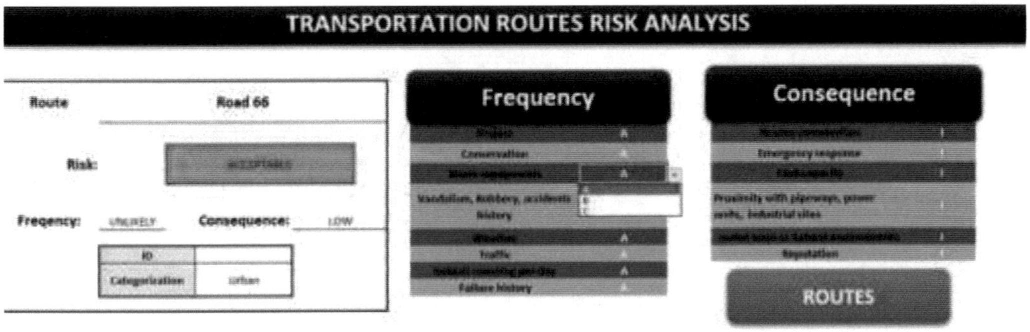


A total of 22 routes were analyzed. It is expected that teams can perform a much higher number of analyses with a minimum of tutoring, given the smooth learning curve for using the tool. Improvement points "appear" from the analysis of the result graphs, allowing recommendations to be properly addressed, as follow:

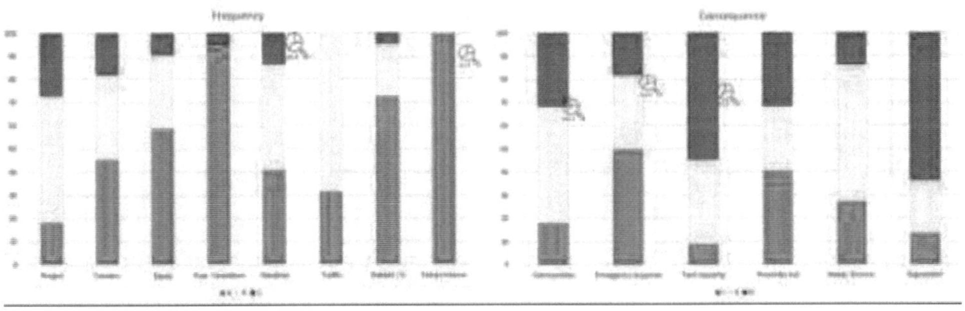

## 3. CUSTOMER RISK ANALYSIS

The 3rd stage of the project (in progress) will be the application of a bow tie diagram to the customer facilities.

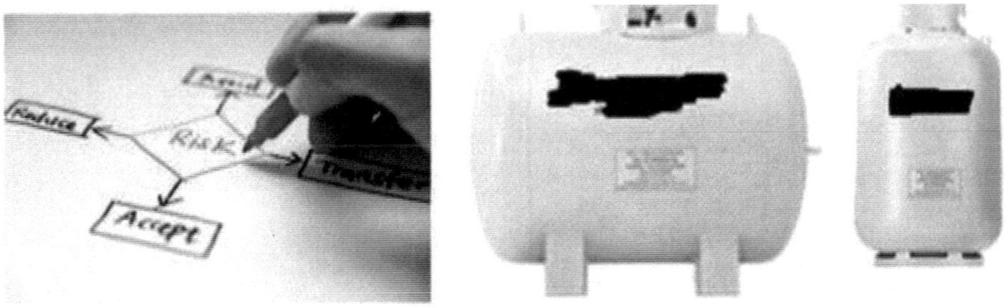


The bow tie will facilitate the process of identifying barriers for subsequent analysis.

## 4. BARRIER AUDIT

Step to be performed as follows below:

- Definition of critical risk scenarios for the preparation of protocol d andaudit;
- Preparation of audit protocols of critical scenarios and preventive and mitigating barriers;
- Audit of barriers / critical controls of the scenarios defined in the 1st step
- The systems to be analyzed are those of customer facilities. CCPS Bow Tie Approach was the chosen method.

## 5. IMPLEMENTATION OF PSM

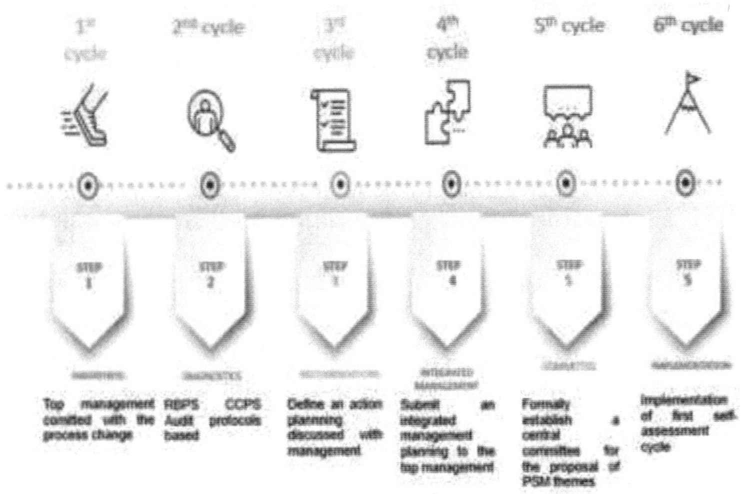


The PSM system implementation strategy is structured in the following steps:

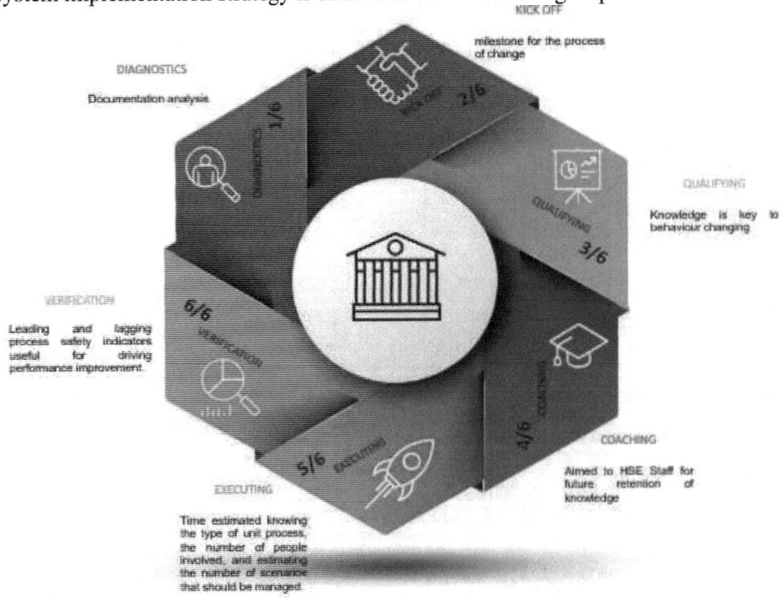

## Conclusion

The area of transport of dangerous products, specifically the transport of LPG in large cities, presents many positive contributions to society, however this logistical process has many risk scenarios with high severity in all stages, from loading to the process bases, distribution of the LPG via tank trucks, and refueling at customers' different facilities. This Project allows the organization to not only make a technical diagnosis of the risks of the entire process, but also to set up a Risk Management System that, through the implementation of the requirements of the Management System and procedures, will be allowed to achieve a high performance with the elimination of leakages. gas with the potential for catastrophic accidents.

## 2 References (examples given to show format).

[1] J.A. Klein and B.K. Vaughen. “Implement an Operational Discipline Program to Improve Process Plant Safety,” June 2011.

[2] CCPS. *Guidelines for Chemical Transportation, Safety, Security and Risk Management.* Center for Chemical Process Safety, American Institute of Chemical Engineers, New York, NY, 1995.

[3] RSE. RSE Consulting procedure for Risk transportation qualitative and quantitative risk analysis, 2019. Developed by Professor Americo Diniz Carvalho Neto.

# 618466 The Evolution of Emergency & Crisis Management in Petronas

**Abstract:**

Managing emergency and crisis effectively is a paramount aspect in Process Safety Management, especially in the complexity of oil and gas business. Optimum readiness is aimed to efficiently address potential impacts arising from incidents; immaterial of size; to contain the extent of exposure as quickly as possible, with the primary intent of saving lives and protecting the environment. Often we pay too much attention on the left hand-side of the bowtie (control barriers) that we often overlook the right hand-side of the bowtie (recovery barriers). The objective of this paper is to share the evolution of emergency and crisis management underwent by PETRONAS, with the ultimate aim to ensure our interests (people, environment and asset), and reputation are protected.

PETRONAS operates globally in more than 50 countries, hence effective establishment and implementation of emergency and crisis plans are vital. Historically, emergency and crisis management and preparedness has been a reactive science. We had our own share of incidents, and multiple lessons learnt had been consolidated to further improve our systems and barriers. The integration of emergency and crisis management as part of the HSE Management System (HSE MS) and Mandatory Control Framework (MCF) built a strong foundation and governance, and ensures continual improvement, taking into account all the internal and external lessons learnt. Conducting and monitoring the number of emergency drills and exercises, specifically based on the Major Accident Hazard (MAH), is imperative to ensure effective response in case of real emergency and crisis. Learning from real cases raised the vital need of coordinated response and putting the rights interventions and controls in place to minimize the impact of incident. Effective emergency and crisis management is crucial to support process safety implementation throughout the asset life-cycle covering design, operation and maintenance stage.

Throughout the years, we are also continuously enhancing our capability in terms of resources and people. In a nutshell, this paper will talk about the evolution of emergency and crisis management in PETRONAS, where it started with system establishment and capability building, whilst moving towards digitalization to better manage emergency and crisis. This paper will share some real incident cases and how we learn from these incidents to further intensify our emergency and crisis management.

# 617018 Pressure Relief System Force Balance Assessment – a Review of Its Effectiveness

**Abstract:**

Design documentation requirements for spring-loaded pressure relief systems are very detailed, based on the definition in API Standard 521, 6th edition. The Installation Documentation section of API Standard 521 recommends provision of inlet and outlet piping pressure drop calculations as part of a complete analysis.

Inlet pressure drop is an important factor, since it is considered that excessive inlet pressure drop will affect pressure relief valve performance, and potentially result in chatter which may damage the valve and piping. In situations where inlet pressure drop exceeds 3% of set pressure for spring loaded pressure relief valves, previous editions of API Standard 520 Part 2 have allowed for an "engineering analysis" to be conducted for these systems - however the content of the engineering analysis has never been well defined.

The latest version of API Standard 520 Part II provides a more detailed description of an engineering analysis and allows a Force Balance Assessment to be conducted for spring loaded pressure relief valves exceeding 3% inlet pressure drop. This method can predict valve stability during relief flow, and identify whether chatter, flutter or stable flow is likely to occur. Valve characteristics such as valve opening and closing times are critical input parameters for this analysis. If stable operation can be demonstrated, then expensive mitigation efforts, such as modifications to piping or the valve itself, can be avoided.

This paper presents an overview of the force balance assessment in a step-by-step manner. Additionally, a sample of spring-loaded pressure relief systems which exceed the 3% inlet pressure drop criteria is reviewed. The aim of this review is to determine the effectiveness of conducting a force balance assessment, and to identify which systems show stable operation, and which systems were identified as likely to chatter.

# 618390 Chemicals Management inside a Refinery - Does Everyone Knows the Hazards of the Chemicals on Site?

**Abstract:**

Chemicals are widely used in Oil&Gas industries. Each of them has different phsyical and chemical properties and these differences make it hard to store and use them in a proper way.

In the refineries, those widely used chemicals require a good storage and transportation standards. In order to create a safe environment for those areas, Tüpraş has created its own "Chemical Interaction Matrix". In this matrix, various aspects of chemicals are handled and each Unit has its own matrix.

As a starting point, each of the chemicals are checked as per their stability and reactivity properties from their safety data sheets. The chemicals that do not have detailed information are reacted to each other in the laboratory in order to see if there is a dangerous reaction.

Chemicals that may cause a dangerous reaction to people and environment are stored seperately and the properties of each are shown in the matrixes. These matrixes are hanged on the unit boards so that everyone can see and check if the storage is provided correctly.

It is important to keep these matrixes up-to-date. The matrix is updated when a new chemical arrives in the unit and everybody is informed about update. The update and information is achieved wit the actions that we tracked via our MoC (Management of Change) system.

# 617653 Operational Risks Involved on Integrating Flare Gas to Fuel System


**Abstract:**

Flaring is a safe industrial process to dispose waste or excess gas from a plant. But, it has limitations in energy conservation and environmental loss. Flare gas integration to cogeneration plant benefits from utilizing high calorific value waste gases as supplemental fuel to cogeneration plant or gas turbines. Earlier studies on flare gas integration have shown environmental incentive and economic viability. However, the impact of flare gas quality on the operation of cogeneration plants are yet to be identified.

This study delves into the operational hazards and risks arising from the variability of flare gas stream. A novel framework is proposed which can be applied to new or integrated processes which has inadequate information on its operational performance and associated hazards. The study identifies the technical hazards through design data collection, process simulation and sensitivity analysis. Followed by a systematic approach to evaluate the critical operational event occurrences and their impacts through scenario development and quantitative risk assessment.

An important outcome from this study is the identification of critical flare-fuel gas parameters affecting the cogeneration operation through process simulation. High hydrogen content and rich fuel-air ratio in the boiler can affect the boiler operating conditions. Increase in the hydrogen content in fuel system can increase the boiler gas temperature, combustion mixture and flame stability inside the firebox. Risk analysis through Bayesian Network showed a significant risk escalation. With a presumed frequency of 12 hrs/year of high and medium hydrogen gas flaring, there is a substantial rise in the occurrence of boiler gas temperature exceeding design limit and rich fuel mixture formation in the firebox. These top events has major influence on the operational time as well as loss control events - flame impingement and tube rupture.

# 618388 Process Hazard Analysis Methodologies and Refinery Applications

**Abstract:**

Process Safety Management System starts with managing the process related risks and from the first step of establishing a new process unit that is needed to evaluate the risks and define the necessary measurements. Process Hazard Analysis methodologies are the systematic tools which help to evaluate the risks in a process. For different needs there are several methodologies like DOW FETI, HAZOP, What if, HAZID, Fault Tree and Event Tree Analysis, Scenario Modelling, Facility Siting etc. Some of these methodologies are semi quantitative some are quantitative methodologies. In order to define and prevent major industrial accidents in our refinery we use DOW FETI, HAZOP, FTA and ETA and Scenario modelling. Moreover for some kind of projects we may also apply What if and HAZID analysis instead of HAZOP depending on the complexity. Facility Siting is another tool for risk related studies which is used to evaluate the risks for buildings in order to protect people inside the buildings.

The workflow for a full unit PHA study starts with DOW FETI calculation which is used for risk ranking of al static equipment inside the unit. DOW FETI calculation is done in accordance with the equipment temperature, pressure, hold-up amounts, components and also reaction types if there is any. Then risk assessment process continues with HAZOP studies which is for scenario identification. In HAZOP, first step is chopping the unit into small parts to be evaluated which is called nodes. After that all nodes are evaluated using P&IDs and PFDs by keywords like temperature, pressure, flow, level etc. HAZOP helps to understand what kind of scenarios are possible inside the unit and these critical scenarios are taken through FTA and ETA studies. HAZOP is a semi quantitative risk assessment method where FTA and ETA are quantitative methods. FTA is applied for critical scenarios in order to find out the probability of top event then ETA is then applied in order to find out the probability of scenarios like fire, explosion, toxic dispersion etc. caused by this top event. FTA and ETA analysis all use the statistical PFD and failure rate values of all initiating events and barriers defined in HAZOP for the related scenario.

Scenario Modelling is the following step which is used for understanding the effect of the probable scenario. The most probable scenarios coming from ETA is modelled to see the effect zones. These effect zones are used to plan Emergency Response Studies for all these potential major accidents.

Facility Siting is another check point in a unit PHA study all potential scenario modelling results are used to evaluate the risks for the buildings and also if there are any in the effect zones of fire, explosion or toxic release then there should be actions defined.

# 619040 Identification of the Process Hazards By Way of Technical Integration:


**Abstract:**

Undisclosed threats can lead to **death** and **devastation** in the process industry. Right and constructive hazard recognition supports security procedures, employment health and risk control techniques. It thereby is essential to the health and security of all company employees and to effective business practices.

Without understanding of process hazards, risk cannot be controlled and risk assessment is compulsory for incident prevention. Managers want their workers to be able to recognize not only apparent but evolving threats effectively. Risk assessment needs various **strategies**, **skilled** and testing teams to identify risks and create obstacles to mitigation in their operating environment. Different approaches can enhance the detection of hazards to the risk assessment process.

This paper explains the approach used in the implementation of combined **traditional** and **non-conventional**, **qualitative** and **quantitative** approaches, to identify process hazards, risk assessment and risk management. An insight on the non-conventional techniques *(Temporary repairs reduction & effective permit to work system)* will enable conference audience from all sectors of process industry, in developing integrated process hazard identification and evaluation

# 617089 Keep the Lion inside the Cage


**Abstract:**

*Keep the Lion inside the Cage*

All Chemicals, that we utilize to produce our final Products, are considered as Lions. We have to maintain their healthiness in good condition & under control inside the Plant (Cage) elements (Pipe, Equipment, Vessel, Tank). Process Safety Engineer's Career is an adventure where all safeguards, Barriers & mitigations shall be designed, constructed & operated safely in order to achieve the Process Safety ultimate Goal: "Prevent Loss of Containment". When the chemical leaks, the Lion is free to have high Consequences. With integration for all Reliability, Process Safety & Asset Integrity, we can keep the Lion inside the Cage & eliminate all Probability to have any Consequence whatever its Risk Level.

In order to achieve our ultimate goal, several aspects shall be managed & steward-shipped by the own Organization. There are many Process Safety pillars starting from "hazards identification" till removing the Risk or reducing it to insignificant level; including The Process Safety Competency Development for all Organization levels, Process Safety Knowledge Enhancement, System Support, Technical Guidelines, Risk Management, Process Safety Performance Monitoring and end with the Emergency Preparedness & Response Capability.

The beauty of the Process Safety Concept is the alignment with the Emotional Intelligent that we feel during the dealing with Chemicals (Lions). They shall be always contained & maintained inside their containments. In the same manner of having a specific Process Safety Competency Development Plan for All members of Organization to maintain the Chemicals inside the containment, it shall be available for all members in order to have the Lions inside the Cage. Our ultimate goal is to complete the Journey in safe manner & to maintain the Healthy condition ... in order to keep "Chemicals" inside the pipe & to keep the Lions inside the cage.

SPRING21
+17TH GCPS
A Joint AIChE and CCPS Meeting

# Human Factors Elements Missing from Process Safety Management (PSM) Systems

**William Bridges**
**President**
**Process Improvement Institute, Inc.**
**1321 Waterside Lane,**
**Knoxville, TN 37922 USA**
**wbridges@piii.com**

**William C. Rhodes II, P.E.**
**Process Improvement Institute, Inc. (PII)**
**82 Via Santa Maria**
**San Clemente, CA 92672**
**wrhodes@piii.com**

PROCESS IMPROVEMENT INSTITUTE


Prepared for Presentation at
American Institute of Chemical Engineers
2021 Spring Meeting and 17th Global Congress on Process Safety
Virtual
April 18 - 22, 2021

# Human Factors Elements Missing from Process Safety Management (PSM) Systems

**William Bridges**
**President**
**Process Improvement Institute, Inc.**
**1321 Waterside Lane,**
**Knoxville, TN 37922 USA**
**wbridges@piii.com**

**William C. Rhodes II, P.E.**
**Process Improvement Institute, Inc. (PII)**

**Keywords:** human factors, process safety, PSM, human error, risk-based process safety, RBPS, conduct of operations, operational discipline

**Abstract**

Process safety is about controlling risk of failures and errors; controlling risk is primarily about reducing the risk of human error. All elements of Risk-Based Process Safety (RBPS) and alternative standards for process safety (such as US OSHA's standard for Process Safety Management [PSM] or ACC's Process Safety Code™ [PSC]) have many elements, and each of these in turn helps to reduce the chance of human error or else helps to limit the impact of human error. But each process safety standard has some weakness in the control of human error.

This paper presents an overview of human factor fundamentals, discusses why many PSM systems are weak on human factors and outlines a comprehensive process safety element on Human Factors. It describes what belongs in each category within the Human Factors element and explains the intent, content, and the benefit of each category. The paper also presents examples of Human Factors' deficiencies and selected examples of industry practices for human factors control are provided. This paper builds on earlier papers, starting from 2010, on the same topic.

## 1 Introduction

All accidents (or nearly all, if you consider that there are some natural phenomena that we either cannot guard against or choose not to guard against) result from human error. This is because humans govern and accomplish all of the activities necessary to control the risk of accidents. Humans influence other humans in the process – not only do humans cause accidents (unintentionally) by making errors directly related to the process itself, but they also cause errors by creating deficiencies in the design and the implementation of management systems (i.e., we make errors in authorities, accountabilities, procedures, feedback, proof documents, continual improvement provisions). Ultimately these management systems govern the human error rate directly contacting or directly influencing the process. The process-related activities where errors have the most influence include:

- Designing of a process
- Engineering of a process
- Specifying the process components
- Receiving and installing equipment
- Commissioning
- Operating the process
- Predicting safeguards necessary to control the risk at an acceptable level and sustaining these safeguards for the life of the process
- Maintaining, inspecting and repairing the process
- Troubleshooting and shutting down the process
- Managing Process Changes

**99% of accidental losses (except for natural disasters) begin with a human error** ***(supported by data from more than 1500 investigations)***

**Root causes of accidents are management system weaknesses**
***(Center for Chemical Process Safety, American Institute for Chemical Engineers, "Guidelines for Investigating Chemical Process Incidents", 2003) – OSHA agrees***

**Weak Management Systems → Human Error →Accidents**

**Figure 1: Relationship between Accidents and Management Systems**

Recent major accidents have highlighted the need for increased focus on Human Factors. The US Chemical Safety Board (CSB) cited human factor deficiencies as one of the main contributors of the catastrophic accident at the BP Texas City Refinery in March 2005. The human factor deficiencies included lack of control of worker fatigue, poor human-system-interface design, poor communication by radio/phone, out-of-date and inaccurate operating procedures, and poor (no) communication between workers at shift handover. The CSB cited similar issues from many other accidents, including the Chevron Richmond Refinery fire in 20212, and has urged industry and the US OSHA (the regulator) to pay much more attention to human factors. This led to the OSHA National Emphasis Program for Refineries to include human factors as one of the 12 core elements it reviewed in detail across many of the 148 oil refineries in the USA around 2010.

Implementing human factor engineering and policies to prevent accidents is not a new concept. Nearly all (or all, from a more complete perspective) of the causes and root causes of major accidents in the past 40 years have been the result of poor control of human factors. This has been cited in many root causes analysis reports and papers concerning these major accidents.

Process Safety Management (PSM) systems based on OSHA's PSM standard are likely lacking the fundamental human factor standards that if applied across the applicable PSM elements, would work together to reduce human error. The *Risk Based Process Safety (2007)* industry standard from the CCPS/AIChE does contain the human factor standards, but these are presented under at least six (6) of the PSM elements instead of under a stand-alone human factor element. This guideline does not provide a needed road map to help companies transition to RBPS from the minimum PSM systems defined in OSHA's PSM standard. A starting point for this transition should be implementing a human factor element comprised of the human factor categories missing from most PSM systems (especially from those based on the OSHA PSM standard).

***Definitions for Use in This Paper***

- This paper will use the term **Human Error** to mean the errors that are made during direct interface or direct influence of the process.
- **Human Factors** are those aspects of the process and related systems that make it more likely for the human to make a mistake that in turn causes or could cause a deviation in the process or could in some indirect way lead to the increased probability of an accidental loss.
- **Management systems are** the administrative controls an organization puts in place to manage the people and workflow related to the process under consideration, and so these inherently attempt to control human factors

## 2. Human Error Fundamentals

### Types of Human Error

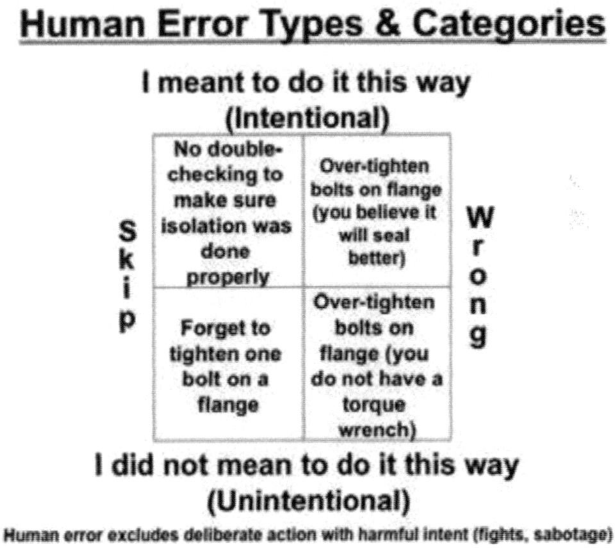


**Figure 2: Types and Categories of Human Error**

In simplest terms, there are only two types of human error: *Errors of Omission* (someone skips a required or necessary step) and *Errors of Commission* (someone performs the step wrong). But in addition, these errors occur either inadvertently (unintentional error) or they occur because the worker believes his or her way is a better way (intentional error, but not intentional harm). Intentional errors can usually be thought of as errors in judgment. Some believe a "lack of awareness of the risk" causes these errors, but in actual practice, the worker commits an intentional error may wellbe aware of the risk. They instead believe they know a better way to accomplish a task or they believe there are already too many layers of protection (so bypassing one layer will not cause any harm).

Regardless of type or category of human error, the organization can and should exert considerable control of the errors.

***Controlling Human Error through Management Systems***

A process is a combination of the utilities, raw materials, and human actions (direct actions and those actions involving programming the process to accomplish automatic functions). If anything goes wrong with these Inputs, or if there are basic design flaws or basic fabrication flaws in the process, then the Outputs will not be desired. The desired output is acceptable (or high) production rates at acceptable or higher quality factors with no harm to the humans (long term or short term), no harm to the environment, and with acceptable (or higher) life of the process components. The negative outcomes resulting from humans failing to control the raw material quality, failing to control the utility levels consistently, making errors directly related to the operation of the process, or making errors in the care of the process (such as maintenance) will result in lower production, lower quality, higher number and severity of safety-related accidents, and more negative impact on the environment. The potential (probability) of the negative outcomes is collectively referred to as ***business risk*** – more precisely, the risk is a product of the likelihood of one or more of these negative outcomes and the severity of each outcome.

A critical concept is therefore:

***If an organization does not directly control risk, the organization cannot directly control quality, safety, environmental impact, or production to acceptable levels. An organization must sustainably control human error to manage the risk of accidental losses that impact quality, safety, the environment, production, or assets.***

In order to sustainably control the risk of a complex process (such as an oil/gas operation, refinery, chemical plant, steel plant, automobile manufacturing, aircraft manufacturing, etc.), the organization must design and implement management systems to:

- **Understand the Risk** – This involves predicting problems; which in turn includes predicting the risk of possible accident/loss scenarios, establishing the appropriate design and the right layers of protection to control risk to a tolerable level
- **Control Risk Factors Day-to-Day** – This involves controlling the original design by maintaining the established layers of protection and managing changes to the design using integrated management systems
- **Analyze Actual Problems and Determine Weaknesses in the System** – This involves identifying weaknesses in designs and management systems and weaknesses in risk understanding through root cause analysis of actual problems (losses and near-losses)

These three elements are illustrated in the following figure:

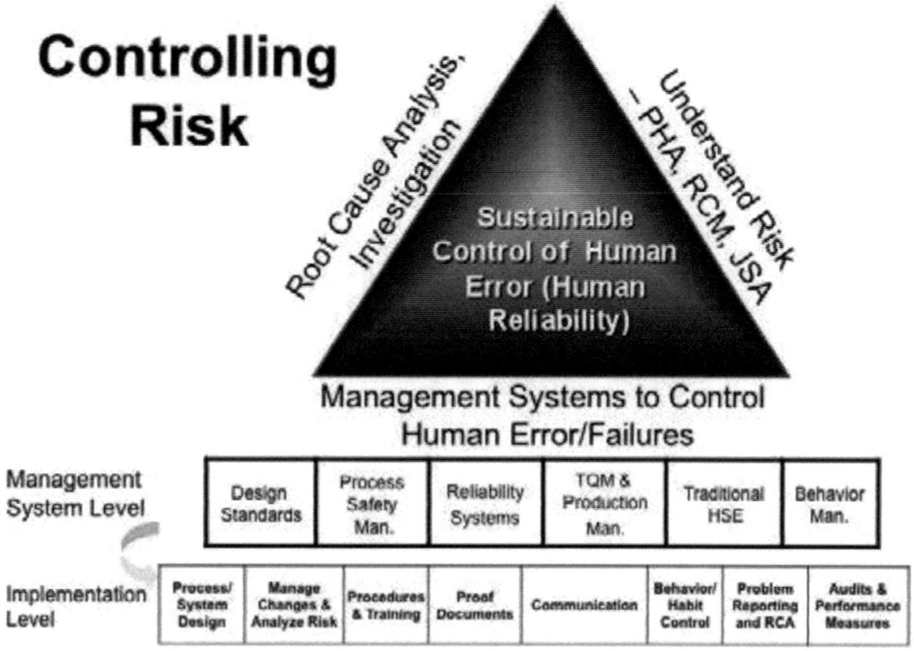


*© Process Improvement Institute, Inc. (2004-2021)*

**Figure 3: Controlling Risk through Human Reliability**

Management systems control the interaction of people with each other and with processes. They are the high level procedures we use to control major activities like conducting PHAs, management of change, writing operating procedures, training employees, evaluating fitness for duty, conducting incident investigations, etc. If management systems are weak, then layers of protection will fail and accidents will happen.

To reiterate, accidents are caused by human error. In general, Process Safety Management (PSM) is focused on maintaining these human errors at a tolerable level because:

- All accidents happen due to errors made by humans; including premature failure of equipment. There are a myriad of management systems to control these human errors and to limit their impact on safety, environment, and quality/production
- When these management systems have weaknesses, near misses occur
- When enough near misses occur, accidents/losses occur

## 3. Human Factor Categories and Typical Impact of Each

To minimize human error, process safety systems should address the **Human Factors Categories** (see various US NRC and US DOE standards from 1980s and 1990s)[1,2]. The table on the next page lists the key human factor categories along with multiplication factors that poor human factors can have on the base human error rates. If all of these human error rates are controlled very well, then the optimized human error rates listed in the next section are achievable.

## TABLE 1: SUMMARY of 10 HUMAN FACTOR CATEGORIES

***Based in part on:*** Gertman, D.; et. al., *The SPAR-H Human Reliability Analysis Method*, NUREG/CR-6883, U.S. Nuclear Regulatory Commission, Office of Nuclear Regulatory Research, Washington, DC, August 2005. PII has modified the list slightly to account for general industry data and terminology and to incorporate PII internal data.

| Human Factor Category | Human Factor Issue/Level | Multiplier for Cognitive & Diagnosis Errors |
|---|---|---|
| **Available Time** | Inadequate time | P(failure)=100% |
| (includes staffing Issues) – ***for responses only*** | Barely adequate time (≈2/3 x nominal) | 10 |
| | Nominal time (1x what is expected) | 1 |
| | Extra time (at least 2x nominal and >20 min) | 0.1 |
| | Expansive time (> 4 x nominal and > 20 min) | 0.01 |
| **Stress/Stressors** | Extreme (threat stress) | 5 |
| (includes staffing issues) | High (time pressures such as during a maintenance outage; issues at home, etc.) | 2 |
| | Nominal | 1 |
| **Complexity & Task Design** | Highly complex | 5 |
| | Moderately complex (requires more than one staff) | 2 |
| | Nominal | 1 |
| | Obvious diagnosis | 0.2 |
| **Experience/Training** | Low | 10 |
| | Nominal | 1 |
| | High | 0.5 |
| **Procedures** | Not available in the field as a reference, but should be | 20 |
| | Incomplete; missing this task or these steps | 8 |
| | Available and >90% accurate, but does not follow format rules ***(normal value for process industry)*** | 3 |
| | Good, 95% accurate, follows >90% of format rules | 1 |
| | Diagnostic/symptom oriented | 1 |
| **Human-Machine Interface (includes tools)** | Missing/Misleading (violates populational stereotype; including round valve handle is facing away from worker) | 20 |
| | Poor or hard to find the right device; in the head calc | 10 |
| | Some unclear labels or displays | 2 |
| | Good | 1 |
| **Fitness for Duty** | Unfit (high fatigue level (>80 hrs/wk or >20 hr/day, no day off in 7-day period; or illness, etc.) | 20 |
| | Highly degraded fitness (high fatigue such as >15 hr/day, illness, injury, etc.) | 10 |
| | Degraded Fitness (>12 hr day and >72 hr/wk) | 5 |
| | Slight fatigue (>8 hr per day; ***normal value for process industry***) | 2 |
| | Nominal | 1 |
| **Work Processes & Supervision** | Poor | 2 |
| | Nominal | 1 |
| | Good | 0.8 |
| **Work Environment** | Extreme | 5 |
| | Good | 1 |
| **Communcation** | No communication or system interference/damage | 10 |
| | No standard for verbal communication rules ***(normal value for process industry)*** | 3 |
| | Well implemented and practiced standard | 1 |

If all of these human error factors are controlled very well, then the optimized human error rates listed in Section 5. of this document are achievable. Alternatively, if one or more of these factors are compromised, the human error rate will increase by the values shown in this table. As an example, an individual whose fitness for duty rating is unfit due to the excessive work hours shown in the table will make errors at a rate 20 times greater than an individual whose fitness for duty is normal.

With excellent control of each of the human factors listed above, a company can begin to approach the lower limits that have been observed for human error. These lower limits are about:

- **1 mistake in 100 steps for most procedures-based tasks** (such as starting up a process unit), a little less for a routine (daily) task that becomes almost a reflex

- **1 in 10 chance or a little better for diagnosis and response to a critical alarm**

Excellent control requires superior design and implementation of management systems, which is enabled through a thorough understanding of these factors, as outlined below.

## 4. Incorporating Human Factors into PSM Systems

The basis upon which most companies build their PSM system is their applicable PSM regulation (e.g., the OSHA PSM standard found at 29 CFR 1910.119 in the United States.) Unfortunately, this standard, which was promulgated in 1992 (and has been essentially unchanged since), is devoid of human factor controls with a few exceptions as outlined by Bridges and Tew[3]. The only direct reference to the term "human factors" is mentioned in paragraph (e), Process Hazard Analysis (PHA), which states that the PHA team must consider human factors (presumably in the review of the causes and the quality of the safeguards). The other mention that alludes to human factors, is in Operating Procedures (see paragraph (f)) which states procedures "must be written clearly and understandably." Similar PSM regulations in other countries are likewise silent on human factors.

The next logical place to seek guidance for incorporating human factors into a company's PSM system is industry-specific standards; therefore, some companies may use these to build upon the regulatory framework. In the chemical process industry, such a standard was published in 1985 by the Center for Chemical Process Safety (CCPS), which is a division of the American Institute of Chemical Engineers (AIChE). This standard included a strong emphasis on human factors since Human Factors was one of its original 12 elements. However, in 2007 the CCPS revised their standard as "Guidelines for Risk-Based Process Safety"[4] (RBPS). This revision abandoned the global element on human factors, and instead spread the direct control of human factors throughout 6 elements. The authors believe that this change not only diminishes the importance of human factors, but also reduces the ability of companies to effectively implement human factors throughout their PSM program.

The CCPS partially addressed this shortcoming by publishing a guideline entitled "Conduct of Operations and Operational Discipline"[5] (COO/OD) in 2011. As outlined in Section 8 of this paper, this newer guideline provides a great deal of practical and useful information for implementing *most* human factors; however, the information contained within it is still incomplete. Therefore, to fully integrate human factors into a PSM program, companies should consult the detailed review of each human factor outlined in Section 5 of this paper to develop the necessary tools to fully implement human factors into their PSM systems.

## 5. Details on Incorporating Human Factors Within Existing PSM Systems

### *A. AVAILABLE TIME and STRESS / STRESSORS*

The **Available Time** for the task refers to the time the task is expected to be required for completion. In many cases the time determined for a process is determined using the theoretical times where the conditions are "ideal" and not necessarily realistic. When task and process duration are not realistic people tend to "find the best and easy way" to get it done, ultimately creating latent conditions for error occurrence. **Work Stress** refers to the emotional response that arises when work demands exceed the person's capacity and capability to cope. Some processes have been designed paying little attention to job design, work organization and management systems. Failing to fulfill this requirement eventually results in work stress.

Staffing levels directly impact both factors. Staffing is the process of assessing, maintaining and scheduling personnel resources to accomplish work. An adequately staffed organization ensures that personnel are available with the proper qualifications for both planned and foreseeable unplanned activities. Staffing is a dynamic process in which plant management monitors personnel performance to ensure that overall organizational performance goals are met or exceeded. The result of an effective staffing process is a balance between personnel costs and the achievement of broader organizational goals.

Three key issues must be considered when staffing decisions are made; the first of which is selecting the right staff for a job. Each organization requires the proper amount and type of expertise to safely and competently operate the plant under a variety of conditions. The term "expertise" includes the attributes of talent, effectiveness, knowledge, skills, abilities, and experience necessary to operate and maintain plant systems, structures and components.

The second key issue is avoiding staff overload. Surges in workload, such as during outages, typically require staff augmentation as well as longer work hours for permanent staff. The introduction of contractor personnel or company personnel from other sites may increase the likelihood of errors due to unfamiliarity with the plant, its procedures and hardware, for example. Longer work hours have the potential to increase fatigue, which also contributes to the likelihood of error.

The third key factor is rotating staff every 1 hour or less for tasks that require high vigilance. Humans are inherently unable to remain alert for signals that seldom, if ever, occur. Even a sailor whose life is at stake cannot maintain an effective watch (look-out) for hostile submarines for more than 30 minutes or so. The following figure illustrates the rapid decrease of vigilance with time.

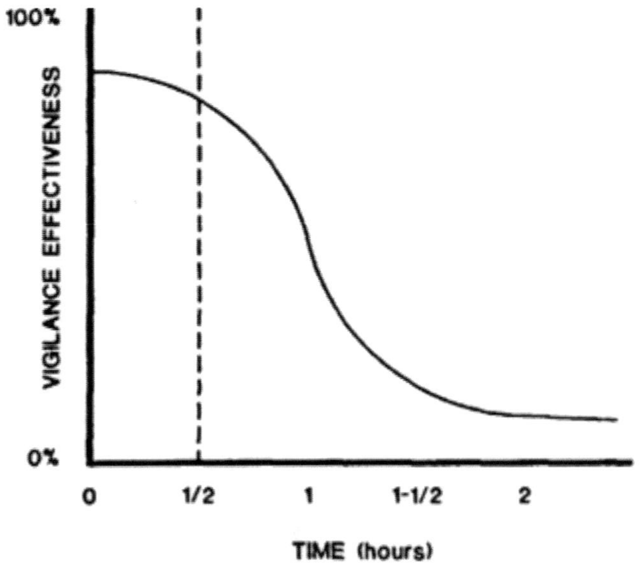


**Figure 4: Vigilance as a Function of Time**

Placing a worker in situations requiring extended, uneventful vigilance may lead to accidents. Therefore, it is important to design control systems in a manner that requires regular operator interaction so that the operator will remain attentive.

## *B. TASK DESIGN*

A task that is designed with the human limits in mind is much more likely to work effectively than one that assumes humans can and will "always" do what is written. The task must consider that humans think and remember and factor in prior data and prior experiences. Factors to consider in assessing **task design** adequacy (i.e., minimizing the likelihood of human error during its completion) include:

- Complexity of the task
- Probability of repeat errors (coupled error) on redundant aspects of the design
- Error Detection and Error Recovery

**Complexity of task** (procedure-based or a call for action) – If the task is too complex, then humans can forget their place in the task, fail to understand the goal of each step or sub-step, or fail to notice when something isn't going right. Task complexity is a function of:

- number of choices available for making a wrong selection of similar items (such as number of similar switches, number of similar valves, number of similar size and shaped cans)
- number of parallel tasks that may distract the worker from the task at hand (leading to either an initiating event or failure of a protection layer)
- number of staff involved (more staff = more complex)
- number of adjustments necessary to achieve the goal
- amount of mental math required (as a rule, NO math should be required in anyone's head when accomplishing a standardized task)

- how much judgment is required to know when you have accomplished each goal within the task

  For most chemical process environments, the complexity of the task is relatively low (one action per step), but for response actions (in which humans serve as the Independent Protection Layer (IPL)), there are almost always other tasks underway when the out-of-bounds reading or the alarm is activated. Complexity is difficult to predict (since it is not known when a human intervention will be needed), but higher complexity can increase error rates by 2 to 10 times.

**<u>Probability of repeat errors</u>** (Coupled errors) – For many maintenance tasks, making a repeat error or "common cause error" or "dependent error" can lead to failures of all backup systems. Such errors have led to MANY airplane crashes and major process safety accidents as well.

Coupling represents the probability of repeating an error (or repeating success) on a second identical task, given that an error was made on the first task. The increased probability of failure on subsequent tasks given that an error has already been made is known as dependence. The list below provides some starting point guidance on values to use:
- 1/20 to1/90 – if the same tasks are separated in time and if visual cues are not present to re-enforce the mistake path. *This error rate assumes a baseline error rate of 1/100 with excellent human factors. If the baseline error is higher, then this rate will increase as well.*
- 1/2 – if the same two tasks are performed back-to-back, and if a mistake is made on the first step of two. *This error rate assumes a baseline error of 1/100 with excellent human factors. If there the baseline error is higher, then this rate will increase as well.*
- 8/10 to 10/10 – if the same three tasks are performed back-to-back and strong visual cue is present (if you can clearly see the first devices you worked on), if a mistake is made on the first step of the two or more
- Two or more people become the same as one person (with respect to counting of errors from the group), if people are working together for more than three days; this is due to the trust that can rapidly build.

These factors are based on the relationships provided in NUREG-1278[1] and the related definitions of weak and strong coupling provided in the training course by Swain (1993)[1] on the same topic, as shown here in Table 2. The following relationship is for errors of omission, such as failing to reopen a root valve or failing to return a safety instrumented function (SIF) to operation, after bypassing the SIF. The qualitative values in Table 2 are based jointly on Swain[1] (1993) and Gertman (SPAR-H, 2005 which is NUREG/CR-6883)[2].

One can readily conclude that staggering of maintenance tasks for different channels of the same SIF or for related SIFs will greatly reduce the level of dependent errors. Unfortunately, most sites PII visits do not stagger the inspection, test, or calibration of redundant channels of the same SIF or of similar SIFs; the reason they cite is the cost of staggering the staff. While there is a perceived short-term higher cost, the answer may be different when lifecycle costs are analyzed.

**Simple Rule: Staggering of maintenance can prevent significant number of human errors in redundant channels. In fact, the US Federal Aviation Administration (FAA) requires staggering of maintenance for aircraft with multiple engines or multiple control systems (i.e.,**

**hydraulics) (*FAA Advisory Circular 120-42A*, as part of Extended Operations (ETOPS) approval).[6]**

**Table 2: Guideline for Assessing Dependence for a within-SIF Set of Identical Tasks (based partially on SPAR-H, 2005 [1,2], and partially on field observations by PII)** Courtesy Process Improvement Institute, Inc., All Rights Reserved

| Level of Dependence | Same Person | Actions Close in time | Same Visual Frame of Reference (can see end point of prior task) | Worker Required to Write Something for Each Component |
|---|---|---|---|---|
| Zero (ZD) | No; the similar tasks are performed by different person/group | Either yes or no | Either yes or no | Either yes or no |
| Zero (ZD) | Yes | No; separated by several days | Either yes or no | Either yes or no |
| Low (LD) | Yes | Low; the similar tasks are performed on sequential days | No | Yes |
| Moderate (MD) | Yes | Moderate; the similar tasks are performed more than 4 hours apart | No | No |
| High (HD) | Yes | Yes; the similar tasks are performed within 2 hours | No | No |
| Complete (CD) | Yes | Yes; the similar tasks are performed within 2 hours | Yes | Either yes or no |

Once the level of dependence is known, the probability of either repeat success or repeating errors on identical tasks can be estimated. For these probabilities, we use Table 3, which is a re- typing of Table 20-17 from NUREG-1278[1] (and the similar table in SPAR-H [Gertman, 2005][2]).

**Table 3. Equations for Conditional Probabilities of Human Success or Failure on Task N, given probability of Success (x) or Failure (X) on Task N-1, for Different Levels of Dependence** *Courtesy Process Improvement Institute, Inc., All Rights Reserved*

| Level of Dependence | Repeating Success Equations (but shown as error probability) | Repeating Failure Equations |
|---|---|---|
| Zero (ZD) | $P_{Success@N} = x$ | $P_{Failures@N} = X$ |
| Low (LD) | $P_{Success@N} = (1+19x)/20$ | $P_{Failures@N} = (1+19X)/20$ |
| Moderate (MD) | $P_{Success@N} = (1+6x)/7$ | $P_{Failures@N} = (1+6X)/7$ |
| High (HD) | $P_{Success@N} = (1+x)/2$ | $P_{Failures@N} = (1+X)/2$ |
| Complete (CD) | $P_{Success@N} = 1.0$ | $P_{Failures@N} = 1.0$ |

**Error detection and error recovery** – Is there enough feedback in the process to allow the worker to realize (in time) that they made a mistake? Have they been trained on how to reason through how to recover from mistakes they or others make? (Sometimes, doing a step too late is far worse than NOT doing the step at all.) Is there enough time available for the type of intervention necessary?

## *C. EXPERIENCE/TRAINING*

**Training** is necessary for general functioning of management systems and for task-specific skills such as how to start up a compressor, repair a pump, lead a root cause analysis, perform a proper Lock-Out/Tag-Out, etc. Organizations normally do a good job of training, including hands-on training. Training systems can be weak (inadequate) however if the training system does not adequately:

- Address how to troubleshoot the process (i.e., handle process deviations or upsets).
- Provide workers with a mental model(s) of the process so they can perform the assigned tasks correctly, diagnose process upsets properly and quickly, and understand the consequences of their actions.
- Provide workers with enough practice of critical tasks

Ironically, training is many times listed as the "cause" or "root cause" of an accident, when in fact the training is adequate, and some other human factor is the major cause.

## *D. PROCEDURES*

Many procedures do not follow best practices for controlling human error, and so the written procedure actually "contributes" to increased error rates. Further, many organizations are missing guides on how to troubleshoot (what to do when process deviations occur). The best practice rules for writing and validating procedures have been published for many years ***(see Bridges, 1997-2010)***[3, 7, 8]. Below is a checklist based on the current set of best practice rules for developing operating, maintenance, and other work-instructions (procedures):

**TABLE 4: PROCEDURE QUALITY CHECKLIST (courtesy PII, 2008)**

| # | Issue | Response |
|---|---|---|
| | **Procedure Content Checklist** | |
| 1 | Is the procedure drafted by a future user of the written procedure? (Engineers should not author procedures to be used by operators or maintenance staff.) | |
| 2 | Is the procedure validated by a walk-down in the field by another future user of the procedures? | |
| 3 | Is the procedure reviewed and commented on by technical staff (engineers or vendors)? | |
| 4 | Is the procedure checked versus the Page and Step format rules below? | |
| 5 | Is a hazard review of step-by-step procedures performed to make sure there are sufficient safeguards (IPLs) against the errors that will occur eventually (when a step is skipped or performed wrong)? | |

| # | Issue | Response |
|---|---|---|
| 6 | Is the content measured using "newly trained operators" to judge the % of steps that are missing, steps that are confusing or wrong, and steps that are out-of-sequence? (A score of 95% accuracy of content is good.) | |
| | **Page Format Checklist** | |
| 1 | Is the title of the procedure the largest item on the page? | |
| 2 | Is the procedure title clear and consistent with other titles, and does it uniquely describes the topic? | |
| 3 | Are the document control features the smallest items on the page? | |
| 4 | Are temporary procedures clearly identified? | |
| 5 | Is white space used effectively?<br>• Is there one blank line between each step?<br>• Does indentation help the user keep their place?<br>• Are the margins large enough to reduce page congestion? | |
| 6 | Is type size 12 point font or larger? | |
| 7 | Is serif type used (rather than sans-serif)? | |
| 8 | Is mixed case used for words of steps, with ALL CAPS used only for special cases (such as IF, THEN, AUTO, and WARNING)? | |
| 9 | Is the step number very simple (single level of number used)? Only an integer? | |
| 10 | Have sections or information not necessary to performing the steps been moved to the back or to another part of the manual or training guide? | |
| 11 | Are section titles bold or larger than the text font? Do sections have clear endings? | |
| 12 | Is the decision on electronic presentation versus hard copy correct? Are electronic linkages to procedures clear and accurate and easy to use? If paper is chosen, is the color of the paper appropriate? | |
| 13 | Is the overall page format (such as Outline format or T-Bar format) appropriate to the use of the procedure? | |
| 14 | Are play script features added for tasks that must be coordinated between two or more users?<br>• Play script is normally used when there are two or more hand-offs of responsibility for steps. | |
| 15 | Are rules followed for formatting of Warnings, Cautions, and Notes? (See annotated rules, such as Warnings are for worker safety and Warnings must clearly standout from rest of page.) | |
| | **Step Writing Checklist** | |
| 1 | Is each step written as a command? | |
| 2 | Is the proper level of detail used throughout? This is judged based on:<br>• *Who* will use the procedures<br>• Same level of detail used in similar procedure steps | |
| 3 | On average, is there only one implied action per instruction? Best practice is to average 1.2. | |
| 4 | Does the procedure indicate when sequence is important?<br>• If sequence matters, each step should be numbered (with an integer or letter)<br>• If sequence does not matter, bullet lists should be used | |
| 5 | Are only common words used? Apply "education" level test (5 grade reading level is best) | |
| 6 | Do all acronyms, abbreviations, and jargon aid understanding?<br>• Develop a list of such terms for use in procedures *and* communication.<br>• Use terms that users use (within reason) | |
| 7 | Is each step specific enough? No room left to guess/interpret:<br>• The meaning of a word or phrase (Check vs. Make sure)<br>• The intent of a step or series of steps<br>• A desired quantity or value<br>• To what equipment the step applies | |
| 8 | Is the procedure free of steps that require in-your-head calculations?<br>• Values expressed as ranges rather than targets with error bands<br>• Conversion tables, worksheets, or graphs provided where needed | |

| # | Issue | Response |
|---|---|---|
| 9 | Are graphics to the user's advantage?<br>• No explanatory paragraphs or lengthy instructions that could be replaced by a picture<br>• No impressive graphics that provide no real advantage | |
| 10 | Are references to the user's advantage?<br>• No lengthy explanations or instructions that could be replaced by branching to a reference<br>• No references to a procedure that references still another<br>• No gaps or overlaps between this procedure and a referenced document<br>• If branching, must branch to a procedure, not to a specific step in a procedure | |
| 11 | Are rules followed for writing warnings, cautions, and conditional steps?<br>• No more than 2 per page<br>• No actions within a warning or caution (actions must always be numbered steps)<br>• Warnings and Cautions contain descriptions of potential consequences | |

For procedures to be effective in ensuring that tasks are performed correctly, they must be used. There are many reasons that workers may not use procedures. ***Deficient Procedures*** are the most prevalent problem in process industries since procedures have not traditionally been developed from the perspective of optimizing human factors; instead, procedures have been traditionally developed to meet a compliance requirement to have written procedures. Examples of procedure deficiencies (inaccuracies) include:

- Incorrect/incomplete/nonexistent. Most procedures we have audited have been only 70-85% accurate; the inaccuracies include missing critical steps, steps as written are not what needs to be done, or the steps are out of sequence
- No/misplaced warnings. For example, a warning should never **contain** the action to take; it should instead **emphasize** the action to take
- Poor format and presentation rules

Other reasons workers may not use procedures include:

- Procedures are out of date
- No procedure has been written for the task
- Users cannot find the procedure they want to use
- Users don't need a procedure because the task is simple
- Users need more information than the procedures contain
- Users see procedures as an affront to their skill
- Procedures are difficult to use in the work environment
- Procedures are difficult to understand

So, in addition to the rules for writing procedures that are shown in the table above, the organization must also address the reasons that cause the worker not to use the written procedure.

## *E. HUMAN-MACHINE INTERFACE (INCLUDING TOOLS)*

The **Human-Machine Interface** (also known as the Human-System Interface, or HSI) is defined as the technology through which personnel interact with plant systems to perform their functions and tasks. The major types of HSIs include alarms, information systems, and control systems.

Each type of HSI is made up of hardware and software components that provide information displays, which are the means for user-system interaction, and controls for executing these interactions.

Personnel use of HSIs is influenced directly by (1) the organization of HSIs into workstations (e.g., consoles and panels); (2) the arrangement of workstations and supporting equipment into facilities, such as a main control room, remote shutdown station, local control station, technical support center, and emergency operations facility; and (3) the environmental conditions in which the HSIs are used, including temperature, humidity, ventilation, illumination, and noise.

There are three important goals to be achieved in the design and implementation of the HSI. These are:

- **Design for operability** refers to designing the HSI to be consistent with the abilities and limitations of the personnel who will be operating it. Weaknesses in the design processes can result in an HSI that is not well suited to the tasks that personnel must perform to ensure plant safety, resulting in increased workload, decreased performance by personnel, and an increased likelihood of errors.

  The wide-spread adoption of distributed control systems has provided many advances in process control. However, if implemented improperly, these systems can greatly reduce the operability of the facility. Although they were unable to provide as much information, old analog panels were typically laid out in a manner such that experienced operators were able to monitor the entire process and effectively identify conditions meriting attention. Due to space limitations, and their cost (e.g., hardware and installation), alarms were typically limited to those deemed to be associated with a significant process upset.

  The advent of distributed control systems has eliminated the space limitations associated with alarms, and greatly reduced their costs (in many cases, alarms can be added merely through programming). Without proper alarm configuration structure and discipline, this can lead to "alarm flooding" (an excessive number of simultaneous alarms) during abnormal operating conditions. The human error of failing to respond properly to an alarm increases by a factor of over 1,000 if ten alarms activate simultaneously compared to a single alarm activation.

  Therefore, it is critical that control and alarm system designers in facilities with distributed control systems properly prioritize information so that operators can respond appropriately to abnormal conditions. Further, trouble-shooting guides should be readily accessible via the distributed control system to reduce human error in alarm response.

  Beyond the topic of alarms, the proper design of the visual displays can also reduce human error. Certain color combinations, such as a light gray background with darker gray equipment and black lines coupled with a flashing red indication for those components in an alarm state effectively draw the operators' attention to the relevant part of the process requiring attention. Further, an effective display should include:

  - Safety systems that are bypassed, impaired, or out of service
  - Process alarms that are disabled
  - Process conditions that are not within established limits
  - Backup systems that are unavailable
  - Special maintenance or testing activities that are currently in progress

- **Design for maintainability** refers to designing the HSI and associated plant equipment to ensure that personnel are able to perform necessary maintenance activities efficiently. Weaknesses in the design process can result in systems that impose excessive demands on personnel for maintenance and, therefore, are prone to maintenance errors or problems with reliability and availability.

- **Design for flexibility** refers to the way that changes, such as upgrades to the HSI, are planned and put into use. A new HSI component may require the user to perform functions and tasks in new ways. Skills that the user developed for managing workload when using the former design, such as ways for scanning information or executing control actions, may no longer be compatible with the new design. The new HSIs must also be compatible with the remaining HSIs so that operators can use them together with limited possibilities for human error. Also, HSI modifications may not be installed or put into service all at one time, causing the user to adapt to temporary configurations that are different from both the original and final configurations. Weaknesses in HSI implementation can increase operator workload and the likelihood of errors.

**Tools** are a special category of HSI, which typically refers to hand tools or devices that are generally designed with the user in mind. A big part of human factor consideration is how to make the equipment and process operation mistake-proof (to prevent errors as much as possible). The following are some factors to consider when error proofing in designs:

- **Design for unambiguous assembly.** Design the product or device such that the assembly process is unambiguous (by designing components so that they can only be assembled one way); i.e., design matching parts that are easy to insert and align. For example, use notches, asymmetrical holes, and stops to mistake-proof the assembly process. Products that go together in only one way require less worker training, perform more reliably, and can be repaired more quickly.
- **Consult workers.** Operators, technicians, and maintenance personnel can pinpoint the most troublesome areas.
- **Avoid symmetry.** When a particular orientation is critical to the design, avoid symmetry. For example, use nonsymmetrical hole patterns.
- **Use labels sparingly.** Labels tend to come off equipment too easily and often are wordy.
- **Review the environment.** Environmental problems that encourage mistakes include poor lighting; high/low heat; excess humidity, dust, and noise — anything that distracts workers.

The following are some factors to consider when error proofing in process/operations:

- **Error-proof mechanisms.** Error-proof mechanisms are very powerful in improving system reliability when incorporated into the design. These mechanisms, by design, will not allow a user to perform an illegal operation. For example, if a user enters a value that is outside the accepted range of operation, the control logic will not accept the value.

- **Automatic alerts.** Automatic alerts immediately inform the user of an illegal operation to prompt corrective action. These alerts are particularly useful in critical operations that allow time for corrective action before some adverse consequence occurs. For example, a piping manifold may be designed such that a warning alarm sounds if valves are not opened/closed in the proper, critical sequence, or if a critical valve is left opened/closed.
- **Automatic system shutdown.** Automatic system shutdown should be incorporated into the design when an illegal action is performed during a critical operation and no time is available for corrective action. For example, if an operator uses the wrong sequence in charging a reactor, the reactor will shut down before other materials are charged that may lead to critical temperatures and pressures.

## *F. FITNESS FOR DUTY*

Successful task performance requires that the capabilities that workers bring to the task fall within an expected range. **Fitness for Duty** issues include reduction in an individual's mental or physical capabilities due to substance abuse, fatigue, illness or stress, which increases the likelihood of errors. Types of possible impairments include:

- Physical attributes – strength, reach, eye-sight and color acuity, hearing,
- Mental attributes – drug and alcohol (abuse), mental stress (on and off the job);
- Fatigue – issues from on the job and off the job (especially control of hours per work-day and per work-week)

Safeguards to prevent fitness-for-duty-related errors include company programs for the detection and prevention of potential or actual impairment, as well as the individual responsibility of workers to decline assignments if they are impaired for any reason. The latter safeguard is a weak one, however, because humans are generally over-confident of their capabilities when under the influence of drugs or alcohol, or are stressed, fatigued or ill. Other factors that may discourage self-reporting include the fear of poor performance reports from supervision or not receiving pay for extra overtime.

Company programs that may be implicated in errors caused by personnel impairment include:

- **Fitness-for-Duty Program** – Company fitness-for-duty programs are primarily responsible for detecting and preventing impaired personnel from performing tasks that may affect public health and safety. Medical evaluations of personnel, behavioral observation programs, employee assistance programs and drug and alcohol testing are used to detect impairment. Impaired workers can be prevented from performing tasks by establishing protocols for instances in which a worker is believed to be unfit for duty, training for supervisors on detection of and response to fitness for duty issues, and employment guidelines for personnel who voluntarily enter drug or alcohol treatment programs. Weaknesses in a Fitness for Duty program may allow impaired personnel to have access to vital areas in a plant where they could commit errors. One excellent starting point for a Fitness for Duty sub-element is the guidance provided in ***US NRC's 10 CFR 26 (2005).***

- **Overtime Policies and Practices** – Most companies establish limits for work hours to reduce on-the-job fatigue. It has been shown that 17 hours of work without a break results in the same error rate as being legally drunk. And, at 10 days straight of 12 hours work-days, the error rates for non-routine tasks such as startup of a continuous unit can increase to 1 mistake in 5 to 10 steps (as opposed to the target of 1 mistake in 100 steps). Routine authorization for work hours in excess of those recommended may result in fatigued workers. Further, a practice of excluding training or meetings that occur outside of an individual's normal work schedule from work-hour limitations will also contribute to fatigue. In ***US NRC's 10 CFR 26 (2005)***, the guidance given for control of overtime hours is no more 72 hours of work per 6 day period, no more than 16 hours of work in one day within that period, and a minimum of 24 hours contiguous hours away from work within a 7 day period. The US DOT has even more stringent rules on limiting work hours and establishing required hours for recovery from fatigue.
- **Shift Scheduling** – Shift scheduling may also affect the likelihood that personnel will show performance decrements due to fatigue. A change in the assigned shift or a rotating shift schedule will disrupt circadian rhythms and may increase the likelihood of errors.
- **Safety Culture** – The effectiveness of self-reporting and behavioral observation programs depends greatly upon the safety culture at a site.

## *G. WORK PROCESSES AND SUPERVISION*

**Supervision** is the process by which work is directed and overseen by first-line management. Successful supervision requires a combination of leadership skills and technical competence. Supervision differs from peer checking or quality control because a supervisor has line management responsibility for the worker(s) as well as responsibility for the work activity.

Supervision is more than the moment-to-moment direction of a work activity. Successful supervision requires the assessment and shaping of worker attitudes and motivation, communication and implementation of management expectations for performing work, the assignment of the best-qualified workers to various tasks, as well as the technical competence to identify incorrect actions and stop improper activities before an error is committed. Effective supervision involves directing the work, overseeing how it is performed, and leadership.

Organizations are typically structured to have sufficient supervision of the job, but many times the delineation between the trainer and the supervisor roles is blurred. Supervision can and normally does play a key role in selecting of the right worker for the job, scheduling of workers to match the required tasks for the day/week, and generally overseeing the task execution to ensure policies and procedures are followed. Supervisors are not always trained on all of their key roles in support of control of human factors, such as detecting issues in workers related to fitness for duty or fatigue.

## *H. WORK ENVIRONMENT*

The **Work Environment** refers to the physical conditions in which work is performed. Environmental conditions that can affect performance include excessive vibration and noise, temperature extremes, and insufficient lighting, as summarized below:

- There are two types of **vibration** that may cause errors. The first is whole-body vibration, in which vibration is transferred to the worker from standing or sitting on a vibrating surface. The second is object vibration, in which a stationary worker interacts with a vibrating object in some fashion. The effects of vibration depend upon its frequency and acceleration. Frequency is the number of oscillations (cycles) that occur in one second. Acceleration is the force, or intensity, of the vibration.
- **Noise** is unwanted sound. Noise can cause errors in several ways. It may disrupt communications, affect the ability to perform tasks and annoy personnel. The effects of noise on communications are complex. Even relatively low levels of noise can mask speech, but only under some circumstances. For example, speakers naturally raise their voices when there is background noise and may be able to overcome some of its effects on communication. Being able to see the speaker's face or using standardized phrases also improves communication in a noisy environment. The type of background noise also affects communication. It is easier to communicate over noise that is steady and uniform than noise that includes sharp tonal peaks, such as background speech.
- **Heat exposure** is a common problem in many areas of a plant, such as the turbine building when the plant is operating. The extent to which workers will be affected by heat depends on many factors. These include physical characteristics, such as age, weight, acclimation to heat, physical fitness and dehydration. Other factors that determine the effects of heat on performance include airflow, humidity, clothing and level of physical activity.
- **Exposure to cold** affects the performance of manual tasks. Decreases in the ability to control hand movements begin at an air temperature of approximately 54° F. The fingers may become numb to pain at this temperature and touch sensitivity is reduced. Performance of gross manual tasks, such as those involving the arms and legs is also degraded at 54° F. The speed at which manual tasks can be performed is affected by the rate of cooling. Slow temperature drops have a greater negative impact on manual dexterity than rapid temperature decreases, during the initial exposure period.
- **Adequate lighting** is required for accurate performance of nearly every task in a unit operation.

The organization must have engineering controls to help control each factor, but sometimes there is no other choice but to rely upon administrative controls.

## *I. COMMUNICATION (most importantly, verbal communication)*

**Communication** is the exchange of information while preparing for or performing work. Verbal communication occurs face-to-face, by telephone, sound-powered phones or walkie-talkies, as well as over public address systems. Written communication occurs, for example, through policies, standards, work packages, training materials, and e-mail.

Communication involves two sets of behaviors: (1) creating and sending messages and (2) receiving and interpreting them. Communication always involves at least two individuals, the sender and the receiver, and occurs:

- Between individuals

- Within and among work groups
- In meetings
- In pre-job or pre-evolution briefings
- During shift turnover

Successful communication requires several steps. The sender first develops the intention to communicate **either** verbally or in writing. The sender then composes a message that presents the meaning as clearly as possible. The receiver must pay attention to the message and then interpret its meaning. If the communication is successful, the receiver interprets the message consistently with the sender's intended meaning. However, there are many potential errors in both sending and receiving the communication:

- **Sending Errors**
    - *Content wrong*
    - *Content inconsistent*
    - *Content inappropriate for the job*
    - *Content inappropriate for the receiver*
    - *Standard terminology not used*
    - *Familiar terminology not used*
    - *Message production inadequate or interfered with*
    - *Necessary information not sent*
    - *Wrong place or person*
    - *Wrong time*
    - *Sending verification failure*

- **Receiving Errors**
    - *Information not sought*
    - *Information not found*
    - *Information not used*
    - *Receiving verification failure*
    - *Message misunderstood.*

Communication errors may be reduced in the following ways:

- Structuring messages in a standard format, which alerts the receiver if important information has been skipped
- Providing written instructions or procedures for complicated tasks
- Establishing protocols to repeat key parts of verbal communications
- Preceding special or nonroutine activities with a pre-job briefing
- Using structured protocols, checklists and logs to supplement written instructions

One key opportunity for communication errors is when a new shift takes over responsibilities from the previous shift. An effective tool to reduce these errors is a properly designed shift handover process that includes appropriate checklists and logs of activities, and a structured and consistent method for verbally reviewing this information. Additionally, the logs and discussions need to include:

- Routine operations
- Transition operations (e.g., changes to produce a different product or switch to a different raw material source)
- Nonroutine operations (e.g., maintenance activities or temporary operating conditions)

## 6. Beyond Optimizing Human Factors: Best Remedies for Human Error

Once an organization optimizes human factors, they must recognize that human errors will still occur – albeit at a lower frequency and likely with a smaller adverse impact than before. Therefore, in cases in which the consequence of a human error is unacceptably high, the organization should implement solutions that will lower the residual risk further. These solutions can be designed to either help prevent the error or compensates for the error, but **they must be independent of the human error initiating event**.

Bridges and Bridges [25] have compiled the following list of the best remedies for human error for these situations:

- Captive Key. Applied to the valve handle hub. PFD = 0.01. Requires placing one component, such as a valve or door, into an open or closed position before releasing the key needed to move another component into a potentially unsafe position. Can be used to make a sequence of steps for certain tasks difficult or impossible to skip or perform in the wrong order.
  - Special consideration: only one person at the site should have access to the machine that produces keys; only this person should be authorized to replace a key; copying of keys should not be tolerated and strictly enforced.

- Limit switches on valve position. Ensures that valves are in the correct position, potentially depending on mode or compared to another valve. The concept is similar to captive key, but the control is by limit switches instead. Switch would trigger shutdown or otherwise prevent misalignment. PFD = 0.1 to 0.01, with 0.01requiring strenuous maintenance practices for switches that is not typical in many plants.
  - Typical limit switches are used for many permissives and mode change interlocks. Maintenance practices and control strategies must be well established to bring the probability of jumpering, knifing, or otherwise defeating a limit switch to a very low probability.
  - Some manual valves can be modified to have feedback/indication (though these are harder to keep working); giving ability to verify if systems are isolated from DCS panel.
  - Not as robust as captive key systems.

- Mechanically coupled valves: 2 valves that are hydraulically or pneumatically or mechanically/physically linked. PFD = 0.1 to 0.01, with 0.01 requiring feedback/ indication on both valves.

- Typically used to ensure that mode switch will be successful, such as when backwashing filters or entering regeneration mode for driers, i.e. preventing only one of two valves from operating.
- Special consideration: Limit switches still needed to verify position, as these valves are difficult to maintain due to their more complex drive or pneumatic systems (can't operate on output alone)

- Instrumented Permissive. Such as an interlock/permissive to prove the pressure is in the right range before allowing an XV to open. Such as where pressurization with a gas is supposed to be done manually before opening a valve for a cryogenic liquid to enter the system, given that the materials of construction experience brittle fracture potential at the temperature (boiling point) of the flashing liquid. PFD = 0.1 to 0.01, depending on configuration as a SIL 1 or SIL 2 preventative SIF.

- Hand-held device with bar code reader. To verify the user has been to each device in the right sequence. PFD = 1 to 0.1 but could be stronger if tied back to the BPCS which would then have the computer, not computer operator, watching the activity to confirm a critical step is performed.

- RF tags and matching of a pair. Similar to Captive Key in concept, but for hose connections, and perhaps other situations. PFD = 0.01

- Unique connections. Unique size, coupling type, thread pattern, etc. to reduce the chance of a wrong connection. PFD = 0.01

- Spring closing lever valves. Manual valve that the human has to hold open by use of the lever handle (typically) that will automatically close when the lever is released. Also called a dead-man valve. PFD = 0.1 to 0.01 (but usually PFD = 0.1 maximum). Such arrangements are applicable to small valves to prevent leaving a valve unattended, such as during manual draining or loading. Can be defeated by tying the handle in the open position.

- Swing Elbow. The elbow and associated piping is designed to that the process can only be lined up in *One* direction at a time. PFD = 0.001. This arrange works well for switching to process modes, such as regeneration of desiccant driers or catalytic reactor beds with very hot air or steam, but when the normal flow alignment is to a hydrocarbon process.

- Unique SIF to compensate for, rather than prevent, a Human Error. PFD = 0.1 to 0.001. The example provided earlier was for an instrumented permissives. This remedy to compensate for a unique human error scenario includes installing SIFs to shut down a process or block or vent a line, such as to eliminate an overpressure scenario that is too large for the current pressure relief system (the scenario is larger flow than any practical PSV or rupture disk system can handle).

- Increase PSV size for a scenario unique to startup or online maintenance. As needed; to achieve the full value of the PFD for the PSV configuration.

- Upgrade materials: Account for scenarios not considered in the original design by upgrading materials enough that the consequences from the human error are no longer possible, reduces likelihood by 2 to 4 orders of magnitude. See the "recommendation" in the Instrumented Permissive for one such example.

- Change design of at-risk components/system: Design out the need for components at risk, or change to a different strategy for certain unit processes. This is the inherently safer approach, and it can include some of the design considerations mentioned above.

Each organization will want to develop such a list with examples including (1) when to use each remedy, (2) what value is risk reduction is gained, (3) how to quickly estimate the cost of the remedy, and (4) an example or two of the remedy to help with clarification.

## 7. Controlling Habits

The direct control of human factors as outlined above is the most effective means to control human error. However, about 15% of human error is due to acquired habits. Some call these Behaviors, but that normally carries negative connotations. Many companies have effective systems for combating bad habits; these systems normally involve peer-to-peer observations and feedback (coaching), and these are many times labeled behavior-based safety management (which is a trademark phrase belonging to BST), or behavior-based reliability, or simply job-observations. A peer-observing-peer system, such as these, can reduce habit-based errors by about 70%.

## 8. Applying Human Factors to Risk Reviews and Root Cause Analysis

It is important to note that while not directly controlling human factors, the PHA/HAZOP (hazard identification and risk assessment) and the Incident Investigation management systems help to identify where and how human factor practices are necessary to control certain risks and help to identify weaknesses in human factor practices and implementations.

**Risk Reviews (PHA/HAZOP, HIRA, JSA, etc.)** – Addressing Human Factors during risk reviews can help identify gaps in all human factor categories since human factor categories are types of safeguards. This is fundamental for predicting where and why humans might make mistakes and for determining (qualitatively, at first) if the protection layers are sufficient if such errors occur and if not, what else is needed. Unfortunately, many organizations do not fully analyze for errors during all modes of operation, and so many (in some cases, most) of the accidents that start with a human error are not predicted, as they should be ***(See Chapter 9.1 of CCPS/AIChE, Guidelines for Hazard Evaluation Procedures, 3rd Edition, 2008[9]; and paper by Bridges, LPS/AIChE, April 2009, Optimizing Hazard Evaluations[10].)*** This practice ultimately strengthens SOPs, human system interface, and most importantly ensures there are sufficient IPLs for scenarios unique to non-routine modes of operation, such as startup, shutdown, and online maintenance.

**Incident Investigation/Root Cause Analysis (II/RCA)** – II/RCA is necessary to learn from mistakes. Although all companies have an II/RCA system, many companies are lacking in the awareness of where human factors fit into an accident sequence. This leads to the II/RCA missing the human factor weaknesses that led to the human error and so the II/RCA may stop at the cause (and even root cause) being "operator error." Many companies also do not get nearly enough near misses reported (the ratio should be about 20 to 100 near misses reported per actual loss/accident ***(See CCPS/AIChE, Guidelines for Investigating Chemical Process Incidents, 2nd Edition, 2003[12]; and paper by Bridges, 8th conference, ASSE-MEC, 2007, Gains in Getting Near Misses Reported[13].)***

## 9. Closing

An organization implementing PSM (or similar programs such as, SMS or OSHA's 18001 for occupational safety) must develop management systems for optimizing human factors to control human error rates. As noted above, three potential sources for guidance for the chemical process industry are regulatory standards, such as the OSHA PSM regulation, the CCPS RBPS reference, and the CCPS COO/OD guideline. As a matter of closing this analysis, it is helpful to compare how well each of these documents address human factor implementation. The following is a rating scale that the authors have developed to gauge how well human factors are incorporated in these sources.

**TABLE 5: DEFINITION OF HUMAN FACTOR RATING SYSTEM**

| Rating | Description |
|---|---|
| 1 | Human Factor is not specifically addressed |
| 2 | Human Factor is nominally addressed |
| 3 | Human Factor is partially addressed |
| 4 | Human Factor is mostly addressed |
| 5 | Human Factor is thoroughly addressed |

The following table summarizes each document's (management system's) ratings for each human factor:

**TABLE 6: HUMAN FACTOR RATINGS FOR PSM REFERENCE DOCUMENTS**

| Human Factor | OSHA PSM | RBPS Guideline | COO/OD (Component of RBPS) |
|---|---|---|---|
| Available Time | 1 | 1 | 2 |
| Stress/Stressors | 1 | 1 | 2 |
| Complexity & Task Design | 1 | 1 | 2 |
| Experience/Training | 2 | 3 | 5 |
| Procedures | 2 | 3 | 2 |
| Human-Machine Interface | 1 | 1 | 4 |
| Fitness for Duty | 1 | 3 | 4 |
| Work Processes & Supervision | 2 | 4 | 4 |
| Work Environment | 1 | 1 | 1 |
| Communication | 1 | 3 | 4 |

As stated above, the OSHA PSM standard does not rate well for incorporation of human factors. It partially addresses them in the area of training, but only nominally at best in the other areas.

The RBPS guideline improves upon the coverage of human factors significantly compared to the US OSHA PSM regulation, particularly in the areas of:

- **Fitness for Duty (FFD)** – specifically by introducing the concept of sleep discipline and overtime limitations, but without specific guidance on limits and without recommended practices for controlling FFD
- **Work Processes & Supervision** – specifically by providing a great deal of emphasis in the areas of process safety culture and workforce involvement
- **Communication** – specifically by providing some (but not all) key communications rules to reduce human error, and by discussing the importance of shift turnover logs, but without enough detail on their content
- **Experience/Training** – specifically by providing a strong structure of the desired training program, albeit at a fairly high level
- **Procedures** – specifically by including most, but not all, of the best practices identified in this paper; the key items missing are the rules to have procedures written by future users of the procedures, and to conduct a hazard analysis of the consequences of human error in following the procedures

The COO/OD guideline further improves upon its parent book, RBPS, in most areas:

- **FFD** – specifically by addressing nearly all the key elements outlined in this paper. One main shortcoming is that it recommends API RP 755 as a guide for Fatigue Management, but this standard is terrible in that it fails to control fatigue that can occur during extended maintenance outages. Further, it does not cover all FFD issues, but rather focuses on fatigue. It would have been better to omit reference to the API RP 755 and instead reference the US NRC standard for FFD, 10 CFR 26. One major advantage of the NRC standard is that it includes the requirements for supervisors to be trained in and search for FFD deficiencies in staff coming onto the shift. A thorough management system for detection and the initial actions and follow-on actions is needed; 10 CFR 26 provides a lot of detail on what is needed.
- **Human-Machine Interface** – specifically by highlighting key DCS design features; however, much of the other factors described in this paper are not discussed.
- **Communication** – specifically by again including key (but not all) communication rules, and by providing specific details about the shift handover log.
- **Experience/Training** – specifically by providing a thorough, and properly detailed, overview of a strong training program.

Taken together, these two CCPS guidelines provide valuable advice on how to incorporate six (6) of the human factors into a PSM program, with notable exceptions in the areas of:

- *Available Time*
- *Stress/Stressors*

- *Task Design and Task Complexity*
- *Work Environment*
- *Procedures - poor/incomplete coverage of best practices*
- *Some gaps within the factors RBPS and COO/OD do cover*

This paper provides sufficient detail for implementing all 10 human factors so that companies following its guidance can be well on their way to building strong human factors elements within their PSM program.

## 9. References:


1. Swain, A. D., Guttmann, H. E., *The Human Reliability Handbook - with Emphasis on Nuclear Power Plant Applications*; NUREG/CR-1278, U.S. Nuclear Regulatory Commission, Division of Facility Operations, 1983.

2. Gertman, D.; Blackman, H.; Marble, J.; Byers, J. and Smith, C.; *The SPAR-H Human Reliability Analysis Method,* NUREG/CR-6883, U.S. Nuclear Regulatory Commission, Office of Nuclear Regulatory Research, Washington, DC, August 2005.

3. Bridges, W., and Tew, R., *Human Factors Elements Missing from Process Safety Management (PSM),* 6th GCPS, AIChE, March 2010.

4. *Risk-Based Process Safety*, CCPS/AIChE, John Wiley & Sons, Inc.; Hoboken, NJ, 2007.

5. *Conduct of Operations and Operational Discipline*, CCPS/AIChE, John Wiley & Sons, Inc.; Hoboken, NJ, 2011.

6. *US FAA Advisory Circular* 120-42A (as part of ETOPS approval)

7. Bridges, W. G., Williams, T. R., "Effective Procedure Writing – A Two-Part Series (Part 1: Reducing Risk of Worker Errors; Part 2: Addressing Process Safety Management Requirements for Operating Procedures)," *Chemical Engineering Progress*, 1997.

8. Bridges, W., and Collazo-Ramos, G., *Human Factors and their Optimization,* 8th GCPS, AIChE, April 2012.

9. *Guidelines for Hazard Evaluation Procedures*, 3rd Edition, CCPS/AIChE, John Wiley & Sons, Inc.; Hoboken, NJ, 2008.

10. Bridges, W., *Optimizing Hazard Evaluations,* LPS/AIChE, April 2009.

11. Bridges, W., Lorenzo, D., and Kirkman, J., *Addressing Human Errors During Process Hazard Analyses,* CCPS Conference, AIChE, 1991 (also published in Chemical Engineering Progress May 1994).

12. *Guidelines for Investigating Chemical Process Incidents, 2nd Edition,* CCPS/AICHE, John Wiley & Sons, Inc.; Hoboken, NJ, 2003.

13. Bridges, W., *Gains in Getting Near Misses Reported,* 8th Conference, ASSE-MEC, Bahrain, 2008.

14. Della Rocco, P.S., et al., *The Effects of Napping on Night Shift Performance,* Federal Aviation Administration, U.S. Department of Transportation, 2000.

15. *Fitness for Duty Programs, Preamble, 10 CFR 26, U.S. Nuclear Regulatory Commission, Revised 2005.*

16. US Chemical Safety and Hazard Investigation Board (CSB), *Anatomy of a Disaster: Explosion at BP Texas City Refinery,* published 2006.

17. University of South Australia (UniSA), Adelaide Centre for Sleep Research, *SDA Fact Sheet - AT03*, 2006.

18. Hallbert, B.; Whaley, A.; Boring, R.; MCabe, P.; and Chang, Y.; *Human Event Repository and Analysis (HERA)*, NUREG/CR-6903, U.S. Nuclear Regulatory Commission, Office of Nuclear Regulatory Research, Washington, DC, 2007.

19. Bridges, W., and Clark, T., *LOPA and Human Reliability - Human Errors and Human IPLs,* 7th GCPS, AIChE, March 2011.

20. Bridges, W., and Thomas, H., *Accounting for Human Error Probability in SIL Verification Calculations,* 8th GCPS, AIChE, April 2012.

21. Bridges, W., *The Impact of Human Factors on LOPA,* 9th GCPS, AIChE, April 2013.

22. Bridges, W., and Tew, R., *Best Practices for Writing Operating Procedures and Trouble-shooting Guides,* 13th GCPS, AIChE, March 2017.

23. Bridges, W., *Proven Approaches to Ensuring Operators Can Respond to Critical Process Deviations in Time (Human Response IPL),* 13th GCPS, AIChE, March 2017.

24. Bridges, W. and Rhodes, W., *Everything you need to Know about Human Reliability and Process Safety,* 17th GCPS, AIChE, March 2019.

25. Bridges, S. and Bridges, W., *Abnormal Modes of Operation – A risk-based focus,* 16th GCPS, AIChE, March 2020

# 617835 Beware of Blind Risk -What You Do Not Know Can Hurt You


**Abstract:**

This paper discusses the concept of blind risk, how it arises, its dire consequences, and how to best avoid it.

Blind risk is most often due to incomplete, inconsistent, or out-of-date asset and process safety information. Most often, these information quality issues are not obvious or visible, and as such they are a blind risk. They lie dormant, like a beartrap in the undergrowth of a forest until someone inadvertently steps on it. They become gaps in our "layers for protection" -the systems and processes we trust to stop incidents before they happen. But blind risk is also a choice, and it is a measure of a company's tolerance for risk.

Safety, engineering, capital, and operating decisions are only as good as the information they are based on. Most companies may not ***objectively*** know how good (read reliable and trustworthy) the quality of their underlying asset and process safety information is. It does not have to be this way.

There are two things you can do today that will have a significant impact now and into the future:

- Use standardized information quality metrics and tools to measure, monitor and report the true state of asset and process safety information across multiple repositories, applications, and formats. Let all stakeholders including executive management know the good, the bad and the ugly. Use this as a baseline to measure your improvement against.
- Monitor and quantify the cost of poor asset and process safety information. Use this information to justify funding for information quality improvement initiatives.

# 619136 Relief Systems Work Is a Money Pit: Change My Mind

**Abstract:**

As a relief system engineer, I revalidate and correct pressure relief system designs daily. Many times, the owner/operator views the pressure relief systems efforts as a 'bottomless money pit' that never returns value and is seemingly never complete. This presentation will dispel this notion and explain how to ensure your pressure relief systems efforts turn into cost-avoidance measures and high-value money-makers.

This three-part presentation will summarize the business purposes of pressure relief system efforts, describe how good relief systems design bases are converted into cost-avoidance measures, and finally share how targeted relief systems efforts can turn into net-positive returns for your facility.

Example content includes safety and regulatory obligations, keeping pace with industry, insurance impact, risk reduction, reputation, safe operating limit changes, and debottlenecking.

**ISO 9001-2015 and Risk Mindedness in the Management of Heat Exchangers in Ammonia Synthesis Process**

Festus Addo-Yobo, Adrian Lutchman, Sandra Maharaj, Devon Indar
The University of Trinidad and Tobago, Point Lisas Campus, Esperanza Street, Brechin Castle,Trinidad and Tobago.

## Introduction

The ammonia synthesis process is one of the most important in the manufacturing Industry because ammonia is a feedstock for diverse products including fertilizer. This process is an interconnected system of production units which include: generation of hydrogen, separation of air to produce nitrogen, ammonia synthesis using catalysts, ammonia condensation and recycling. In the twenty first century efforts were concentrated on the optimizing of the processes either independently or as groups of processes providing inputs to various stages of the processing plant (CEP 2015). As integrated design has been largely successful in optimizing the production plant that is delivered to the producer, attention is now focussed on running the plant so as to optimize its performance. Plant safety, operational efficiency, effectiveness of operation, utilization of resources and productivity are desired outcomes. To achieve these ends, Process Control Engineers must necessarily have conversations with custodians of Quality Management Systems (QMS) in order to optimize the plant performance at a higher management level.

Bayer Technology Services offer Online Optimization & Operational Excellence solutions . Others utilize the TQM and Six Sigma quality assurance models. Nowadays, most plants are run by Distributed Control Systems (DCS) and have modelling software such HYSYS for carrying out such tasks. Thus the opportunity to optimize plant performance online has become feasible. BASF Services have teamed up with Haldor Topsoe to offer Ammonia producers new optimization opportunities online (Process Worldwide 2020).

Ammonia plants have optimized energy recovery through design. Waste heat in the process stream exiting the front end, exiting the secondary reformer, is recovered via heat exchange to produce high pressure steam. Similarly the heat produced by the highly exothermic conversion reaction in the backend is also recovered to produce high pressure steam. The heat exchangers used in these sections (shown in red boxes in the adjacent figure) operate under high temperatures (>400 ºC). They thus are susceptible to failure mechanisms not usually encountered in most petrochemical operations.

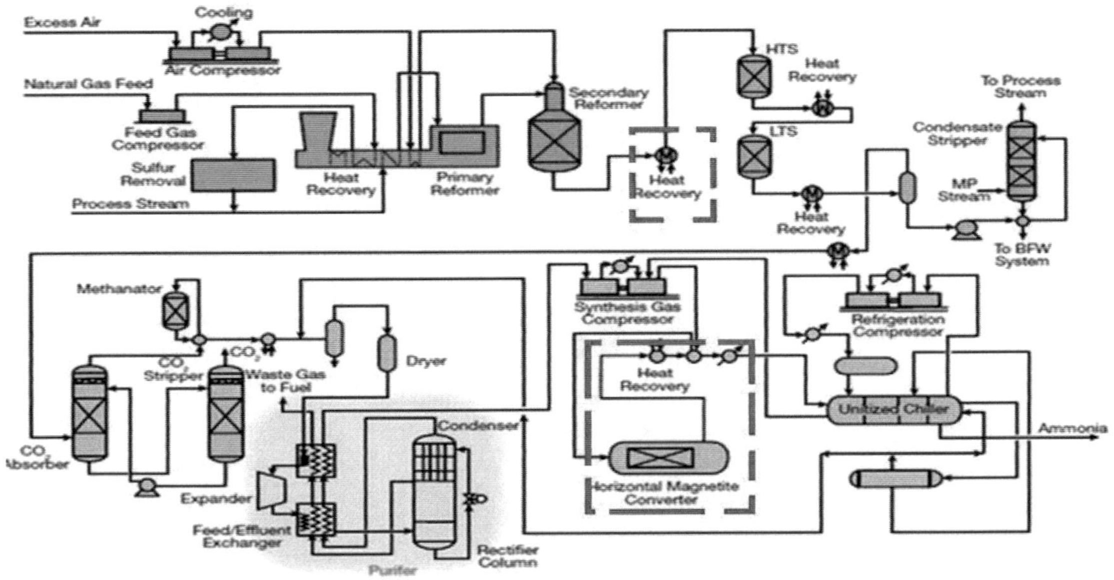

www.aiche.org/cep September 2016 CEP Copyright © 2016 American Institute of Chemical Engineers (AIChE)

## The Quality Management System

An important development in the standard to which quality management systems for production processes must conform is the revision of the ISO 9001 Standard to the 2015 version which makes it a requirement to address risk in all aspects of the QMS and provide anticipatory remedial actions. The standard places the responsibility on top management. Addressing risk is identified in the Standard as having the potential to reduce cost and also providing opportunities. One important way to operationalize risk-mindedness as advocated in the standard is to model the plant performance or the performance of relevant components and investigate scenarios to identify the risks. Heat exchanger tube thinning In the nuclear power generation industry is well researched and reported (Gao et al – 2017, Jonas – 2004, Snachez-Carrera – 1984). Remedial action in the nuclear industry includes scheduled inspection and replacement of heat exchanger tubes prior to their rupture according to protocols using Computer Programmes e,g,CHECWORKS™ (Smith & Horowitz – 2014). In the Ammonia Industry, non-destructive testing can identify tubes that have thinned significantly, which are then blocked off as a preventive measure to avoid water leaking from them to the process stream.

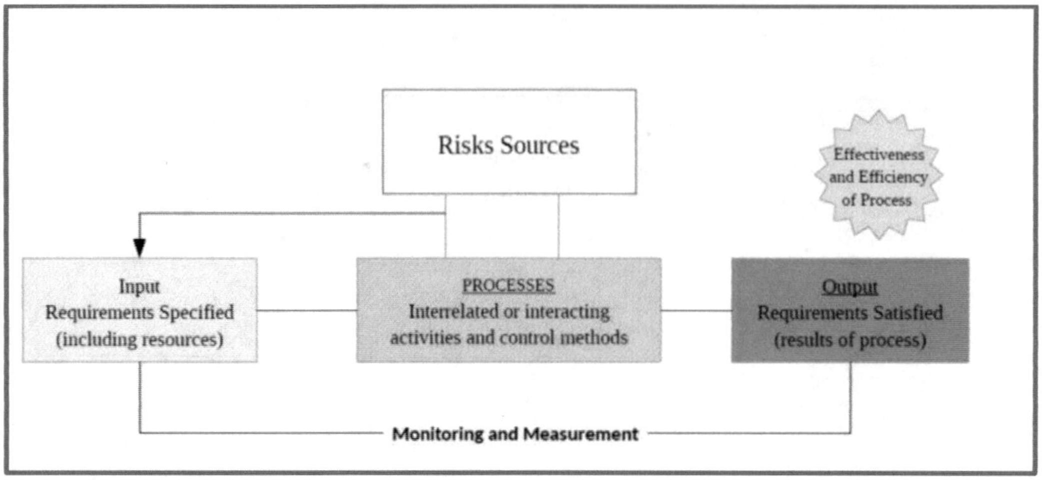


Juxtaposing the great effort and successes in ammonia process optimization to the experience of plant operators, demonstrates the existence of a gap. Even when plants are controlled at desired set points, heat exchanger tubes can and do fail leading to days and weeks of unplanned downtime. Given the high throughput of the plant (600 tonnes/day), this represents significant reduction in plant productivity. Such failures are risks that can be expected, though the events cannot be predicted with accuracy. Heat exchanger networks form an important part of the ammonia manufacturing process. Therefore, these risks should be managed by modelling the operation of the networks. In this work we use HYSYS to model the performance of the heat exchanger network used in typical Ammonia Plant (Chemical Engineering Progress 2015) before and after heat exchanger tubes are blocked off. We investigate numerically, the influence using Boiler Feed Water (BFW) to manage four of the interacting process variables which influence the risk of tube failure: process gas temperature, the velocity of the steam-water mixture in the tubes, the average temperature and the quality of the steam-water mixture in the boiler tubes.

**Identification of Risk Factors**

The mechanism of platelet erosion has been identified as leading to the thinning of ductile stainless such as the high carbon steel used in the nuclear industry. Abrasive particulates scrape the surface and take off the protective oxide which is swept away by the current of fluid flowing in the tubes, leading to further corrosion. This is known as flow assisted corrosion. Levy (1986) shows that above about 250ºC, temperature speeds up tube thinning process exponentially. Other factors which contribute to Flow Accelerated Corrosion (FAC) are the velocity of the particle, their hardness, the shape and the impinging angle. The geometry of the pipeworks is very important because bends and fittings serve as targets for high velocity dislodged protective oxide.

The HT exchanger tubes which carry BFW for cooling process gases are made of Stainless steel 304L whose mass fraction of chromium are as high as 18-19%. Chromium forms an oxide layer up to 300nm which protects the steel. Should this layer be removed, chromium would react with dissolved oxygen in the feed water to form Chromium Oxide or an even more protective layer comprising a solid solution of Fe, Ni, Cr oxide. Even though the BFW is purified by ion-exchanged to remove chlorides, the possibility of chloride contamination by downstream cooling seawater cannot be ruled out. Chromium can be leached from the protective layer by presence of alkali chloride, which could react with oxygen to release chlorine which further attack chromium oxide and leach out the chromium at temperatures above 500ºC. At lower temperatures, it is postulated that the chloride could react with the oxygen and chromium oxide according the eqn below.

$Cr_2O_3$ (s) +4KCl(s/l)+$2H_2O$ +1/2 $O_2$ ↔ $2K_2CrO_3$ (s/l) + 4HCl.

The chromate formed would either remain in solid phase in the absence of liquid water or dissolve in the hot BFW water and be leached away. In either case, no contiguous layer of protective chromium oxide layer would remain and the iron underneath would rust. This mechanism would lead to thinning. Thus, the risk factors are: the contamination of the BFW with chloride, high temperatures and the oxygen levels in the BFW. The steam weight would also constitute a risk factor. The presence of nitrates is a risk factor as it known to corrode SS304L. These two mechanisms would apply to all boiler tubes in all the heat exchangers and thus all the tubes would be equally exposed to this risk.

In an effort to increase heat transfer coefficients in heat exchangers in the ammonia process, "static" mixers are often inserted in boiler tubes. The increase in the heat transfer coefficients is accompanied by the disruption of the laminar boundary sub-layer which also leads a higher mass transfer coefficient, thus increasing the potential for leaching chromium and iron ions from the metal. The additional risk factor associated with presence of static mixers in tubes is the higher velocities in the pipes they cause by occupying space.

The tube inserts pose additional risks introduced at the design stage. They are pendula bodies which will undergo complex vibrations and translational motions during plant operation. The amplitude of the vibrations will depend on the flexural rigidity of the metal. Their motion pose two additional risks of tube thinning: they can impose forces normal at the tube wall which could disrupt the formation of replacement protective oxide layer (F. Addo-Yobo Unpublished work); they could also make contact with the tube wall, peel off existing protective layer and set off vibrations in the walls which can continually weaken the replacement oxide layer and increase its porosity to migrating metal ions thus accelerating erosion and corrosion thinning of the steel underneath.

Thus the operational risk factors are: steam Weight, temperature in boiler tubes, velocity in the boiler tube. Design risk factors are: presence and motion of inserts for promoting turbulence, vibrations; and Utility risk factors are: contamination of process water with chlorides and nitrates after the ion-exchange stage.

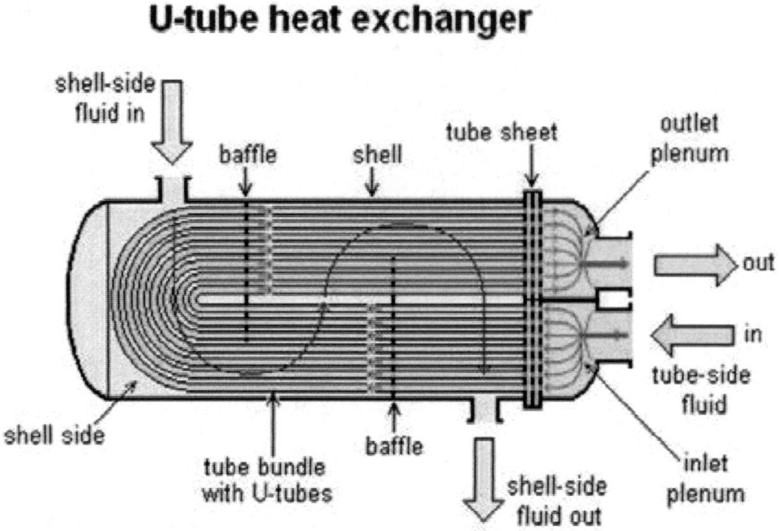


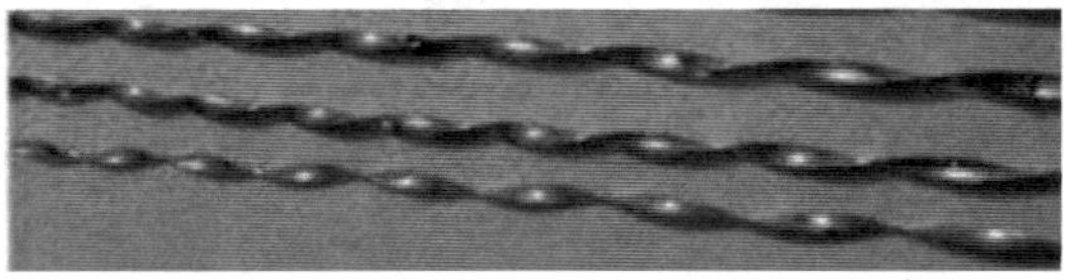

**The Models in Hysys**

he heat exchangers enclosed in the dotted red boundaries in the Process Flow Diagram above are modelled as a network. E2A is immediately downstream of the ammonia converter, E2B is the second in that stream. E111 is the exchanger in the front end stage of the process. The flow of Boiler Feed Water for cooling, in the boiler tubes in the heat exchanger E2A, E2B and E11, commence as single phase then becomes two phase with the generation of steam as heat is transferred from the process gases on the shell side. The dispersed two phase model in HYSYS was adopted to account for this.Whereas standard HYSYS models for heat exchanger could not accommodate finned and unfinned tubes in one exchanger, the rigorous EDR models were modified successfully to include this capability without changing the hydraulic and heat transfer characteristics. We were also able to incorporate variations in the process gas specific heat with composition without much computational expense.

From heat balance, the weight fraction of steam in the tubes ($v_f$) relates the process gas flowrates ($\dot{m}_G$) and temperature drop ($T^G$), BFW flow rates ($\dot{m}_G$) and changes in enthalpies ($H^W$). The steam weight was computed as shown in the adjacent equation.

$$v_f = \frac{\frac{\dot{m}_G}{\dot{m}_w}\left[C_{pc2}(T_1^G - T_2^G) + C_{pc1}\left(T_{1,in}^G - T_{2,out}^G\right)\right] + H_1^w - H_2^w}{H_2^v - H_2^w}$$

Thus confirming that the flow rate of the BFW is a suitable manipulated variable for controlling gas outlet temperatures. As stated above, the velocity of BFW in the tubes is important and its value can pose a risk. Its value varies with the weight of steam. The dispersed two phase model for steam-water mixes in HYSYS was used for modelling the flow and heat transfer to BFW in the tubes of the heat-exchanger. The shell side heat transfer coefficient in HYSYS are predicted by the well-known Dittus-Boelter equation.

The models built in HYSYS accurately predicted the output pressures and temperatures of the process gas entering and leaving the three heat exchangers, under standard conditions, within 3% percent of the expected values, as shown in the Table1. The simulations were carried out under standard operating conditions at 100% plant rate. The table also shows the risk values of these deviations from the TEMA values, which will be discussed later.

**Results and Discussion**

Often process Engineers find the need to increase the throughput of a Plant beyond the design values in order to improve productivity. However, this may carry additional risks of shortening the lifespan of the heat exchangers. In order to protect the heat exchangers and avoid overloading them, the product stream can be spilt and a percentage diverted around exchangers E2A and E2B. Table 2 compares output variables when model plant is operated with higher throughputs than the design levels. Columns 4,5,6 of Table 2 below show that the tubeside temperatures can be maintained within 0.3% of the design values with the bypass arrangement. However, steam weight increases beyond the prescribed TEMA values of 30%, thereby increasing potential operational risk. The weight fraction of steam increases beyond prescribed TEMA values with additional feed rate of 7% to 17%. Columns 5 & 6 of the above table show that at around 17% excess feed rate, increasing the cooling water rate by around 8% reduces the steam weight fraction to values close to the prescribed limit, thereby reducing operational risk and at the same time raising production rate.

The studies show that when 5% of tubes in E2A are blocked to mimic the response of industry to tube thinning beyond a certain level, to eliminate the risk spilling water from burst tubes wetting the product gases, the steam weight remains at reasonable levels close to the TEMA limit. However, the velocity in the tubes increase to values close to the prescribed TEMA limit because of the reduced cross sectional area available to the flow of BFW. The risk values of plugging tubes in exchanger E2A on exchangers E2B and E11 are negligible. Tables 3 summarizes the predictions of output variables caused by the plugging 10% of the tubes in E2A on E2A. The simulations show a reduction in the area available for heat transfer, in exchanger E2A by about 3% accompanied by a reduction in the heat duty of ca 4.3%. The reduced cross-sectional area available for flow on the tubeside results in 11% increase in the mean velocity through the tubeside and a 7% increase in the maximum velocity in the tube. The latter is close to the prescribed TEMA limit thereby increasing the operational risk posed to the plant by the blocking the tubes. The effect of this increase in velocity could be a shortening on the service life of E2A.

**Table 1: Comparison of the predicted plant output variables with the design historic values.**

| | Mass Flow Rate | % Deviation of predicted Temperature from Standard TEMA values | Risk Value | % Deviation of predicted Pressure from Standard TEMA values | Risk Value |
|---|---|---|---|---|---|
| | | | | | |
| EE11 Gas Inlet | Standard | 0.00 | None | -1.85% | None |
| EE11 Gas Outlet | Standard | -0.21% | None | | None |
| EE11 BFW Inlet | BFW rate | -0.39% | None | | None |
| E2A BFW Inlet | BFW rate | 0.18% | None | | None |
| E2A Gas Inlet | Standard | 0.00 | None | 1.31 | None |
| E2A Gas Outlet | Standard | -0.05% | None | 2.62 | None |
| To **** (E2A BFW Outlet) | BFW rate | -0.29% | None | -1.33 | None |
| BFW | BFW rate | 2.53 | | | None |
| E2B Gas Outlet | Standard | 1.57% | Negligible | 2.80 | None |

Table 2: Effect of Increasing the Plant Throughput

| KEY PROCESS PARAMETERS | Design Specification or Historic data | 100% PLANT RATE | 7% increase in feed process gas | 17.4% increase in feed process gas | 16.7% increase in feed process gas |
|---|---|---|---|---|---|

| | | | | | |
|---|---|---|---|---|---|
| **Process Gas Rate (kg/s)** | Standard | No deviation | +7% | +17.4% | +16.7% |
| **Deviation in Water circulation rate (Wt/Wt)** | Standard | No deviation | -17.2% | -21.9% | -15.7% |
| | | | | | |
| **Deviation in steam weight** $\frac{Predicted - Tema\ limit}{TEMA\ limit}$ | -23.3% | -23.3% | +33.73% | +46.8% | +13.4% |
| **Risk Status E2A due to steam weight** | None | None | Highest | Highest | Highest |
| **Deviation in tube-side outlet temperature of E2A** | None | -0.29% | -0.14% | -0.143% | -0.15% |
| **Risk value of tube temperature E2A** | None | None | None | None | None |
| | | Mean/Max | | | |
| **E2A TUBESIDE % Velocity deviation** | No Data | -61%/-20% | -64%/- 3.7% | - 63%/+3.6% | -63%/- 3.7% |
| **Risk Value E2A %Vel deviation** | N/A | No Risk | /Localized Medium Risk | /High risk | /Medium risk |
| **E2B % dev in TUBESIDE VELOCITY** | -51% | -59%/-57% | -66%/-64% | -66%/-64% | -63%/- 61% |
| **Risk value** | No risk | No risk | No risk | No risk | No risk |
| **E11 TUBESIDE VELOCITY (m/s)** | 1.92 | -60/-58 | -66/-64 | -66/-63 | -64%/- 61% |
| **Risk value in E11 tube side velocity** | | No risk | No risk | No risk | No risk |

**Table 3: The Impact of blocking off 5% (10%) of the tubes Exchanger E2A of Performance of E2A**

| Heat Exchanger Characteristics | E2A with no plugged tubes at | E2A with plugged tubes | Percent Deviation |
|---|---|---|---|

| | theoretical conditions | at conditions from DCS | |
|---|---|---|---|
| **Effective Surface Area available for Heat Transfer ($m^2$)** | * | * | -3.07 |
| **Heat Duty (kW)** | * | * | -4.35 |
| **Dev from Standard in Mean Temperature of Outlet of tubeside E2A** | -0.15% | -0.16% | |
| **Risk value** | No Risk | No risk | |
| **Corrected Mean Temperature Difference (°C)** | * | * | 7.34 |
| **Deviation in Heat Transfer Coefficient** | * | * | -8.1% |
| **Mean Shellside Velocity (m/s)** | -56.2% | -56.1% | 0.28 |
| **Maximum Shellside Velocity (m/s)** | -53.2% | -53.2% | 0.00 |
| **Risk value of deviation in maximum shell side velocity** | No risk | No risk | |
| **E2A TUBESIDE % mean Velocity deviation** $\frac{mean\ velocity - Tema\ vel\ limit}{TEMA\ velocity\ limit}$ | -63% | -59% | |
| **Risk Value E2A %Vel deviation** | None | None | |
| **Dev maximum Tubeside Velocity (%)** $\frac{Predicted\ max\ velocity - Tema\ vel\ limit}{TEMA\ velocity\ limit}$ | -6.7% | -0.23% | |
| **Risk value of Predicted maximum velocity** | Medium risk | Medium risk | |

**Discussion, Conclusion & Recommendation for Future work**

The studies have demonstrated the usefulness of building a rational model of the operation of the heat exchangers in the Ammonia process based on the diagrams presented in CEP (2015). They confirm the remedial actions taken by the Industry to improve productivity and maintain safe operation are sound. However, they point to the unfortunate fact that the quest to increase the heat transfer coefficients in the U-tube heat exchangers by inserting turbulent promoters are fraught with design risks that might not be possible to eliminate.

The simulations coupled with the present understanding of the mechanism of corro-erosion of high chromium stainless steel products (Folkeson et. al. 2011) suggest that once tube thinning is allowed to commence, caused by a risk factor introduced at the design stage, e.g. the inclusion of unfastened tube inserts, which increase heat transfer coefficients but which disturb the protective oxide layer, the exchangers may have to be eventually de-commissioned because of the difficulty of laying down a contiguous protective layer as shown in the schematic shown above right.

In order to firm up on the mechanisms that trigger tube thinning by corro-erosion of stainless steel boiler tubes in the Ammonia Process, further research into the risks posed by tube inserts and the careful weighing up of the cost of mitigating this factor as required by the Standard need to be carried out.

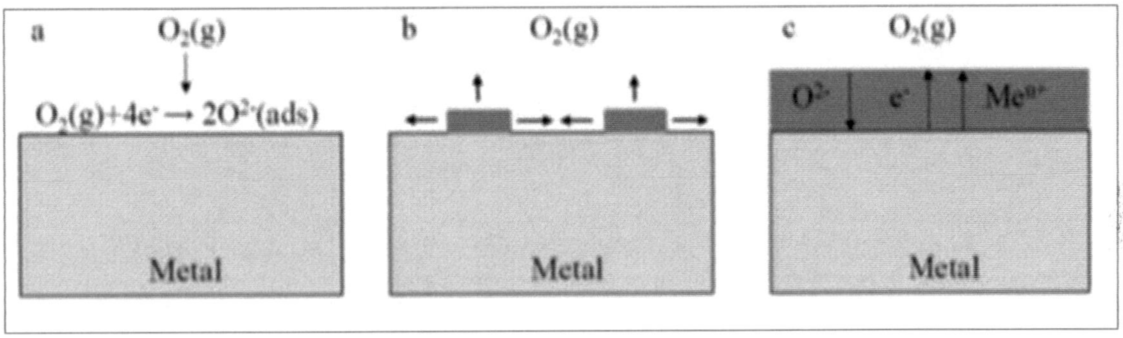


**Schematic of the initial formation of an oxide scale by adsorption of oxygen, oxide nucleation and growth to form a continuous oxide scale (stages a–c).**

## References


1. Venkat Pattabathula & Jim Richardson 2016, Introduction to Ammonia Production, Chemical Engineering Progress pp 69-75
2. Process Worldwide: 12:08:2020 Editor Ahlam Rais. Haldor Topsoe, BASF to Offer Ammonia Producer new Optimisation possibilities. ID 46800904 https://www.process-worldwide.com/haldor-topsoe-basf-to-offer-ammonia-producers-new-optimisation-possibilities-a-955495/
3. The Role of CHECWORKS in an Effective FAC Programme, David Smith and Jeffrey Horrowitz , ICONE 22, 22nd International Conference on Nuclear Engineering, 2014 Prague
4. Gao, Weimin, Li, Yungang and Kong, Lingxue 2017, Numerical investigation of erosion of tube sheet and tubes of a shell and tube heat exchanger, *Computers and chemical engineering*, vol. 96, pp. 115-127, doi: 10.1016/j.compchemeng.2016.10.005.Boiler tube failure in Heat Exchangers for Power generation in the Nuclear Industry

5. L.E Sanchez-Carrera, "Mechanism of Corrosion-Erosion in Steam Extraction lines in Power Station", PhD Thesis Dept of Mechanical Engineering MIT 1984

6. Jonas, O. (2004). FAC and Cavitation: Identification, assessment, monitoring, prevention. *Combined Cycle Journal*.

7. Suez Water Technologies and Solutions. (2020). *Handbook of Industrial Water Treatment Chapter 11- Preboiler & Industrial Boiler Corrosion Control.* Retrieved from https://www.suezwatertechnologies.com/handbook/chapter-11-preboiler-and-boiler-corrosion-control

8. Md. Moniruzzaman Bhuyan1, Ujjwal K. Deb, M. Shahriar, Simul Acherjee 2017, Simulation of Heat Transfer in a Tubular Pipe Using Different Twisted Tape Inserts, Open Journal of Fluid Dynamics, 2017, 7, 397-409, http://www.scirp.org/journal/ojfd

9. Folkeson, N., Jonsson, T. Halvarsson, M. Johansson, L.-G., Svensson J.-E., The influence of small amounts of KCl(s) on the high temperature corrosion of a Fe-2.25Cr-1Mo steel at 400 and 500°C. *Materials and Corrosion*, (2011), 62(7), pp. 606–610

10. F, Addo-Yobo 2021, Motion of Pendula inserts in heat exchanger tubes in the Ammonia Process and their impact on exchanger risk, Unpublished work.

# 617202 Optimal Placement of Gas Detectors Using Dispersion Modeling

**Abstract:**

Historically, gas detector placement was performed by experienced designers using rules of thumb and judgment based on fundamental principles of gas cloud behavior. Unfortunately, designs using these approaches have been inconsistent and, in many cases, poorly performing. Historical data from the UK HSE shows that certain fixed gas detection systems routinely were only able to detect less than 70% of "major" gas releases. In the past decade, advances in computer modeling and 3D graphics has enabled the implementation of chemical process quantitative risk analysis (CPQRA) in the service of gas detection system design. The general process for using CPQRA to develop and verify a gas detection performance metric called coverage, which is essentially the fraction of the all release scenarios that are detected by a gas detector array, has been documented in the International Society for Automation (ISA) technical report ISA TR84.00.07 – Guidance on the Evaluation of Fire, Combustible Gas, and Toxic Gas System Effectiveness. This tutorial provides an overview level explanation of the workflow for the process of gas detection mapping, including the risk-based selection of performance targets and quantitative verification of the achievement of the coverage performance targets using dispersion modeling results and leak rate estimations that are similar to the proven approaches used in CPQRA. The tutorial goes through the entire process of gas detector placement for a sample facility where liquefied petroleum gas products are stored, beginning with risk assessment and going all the way through a completed gas detector layout.

<table>
<tr><th>Presenter Information</th><th>Contact Information</th></tr>
<tr><td rowspan="2">Presenter Name:<br>Bejoy K. Bharatiya</td><td>Contact Name: Bejoy Bharatiya</td></tr>
<tr><td>Telephone No.: +61 413 098 008</td></tr>
<tr><td rowspan="3">Presenter Biography:<br>Bejoy Bharatiya was Process Safety Advisor at DNV GL Offshore and Industrial Services (B) Sdn. Bhd. He has over 30 years of professional experience in the upstream oil & gas, refining, petrochemical, and mineral processing industry. Prior to joining DNV, he held various positions, Process Safety Group Lead at Saudi Aramco Shell Refinery Company (SASEF), Senior Process Safety Engineer at SANTOS (South Australian Oil Search) Pty Ltd. Australia, Senior Operations Engineer with Saudi Aramco Gas Operations, and the Principal Process Engineer of Kellogg, Brown & Root (KBR), Australia.</td><td>Fax No.:</td></tr>
<tr><td>E-mail Address (Required):<br>bejoy2014@gmail.com<br>pinkubejoy@iinet.net.au</td></tr>
<tr><td>Your Company/Department:</td></tr>
</table>

**Abstract (max of 1500 words)**


## HSE Philosophies in Capital Projects of Petroleum Refinery and Petrochemical Complex - A Practical Approach

Bejoy K. Bharatiya

The petroleum industry has long since recognized that certain risks are involved in the major hazard industries. It becomes a major challenge to minimize risk by comprehensive HSE framework in the project execution of a major process facility in the hydrocarbon industry. Over the years, the oil and gas industry has learned lessons from a number of process safety events and various hazards such as loss of control and loss of containment top events that could result in: fires, vapour cloud explosions (VCE), toxic and asphyxiant exposure, which occurred in various parts of the world and are not limited to the country of origin. The design and operation of process facility for the refinery and petrochemical complex have been the subject of major hazard to the oil industry.

It has been the continuous effort of the industry to update the design and operating philosophies to improve the inherent safety of the processing units. The present paper is concerned with defining the major safety considerations of international oil companies, and Industry Standards in designing and operating the refinery and petrochemical process facilities. HSE philosophies are required to guide and drive the design intent of HSE critical systems. The main safety philosophies include: ESD and blowdown, fire and gas detection, toxic gas (H2S) detection, active and passive fire protection (including guidelines for firefighting systems and fire protection), layout HSE philosophy, process safety and containment etc.

This paper provides the basic safety design guidelines for a major project but does not attempt to cover in detail every aspect of safety design. Reference should be made, therefore, to specific discipline philosophies and specifications as appropriate.

The primary environmental philosophies include venting and flaring of hydrocarbons and H2S/SO2, CO2 production, solid waste and effluent wastewater disposal, energy efficiency etc. In the project planning and execution, these philosophies must be effectively embedded in the design through the HSE premises document.

For the major project execution, the HSE philosophies are written in a way to instruct the design functional specifications to be compatible and aligned with the existing facility, but are updated to take into account of new technology, new knowledge and other industry requirements.

HSE philosophy shall be considered throughout all the project development phases, engineering, construction, start-up, and operation via dedicated and well-structured efforts. The safety philosophy, and resulting protection of persons and assets, is combining all provisions undertaking during all project phases.

HSE Philosophy for the detailed engineering phase is prepared with the objective to elaborate the Project design philosophy with respect to identification and elimination of hazards and to promote a design that is inherently safe. Objectives of this HSE philosophy are:

- Provide a safe working environment for personnel
- Minimize the potential for hazardous occurrences
- Avoid exposure of personnel to potential hazard
- Contain and minimize the effects of hazards
- Provide adequate safe access and escape away from hazards
- Protect the environment
- Provide a design that is fit for purpose for field life with the required integrity for safety and environmental protection

**HSE Design Philosophies**

During the detailed engineering, and subsequent construction and commissioning, particular care will be taken by achieving:

- Good engineering practice
- Application of an all-encompassing QA system
- Development and implementation of a safety management system
- Design reviews at 30%, 60% and 90%, layout review, HAZID, Design Safety Review (DSR), accessibility reviews and other reviews.
- Qualitative and quantitative risk assessments and IPF review
- Design of installation for extreme weather conditions
- Execution of an extensive pre-start up review of the plant before introduction of feed (Hydrocarbon, gas) covering both hardware and, control systems and procedures.
- Conformity check with local regulations, project standards and other applicable industry standards and directives.

**Hazard assessment** – Hazard assessment is based upon the hazard of the materials present in the processing facility. Some of the highest-ranking typical hazards included in the risk assessment (to be taken from facility Hazard Register) are:

| Hazard | Top Event | Consequence |
|---|---|---|
| All Liquid Hydrocarbons (HC) including condensed flare gas | Loss of Containment | Fire |
| Toxics / H2S / N2 /BTX (Aromatics-Benzene, Toluene, Xylene) | Loss of Control or Containment | Personal Exposure |
| All HC, including H2 enriched HC | Loss of Containment | Generation of vapour cloud leading to VCE or fire |
| Heavier than air HC | Loss of Containment | Vapour cloud and/or Fire resulting from heaving than air HC |

The design for the active/passive fire protection systems, fire detection systems, gas detection systems and general HSE requirements shall comply with the HSE Philosophy, project standards and referenced codes/standards, specifications, and regulations.

The fire safety of a plant is the degree of reduction in vulnerability to fire and/or explosion incidents in terms of their probability of occurrence and associated potential damage. This degree is maximized by the integrated application of fire prevention, fire protection and fire-fighting measures.

The philosophy is to achieve the desired degree of fire safety primarily by fire prevention measures, which will also result in an optimum availability of the facilities to produce. Where despite the preventive measures potential fire hazards remain, fire protection measures are taken. The basic objective of fire protection is to limit or prevent the escalation of a fire to avoid risk to life and to minimize material damage.

The Project shall be divided into separate fire risk areas of possible fire involvement. This separation may include major pipe racks, roadways, or vacant space. For estimating firewater demand, only one major fire occurring in one fire risk area at any time is to be considered. The fire water system shall be developed to meet the highest firewater demand of the worst-case firewater scenario. Extension of the existing firewater system on the Refinery to cover the new installations will be considered. Harmonization of fire safety measures shall be maximized.

In addition to fire protection measures, the plant fire-fighting organization is available to control and/or extinguish fires that occur despite the fire prevention and fire protection. All shift operators to be trained in industrial firefighting form the first intervention team (FIT) and additional fire fighters will be available from the Fire Station Crew. The combination of FIT and the existing Fire Station Crew will form an effective team, available within minutes after an incident is detected. If required, extra support can be provided through the external fire brigade.

There shall always be more emphasis on fire prevention, fire protection and fixed and semi-fixed firefighting systems than on the capacity of any fire-fighting organization.

The existing fire safety facilities in terms of mobile fire-fighting equipment and fire brigade organization of the refinery complex will be used to a maximum extent.

Furthermore, as the refinery hydrotreating and Sulphur Recovery Units process/handle large quantities of $H_2S/SO_2$, special emphasis shall be made during the design phase for the installation of suitable system for $H_2S$ service such as high mechanical integrity (e.g. materials, pipe types, double pumps seals etc.), safeguarding, $H_2S$ detection and automatic isolation/shutdown system. Design of this system shall be as per project standards and actual HSE philosophy for $H_2S$ detection system.

Safety relief venting shall be provided for equipment, which is subject to internal pressure build-up from an external heat source. Pressure relieving devices shall be designed in accordance with project standards and/or industry standards.

Plant layout and spacing shall be developed based on project standards and industry practices, e.g. API RP/STDs, General Electric GAP Services (GE GAP) 2.5.2 "Oil and Chemical Plant Layout Spacing. Safety distances will be optimized to reduce the area utilized once more detailed design is progressed and risk studies have been completed (e.g. Physical effect modeling).

*This paper discusses industry experience and summarizes the salient design considerations and risk assessment to improve safety integrity in the execution of capital projects for refinery and petrochemical complex. An effort has been made to highlight the HSE methodologies adopted by major oil companies using project standards, and industry best practices and standards like API, ASME, NFPA, IEC, ISO etc. Emphasis has been made to increase the awareness among the discipline design engineers and process safety engineers to develop safer designs of the refinery and petrochemical facilities.*

# 619065 Addressing Aging Processes and Equipment

**Abstract:**

Number of the fertilizers plants across the globe have been installed and are in operation since many decades. The operation with the old equipment and conventional process techniques is being managed coupled with budget constraints. Hence the old aged plant equipment have high risk of failure under the shed of outdated processes, and may cause severe damage to the plant assets as well as to human. Consequently, one has to bear more and repeated maintenance cost and regular replacements of equipment.

Fatima Fert Plant (Previously Daud Hercules) also commissioned in 1969 and is in operation to date. Dealing with the challenge of old aged equipment and process techniques, very strict and robust reliability plans have been established and implemented to inspect the equipment with the advanced ultrasonic techniques and to make sure that the asset is in operation under controlled parameters. Based upon the finding of inspections, 10 year rolling plan have been developed to replace the equipment that are prone to failure.

The paper will be providing the details of the challenges and risks faced due to old age equipment and the way forward to mitigate and control that risk with optimized budget utilization.

# 619392 Why Is There a Regularity of Process Incidents Occurring, or Is It All Just a Re-Occurrence of the Past?

**Abstract:**

We attend conferences to hear the experiences of others, learn lessons from past events and take that experience and learning back to our place of work. Yet a simple analysis of suggest there is a plateau in the progress to reducing the frequency of process safety incidents.

So perhaps the question to address is, "Why are lessons not being learnt"?

This paper will highlight various data demonstrating the "plateauing" of performance; highlight particular recurring indicators that need to focused on and look at how learning may be distributed to those that need to increase their knowledge of PSM. There will be a particular emphasis of the use of case studies in this approach.

The paper will particularly interest those still on the learning journey in PSM and those questioning alternative approaches.

# 618397 How You Manage the Actions Coming through Process Safety Management Systems?

**Abstract:**

For improving process safety in the plant, we are putting great afford managing task of operators. Alarm management is important for control operator's tanks. Quantity of the alarms should not be more then control operator can manage.

However, there is gap in engineer's tasks management. Workload of an engineer is an important parameter in complex plants. This workload can be rapidly increase with HSE related tasks. As a result of Management of Change, Incident investigations, ... most of the tasks are given to operations engineers, projects engineers, maintenance engineers.

If the workload is too high it is easy for the engineer to miss the important actions. With HSE Action Portal, we aim to give management a view of task load for each department. The department responsible can focus on the overall tasks of the department

HSE Action Portal make possible to reach actions and tasks coming through PSM system elements such as Management of Change, Pre-Start up safety review, incident/accident investigations and process hazard analysis in the same platform. Users are able to analyse their action/tasks in terms of number and completion percent. In this way, users are triggered to follow closely their actions and complete them in a correct way.

It is an useful tool also for top management in order to collect data and create statistics periodically. It can be seen actions/task situations for each refinery and department.

# 617997 Guidelines for Risk Ranking Hazardous Scenarios When Conducting PHAs

**Abstract:**

Most of the companies in the process industry use corporate guidelines, or dedicated criteria and principles for conducting Process Hazard Analysis (PHAs). Examples of factors addressed are the following: nodes selection, process deviations to be brainstormed, level of detail when writing causes/consequences, safeguards categorization, risk matrix to be used, and how to risk rank the identified scenarios.

This paper collects different approaches from leading PHAs at several companies. An analysis comparing these approaches is provided and strengths and weaknesses are identified. Finally, based on safeguards and consequences categorization, guidelines are proposed for consistently risk rank hazardous scenarios.

# 619107 Practical Guide to Operationalize & Visualize Cumulative Process Safety Risks Perspective

**Abstract:**

The paper describes how Process Safety Management can be seamlessly integrated with Operational Risk Management. It demonstrates how operational processes like Permit to Work, MOC, Job Hazard Analysis, Incident Management and Process Safety Management can be brought together to support day-to-day decision making based on the actual cumulative risk status of the facility.

Key thinking is developed to answer questions such as *How do we make Process Safety Risk as visible as HSE* and *How can we prevent previously identified hazardous sequences of events from reoccurring?* , thus increasing the entire system safety.

This paper presents practical approaches in overcoming the challenges in collecting data from disparate sources and learnings on realising Dynamic Operational Risk Management (ORM) from various client implementations. In addition, it touches upon topics such as *How to analyze the data* and *Use of machine learning to predict future risk* .

The paper presents a risk calculation methodology that calculates barrier status and visualizes cumulative risk:

- in a standardized barrier model, based on IOGP 544 standard, in Swiss-Cheese view
- geographically by site and area/location using site graphics
- per critical scenario in a BowTie view

It connects and visualizes process safety risk by automatically collecting data for Tier-1,2 and 3 safety indicators that are covered by various processes spread across the organization.

Furthermore, it shows how the cumulative risk status can be used in operational processes, potential conflict detection and planning of activities to manage and minimize the actual risk.

*Keywords: Barrier Management, Process Safety Management, BowTie Methodology, Swiss-Cheese Model, Operational Risk Management, Project Implementation Methodologies, permit-to-work, MOC, JHA/RA, Machine Learning*

## ENGINEERING & OPERATIONS

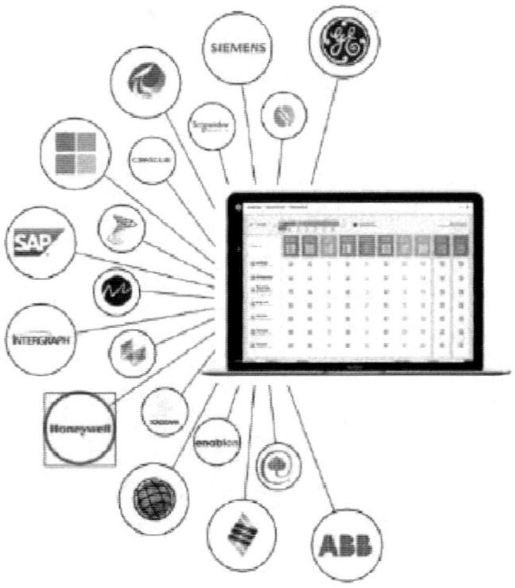


Wolters Kluwer Enablon Proprietary and Confidential

## BARRIER MANAGEMENT INTEGRATION DATA POINTS

**Integrity Data:** CMMS, Audits & Assurance, Competence Management, Inspections, Engineering data.

**Operational Intervention:** Process Control System / Historian, Shift Logs, Operator Rounds, Permits, Temporary defeats.

**Barrier Challenges & Failures:** Incident Management, Management of Change, Process upsets, PSM events.

## BARRIER MANAGEMENT – PROCESS SAFETY MANAGEMENT

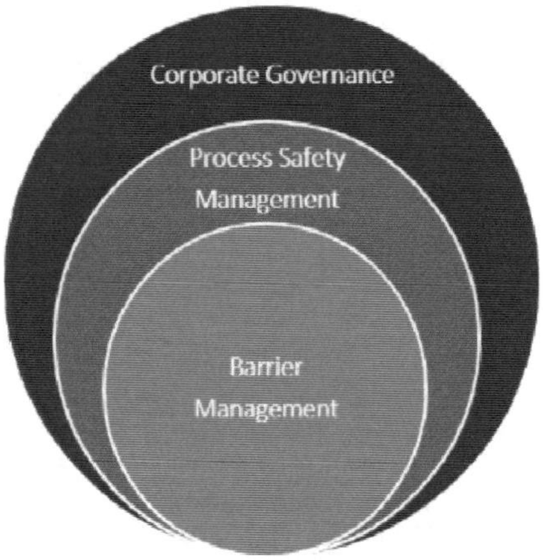


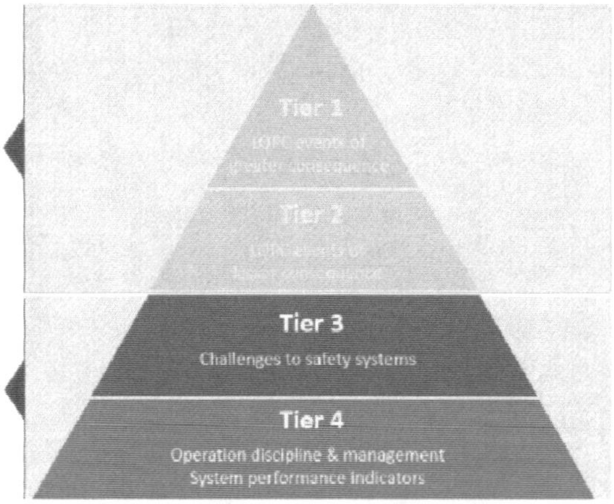


The focus of Barrier Management

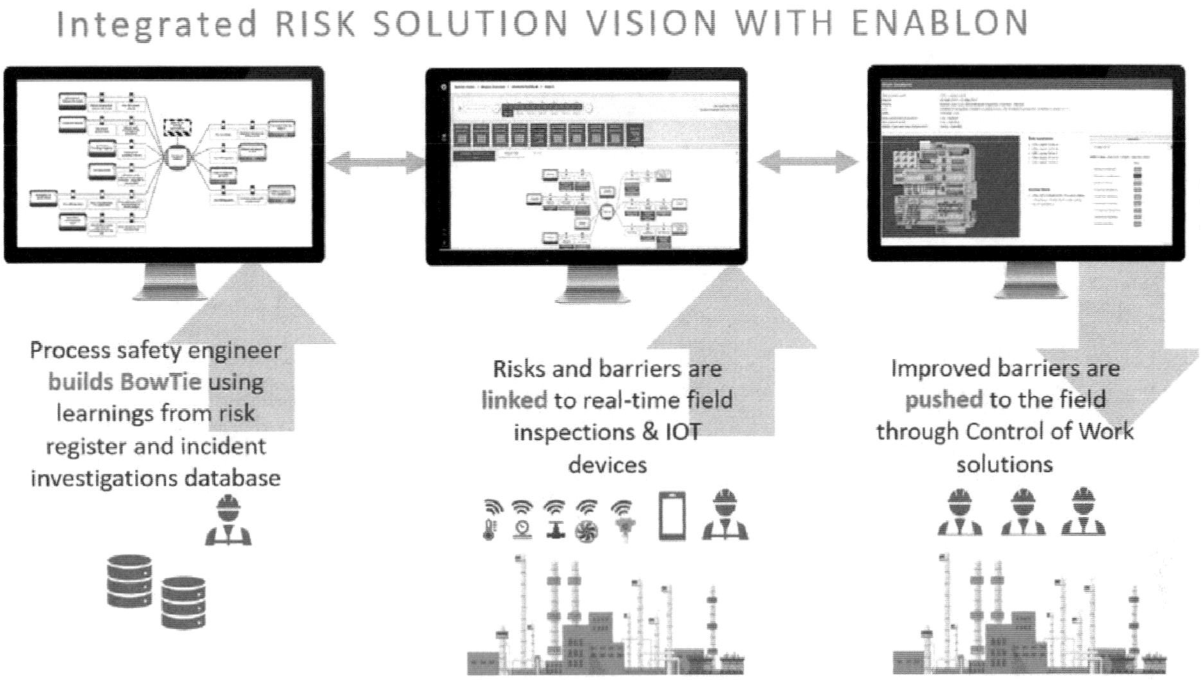
Integrated RISK SOLUTION VISION WITH ENABLON
Process safety engineer builds BowTie using learnings from risk register and incident investigations database
Risks and barriers are linked to real-time field inspections & IOT devices
Improved barriers are pushed to the field through Control of Work solutions

# 619126 Controlling Human Performance Between Both Unplanned and Planned Tasks within Abnormal Operation Mode

**Abstract:**

There is confusion in terminology used in the chemical-related industry for the class of procedures commonly referred to as Abnormal Mode of Operation and Abnormal situation management. This paper provides a clear definition of each mode of operation and gives examples of how the human performance is controlled for each.

1. Normal – either a continuous mode or a normal batch mode of operation.
2. Planned Non-routine (Non-Normal)– Startup, shutdown, are online maintenance are the main non-routine or non-normal modes of planned operation. But planned Temporary procedures, with time limits, are also part of these.
3. Abnormal– These include those activities covered by generalize procedures or guides (planned in a general sense) the activities that don't have written, step-by-step procedures (but noting that many companies have guides for most of these abnormal situations). A further breakdown is possible:
    1. Response to upsets using a Trouble-shooting Guides(TSG) for handling deviations from the operating window (these or normally triggered on an alarm); these may ultimately lead to a shutdown, safe park, or emergency shutdown (all of which are also proceduralized), if the deviation cannot be corrected in time.
    2. Response to failures using a Temporary procedure - such as how to run in bypass mode if the flow controller fails and you want to keep running in manual mode.
    3. Response to Unanticipated Events - The FIRST occurrence of this is handled by what is generally referred to as Emergency-MOC, which is really saying we will make a change immediately, and do the risk review of the change later, and then learn from this one case and proceduralize into a Temp Procedure or TSG for the next time it comes up.
4. Emergency Operations- a diminished or reduced operating plan; normally a Temporary procedure (similar to 3.b. above)
5. Emergency Shutdown - a TSG that fails to resolve the issue in time will go to this; or for some events such as a sudden loss of containment, we go straight to these
6. Emergency Response- normally in conjunction with 4 or 5, but focused on the protection of people, assets, environment, given the release or other imminent harm is in play.

This paper will provide a framework to ensure no mode of operation is overlooked and it will help sites understand what is needed to control risk during each mode of operation. Most of the time in the paper and presentation will be focused on classification 3, Abnormal Mode of operation.

# 619218 Proactive Lopc Indicator Critical Analysis through Power BI Dashboard

**Abstract:**

This work applies Power BI dashboard in order to proceed a critical analysis of proactive loss of primary containment indicator so that possible process problems can be detected before greater events occur by prognoses and decision-making to direct resource management in a more assertive way.

A process safety event is an unplanned or uncontrolled release of any material including non-toxic and non-flammable materials from a process, or an undesired event or condition that under slightly different circumstances could have resulted in a release of material (API 754). The pyramid model can represent the predictive relationship between lower and higher events.

It is important to guarantee the good register of pyramid base events (Tier 3) to achieve enough information, identify the fail installation's weakness, avoid higher process safety events, reduce losses and improve the facility performance. Nevertheless, the only register is not enough to achieve these improvements. Its's necessary to quantify and analyze the data to guide the decision-making. A critical analysis involves the following parameters: leaked product, fail causes, equipment that failed, engineer discipline, LOPC location, consequences.

The first step to make the critical analysis is having a reliable database. This step involves training the operational team in leading and lagging process safety indicators and process safety management. The communication must be direct and clear. Furthermore, it is necessary to provide a satisfactory registration platform with the necessary fields to carry out the subsequent critical analysis.

After that, it is possible to extract all the information to proceed the proactive LOPC indicator critical analysis. At this point Business Intelligence (BI) presents itself as an adequate tool. Power BI, from Microsoft, is one of the most prominent BI tools. It is a collection of software services, applications and connectors that work together to consolidate, make information from different data sources coherent, visually engaging and interactive (from an Excel spreadsheet to a cloud-based data collection).

Through the LOPC critical analysis using Power BI dashboard it is possible to evaluate the Tier 3 incidents, focused on developing a learning approach. Therefore, it is possible to prevent and monitor failures trends and effectiveness of implemented actions. The critical analysis permits to direct the resource management by identifying the equipment and systems with recurrent failure events. It is worth mentioning the effective contribution for Process Safety Culture, since the leadership, operation and related areas could see the cause-effect relationships between the minor events and the daily operations.

# 619342 A Case Study on Ensuring "ALARP" Demonstration to Justify Continued Operation with Respect to "Mah" By Integrating Process Safety Techniques for Existing Process Facility

**Abstract:**

A lot have been written on process safety. However, practical applications of all these theories & studies has its own challenges. At the end, after overcoming the challenges as well as developing, implementing & monitoring strategies, a risk to existing facilities can be reduced to ALARP/ acceptable level.

This paper will throw light & discuss about requirements to integrate different process safety strategies, challenges during implementation & their solutions to demonstrate that existing oil & gas facility is safe for operation, through a practical case study of one complete process train. The methodology includes;

- Identifications of process trains
- Data collection
- Data Review
- Review past changes & their implementation
- Update records & P&IDs
- Carry out/ Update Integrated PHA with process simulation
- Development of Failure Scenario
- Risk Calculation considering existing safeguard (preventive and mitigating measures)
- Risk submission and ALARP demonstration
- Compare risk
- Identification of safety critical elements (SCEs) & Safety Critical Tasks (SCTs)
- Development of Performance Standard of SCEs
- Developing management system of PLAN-DO-CHECK-ACT
- Integrating the same in Audit/Inspection
- Develop training through simulations

Challenges faced by team for existing facility to achieve optimum process safety & ensure ALARP level of risk, and their solutions will be discussed.

## Conclusion

This exercise provides evidence that wherever practical, the design and operations of the existing oil & gas facility has attained a level of risk that Justifies Continued Operation with respect to major fire and explosion events. This demonstration of Justification for Continued Operation is based on the number of safety measures inbuilt at design stage, Quantitative Risk Assessment and various other process safety studies revalidated during operational phase of the oil & gas process facility as well as after implementing changes through MoC. It must be noted that a separate process of action tracking is in place to prioritize and monitor implementation of recommendations provided to reduce the risk further to As Low As Reasonably Practicable (ALARP). A systematic approach of Plan-Do-Check-Act is developed to ensure that, safety critical tasks are carried out in timely manner by competent personnel, to ensure safety critical elements are in place and functional in line with performance standard.

# 616668 Towards Management Systems Perfection – Successes with Implementing a Layer of Verification Analysis to Continually Treat Process Safety Risks

**Abstract:**

Trust, but verify, by continually checking everything! This paper will demonstrate the importance of verification as a critical step in continually improving the treatment of process safety risks, including implementing inherently safer designs and operating practices, and controlling residual risks.

A well-established design concept for management systems is the continual improvement lifecycle involving the steps: Plan-Do-Check-Act (PDCA). Similar to standards for managing quality, environmental, occupational health and safety, and security, models for effective process safety management (PSM) systems also utilize the PDCA cycle. At a higher level, the PSM elements of Mechanical Integrity, Incident Investigation and Compliance Audits address verification in the "Check" step; while every management process of the PSM System should incorporate a "Check" step with verification processes and procedures to monitor and assure compliance and performance. This paper will present a taxonomy of activities for meticulous verification against policies, objectives, plans and other requirements for a vibrant PSM system, showing the importance of examining, analyzing, assessing, confirming, inspecting, testing, and reviewing planned activities, desired conditions and achieved results.

Attendees will benefit from practical takeaways of how each management system element can be developed with "verification designed in" and a verification culture created to treat process safety risks at three opportune times: before the incident (prevention), during the incident (emergency management) and after the incident (incident investigation, root cause analysis and lessons learned).

A comprehensive case study with practical examples will be presented to demonstrate a proven technique of Layers of Verification Analysis (LOVA) as a best practice integrated into the conduct of a process hazard analysis (PHA). The case study will show the audience "how to" enhance the PHA facilitation wearing an "auditor's hat" to raise questions and verify the effectiveness of existing safeguards, all of which are managed by the PSM System. Borrowing from the basic principles of management systems auditing, LOVA asks these questions for each hazard scenario: are systems, processes and procedures associated with causes (of human errors and equipment malfunctions), with consequences (of physical phenomena and resulting impacts), and with safeguards (for both engineering and administrative controls) *(1) in place?, (2) adequately designed? and (3) effectively implemented?* LOVA has also been used for supplemental analyses within the PHA to address human factors, previous incidents, facility siting and changes (e.g. MOCs). Recommendations from the LOVA in PHAs can be used as lessons learned for improving the implementation of the PSM system on a daily basis, without having to wait on the next periodic audit or PHA revalidation. The LOVA technique has been shown to be equally suitable for any stage of the process lifecycle, including inherently safer design evaluations.

The paper will also present three incident case histories to show how missing, inadequately designed or ineffectively implemented verification processes and activities have contributed to process safety incidents.

This paper will draw upon the authors' extensive experiences in designing, developing, implementing and auditing PSM systems, as well as conducting PHAs, building siting evaluations, and incident investigations.

# 617112 Study and Experimental Validation of CO2 Blowdown at Cryogenic Conditions

**Abstract:**

$CO_2$ solidification is one of the critical issues being identified in cryogenic technology process, particularly during the blowdown process. This poses critical challenge in developing the technology for separation of $CO_2$ from natural gas as there are limited studies and data to support the design and simulation of the process. Depressurization of high $CO_2$ fluid at cryogenic temperature would lead to possibility of CO2 solid formation, hence potential blockage of process equipment and venting line.

The new cryogenic technologies are currently being tested for the proof of concept. Hence, a pilot plant, which is a scaled down version of the technology was developed. One of the major challenges faced during the pilot plant testing is the emergency depressurization philosophy as the process involves $CO_2$ solids handling which is uncommon to the industry standard. Lack of experimental data of $CO_2$ blowdown at these conditions presented a technical gap which needs to be addressed to properly predict the solidification and mitigations, hence avoiding over-design of the plant and chances of accidents and unscheduled downtimes can be minimized.

Therefore, this paper will focus on the experimental study which have been done on $CO_2$ blowdown scenarios at a pilot plant. The pressure and temperature profile provide real insight of the solidification of CO2 during several blowdown scenarios. Finally the paper would validate a simulation data conducted using the experimental data as one of the potential prediction tools for CO2 blowdown process.

# 617588 The Supervisors Next Big Challenge: How to Ensure Operational Discipline on the Work Fronts

**Abstract:**

### The Supervisors Next Big Challenge:

### How to Ensure Operational Discipline on the Work Fronts

There is a problem that the petrochemical and oil and gas industries have not solved effectively: how to transform work fronts in environments based on operational discipline.

All the commitment and effort that an organization mobilizes in its risk management, finally, is updated in what happens between supervisors and operators on the work fronts. That is why it is so necessary to work on the alignment of the workforce of operating companies and contractors.

In the last decade, Senior Management became more aware of its role: defining what were tolerable risks and assuring resources. Process Safety, Integrity and Environment ceased to be discursive themes, to effectively have a genuine place at the small table where strategic decisions are made. The biggest challenge has been getting it to operate according to those definitions.

Supervision is carried out on the work front, the true executive arm of the organizations in the field. And for this organizational segment to ensure a disciplined operation, they need to be able to manage the culture.

A common mistake in the risk management approach has been to think of the lack of operational discipline of supervision as a problem of isolated individuals, and not of the context in which these people work, that is, of the context that the activities provide. organizations, from the context produced by culture.

When analyzing the supervision segment, one should not think of undisciplined people, but cultures that tolerate them, or that reinforce risky behaviors and shortcuts to the established norm. If supervision does not ensure operational discipline on the work fronts, it is because the organization does not provide the conditions for this to occur. People are the reflection of the culture in which they operate. Individuals do not impose a personal style of managing risks in isolation, but adapt to group agreements, displaying safe or unsafe behaviors depending on what is collectively tolerated. Field supervisors only take risks that the culture allows them to take.

The change of mindset in the supervision segment should not be built on the basis of shortcuts or "crash plans", but rather in avoiding in a progressive and lasting way the repetition of incidents, through a transformative process: generating a culture of discipline, learning and mutual trust in the work fronts, based on group agreements between operators and contractor companies, managed by a trained supervision to make decisions based on risks, combat the normalization of deviance and keep in mind the sense of vulnerability.

The key is to generate the conditions for effective learning. In addressing group problems, and not individual ones. The challenge is to create the conditions so that operational discipline is not a declaration of principles but a reality in the field. For this they do not need new knowledge, but to deepen what they already know, in the right context.

The pandemic, paradoxically, provided an opportunity to improve the quality of supervision development, personalizing it. Although online self-training has been around for a long time, it is not the most appropriate tool to generate changes in operations. For that, culture needs to intervene.

The next decade seems to keep a central challenge: transforming supervisory teams into custodians of culture and guarantors of risk management, through work fronts based on operational discipline.

| Presenter Information | Contact Information |
|---|---|
| **Presenter Name**:<br>Bejoy K. Bharatiya | **Contact Name**: Bejoy Bharatiya<br>Telephone No.: +61 413 098 008 |
| **Presenter Biography:**<br>Bejoy Bharatiya was Process Safety Advisor at DNV GL Offshore and Industrial Services (B) Sdn. Bhd. He has over 30 years of professional experience in the upstream oil & gas, refining, petrochemical, and mineral processing industry. Prior to joining DNV, he held various positions, Process Safety Group Lead at Saudi Aramco Shell Refinery Company (SASEF), Senior Process Safety Engineer at SANTOS (South Australian Oil Search) Pty Ltd. Australia, Senior Operations Engineer with Saudi Aramco Gas Operations, and the Principal Process Engineer of Kellogg, Brown & Root (KBR), Australia. | Fax No.:<br>E-mail Address (Required):<br>**bejoyb@hotmail.com**<br>**bejoy2014@gmail.com**<br>Your Company/Department: |



**Abstract (max of 1500 words)**

## Project Assurance Review and Audit Framework for Capital Project of a Petroleum Refinery - A Process Safety Approach

Bejoy K. Bharatiya

The petroleum industry has long since recognized that certain risks are involved in the major hazard industries. It becomes a major challenge to minimize risk by **comprehensive project assurance/audit framework** in the project execution of a major process facility in the hydrocarbon industry. Over the years, the oil and gas industry has learned lessons from a number of process safety events and various hazards such as loss of control and loss of containment top events that could result in: fires, vapour cloud explosions, toxic and asphyxiant exposure, which occurred in various parts of the world and are not limited to the country of origin. The design and operation of process facility for the refinery and petrochemical complex have been the subject of major hazard to the oil industry.

It has been the continuous effort of the industry to update the design and operating philosophies to improve the inherent safety of the processing units. The present paper is concerned with defining the major safety considerations of international oil companies, and Industry Standards in designing and operating the refinery and petrochemical process facilities. Inherently safe design and structured project assurance and audit program are mandatory to guide and drive the design intent of HSE critical systems. The main safety philosophies include: ESD and blowdown, fire and gas detection, toxic gas ($H_2S$) detection, active and passive fire protection (including guidelines for firefighting systems and fire protection), layout HSE philosophy, process safety and containment etc.

This paper provides the **assurance/audit framework** for a major project but does not attempt to cover in detail every aspect of safety design. Reference should be made, therefore, to specific discipline philosophies and specifications as appropriate. (Refer to **Figure 1**).

Project audit or assurance program shall be applied throughout all the project development phases, engineering, construction, start-up, and operation via dedicated and well-structured efforts. The safety philosophy, and resulting protection of persons and assets, is combining all provisions undertaking during all project phases. (Refer to gated process, 1-6 **Steps** explained below).

### HSE Design Philosophies

During the detailed engineering, and subsequent construction and commissioning, particular care will be taken by achieving:

- Good engineering practice
- Application of an all-encompassing QA system
- Development and implementation of a safety management system
- Design reviews at 30%, 60% and 90%, layout review, HAZID, Design Safety Review (DSR), accessibility reviews and other reviews.
- Qualitative and quantitative risk assessments and IPF review
- Design of installation for extreme weather conditions
- Execution of an extensive pre-start up review of the plant before introduction of feed (Hydrocarbon, gas) covering both hardware and, control systems and procedures.

- Conformity check with local regulations, project standards and other applicable industry standards and directives.

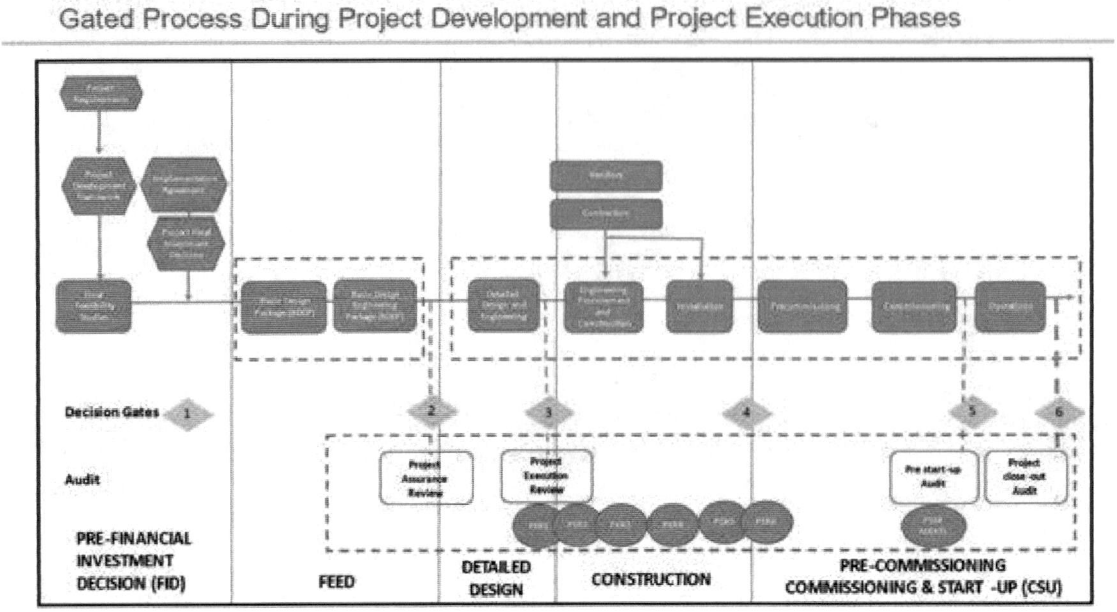


**Figure-1** Gated Process for Project Assurance/Audit Framework for Project Execution

**Project Assurance Review/Audit** is required at all stages of project execution.

**1. Pre-Financial Investment Decision (FID) stage:**

1) Management: review critical contractor's capability, business license, & project experiences, PEP, schedule.
2) QA/QC: review critical contractor's ISO 9000 certificate, QA/QC plan, standards/policies/procedures (**SPP**), sample check of work records/reports etc. (see Table-1 International Codes and Standard)
3) HSE: review critical contractor's ISO 18000 certificate, project HSE plan, SPP, sample check of work records/reports etc.
4) Technical: review Feasibility Study, Environmental Impact Assessment (EIA) reports., compare critical design area engineering standards against international standards as well as best industry practices, etc.

**Deliverable**: review comments or audit findings

**2. Front-end Engineering Design (FEED) Stage:**

1) Management: review critical contractor's capability, business license, & project experiences, PEP, schedule
2) QA/QC: review critical contractor's ISO 9000 certificate, QA/QC plan, SPP, sample check of work records/reports etc.
3) HSE: review critical contractor's ISO 18000 certificate, HSE plan, SPP, sample check of work records/reports, including **PHAs/HAZOP/QRA** etc.
4) Technical: review FEED engineering design, & risk assessment e.g. HAZID, HAZOP, QRA, etc.

**Deliverable**: audit findings on Statement of Review for FEED, Statement of Review for FEED Safety Study.

**3. Detail Design Stage:**

1) Management: review critical contractor's capability, business license, & project experiences, PEP, schedule

2) QA/QC: review critical contractor's ISO 9000 certificate, QA/QC plan, procedures, standards/policies/procedures (**SPP**), sample check of work records/reports etc.
3) HSE: review critical contractor's ISO 45001 certificate, HSE plan, standards/policies/procedures (**SPP**), sample check of work records/reports etc.
4) Technical: review Detail Design engineering design, design variation, & risk assessment e.g. closeout of HAZOP recommendations, etc.

**Deliverable**: review comments or audit report on Statement of Review for Detail Design, Statement of Review for Detail Design Safety Studies.

**4. Procurement Stage**:

1) Management: review critical contractor's capability, business license, & project experiences.
2) Technical: review specifications of safety critical equipment and QA/QC process.
3) Engineering requirement e.g. specification, database, drawings etc.
   - Manufacturing documents e.g. drawings, inspection & test plan, witness Factory Acceptance Test (FAT), review FAT report & Manufacturing Databook/Spare parts, etc.
   - Inspection & Test Organization documents e.g. Inspection Report, 3rd party certificate etc.

**Deliverable**: review comments or audit findings.

**5. Construction Stage:**

1) Management: review critical contractor's capability, business license, & project experiences, PEP, schedule, etc.
2) QA/QC: review critical contractor's ISO 9000 certificate, QA/QC plan, standards/policies/procedures (**SPP**), sample check of work records/reports etc.
3) HSE: review critical contractor's ISO 45001 certificate, HSE management system, plan, procedures, standards/policies/procedures (**SPP**), sample check of work records/reports etc., construction environmental monitoring, waste treatment plan, emergency response plan
4) Technical: compare critical design area project engineering standards against international standards, review construction methodologies, check list, punch list, site construction/safety/environmental/QA/QC activities, design variation, mechanical completion, handover report etc. (**see Mechanical Completion Certificate Checklists** for conducting site audits

**Deliverable**: review comments or audit report, Statement of Review for Construction

**6. Pre-Commissioning & Commissioning Stage**:

1) Management: review pre-commissioning & commissioning plan schedule, integration plan etc.
2) Technical: review pre-commissioning & commissioning plan, isolation list, punch list, Site Acceptance Test (SAT), red-line markup drawings, design variation, witness pre-commissioning & commissioning activities, Prestart-up Safety Review (PSSR), ready for startup inspection etc.

**Table – 1 International Codes and Standards**

| | |
|---|---|
| IEC 79 | Electrical Apparatus for Explosive Gas Atmospheres |
| API RP 520 | Recommended Practice for the Sizing, Selection and Installation of Pressure Relieving Devices (Part 1 and 2) |
| API RP 521 | Recommended Practice for Pressure Relieving and Depressurizing Systems |
| NFPA 10 | Portable Fire Extinguishers |
| NFPA 11 | Low Expansion Foam |
| NFPA 13 | Installation of Sprinkler Systems |
| NFPA 15 | Water spray Fixed systems for Fire Protection |
| NFPA 16 | Deluge Foam Water Sprinkler and Foam Water spray systems |
| NFPA 17 | Dry Chemical Extinguishing systems |
| NFPA 24 | Installation of private fire service mains and their appurtenances |
| NFPA 30 | Flammable and Combustible Liquids Code |
| NFPA 72 | National Fire Alarm Code |
| NFPA 101 | Life Safety Code |
| NFPA 2001 | Clean Agent Extinguishing System |
| NFPA | Fire Protection Guide to Hazardous Materials |

| | |
|---|---|
| IP Model Code of Safe Practice Part 3 | Refinery Safety Code |
| IP Model Code of Safe Practice Part 19 | Fire Protection at Petroleum Refineries and Bulk Storage Installations |
| BS 5839 | BS 5839 Part 1988 for Fire Detection and Alarm Systems for buildings |
| IEC 61511 | Functional Safety, Safety Instrumented Systems for the Process Industry |
| ISO 14692 | Petroleum and Natural Gas Industries – Glass Reinforced Plastics (GRP) piping |

*This paper discusses industry experience and summarizes the salient design considerations and risk assessment to improve safety integrity in the execution of capital projects for refinery and petrochemical complex. An effort has been made to highlight the* ***project assurance/audit framework*** *with HSE methodologies adopted by major oil companies using project standards, and industry best practices and standards like API, ASME, NFPA, IEC, ISO etc. Emphasis has been made to increase the awareness among the discipline design engineers and process safety engineers to develop safer designs of the refinery and petrochemical facilities*

# 619336 Case Studies – Beware of Common Errors during Risk Assessment of Powder Handling


**Abstract:**

**Case studies – Beware of Common errors during Risk Assessment of Powder handling**

"Single size shoe does not fit all," sounds so much in synchrony when we want to assess risk related to powder handling. Powders, unlike liquids have varied challenges right from it's transportation, receipt storage and handling.

Dust handling can have primarily 2 major hazards:

1. Dust explosion
2. Dust thermal decomposition

Few of the important parameters to understand for dust explosion are:

1. Fuel characteristics of the powder:
2. MIE: Minimum energy that can ignite dust cloud in the air.
3. MIT: lowest temp of a hot surface that will cause dust cloud (not dust layer) to ignite.

- Particle size: Smaller size poses higher risk

1. Kst: Max Rate of pressure increase. lower the kst number means the lesser severity 0 bar m/s means no risk while >300 bar m/s means high risk and bigger explosion severity.
2. Ignition source:
3. Hot Surfaces
4. Mechanical sparks/friction

- Electrical motors/contacts

1. Static electricity

Few important parameters to understand and control thermal decomposition in powders is:

1. Onset temperature of decomposition
2. Ageing of powders in vessels that the in process powders are handled such as Dryer, pipelines etc
3. Bulk storage stability of powders

Powders have challenge in terms of dissipation of heat, sampling of representative samples and uniform temperature indication and thus control.

The case studies will focus more on common errors during defining of controls for powder handling and will focus on how preventive controls can be built in to avoid major incidents.

| Sr.No | Case Study Title | Controls in brief |
|---|---|---|
| 1 | Drying of Powders with detonation property | Selection of right type of Dryer to ensure metal tramp is not an issue<br><br>Inert atmosphere |
| 2 | Aging of powders and thus lowering of thermal decomposition onset | Thermal studies post commissioning to ascertain periodic cleaning of process vessels such as dryers to avoid decomposition at lower temperatures |
| 3 | Handling of shock sensitive powders | Handling the powders in situ in inert atmosphere<br><br>Ensuring that all process lines such as filter to dryer are flushed every batch<br><br>Higher authority control for permit to work on such systems<br><br>Avoid use of agitators for such processes. |
| 4 | Handling low MIE powder formulations including drumming out | Static charge generation is a concern. Some preventive controls to be considered are:<br><br>Ensuring that drop height of powder is less than 3 mtr.<br><br>Use of earthing continuity stem such as earthrite system.<br><br>Ensure personnel use antistatic body protection and shoes.<br><br>Use of conductive FIBC or drums |

Good housekeeping

All the above studies are based on Author's 10 years of experience in process hazard analysis for bulk chemical manufacturing and is expected to benefit the audience at large.

# YES You Can: Case Study of Implementing a Center for Chemical Process Safety Risk Based Process Safety (CCPS RBPS)

**Nestor Humberto Sposito**
PPR Consultora
Bahia Blanca, Buenos Aires, Argentina
pprconsultora@gmail.com
**Juanita Corredor**
Process Safety Leader Brinsa
Cajicá, Colombia
juanita.corredor@brinsa.com.co
**John Murphy**
Process Safety Services
Punta Gorda, FL
hamjfm@embarqmail.com
**John Diaz**
Quality Director, Brinsa SA,
Cajicá, Colombia
john.diaz@brinsa.com.co



Prepared for Presentation at
American Institute of Chemical Engineers
2020 Spring Meeting and 17th Global Congress on Process Safety
Virtual
April 18 - 22, 2021

# YES You Can: Case Study of Implementing a Center for Chemical Process Safety Risk Based Process Safety (CCPS RBPS)

**Nestor Humberto Sposito**
PPR Consultora
Bahia Blanca, Buenos Aires, Argentina
pprconsultora@gmail.com
**Juanita Corredor**
Process Safety Leader Brinsa
Cajicá, Colombia
juanita.corredor@brinsa.com.co
**John Murphy**
Process Safety Services
Punta Gorda, FL
hamjfm@embarqmail.com
**John Diaz**
Quality Director, Brinsa SA,
Cajicá, Colombia
john.diaz@brinsa.com.co

**Keywords:** Process Safety, RBPS,

## Abstract

This paper discusses the challenges and opportunities that arise in the implementation of a risk-based process safety management system in a hazardous materials facility in the chlorine-alkali industry, starting from a relatively low level of compliance with standards, rules of good art, and a general lack of knowledge of the basic principles of process safety.

The motivation to begin the implementation of risked-based process safety (RBPS) came from an external audit requested by the company in November 2017 to evaluate its compliance with several pamphlets from the Chlorine Institute. At that time, the organization did not have a process safety sector and safety efforts were oriented towards personal safety. The final report included a long list of recommendations, each of them associated with good practice from the CI pamphlets. These recommendations were also associated with one of the 20 elements of the CCPS RBPS.

The authors discuss how the implementation plan was developed, the awareness and training of the top and middle levels of the organization's leadership to support the project, the mechanism for selecting priorities, and the process of preparing the schedule including scope, responsibility, and compliance dates.

The paper discusses examples of improvement in facilities, equipment, systems, and procedures in the operational, maintenance, engineering, etc. sectors, the incorporation of systematic and rigorous hazard identification and risk analysis techniques, and the benefits of implementation and monitoring the recommendations suggested in the studies. The strong leadership commitment and the process monitoring methodology by the company's senior management drives the implementation and gives visibility to the management system within the entire organization.

The authors also discuss the economic results of the implementation and quantify the impact of the improvements.

The pandemic unleashed in the world, altered the course of implementation of the process safety management system. The paper discusses the alternative and exceptional measures adopted to mitigate this impact and the results of those measures.

While it is true that the progress toward total implementation of RBPS is very important, it also important to understand what is missing to achieve the full implementation of the 20 elements of RBPS, the appropriate level of maturity necessary in the culture of the organization to achieve continuous improvement and what is the plan to address the challenges to reach each stage of implementation.

The authors encourage all organizations that have not yet implemented a process safety management system to understand that it can be done, even in less than ideal conditions, with very old facilities and a poor or non-existent process safety culture, if there is a strong commitment from the top leadership of the organization.

# 1 Introduction

A process safety management system that is robust and reliable is necessary to ensure the safe operation of a facility involving highly dangerous products and operating conditions, such as the ones found in a chlor-alkali chemical plant.

Below, there will be a brief introduction of the installation's, which was the subject of this study, most important characteristics.

Brinsa is a Colombian company, leading in salt chemistry, with 25 years of experience. They extract and refine the salt, and from this raw material, they develop the chemistry for the chlor-alkali which supplies the whole industry. The company has manufacturing facilities in Colombia, Costa Rica, and the Dominican Republic, as well as a commercial presence in Latin America.

The chemical plant has been increasing the historical monthly production record of chlorine-soda, through electrolytic cell technology, in order to achieve the separation of the sodium chloride molecule. This is cutting-edge technology that improves process efficiency as well as environmental control.

There are inherent risks involved in the chlor alkali processing because of the hazardous substances involved in the process. Some are classified as corrosive, toxic, and flammable.

The chemical plant has an installed capacity of 100 ton/day of liquid chlorine, 225 ton/day of 50% soda, 160 ton/day of hydrochloric acid, 180 ton/day of 15% sodium hypochlorite (batch production), and 260 ton/day of 5.25% sodium hypochlorite (continuous production). Additionally, there is a storage capacity for 1540 tons of sodium hydroxide, 503 tons of hydrochloric acid, 421 tons of 15% sodium hypochlorite, 302 tons of 38% sodium chloride, 327 ton of lineal sulphonic acid, and finally, and 141 ton of chlorine as a finished product.

The plant has 4 boilers that supply steam at 400 PSIG and 400 °C. Part of the steam that is produced is used in the plant's turbo generators to produce up to 8.5 MW, achieving energy self-sufficiency. There is also a sulfonation plant, which produces raw material for the manufacturing of the cleaning products, and a blowing plant that fabricates packages and lids.

The company has a total of 1400 workers, between fixed and temporary workers and 400 permanent contractors.

## 2 External Audit. Trigger for the Process.

In November 2017, the company's leadership, decided to conduct third party audit of the chlorine production plant, including all of its stages (membrane, compression, storage, packaging, shipping, and distribution) to access the level of the compliance to the Chlorine Institute's Pamphlets applicable to chloro alkali production.

A protocol was developed (a checklist), based on the guidelines from the Chlorine Institute and the industry's experience. The list was used as an evaluation tool during the 5-day visit to the chlor-alkali process plant.

The evaluation activities in the field included:

- A tour of the facilities
- Review of the following programs with the plant's personnel:
  - Incident reports
  - Emergency preparation and response plan
  - Standard Operating Procedures
  - Plan for personnel training and development assurance
- Review of the pertinent Chlorine Institute checklists, with the plant's personnel.
- Presentation and discussion of the CCPS Risk-Based Process Safety program to the plant's personnel and leadership.

### 2.1 Findings and Priority Recommendations.

Below, are itemized the most recommendation that was given, as mandatory, by the auditors:

- Develop a written program for the investigation and report of process safety incidents. (1)
- Develop a written program for the involvement of concerned parties, including information on the process dangers, the preparation and emergency plan, self-protection measures, the alarm system, medical treatment, etc.
- Develop a plan for the performance of a periodical risk identification and analysis for the high-risk process areas.
- Develop a written plan on the reliability and mechanical integrity of the plant's assets. Define the critical equipment and an inspection and maintenance plan.
- Develop and implement a Management Of Change program (MOC)
- Develop plans of electrical areas classification in accordance with international standards
- Develop a solid culture of process safety. Establish process safety as a core value
- Provide solid leadership from all levels, starting with the board of directors (lead by example)
- Improve the operative procedures management system
- Establish a Process Safety role

The audit results were presented and shared on the last day with the facility leadership team. The team showed itself very willing to implement all of the recommendations, especially those cataloged as mandatory, and to launch a project to implement a process safety management system that didn't exist at that time in the organization.

# 3 Awareness & Planning

The company decided to hire the services of a process safety specialist consultancy to advise and monitor the implementation of the Risk-Based Safety System

The concept of process safety was new for all of the organization levels. Initially, a broad training and awareness plan was launched on process safety topics using different courses and workshops encompassing the different organization levels.

The plan was designed to bring them the basic process safety knowledge and to achieve the commitment from all the people to the future program implementation.

Workshops were designed with different scopes and approaches in the function of the target audience, starting with the management and leadership levels.

The first workshop was targeted at leaders from different sectors of the organization. During the 3 days training, after 24 training hours was completed, all participants were summoned to demonstrate their commitment to the system by answering the following questions:

> From your role, what is your commitment to Brinsa's Process Safety implementation?
> What strategy will you use for the implementation?

Indicate which are the 5 activities you will carry out immediately to improve the Process Safety development in your area or sector

All of the resulting material was collected in a report which was subsequently raised to general management and was a significant contribution to the further planning and implementation

Additionally, at this first stage, a 4-hour workshop was developed for the top management of the company, where top management members took part, including the company's CEO. This was key for having the commitment of the highest authorities in the organization on the management system implementation.

Table 1 shows the training that was carried out covering a total of more than 2000 men-hr of training in safety process topics by the end of 2020. During 2021 it is intended to train operators and personnel in sectors not associated with operations such as contracts, human resources, and purchases through specially designed courses for these groups.

**Table 1. PS Training Course**

| Course Name | Top Management | Directors | Senior Management | Supervisors | Operators | Total Men-hours |
|---|---|---|---|---|---|---|
| RBPS Introduction | | | 168 | 240 | | 408 |
| PSM Audit | | 24 | 168 | 192 | | 384 |
| PSM Audit Refresh for Auditors | | 8 | 56 | 48 | | 112 |
| RBPS Basic - Virtual | | | 40 | 32 | | 72 |
| Process Safety Advanced | | 24 | 336 | 288 | | 648 |
| Inspiring Leadership | 12 | 12 | | | | 24 |
| MOC Sensitization | | | 46 | 40 | 180 | 266 |
| MOC Implementation | | | 46 | 40 | 180 | 266 |

| | | | | | | |
|---|---|---|---|---|---|---|
| Total | 12 | 68 | 860 | 880 | 360 | 2180 |

At the end of the first training, an 8-hour workshop was held to determine the planning for the implementation of the 4 pillars and the 20 elements of the Process Safety management system. The managers, chiefs, and leaders who participated were divided into multidisciplinary working teams.

The consultant developed a tool that contemplated an assessment for each element the following

- Element's impact on the organization's risk level
- Element's level of development at the organization
- Resources availability

| Rating Guide | | | | |
|---|---|---|---|---|
| Very High - The implementation will immediately have a very strong impact on the prevention and mitigation of the risks of the installation (Weighted as 4) | | | | |
| High - The implementation will immediately have a considerable impact on the prevention and mitigation of the risks of the installation. (Weighted as 3 points) | | | | |
| Medium - The implementation will immediately have a moderate impact on the prevention and mitigation of the risks of the installation. (Weighted as 2 points) | | | | |
| Low - The implementation will immediately have a low impact on the prevention and mitigation of the risks of the installation. (Weighted as 1 point) | | | | |
| **Impacto del Elemento en el Nivel de Riesgo de la Organización** | | | | |
| **Process Safety Commitment** | Very | Hig | Mediu | Lo |
| Proces Safety Culture | X | | | |
| Standar Compliance | X | | | |
| Proces Safety Compdetencies | | | X | |
| Workforce Participation | | X | | |
| Disclosure to Stakeholders | X | | | |
| Current score count by ranked category | 3 | 1 | 1 | 0 |
| Percentage of current scores by ranked category | 60% | 20% | 20% | 0% |
| Average Score | 3.40 | | | |
| **Understanding Hazards and Risks** | Very | Hig | Mediu | Lo |
| Knowledge Management Process | X | | | |
| Hazard Identification and Risk Analysis | X | | | |
| Current score count by ranked category | 2 | 0 | 0 | 0 |
| Percentage of current scores by ranked category | 100% | 0% | 0% | 0% |
| Average Score | 4.00 | | | |
| **Risk Management** | Very | Hig | Mediu | Lo |
| Operative Procedures | X | | | |
| Safe Work Practices | X | | | |
| Asset Integrity and Reliability | X | | | |
| Contractor Management | | X | | |
| Guarantee Training and Performance | X | | | |
| Management of Change | X | | | |
| Operational Availability | X | | | |
| Conduct in Operations | | | X | |
| Emergency Management | X | | | |
| Current score count by ranked category | 7 | 1 | 1 | 0 |
| Percentage of current scores by ranked category | 78% | 11% | 11% | 0% |
| Average Score | 3.67 | | | |
| **Learn from Experience** | Very | Hig | Mediu | Lo |
| Incident Investigation | X | | | |
| Measurements and Metrics | | X | | |
| Audit | | X | | |
| Management Review and Continuous Improvement | X | | | |
| Current score count by ranked category | 2 | 2 | 0 | 0 |
| Percentage of current scores by ranked category | 50% | 50% | 0% | 0% |
| Average Score | 3.50 | | | |

**Figure 1**. Element prioritization for PSM

The teams had to classify each elements:

- High
- Medium
- Low

Then the tool performed a weighted calculation to determining the implementation priority of the each element of the management system.This was done to achieve the greatest possible impact in reducing the risk of the installation, to assure short-term implementation progress results and to maintain a very strong commitment of the organization. The workshop results and recommendations were presented to the company's management team and finally, after a few requested changes, it was approved. The plan contemplated 3 stages with priorities 1, 2, and 3 for the implementation of the different elements. The elements Process Safety Culture, Management Review, and Continuous Improvement (Stage 0) they would start in stage 1 and continue to develop during the entire implementation process. It was estimated that all the stages would be completed in 3 years. Subsequently, based on the experience, the maturation of the program, and the impact of the pandemic, it was updated extending to 5 years.

# 4 Organizational Strategy for Implementation.

A Company has to strive to make sure that Process Safety activities are completed in such a way that it minimizes the risk for employees, the public, and the environment, whichever the risk level. The activities that have high-risk levels could require higher attention levels from Management and higher resources assignment. The RBPS methodology intentionally guides the companies to direct efforts to the highest-risk levels sectors.

In compliance with the RBPS strategy, the organization decides to start with the Chemicals Plant, due to the high risk of the process and the results obtained from the diagnosis assessment. Likewise, a specific task force was created, in accordance with the auditor's recommendations and the industry best practices.The focus of the task force would be to develop the implementation methodologies for each of the elements of the RBPS system, follow up on the work plan, ensure interdisciplinary participation from the different sectors in the organization and ensure the system sustainability. The "Process Safety" team was created under the Quality Management integrated by a Chief, 3 professionals, and 1 analyst.

In parallel, the maintenance area team designed its preventive maintenance model and strategy. The Assets Integrity and Reliability task force, which was assigned to the Engineering and Maintenance Management, composed of 1 chief and 5 engineers. It had specific focus on preventive and predictive maintenance. The Technical Documentation center attendants were included in this work team,.They are responsible for centralizing and managing all the organization's projects and processes information.

With the emphasis on the implementation of the elements, a work plan was established with different activities, responsible and execution dates to strengthen the implementation process and to follow an effective realistic schedule.

The Process Safety Area, in alignment with the other company areas, will follow up on the progress and closure of the assigned activities. In Figure 2 a general schedule is presented of each of the elements with its implementation general plan. It should be noted that the Process Safety Culture, Management Review and Continuous Improvement elements are to be addressed the entire 5 years of implementation. Other elements started in 2018 will continue their work strategy with assistance of the consulting firm.

| Elements | Start | End | % Projected to Dec/20 |
|---|---|---|---|
| Process Safety Culture | 02/03/2018 | 20/12/2022 | 73.30% |
| Management review and Continuous Improvement | 02/03/2018 | 20/12/2022 | 73.30% |
| Contractor Management | 02/04/2018 | 20/12/2021 | 73.30% |
| Operative Procedures | 05/03/2018 | 30/06/2020 | 100.00% |
| Knowledge Management Process | 05/03/2018 | 20/09/2020 | 100.00% |
| Operations Conduct | 02/05/2018 | 20/04/2021 | 86.10% |
| Standards Compliance | 01/06/2018 | 28/12/2021 | 72.10% |
| Hazards identification and Risks Analysis | 01/06/2018 | 30/01/2020 | 100.00% |
| Reliability and Assets Integrity | 1/10/2018 | 20/12/2022 | |
| Emergency Management | 01/06/2018 | 20/12/2019 | 100.00% |
| Training Assurance | 01/06/2018 | 20/12/2021 | 72.10% |
| Preparation for Operation | 01/07/2018 | 20/09/2019 | 100.00% |
| Labor Force Participation | 05/03/2018 | 28/02/2020 | 100.00% |
| Management of Change | 02/04/2018 | 28/12/2021 | 55.60% |
| Incidents Investigation | 05/03/2018 | 30/06/2019 | 100.00% |
| Audit | 02/04/2018 | 20/12/2021 | 71.10% |
| External Audit | 2/11/2023 | 6/11/2023 | 0.00% |
| 2019 Internal Audit | 3/12/2018 | 28/02/2019 | 100.00% |
| 2020 Internal Audit | 20/12/2019 | 20/11/2020 | 100.00% |
| 2021 Internal Audit | 20/02/2021 | 20/11/2021 | 0.00% |
| 2022 Internal Audit | 20/02/2022 | 20/11/2022 | 0.00% |
| Safe Work Practices | 05/03/2018 | 20/04/2021 | 89.50% |
| Measurements and Metrics | 05/03/2018 | 20/12/2020 | 100.00% |
| Process Safety Competency | 1/10/2018 | 20/12/2021 | 69.20% |

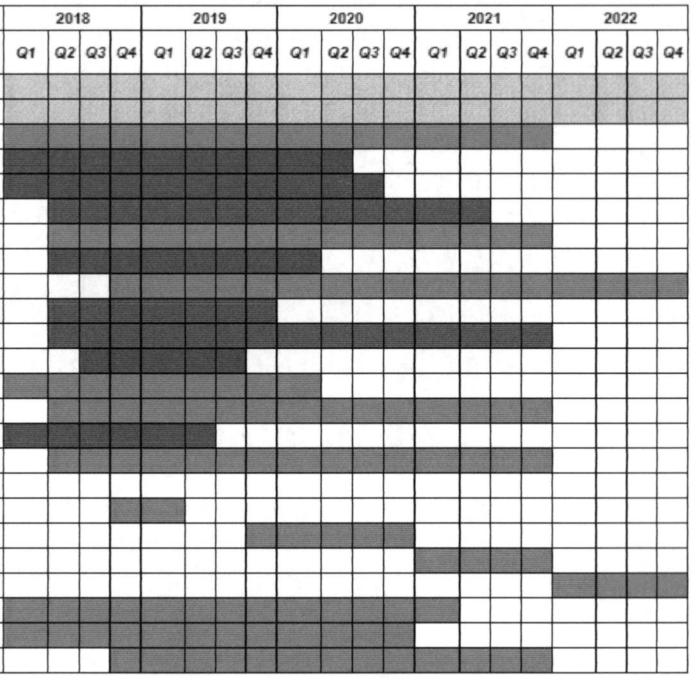


Stage 0 Stage 1 Stage 2 Stage 3

**Figure 2**: Management System Work Plan Implementation schedule

Next, the results of some elements of the system are presented, and the advances obtained during their development and implementation.

## 4.1 Hazard Identification and Risk Analysis.

The second pillar of the RBPS system "Understanding Hazards and Risks" is supported by two elements:

- Knowledge management
- Hazard identification and Risk Analysis

This pillar was identified at the diagnostic stage as High Priority(See Figure 2). Thus both elements start to be developed simultaneously, given the interdependence that exists between them. The organization decided to adopt HAZOP as an evaluation tool, complementing the company's risk matrix. HAZOP studies were developed for 10 chemical plant processes, with the participation of the entire interdisciplinary team. This made it possible to identify the current situation of the plant and assess the hazards according to the organization's risk matrix. The study was divided into two stages. Onstage implemented in 2018 included the Chlorine Circuit, Electrolysis, Hydrogen, Hydrochloric

Acid, and Sulfonation processes resulted im 162 recommendations categorized as very high and high risks The second stage started in 2019 included the CEDI (Chemical Distribution Center by its Spanish acronym meaning), Sodium Hypochlorite, Caustic Soda, Calcium Chloride and Demercurization processes resulted 176 very high and high risks recommendations.

In general, there's a high risk at the analyzed plants. 72% of the identified scenarios are at extreme or high risk. The recommendations generated by the HAZOP team when implemented significantly reduce the risk. They mainly focused on strengthening the designs with engineering measures to prevent and mitigate incidents.

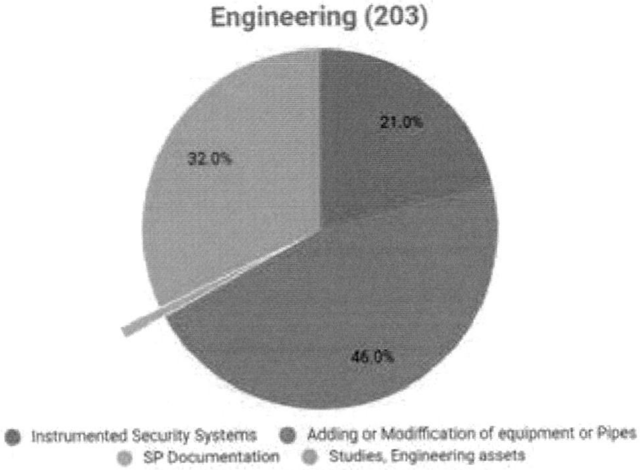


**Figure 3:** Engineering Recommendations

About 60% of recommendations are engineering improvements, which include among others, instrumented safety systems installation (21%), modification or adding new equipment or piping (46%), engineering studies (32%).

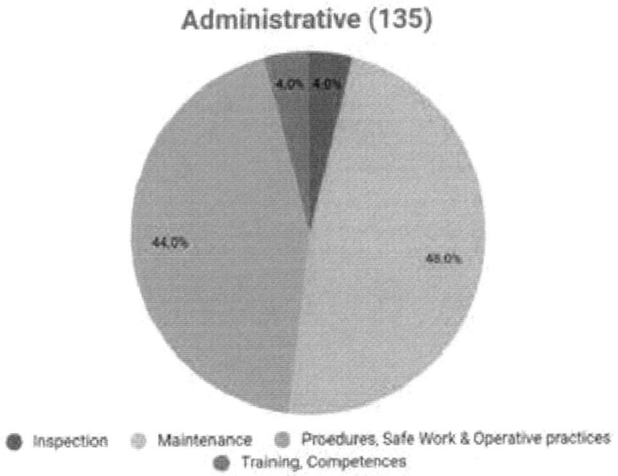

**Figure 4:** Administrative Recommendations

40% of remaining recommendations were administrative barriers improvements such as inspections, maintenance, training and qualification, procedures, operational and safe-work practices, and in general, process safety standards (Figure 4).

A 3-year work plan was defined for closing gaps, with budgetary assignments and defining specific task forces for gap closing.

## 4.2 Operating Procedures and Performance Assurance.

The RBPS element that ensures adequate development, timely maintenance, and the consistent use of operative procedures (procedures) is one of the 9 elements at the RBPS pillars: Risk Management.

Good operative procedures, among other things, describe the process, the hazards, tools, protective equipment, control systems and also provides troubleshooting instructions when a non-expected behavior emerges. They also address abnormal situations such as emergency shutdowns and temporary operations.

Operative procedures at Brinsa didn't comply with many of the requirements outlined above. Therefore, a standard is generated for drafting operative procedures according to the CCPS recommendations. This way, the participation of the operators in drafting procedures, created synergy within the personnel expertise and the information related to the operation manuals, the review by the supervision, and the coordination and approval by the Plant management. This procedure ensures that the input of different management levels is guaranteed and the documentation is appropriate. The drafting of all operative procedures required about 2 years. The final result was much more user friendly procedures, with adequate language for easy comprehension by users. Pictures were also included and unambiguous instructions of the steps to follow during the operation.

By the end of 2019, after the information and procedures updated and approved for the processes of Compressors, liquefaction, drying and Chlorine storage, de-chlorination and brine treatment, sodium hypochlorite production, calcium chloride and hydrochloric acid, a personnel retraining program was implemented. It included the technical skills assessment of the operators. it included 3 parts:

- Performance list: It is an evaluation of the knowledge of operational practices in the field.
- Theoretical evaluation of the process: The questions are asked taking into account the operating procedures.
- Cause-Effect evaluation: It measures the ability of the operator to understand the main risks that can be produced by the process and his role to minimize, delete or correct the risk.

After completing the first evaluation, feedback on weaknesses was carried out and subsequently, training was given to each operator. Finally, a second evaluation was carried out, emphasizing the non-conformities. When the training was completed all of the operators had a rating above 90%.

After updating the operative procedures, the first phase of generating the Pre-startup Safety Review (PSSR) lists was started, with the same creation, review and approval flow. In this first phase, the lists only contemplate start-ups after programmed stoppage or non-programmed stoppages without changes or modifications of the process were included. The documentation was developed during 2020. A second phase is planned for 2021that will ensure that the RFSR lists will be also mandatory for new projects.

## 4.3 Incident Investigation.

Developing, sustaining, and enhancing the organization's incident investigation competency is one of four elements in the RBPS pillar of ***learning from experience.*** (1)

In accordance with RBPS methodology, the company implemented a process for reporting, tracking and investigating incidents that include: 1. a formal process for investigating incidents, including staff provision, conducting, documenting and tracking process safety incidents & 2. Incident and incident investigation data trends to identify recurring incidents. This process also manages the resolution and documentation of recommendations generated by the investigations.

From the implementation of the process, a culture of reporting process safety incidents, including near misses, gradually began to emerge. Reports include, but are not limited to, chlorine leaks, spills (or drips) of liquid and corrosive substances, safety valve activation, fire outbreaks, among others. These events are investigated by interdisciplinary teams (mechanical, electrical maintenance, reliability, operation, process safety) through different investigation methodologies, with the use of Tap Root® being preferred in the organization.

The investigations involve the socialization of the lessons to be learned from the events that occurred at all levels in the operation, through daily operation meetings (DOM), and the lessons are left at different points of the plant for a free consultation. In Figure 5, the lesson-to-learn model currently being socialized

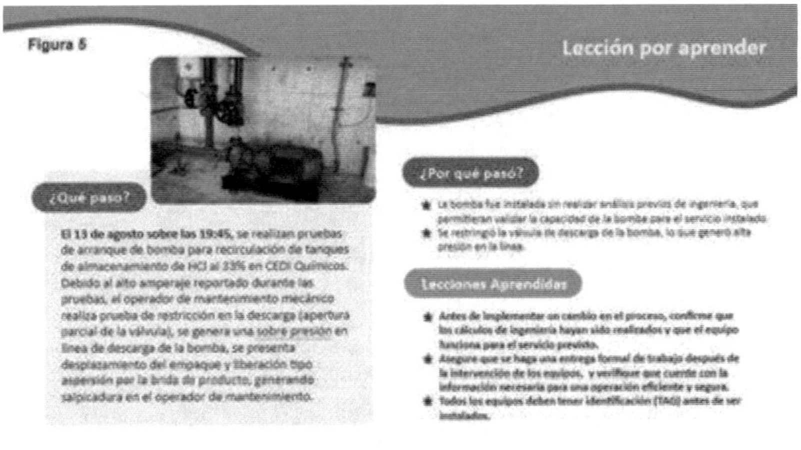


**Figure 5:** Lesson-to-learn model being socialized

With the implementation of this element, different levels of the organization have been reached, a culture of both reporting and interdisciplinary understanding of the events is being generated.

In Figure 6, the increase in process safety incidents reported by the Chemical Plant is evidenced. The conclusions and recommendations of these investigations led to substantial improvements in the processes, increasing safety and operational reliability.

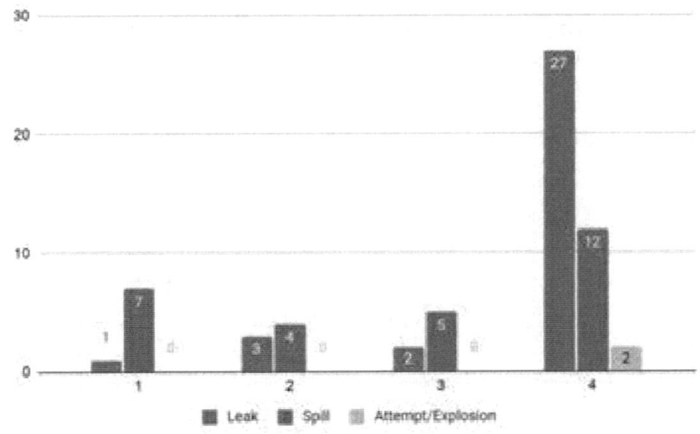


**Figure 6:** Process Safety reported incidents

During the first 3 years of implementation, activities of all the main elements that have allowed a systematic advance in the implementation have been developed. This has also made it possible to unite efforts from different fronts, create a comprehensive system and receive genuine support from the senior management team.

## 4.4Management Review and Continuous improvement

Since 2018, monthly monitoring is carried out by the management team and operations leadership to review the progress in the implementation of the process safety management system, opportunities for improvement, delays, budget execution and redefinition of strategies that leverage the system implementation. In these evaluation meetings, the gap closures identified in risk analysis and incident investigation are also reviewed. The results and conclusions of these evaluations are presented and analyzed every 4 months at the company's shareholders' meeting.

# 5 Investment

The RBPS implementation requires allocating human, material and financial resources to implement a robust, consistent and sustainable system over time.

The leadership of the organization, understanding its responsibility, approved for the first 4 years of implementation an approximate budget of 10,000,000 USD. This budget was assigned, among other things, to Training and performance assurance, incorporation of resources, developing the procedures and policies required by the RBPS and implementing improvements in the plant arising from the risk studies, the requirements of the management system and the conclusions of the investigation of process safety incidents.

Some improvements and changes involve large projects, and others are "minor" improvements that seek to reduce the human factor as a source of risk.

Below we will describe some of the main improvements made to the installation:

Around 30 level transmitters were installed in different process tanks, to eliminate the level glasses with which the level visualization was conducted (Figure 7).

| BEFORE | AFTER |
|---|---|
| **15% Sodium Hypochlorite Preparation tanks 688 / 689.** Level visualization through a sight glass, without an alarm system to alert the overflow of the tanks. | Installation of level transmitters with alarms, giving the operator greater control over the operation and identification of substandard conditions, both due to overflow and lack of power. |

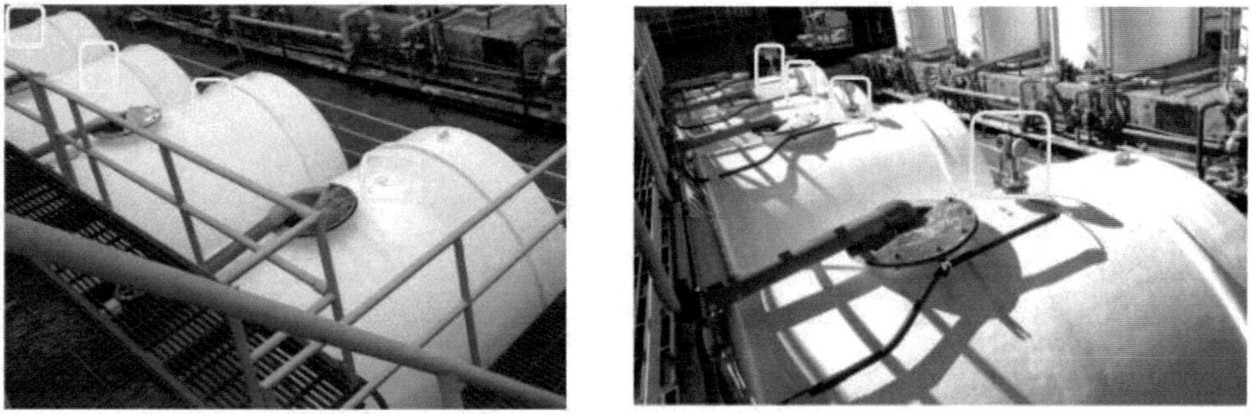

**Figure 7:** Level Transmitter, 15% Sodium Hypochlorite Preparation Tanks

A project was created for the installation of new liquid chlorine storage tanks, meeting the minimum requirements defined by the Chlorine Institute pamphlets, the EUROCHLOR technical definitions, and the available ASTM and ASME construction standards, in replacement of existing ones that did not comply with the RAGAGEP standards mentioned above. The storage system will start operations in the second half of 2021 (Figure 8).

**Figure 8:** Liquid Chlorine Storage Tanks

The implementation of the safety management system allowed the organization to know in greater depth the hazards associated with the handling of hydrogen and for this reason it was decided to carry out a specific HAZOP study for the system with the advice of a hydrogen expert. The study incorporated the risk rating using the company's matrix and based on this, approximately 40 recommendations for improvement were obtained. Most have already been implemented, especially those associated with the elimination of equipment and systems that were not operating after the change of technology from mercury cells to membrane cells.

**BEFORE**

Equipment in the out-of-service area of the hydrogen filtration system, excessive obstacles in the area.

**AFTER**

Dismantling of equipment, cleaning the area, facilitating access to operational areas. Image was taken on May 30/20

**Figure 9:** Hydrogen Filtration System dismantling

In the HAZOP study carried out on the hydrogen system, it was identified that in the hydrochloric acid receiving tanks there was an imminent risk due to the presence of an explosive atmosphere (accumulation of hydrogen gas). This recommendation was verified with the technology provider, and an engineering study was started to implement the improvement. Tank blanketing, or padding with N2 was installed.

| BEFORE | AFTER |
| --- | --- |
| Mersen Recommendation (HCl Furnace Supplier): Ensure nitrogen purge in the process tanks with minimum flow control to avoid an explosion due to hydrogen accumulation. In previous years one of the tanks had an explosion with the expulsion of part of the tank.<br>Rec. 315. Consider adding nitrogen as a blanketing gas to these Receiver Tanks. | Connections and adjustments are made in the receiver system and gas scrubbing to maintain a permanent purge to both receiver tanks of HCl guaranteeing the displacement of remaining gases and avoiding any accumulation of hydrogen inside. |

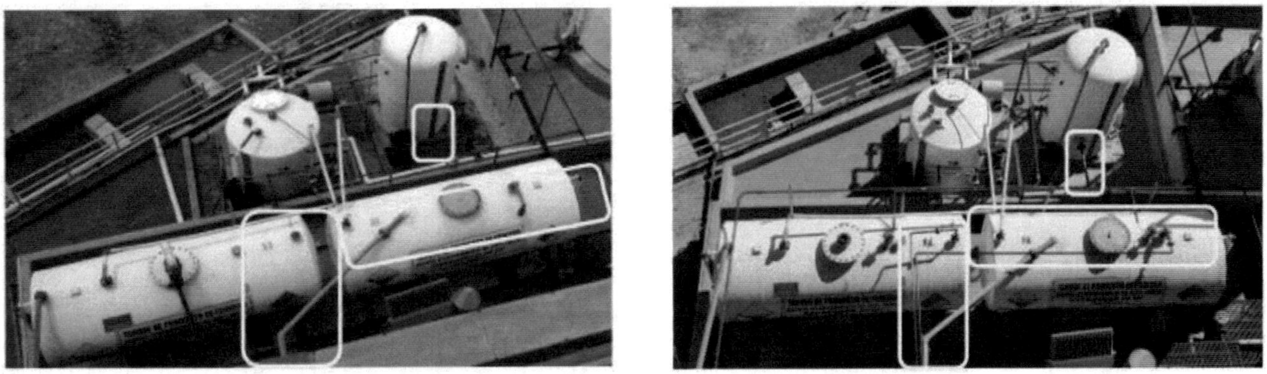

**Figure 10.** Blanketing gas to the hydrochloric acid receiving tanks.

Several processes in the facility still have a highly manual operation. During the HAZOP study, several recommendations were made associated with the automation of some processes. The figure shows the implementation of the automatic valve to guarantee the start-up of the backup fan of the sodium hypochlorite fan, to avoid a plant shutdown and a chlorine leak due to the pressure increase in the chlorine collector.

| BEFORE | AFTER |
| --- | --- |
| Rec. 20. Implementation of the automatic start of the spare fan in case of fan failure in service.<br>Risk: Loss of vacuum in the absorption tower and | Automatic valve installation and control loop programming was performed. |

loss of receipt of Cl2 with stoppage of the chemical plant due to increased pressure in the Cl2 receipt lines and subsequent risk of Cl2 leakage into the atmosphere with risk to people of the area.

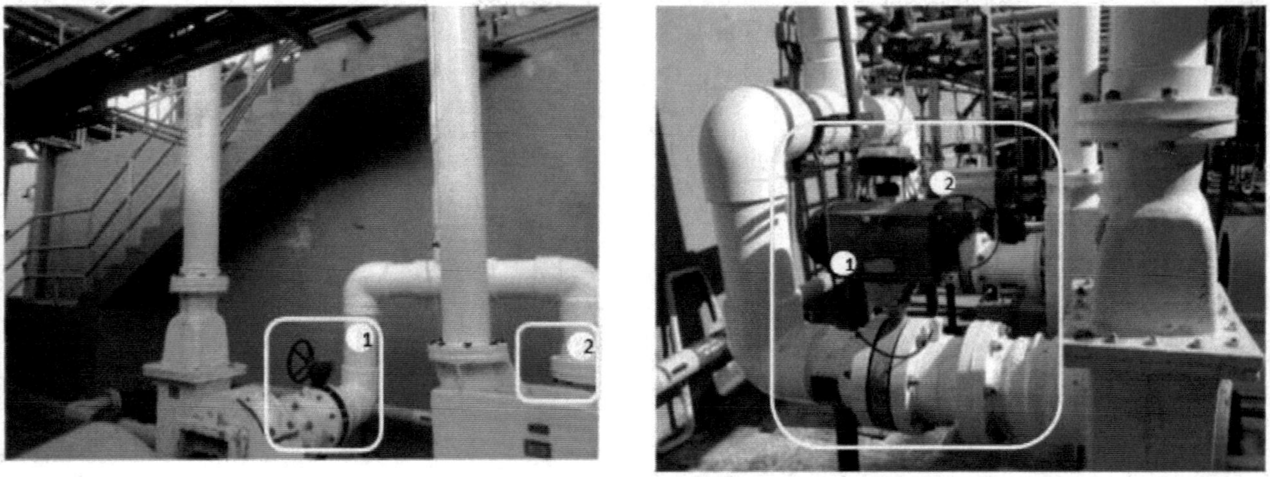


**Figure 11.** Automatic startup of the sodium hypochlorite backup fan.

Consideration of the "human factor" is a key aspect of a process safety management system. This discipline takes into account, in the work environment, the interaction between people, facilities and equipment, and management systems. (2) Based on these concepts, the organization decided to incorporate them into many aspects of the management system, for example, to improve the consistency of all technical documentation, including operating procedures and PSSR checklists, which includes updating of P&IDs, datasheets, among others, a program was launched to improve the identification of all equipment, instruments and pipes in the plant. (See Figure 10, Figure 11, Figure 12).

| BEFORE | AFTER |
| --- | --- |
| Unidentified valves can lead to their misalignment and therefore a mixture of chemicals that can give off dangerous gases Cl2, H2S, generating risk to people in the area. | Valves are tagged, minimizing operational errors due to confusion. |

**Figure 12.** Valves tagging

| BEFORE | AFTER |
| --- | --- |
| Tanks 769 and 779 unidentified. | The identification of the auxiliary storage tanks of sodium hypochlorite at 15% is carried out. |

**Figure 13.** Tanks tagging

| BEFORE | AFTER |
| --- | --- |
| Unidentified pipe, an event occurred due to mixing of raw materials due to misalignment of valves. | The identification of the pipes is carried out to minimize errors in the direction of fluids. |

**Figure 14.** Pipes tagging

## 6 AUDIT

The only way to determine that a PSM program is working properly is to thoroughly examine both its design and implementation on a regular basis. Therefore, a PSM audit represents a way to measure the effectiveness of the PSM program and also allows a comprehensive comparison with predetermined PSM metrics. If the PSM program is not measured carefully, it cannot be controlled or improved. (3)

A successful PSM program audit begins with a commitment by the Company's senior and facility management to conduct periodic audits of these programs. This is essential to allocate appropriate resources to conduct the audits, ensure that findings and recommendations are actively and carefully addressed promptly, and ensure that on-site activities allow a successful development of the audit.

(Successful PSM program auditing according to CCPS begins with a commitment by the senior company and site management to perform periodic audits of these programs, to allocate the appropriate resources to perform the audits, to ensure that the findings and recommendations are actively and carefully addressed in a timely manner, and to ensure that the activities on-site during the audit are arranged to support the audit to the maximum extent possible.)

These principles were strongly adopted by the leadership of the organization and based on this, the process safety audits program was established to accompany the implementation of the system in order to control its effectiveness and make adjustments and corrections promptly. The plan first contemplated an internal training process and the development of audit protocols for the 20 elements of the RBPS. All this material was based on the CCPS audit book

Then it was decided to start with internal audits for partial elements of the management system. So far, 2 internal audits have been carried out, one in person in 2019 and one in 2020. Despite the challenges posed by the global situation due to the pandemic, the company decides to continue with the internal audit plan in 2020, counting with remote assistance from the consulting firm and blended participation by the audit team and the audited team, seeking to guarantee the biosafety protocols defined by Brinsa.

The audited elements in 2020 are in an advanced development process with partial implementation; a greater understanding of the spirit and scope of the audit and a more agile development were observed.

Ultimately, and beyond the qualification and findings, there is evidence of growing knowledge and commitment of most sectors of the organization with the process safety subject.

Procedures and tools have been incorporated that are strengthening the entire process safety management system. The Asset Integrity and Reliability element shows a significant improvement in its rating (2.44 in 2019 vs. 3.19 in 2020) derived from the important incorporation of procedures and tools for planning and monitoring the reliability and maintenance tasks of the equipment to be throughout their lifecycle.

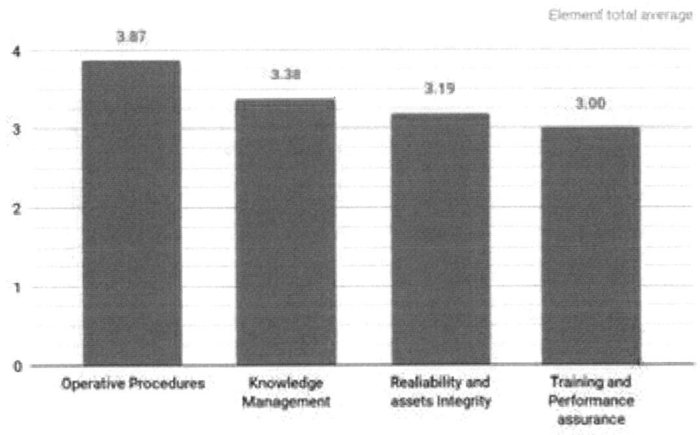


**Figure 15:** Qualification of audited elements 2020

The "operating procedures" element has notably improved in the second audit compared to the first one. It is worth highlighting the existence of a procedure that establishes the scope, format, responsibilities, etc. for the development of the operating procedures, and promoting the active participation of the operators in carrying out / reviewing them.

## 7 CONCLUSIONS

The most important conclusions of this case study are:

- An RBPS can be applied with very good results, in an organization with many years of operation and a culture focused only on personal safety.

- Awareness and training in the concepts of process safety before the implementation of the management system, at all levels of the organization, especially leadership, are essential to achieve understanding and awareness of its importance.

- The support and commitment of the organization's leadership is key to implement a process safety management system in organizations that are unaware of the issue and their safety concern is exclusively personal safety. Leadership must provide the human and material resources and make the necessary investments for a correct implementation as happened in this case.

- Implementation poses a huge challenge when starting from scratch. However, it can have very good results in the short and medium-term if it is properly planned and with the active participation of all levels of the organization in such planning.

- The proper selection of the elements of the management system on where to start the implementation is essential to try to quickly reduce the risks of the installation and at the same time show significant progress that maintains the interest and commitment of all levels of the organization, especially the leadership.

- The implementation requires a specific sector of Process Safety to lead it, monitor progress and solve limitations and challenges during the process.

- The implementation of the management system introduced benefits in other areas of the organization such as production by achieving increased reliability and operational continuity through the creation of a previously non-existent asset integrity group.

## 8 Bibliography

1. **Center for Chemical Process Safety, American Institute of Chemical Engineers.** *CCPS. Guidelines for Risk-Based Process Safety.* New York: Wiley, 2007.

2. **Center for Chemical Process Safety, American Institute of Chemical Engineers. CCPS.** . *Human Factor Methods for Improving Performance in the Process Industries.* New York: Wiley, 2007.

3. **Center for Chemical Process Safety, American Institute of Chemical Engineers. .** *Guidelines for Auditing Process Safety Management System.* New York: Wiley, 2011.

SPRING21
+17TH GCPS
A Joint AIChE and CCPS Meeting

# BS Team: Finding Competence Gaps

**Salvador Ávila**
**UFBA**
**avilasalva@gmail.com**

**Luiz Fernando Lopes de Souza**
**COPENOR**
**lflsouza@globo.com**

**Gunther Josua Costa**
**UNIVASF**
**guntherjosua@gmail.com**

**Lucas Menezes Pereira**
**UFBA**
**lucasmpereira13@hotmail.com**



Prepared for Presentation at
American Institute of Chemical Engineers
2021 Spring Meeting and 17th Global Congress on Process Safety Virtual
April 18 - 22, 2021

# BS Team: Finding Competence Gaps

**Salvador Ávila**
**UFBA**
**avilasalva@gmail.com**

**Luiz Fernando Lopes de Souza**
**COPENOR**
**lflsouza@globo.com**

**Gunther Josua Costa**
**UNIVASF**
**guntherjosua@gmail.com**

**Lucas Menezes Pereira**
**UFBA**
**lucasmpereira13@hotmail.com**

**Keywords:** BS, Competences gaps, Decision models

## Abstract

The academy form competences to work in the routine, without worries, without unknown hazards. The rules and objectives in routine activities are clear and require operational discipline to control the cost of production and product quality. The current educational model does not prepare society and groups for an emergency or even for decisions in a crisis, where resources are scarce mainly because the necessary knowledge about danger and its mechanism of action is lacking.

What really happens in the chain event scenario when the danger is unknown? What are the typical models of action of the emergency leader and his team? What are the required actions in crisis management required for a BS type event considering the scarce resources, the high level of stress and the direction of the danger energy? Which models are indicated for the situation of unknown danger and its intense energy flow with high impacts during these events? What is the appropriate mind map of those who lead and execute the actions considering the low level of visibility in the events and the dynamic of geopolitics in the BS scenario where, depending on the preparation for emergency, in case of intense demand for attention and action, possible modifications may occur in the map mental for decisions?

This work intends to review the discussions already performed about new concepts and tools to be used in the crisis prevention area. Between this works we will understand the effect of the stress level on emergency decisions valuating human error and observing team response using LODA tool. The Human elements designed in the organizational culture, can stop the high danger energy that can flow from some industries. The principal aspects to be constructed and monitored are technology (risk and complexity); management (leadership and stress); and behavior (cooperation, commitment, competence, and communication).

By the other hand, human factors are analyzed to avoid situations which hazard energy carriers during top events, as the effect of cultural aspects to chain reaction until the occurrence of disaster. Decision models for the emergency brigade (leader and team) indicate the motor and cognitive gaps that cause lack of control during contingencies. Finally, the analysis of simulation types, by rotating observation of field simulation allow find the deficiencies of the emergency team in a real situation in the future.
The methodology divided in following steps are dynamic: first, rescue the discussions about stress and the BS events and, in second step, recognize the intense hazard energy flow through failure events, antecedents, elements and human factors. In third step, we try to locate the gaps and environmental conflicts influencing losses from accidents. Finally, we assess the current event of covid-19 pandemics to suggest correct emergency preparedness actions in the preparation of leaders and their teams regarding cognitive and motor characteristics in situations of high demand. When comparing the characteristics already described by TALEB about BS with the event COVID-19, we intend to carry out a case study to indicate the possible causes this global impact crisis, or even, understand the technological disaster causes where uncontrolled events from chemical, nuclear, and oil industry begins chain reactions until the disaster. Unfortunately, a Crisis situation can be caused from several intricated dimensions, making difficult the contingency and mitigation actions.

## 1 Introduction

The academy form competences to work in the routine, without worries, without unknown hazards. The rules and objectives in routine activities are clear and require operational discipline to control the cost of production and product quality. The current educational model does not prepare society and groups for an emergency or even for decisions in a crisis, where resources are scarce mainly because the necessary knowledge about the hazard and its mechanism of action is lacking.
The knowledge learned to graduate the engineer is limited to logic and objective concepts. The students do not discuss different concepts, as social culture, legislation, workstation behavior at the same time, mixing objective aspects with subjective ones. We need to understand Cultural traits, geopolitical aspects, intrinsic variables from worker during routine to discuss about response models for emergencies, necessary to avoid disasters.

*Unfortunately, we only look for new competences after the disaster happen! Why?*

The metrics for Process Safety management are discussed in methods indicated in the CCPS books [1]. Some important issues such as: management and investigation of major incidents [2]; technological changes that cause accidents [3]; lack preventive actions for organizational changes [4] are exemplified and barriers are suggested, so as not to affect the safety of processes with big accidents.
These issues and solutions seek to avoid the accident through management actions, technological systems, and the attempt to control human factors [5] based on the lessons learned by various industries and associations when investigating the events that occurred. This investigation of the past, looking back to the facts, has a limit when external, dynamic environments threaten normality. The crisis scenarios can be very dynamic, not visible and of parallel occurrence, at the same time as those caused by climate change, by social conflicts, by conflicts arising from

degraded geopolitics, and by the effects of multiculturalism in the world of work. and in society - indicating low tolerance behavior.
In 2018, CCPS [6] with the leadership of Eastman, raises the issue of normalization of deviation, indicating repeated situations of acceptance of operational deviations, in the shift routine, are considered normal situations. The lack of research with the indication of bottom-up techniques for the analysis of operational culture and the respective heuristics turns the book into another attempt to guide the occurrence of events looking to the past and without an appropriate view of what is happening in the present. and what will happen in the future, on the factory floor. What are the social emergent that cause deviations to normalize?

*What are the social emergent that cause deviations to normalize?*

In this context, analyzing the internal and external social threats to avoid environments prone to deviations gains importance. Risk aversion, excessive self-confidence, and fear of guilt for human error are considered cultural events that produce underreporting. This phenomenon changes the shape of Bird's pyramid [7] to the Christmas tree as shown in Figure 1 and causes the accident to occur already with the chain reaction triggered. The assessment and actions resulting from the notification (not performed) could avoid the domino effect or reduce the risk of major events.
The accumulation of deviations without treatment and with high level of underreporting, a cultural event, increases the likelihood of Black Swan events. Failure to recognize the Scenario for hazard identification and risk analysis that implies a lack of insight into the operational context and creates uncontrolled situation with natural, technological, political disasters. The Black Swan Event goes beyond the local scale and reaches a global scale, with unknown dangers, or already known dangers, but due to lack of attention, it has the characteristics of unknown danger because it has not been investigated as to the mechanisms of action. We do not know what the appropriate barriers are. Thus, we classify the Black Swan because of totally unknown or operationally unknown dangers.

**Figure 1. Christmas Accident Tree**

## 2 Literature Review

This work intends to review the discussions already performed about new concepts and tools in the crisis area. Between this works will understand the effect of the stress level on emergency decisions valuating human error and observing team response using LODA tool. The Human elements in the organizational design can stop the high hazard energy flow of some industries. The principal aspects to be constructed and monitored are technology (risk and complexity);

management (leadership and stress); and behavior (cooperation, commitment, competence, and communication).

By the other hand, human factors are analyzed to avoid situations in which hazard energy carriers during events, from cultural aspects to chain reaction until disaster. Decision models for the emergency brigade, leader, and team; indicate the motor and cognitive gaps that can result in lack of control during contingencies. Finally, the analysis of simulation types, by rotating observation of field simulation allow find the deficiencies of the emergency team.

Given the great stress present in the geopolitical situation, in which there is a constant struggle for territory, renewable and non-renewable energy sources, in addition to climate changes intensified by anthropic activity, large flows of Hazard Energy (HE) are generated that are commonly not perceived. The unnoticed concentration of this energy gives rise to what is known as latent hazard, and when it evolves to levels that can no longer control or filter its pending of burst and destruction, a megaevent of disaster occurs, which has recently been called the Black Swan.

Due to the lack of a notion of the hazard to which disaster victims are subjected, it is said that the Black Swan is still unknown to the population, and because they are often unprecedented occurrences, they are considered rare and of high impact in multiple respects. For [8], the victims of the twin tower attack on 9/11 were victims of a Black Swan. It was not possible to locate where the hazard energy was concentrated, nor measure it. Gray Swans, according to the same author, are unlikely but impactful events that are outside the normal scenario.

A Black Elephant, according to [9], is a significant threat, risk or issue that is obvious to everyone, but no one wants to address or deal with it, "they get ignored". When it causes high impact, people get surprised and act as if it were an unpredictable black swan.

There are new models of behavior to include high stress and ways to measure this feature through the radar chart. This chart presents the characteristic of presenting multivariable data in the form of a graph.

The managerial, human-social, and technological aspects can create an environment for unsafe behavior, in which they allow the flow of hazard energy across barriers. The human and organizational factors elaborate, by events, the condition for the transfer of the social hazard towards the physical realization of the accident or incident.

In order to prepare people to deal with risk situations and be able to participate in decision-making processes that report to their own lives, it is necessary investment in: 1) Integrated Communication - Training and language competence are really important. Subjective factors such as judgments, culture of guilt and fear of damage to the image cause the worker to omit information that could contribute to the construction of barriers to avoid incidents and accidents; 2) Stakeholders - person or group who have an interest in a particular business, industry or service; 3) Intellectual Capital - Knowledge of preventive and mitigating actions provides people with adequate decision-making; 4) Competence - Understanding the causal nexus of the phenomena involved in the accident, having experience and experience similar to the activities related to the events.

## 3 Models & Hypotheses

The discussion about the competence of Black Swan Team beginning answering questions after observing real situations of emergency. After we feel some gaps and try to answer after investigation about the reasons of emergency team perform this behavior. After each question we construct or explain the scenario and answer with new tools that will compound the methodology.

### *3.1. What really happens in the chain event scenario when the danger is unknown?*

We have nothing to perform about the existence of a disaster from an unknown danger as in actual case of a pandemic. The disaster is a truth, we must reduce its impact, investigate the various dimensions involved with this unknown danger, or unknown operational danger considering the cultural and environmental effects during the isolation and flexibilization.
The study of the origin of the disaster allows us to understand its mechanism of action and possible consequences. The danger that causes the accident-disaster of technological (industry) or natural (climate, virus) origin can be acute, being a fast event with high impact, or it can be chronic, lasting a long time with, probably, immense impacts. Thus, the event becomes a public accident when it affects the population. We classify the danger and consequent events of COVID19 as natural, chronic, and public. This discussion is part of Strategic Intelligence that define the need resources and Hazard Identification, and Analysis of General Risk to define and test the best barriers.
While the danger is at work in an external region or country (not close to our region) we will observe and learn about its mechanisms, we seek to install two instances: The Integrated Management Committee and the Strategic Intelligence Committee. The first will join the necessary resources and leaders to take decision and the second will look for the best actions, benchmark, learning from other countries and regions.
In Integrated Management, Specialists, Managers and Practitioners who are part of Public-Private Institutions meet to define studies that assist in the decision on actions. Some tests are suggested to find the mechanisms of action of the hazard. In this way, barriers are built for known dangers, for the unknown and for threatening environments. Indicators and parameters and variables are mined to understand the impact of the pandemic on the economy and social relations.
Strategic Intelligence intends to identify the source of the main hazard and parallel hazards to project the Future Crisis Scenario. Scenarios are designed to test barriers, variables, resources, and knowledge. We will value to learn from the past, and from abroad experience, sending representatives to recognize the mechanism of these dangers through visits to government and institutions. In this moment a good general knowledge about pandemic situation, technological disaster and natural, impacts on social conflicts and economy, will help do draw an initial description and after, inviting specialists, the crisis leader tries to construct and perform effective actions (corrective, preventive and mitigation actions).
After confirmation that the Scenarios indicate an active Pandemic, the Contingency Committees begin in specific and critical regions. In some cases, such as natural disasters involving an earthquake, breaking of barriers in the mining industry and other technological accidents (fire or toxic spill) we need to install a Contingency Committees. In cases of acute accidents, groups are sent for mitigation action. Therefore, operations and local intervention teams will be installed.
The Black Swan Team at the Strategic (Strategic Intelligence), Tactical (Integrated Management) and Operational (Command and operations) levels need the classification of the type of event and impact for the work of the teams. One of the cases may be the study of technological accidents such as the breaking of barriers in mining or the leakage of toxic gases near the city (similar to Bhopal). The other case can be related to public events of high capillarity about impact, a virus that contaminates quickly and through unknown routes, the current COVID19, or even a series of earthquakes that promote tidal waves and affect the functioning of cities. Or even, the possibility of multiple events occurring involving the technological and the public dimension.

### *3.2. What are the typical models of action of the emergency leader and crisis management required for treat a black swan type event considering the resources, the level of stress and the direction of the danger energy?*

***LODA (PROGRESSIVE STRESS MULTI-LYER MODEL) shows that the mind map change in extremely high stress with wrong decisions.***

Ávila [10], knowing about the possibility of high impact of technological accidents near cities and after confirming the risk to the risk affected by the work team and the population through underreporting. Ávila confirms the occurrence of the phenomenon of cultural normalization of deviation and for the analysis of the consequence of progressive stress, he designed an accident involving an LPG fireball near a city in Brazil.

This event constructed from real events and deviations that occurred, although the top event did not occur, was composed of 65 diversions and successive events, sociotechnical failures in the form of the local routine and complemented with those that occurred in New Mexico. The Accident prepared for the Risk Analysis impacted with 40 fatalities (this simulation was structured in the thesis and published as a tool [10].

The analysis model involved 5 levels of stress in the team, where 1st is uncontrolled process with operational deviations and the last was the crisis scenario in the next city. Progressive stress disperses or disorganizes the mind maps that affect the emergency leader's decisions. In this model it is considered the occurrence of a theater with critical roles where there are communication failures or non-acceptance of the roles in the staging. Observing each phase of the accident and its cognitive stress will indicate social phenomena that represent problems in the treatment of accidents. These problems must be addressed before the actual demand for contingency.

***EMERGENCY PERFORMANCE MODEL indicated that there are characteristics for the leader and the team that are intuitive, affective and cognitive and that it was not found in emergency team in these accidents of the history.***

The studies by Carvalho [11] and Ávila et al [12] indicate that there are specific characteristics that are appropriate for the leader and the emergency team on real demand. In the case of leadership, we can mention perception of real risks (signs and scenarios); competence regarding the danger mechanisms (fire, virus, mechanical impact); energizing the team to act in contingency; a good memory to reinforce the decision; and speed of command since the time of the chain event, until the greatest impact, is very fast, and evacuation measures can save many lives.

In the case of the emergency team, we can mention: the quick cognitive to retrieve documents, a discussion, a phone call and assist the leadership in the decision through opinion and calculations; quick rescue of memory for the analysis of the scenarios, this aid sometimes brings information avoiding the direction of the danger to where there is greater vulnerability, for example, sending of chemical product to empty the burnt tank where there are visitors or maintenance service - the leader does not knows the level of vulnerability of the areas and depends on the team's support for this analysis; emotional balance brings a better quality of the cognitive and motor apparatus; ability to predict future situations through greater knowledge of the event's phenomena; and the ability to act on time, without delay, bringing firmness to field interventions.

Thus, Carvalho [11] and Ávila et al [13] affirm that the preparation of the emergency team mixes past experiences, attitudes, field and memory exercises, team energization, which are often not considered in normal training for teams of emergency. The results of the team's performance must be observed so that there is continuous improvement in performance under real demand.

***COGNITIVE GAP ASSESSMENT IN REAL ACCIDENTS AND SIMULATION where we assess the emergency teams based on public and technologic accidents confirming deficiencies indicated by Carvalho, Ávila e Pichon Riviere.***

This research work analyzed the national and international accidents regarding the characteristics discussed in the models for emergency action [11] [12]. The emergency team was analyzed in Accidents: (a) Chemicals and Oil of high impact in number of fatalities and environment (piper alpha - oil; Horizon - Oil; Union Carbide Bhopal - isocyanate); (b) national accidents of shipwreck in Mar Grande and explosion with fatalities at the FPSO São Mateus in the Brazilian state of Espirito Santo; and accident designed in the form of a simulation for the analysis of fireball risk by LPG in northeastern Brazil.

Specialists working in Risk Analysis, Contingency and Emergency Leadership met to classify the teams in terms of cognitive gaps and emotional control. This analysis was published at LACPS in Buenos Aires and it was concluded, in discussions with Process Safety Managers, that the classic way of training for emergency brigade teams does not meet the need to control real demand. Why? Although the situation presented in the training simulations for the Emergency Brigade, the petrochemical, refining, and oil industries seek typical situations for fire, fire, leaks and environmental impact, there are always organizational factors that affect human factors, there are always social factors that they cause conflicts and affect the quality of communication at work, and there are always false indications, a high backlog in the industry today. Thus, events always happen in multiple dimensions and the possibility of something unknown happening is real. Thus, the greatest value to be given is cognitive-affective-intuitive training for the emergency brigade. Lately, Process Safety Managers work only in Cognitive and without surprises. The level of stress and the style of the Leadership bring a level of comfort that may not be satisfactory for the occurrences of real demands.

***COGNITIVE FAILURE IN DISASTER FIELD SIMULATION where the team is not always integrated and energized in the hot simulation situation, believing only in the obligation to be present. In this case, preparation through the simulation leaves cognitive and compromising gaps for the contingency in the disaster.***

At the CIDEM, Mass Disaster Congress, held in Feira de Santana, UFBA was a rotating observer and investigator during a simulated exercise involving 25 public agencies including Armed Forces (Navy, Army and Air Force), Federal Police, Municipal Guard and State Military Police, SAMU, State Fire Department in addition to the BOPE anti-terrorist squad. Various functions of civil defense, social defense, isolation, and medical care came together in this exercise to prevent events of bomb explosion on buses, tipping of benzene trucks, and kidnapping, actions promoted by terrorist groups, all with an impact on the population and in nature, in the road modal.

The observation of these events indicates that the hierarchy of the agencies is above the recognition of their roles during the simulation and the inter-agency communication. The communication made with noise leads the team to different levels of urgency for mind mapping in the different teams. Some equipment was underused indicating lack of training, therefore, there are human and organizational factors that must be put into discussion to improve the integrated actions of the emergency command. The main characteristic is the lack of common priorities in the agencies promoting a disintegrated management state, where the size of the problem is not known for sure, only when faced with it. We note mainly that communication tools and devices are not used properly. Because it is a cold simulation, prescribed events of common knowledge, many communication failures did not appear. And this is a serious problem when analyzing previous events when municipal, federal, state forces and private governance act around disaster mitigation, for example, by preventing ship collisions. We note that the lack of credit of the great emergency teams on the simulated exercise can induce people to repeat mechanical actions bringing nullity to the Strategic Intelligence team.

### *3.3. Which models are indicated for the situation of unknown danger and its intense energy flow with high impacts during these events?*

***SYSTEMIC FAILURE OR FINGERTIP FAILURE MODEL, where the failure is multidimensional and involves from emotional balance, and knowledge - quick cognitive, we must analyze the connection between these properties to invest in selection and training.***

The systemic failure models that indicate connectivity points of events of different dimensions to cause the top event. The format of the Systemic failure is like the digital of the fault that was published at the European Reliability Congress (ESREL).

***EXERGY OF OPERATIONAL FAILURE AND COMPLEX PROCESSES where the indicators of complexity in the construction of BS are analyzed.***

The discussion on the Exercise of Operational Failure [14] presenting that cognitive losses in the task are related to energy losses in equipment and processes. The presentation of The Investigation on Complex Processes through the tools that connect different sources that cause uncontrol in the process, interpreted through: Connectivity Diagram [15] [16]; of the Materialization Diagram where the failures that transform communication noises from management into a disconnected mental map in execution, are transferred into disturbance in the physical dimension; and finally the logical diagram of the fault indicating regions where to better install the barriers, in sociotechnical functions. These are models discussed at the Brazilian Congress of Chemical Engineering and applied in the Chemical Process Industry (aliphatic and aromatic amines in addition to liquefied petroleum gas).

***NETWORK OF FACTORS AND HUMAN ELEMENTS, BARRIER EFFICIENCY, AND DANGER ENERGY SATURATION where we need to identify the points where danger energy enters saturation to elaborate the best barriers.***

In recent times we have discussed ways to identify the regions in which the greatest flow of danger energy occurs, following the Model of Reason's Swiss cheese [17]. This danger flow circulates among the factors that influence human performance on the sequential, parallel and crossflow route. This multidimensionality of the failure follows the model in a network format of human and organizational factors divided into culture, management, routine and accident. This multidimensionality of the failure also discusses the safeguards of the technological, managerial, and behavioral dimension through the human elements that are design barriers to contain the danger energy that flows until reaching the maximum levels that are the accident and the disaster. Human factors are part of the routine of the operation and human elements are design safeguards. As described in Table 1, in the operation there are different factors that cause incidents and situations of loss. The nature of this event has the following dimensions: Cultural; Managerial; Routine control; and the scale of the accident and disaster. Also, in the same table we present the types of human elements in the project: Management; Technology; and Behavior.

**Table 1 – Human Factors and Elements**

| | |
|---|---|
| **OPERATION** - A. CULTURAL DIMENSION OF HUMAN FACTOR: (1) Social Culture - (2) Social Phenomenon - (3) Organizational and Safety Culture.<br>B. MANAGEMENT: (4) Management - (5) People & Groups.<br>C. DIMENSION OF CONTROL IN THE ROUTINE: (6) Operational Control - (7) Failure Control<br>D. ACCIDENT AND DISASTER DIMENSION: (8) Accident Control - (9) Disaster Control. | **DESIGN** - E. HUMAN MANAGEMENT ELEMENT<br>(1) Stress - (2) Leadership<br>F. TECHNOLOGY<br>(3) Risk - (4) Complexity<br>G. BEHAVIOR<br>(5) Communication - (6) Competence<br>(7) Cooperation - (8) Commitment |

## 4 Methodology

The investigation was carried out following the steps: (1) Recovery from the discussion about stress in the treatment of emergencies, especially in the case of BS type disasters; (2) Recognition of the intense flow of danger energy through antecedent events, human safeguard elements, human factors of industrial operation and consequent events that hinder the return to normality; (3) to locate the environmental gaps and conflicts that influence process losses and the occurrence of accidents, chain reactions and the difficulties of returning to normality, for this we divided the competences in: Strategic Intelligence, Integrated Management, Command and action in Local Operations, and Organizational Resilience. For each competence, possible gaps, deviations, or human errors are discussed, comparing with the expected standard.

Finally, we assess the current event of covid-19 pandemics to suggest correct emergency preparedness actions in the preparation of leaders and their teams regarding cognitive and motor characteristics in situations of high demand.

Here in the methodology, we will indicate the competencies required for the BS team and the mental map of each competency for further discussion on the case of the COVID19 disaster.

### *4.1. Competencies for BS*

As shown in Figure 2, Strategic Intelligence and Organizational Resilience have the purpose, respectively, to identify the danger and to restore the organization to the condition of normal, controlled operation, as it was before the disaster. Therefore, they outline strategies based on investigations of the past, the present, and the future, in addition to comparing local situations with external ones and in other countries. It is important to identify what the danger is and its mechanism of action, map the impacts to project the recovery. The Intelligence group is composed of risk analysts who work with scenarios of different dimensions (health, politics, technology, climate) and who can defragment dispersed information to build Future Scenarios and explain the events, even if the danger is still unknown, through testing and observation.

Another group that will guarantee the application of resources in the right place and that will effectively communicate the current situation involving the population and cities, industry, commerce, and services, is the integrated management group that will develop appropriate indicators for decision making from the disaster impact curve and from the data and information generated by strategic intelligence.

The local operations and command group works to carry out the activities indicated by the integrated and recommended management group, guided by strategic intelligence. The role of the

leader is important to energize the team and the role of the team providing cognitive and motor support in this achievement.
In this way, the Academy, Public Agencies and Risk activities such as the chemical industry and the oil industry need to structure themselves to prepare competencies that know how to act in the investigation of scenarios, in integrated public-private crisis management, in command and in carrying out operations. contingency and the recovery of local and regional activities.
Cold simulations [18] do not bring the expected result to structure contingency teams. Crisis Managers have not been effective in the application of resources, in communication and in the strategies drawn from the decision tools, as presented in the pandemic of COVID19 [19]. The risk management team is not able to design the scenario because it does not identify the danger and its mechanisms and impacts. We have no skills prepared for Organizational Resilience and we lose control of the situation during the emergency period. Thus, we need to develop the skills mentioned in Figure 2 and confirmed from the team's mind maps.
Mind maps are average informal rules accepted by the group after all the list of organizational priorities, group preferences, and regionalities. This map will then show how the team works in each of the four blocks of competence and how the teams should work.
Note that the models do not characterize the relationship of urgency of time, recognition of danger and its mechanisms, also the importance of the discussion on the types of communication (between groups, with the community, with the population or inserted in a political environment) and care in relation to false-incomplete or anticipated information.
The final analysis is a brief discussion of the mind map for BS events in general and for the specific case of the COVID 19 Pandemic.

## 5 Case Study

The Research Group on Operational and Dynamic Risks (GRODIN) has prepared a Master Program for Disaster Preparedness and Crisis Management and has been acting as a critical observer of Government actions and people's reactions to address COVID19, discussing tools in the Strategic, Managerial and of Operations.
Unfortunately, new strains of this virus are emerging, in a transformation of unknown origin, which shows once again that we are during a Black Swan Event. We must deal with this matter coldly to act in the face of the terror of mass death, and to avoid the wrong decisions. The skills presented in Figure 2 must be developed and available to act in this Global Crisis:

**(1) IE Strategic Intelligence** (study of the origin of the danger, its mechanisms, and the construction of future scenarios, where spurious data and warnings about changing the danger line must be valued, investigating the past to understand the present, investigating the future for disaster preparedness.
**(2) Integrated Management and Crisis Project - GIP** (data flow, organization of resources, development of tools for decision appropriate to the mechanisms of action of the danger and its dynamism when the danger is unknown, communication mode and the treatment of the sub- political-economic and technological notification).
**(3) Command of local operations COL** with the preparation of resources for the here and now based on the organization of the Integrated Crisis Management in the projections of the future created by the IE (people, structures), reviewing the Integrated Management and the IE with new data from the present coming from the COL, having a preparation of the leadership and the team to avoid simulating extreme stress.
**(4) Organizational Resilience RO** where, with the control of the situation, seek the causal nexus based on practice and science in order not to lose control of the recovery. Always value spurious data

and information and warnings, classifying the quality of the information. Motivate IE for further investigations about the danger, its mechanisms of action and possible impacts before it happens.

Taleb [8] stated that in the face of an unknown danger, "everyone" will try to explain without knowing for certain what he is talking about (we need IE), demonstrating dominance that does not exist. Others who are comfortable with being a subject of scientific knowledge, virology, are not concerned with the preparation for situations not lived in the past, so we consider that unknown danger = known local non-operational danger. Thus, with chronic and growing impacts, we will have uncontrolled scenarios due to lack of resources, infrastructure, and lack of competent people for action.

The Black Swan has several dimensions that occur at the same time, Health, Economy, Politics, Logistics, Climate and Geopolitics, regardless of the dimension of origin of the danger. This multidimensionality gets in the way of decisions since the leaders of knowledge, public and private management, do not have the habit of thinking, investigating and deciding considering the systemic failure and its multidimensionality.

The low quality of information, underreporting and / or addictive data, undermines the role of management and science, which creates addictive or incorrect interpretations following mathematical models that are not practical. Observation of facts and events must be valued in times of Crisis (Disaster) to enable the endless elaboration of multidimensional scenarios to dynamize the Strategies.

In times of crisis, it is necessary to work with two teams that feed and feed each other on information. The IE and RO must be calm and when analyzing fatality data, review future actions and investment of restricted resources. The GIP and COL that see in the here and now practical needs to control the mitigation teams towards operational control and human errors caused by extreme stress (EE).

The IE and RO must be calm in the construction of Scenarios with weekly average of 4-day moving averages of fatalities to assess the effectiveness and effectiveness of previous actions, seek to explain or approach the truth about the danger. IE and RO's lack of calm in analyzing the scenario can vitiate interpretations in a single dimension of the disaster leading to incorrect projections of the future. It is necessary to change the position of the gaze to see what is hidden. Thus, we have developed guidelines to be followed:

(1) Installation of Crisis Committees for IE, GIC, COL and RO with a multidisciplinary and dynamic team; (2) Choosing-developing easy-to-use and effective statistical tools to support decisions, therefore, the weekly average of the 4-day moving average tries to avoid underreporting and spurious data that does not indicate a trend; (3) Enhance research on spurious data or information or warnings that may indicate a change in the danger or its mechanism and, consequently, its impact; (4) Prepare the team to maintain the quality of the cognitive and also adapt to the "time available" to guarantee the effectiveness of the actions. Extreme stress increases human error from 10% to 70%, bringing incorrect decisions and actions. We have to use tools to preserve the cognitive in extreme stress !; (5) Build a knowledge network platform, develop practical tools based on grounded theory, information flow, integration, and critical observation; (6) Following the film World War Z, it is important to treat spurious data and observe-validate unconventional local solutions that can be the difference in the application of restricted resources; (7) Following the UTOPIA Series, recognizing the political, geopolitical movements that have low visibility, can indicate simulacra of information, fake news. When knowing the fake news and underreporting, the safeguards will be more assertive, the aim is to test the hypotheses to protect the population; (8) Even if the situation is more optimistic, maintain recourse to pessimistic

scenarios, considering the opinion of people who are reliable but who project a different future from the majority as in the film World War Z.
I forgot the following aspect about the Leader to avoid Accident-Disaster Situations discussed by Collins [20]: (1) Not wanting the power and nor having vanities of signatures; (2) Extreme humility in dealing with unknown hazards; (3) Active listening for events, warnings and deviations; (4) Fierce determination in decisions and actions; (5) Recognize the possibility of own and team mistakes in strategy and action; (6) valuing the organizational and family unit.

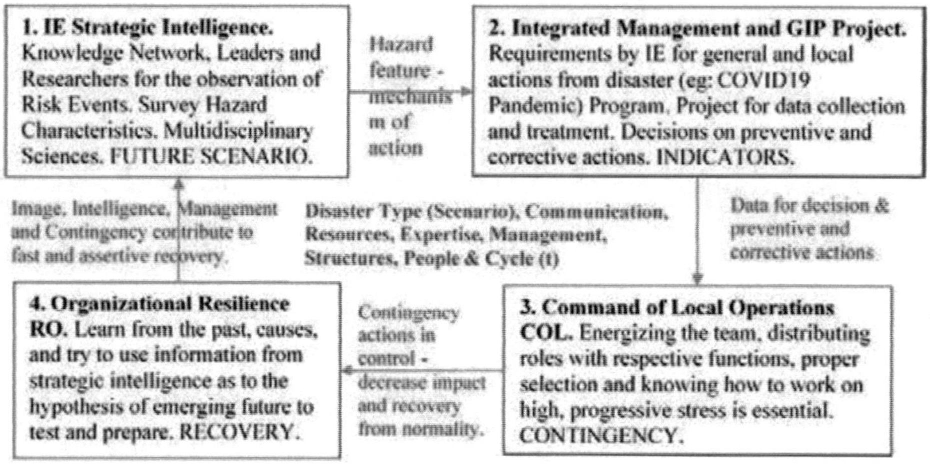


**Figure 2. Competencies for the BS Team: Intelligence, Management and Operation.**

### *5.1. Mind Map for BS and COVID19*

The characteristics already described by TALEB and Avila using the models from research about the dynamic risk we show the principal competences to attend a BS event. In Table 2, we discuss about the generic pattern and a case of COVID19 BS team.
The informal rules that turn into actions follow mind maps by type of function, Intelligence, Management and Operation. Below are examples of expected patterns and deviations observed in different mind maps.

**Table 2. Mind Map for BS**

| Mind Map | Generic standard | Expected standard (COVID19) | Deviations (GENERIC) |
|---|---|---|---|
| Intelligence | - To study the accidents and incidents; | - Study the disease in depth knowing the form of contagion, consequences, forms of treatment; | - Study superficially without exploring for learning; |
| | - In the crisis, seek as much information about the products involved, the team exposed, the area affected; | - In the crisis, seek as much information as possible about the disease, exposed teams, affected areas; | - Search for a solution without knowing what we are facing, people involved; |
| | - Know the risks that are relevant to the business; | - Know the risks that are relevant to the disease; | - Not knowing the extent of the risks and not preparing for it; |
| M an | - Know the resources available to face the crisis; | - Know in depth all the resources available for coping with the disease; | - Have resources that are not available. |

| | | | |
|---|---|---|---|
| | | | - Wait to seek resources in crise; |
| | - Treat accidents and incidents with preventive and corrective improvement actions; | - Predict and seek resources to strengthen the fight against the disease with the creation of scenarios; | - Treat accidents and incidents superficially without seeking a solution; |
| | | - Treat deviations with preventive and corrective improvement actions; | - Treat deviations superficially without seeking a solution; |
| Operations | - Know and study their role in acting during the crisis; | - Know and study their role in acting during the pandemic crisis; | - Facing the crisis in a way that is not consistent with the procedures causing conflict with the team; |
| | - Communicate any changes that are made to the procedures by training the others involved; | - Communicate any changes that are made to the procedures by training the others involved; | - Carrying out activities in a manner inconsistent with the procedures, putting the activity at risk; |

# 6 Conclusion

The current educational model prepares society for routine operations only, always with a focus on the main objective of producing with quality and low cost. The safety of the process is studied in a superficial way, focusing on the clearest and expected risks. BS are situations that arise during an event resulting from the sum of risks that may have been studied separately but never together. This moment requires several skills from the teams that work on the front line and coping and mastering the situation can take more time and we know that in an emergency, time and knowledge are fundamental. The intention of this work was to present a review and discussion about new concepts and tools that can be used in the prevention of human errors during crises (LODA) and thus we were able to understand the effects that the level of stress can influence in the decision making during the crisis and compromise coping with the crisis based on the understanding that the mental map changes completely in the crisis.

The presentation of the methodology with the necessary skills to face the BS and the mental map models are fundamental to create the PDCA of the crisis that always starts with strategic intelligence, integrated management, Operations command, and the organizational resilience that will feed the strategic intelligence in a learning cycle and the search for continuous improvement. The more we study this methodology, we see that although the event is unexpected, the knowledge and the feedback of information for the knowledge of the risks, will make that the mastery of the situation is reached more quickly.

The academy needs to better prepare society for unexpected situations. BS events are becoming more common, given the COVID19 pandemic experienced today and methodologies applied at the time of the crisis can save lives.

# 7 References

[1] T. Overton, M. Broadribb, C.N. Cheung, E. Christie, S. Cowher et al. "Guidelines for Process Safety Metrics", pp. 1-178, 2009.

[2] Center for Chemical Process Safety (CCPS). "Guidelines for Investigating Process Safety Incidents", Third Edition, 2019.

[3] Center for Chemical Process Safety (CCPS). "Guidelines for the Management of Change for Process Safety", 2010.

[4] Center for Chemical Process Safety (CCPS). "Guidelines for Managing Process Safety Risks During Organizational Change", 2013.

[5] Center for Chemical Process Safety (CCPS). "Human Factors Methods for Improving Performance in the Process Industries", December 2006, 248 p.

[6] Center for Chemical Process Safety (CCPS). "Recognizing and Responding to Normalization of Deviance", September 2018, 176 p.

[7] Herbert W. Heinrich, Industrial Accident Prevention: A Scientific Approach, 1931

[8] N. Taleb, "The Black Swan: Second Edition: The Impact of the Highly Improbable". Random House Trade Paperbacks, 2010, 469 p.

[9] P. Ho, "Future Beasts and Where to Find Them". Available at https://www.psd.gov.sg/challenge/ideas/feature/future-beasts-and-where-to-find-them. Accessed on January 22, 2021.

[10] S.F. Ávila. "Dependent Layer of Operation Decision Analyzes (LODA) to calculate human factor, a simulated case with PLG event", 2011.

[11] A.C.F. Carvalho. "Diagnóstico de lacunas nas competências da equipe para enfrentamento de situações fora da rotina". Master's dissertation – Federal University of Bahia, Polytechnic School of Engineering. Salvador, 2016, 103 p.

[12] S.F. Ávila., B. Mrugalska, M. Wyrwicka, M. Souza, J. Ávila, et al. "Cognitive and Organizational Criteria for Workstation Design", 2019.

[13] S.F. Ávila., A.C.F. Carvalho, G.P.J. Portela, C. Costa. "Learning Environment to Take Operational Decision in Emergency Situation". Procedia Manufacturing. vol. 3, pp. 1780-1787, 2015.

[14] S. F. Ávila. "A exergia da falha operacional, uma discussão filosófica com bases práticas". In: Proceedings of 19th Brazilian Congress of Chemical Engineering, 2012, Buzios.

[15] S.F. Ávila. "Análise de falha em processos complexos". In: Proceedings of 19th Brazilian Congress of Chemical Engineering, 2012, Buzios.

[16] C. Perrow. "Normal accidents: living with high-risk technologies". Princeton University Press, Princeton, N.J., 1999.

[17] J. T. Reason. "Human error". Cambridge University Press, 1990.

[18] S. F. Ávila., I. Cerqueira, V. Júnior, J. Nascimento, C. Peres, C. Brunoro, P. Arezes. . Integrated Management in Disaster: Discussion of Competences in a Real Simulation. AHFE 2019.

[19] S. F. Ávila., J.S. Ávila, A. A. Santos, M. C. N. Carmo, L. F. L. Souza. Crisis Decision Tool, fighting against COVID-19. AHFE 2021.

[20] J. Collins. Level 5 Leadership: The Triumph of Humility and Fierce Resolve. Harvard Business. 2005.

**AIChE Paper Number: 23A**

# MEGA CRACKERS: CHALLENGES AND THE ROAD AHEAD

*Prasada Raju Vegesina*
*Manager, Initial Operations*
*Lummus Technology*

Prepared for Presentation at the 2021 Spring National Meeting
vFairs Virtual Platform, April 22, 2021

# MEGA CRACKERS: CHALLENGES AND THE ROAD AHEAD

Prasada Raju Vegesina
Manager, Initial Operations
Lummus Technology

Abstract:

The definition of 'World-Scale' for operating plants has been rapidly evolving, enabled by innovations in process technologies and advancements in equipment design. Robust reliability and safety integrity systems adopted by operating companies are helping sustain these facilities, resulting in increased demand of such capacities. Favorable ROI make them compelling. Nonetheless, the size itself adds complexities at each stage of the project.

Emission/Environmental considerations, increased focus on embracing and sustaining green initiatives by producers, robust regulatory practices and compliance, local community pressures etc. have put added impetus on minimization of flaring and thus optimization and effective execution of commissioning and startup activities. While companies do focus on these activities, the factors affecting the same go beyond these specific activities.

This paper highlights critical challenges to commissioning and startup activities with a focus on mega crackers based on actual experiences. It intends to draw attention towards contributory factors from each stage of the project, viz. design, equipment selection, precommissioning, commissioning and startup. It is true that most challenges experienced in commissioning and startup are a result of design considerations and poor precommissioning and preservation. These happen in spite of the numerous risk control processes and systems that companies follow.

Initial Ops at Lummus Technology is typically involved in internal P&ID reviews, developing Standard Operating Manuals, carry out mechanical checks at site and advice clients in precommissioning, commissioning and startup of facilities. This provides a 360° perspective of the processes involved and the opportunity to do so with multiple world class clients' help understand the challenges and address them effectively. This paper intends to highlight such experiences.

## Introduction

The significant role played by commissioning and startup activities in the overall success of a project had been well recognized by the industry. Procedures and practices have evolved and companies have become increasingly efficient in handling the same. Many Ethylene producers have handled or atleast have personnel who have successfully handled multiple plant startups. Capacities of upto a million ton of ethylene have become common. Several processes and risk systems are in vogue to anticipate and effectively address any potential challenges.

However it is still not unusual that similar if not the same lessons are being learnt in many of these startups. The challenges get further accentuated and augmented with the ever increasing size of the facilities. Having said that, with the hindsight of experience drawn from mega cracker startups, it is evident that these can be addressed with a clear understanding of potential challenges and appropriate planning and execution.

Contributory factors based on actual experiences from each stage of the project viz. design, equipment selection, precommissioning, commissioning & startup and last but not the least, human factor are highlighted in the paper.

## Design Considerations

Design considerations and equipment selection play a significant role in commissioning and startup of the facility. It is necessary that impact of such be analyzed for a smooth and successful commissioning and startup experience. Below are some examples from a few mega crackers.

To be mega in size, most producers have been looking towards multi-feed crackers which often come integrated with refining facilities. During the design phase, most producers remain confident of their ability to maintain the feeds at startup and in a bid to reduce costs the design evaluation is sought only for normal turndown operations. It's only during the startups, the difference between *50% of plant load* and *50% of feed types* is actually realized. The startup becomes challenging and some of the systems remain unstable.

This holds particularly true for multi-feed crackers with high liquid feed content. Often clients procure flux oil (make up quench oil) for the startup with compromised specifications from their own refineries. While this in itself is not a major issue, when the startup is done on gas-only feed or heaters with no feed are turned into Fractionator for prolonged periods of time, the lighter aromatic components within the make-up oil gets stripped off resulting in high viscosity of the circulating oil. Viscosity of upto 50 cps was experienced in a cracker due to unexpected delays in feed in.

If some cases the make-up oil contained excess naphthalene, which tend to accumulate in the gas oil section in the fractionator. In a multi-feed unit, while cracking gas-only feeds to start-with, in absence of heavy aromatics to keep them in fluid state, they tend to solidify and choke the respective system. Once choked it's a tough task to get back to normal. However these could be addressed if the project scope clearly states the possibility of lack of certain types of feeds during startup. Provisions like heat tracing, flushing arrangement etc. can be made as part of the design if these are envisaged at the right time.

Quench oil circulation pump in a liquid cracker are among the larger equipment. Often they come with a continuous external seal flush, typically from pan oil/purge oil which is expected to work well during normal operation. However with these large pumps needing anywhere upto 1m3/hr. of seal flush, any delay in startup would not only necessitate large back up inventory but also lead to increase in Gasoline Fractionator level. While this can be recovered and reused prior to cracking, the same may not be possible after cracking due to presence of cracked components. Selection of the pump seals need careful consideration. While the external seal flush arrangement is not necessarily unwarranted, piping flexibilities may be considered keeping in view the above concern.

Process and equipment safety norms are continually evolving. Enhanced SIL involve multiple shut-off valves. Mega crackers imply very large valves and given the size requirements these tend to be mostly butterfly type. Valves on super high pressure steam to turbines are subject to very high pressure differential when in shutoff position. Failures experienced range from component (positioner related) to shearing of the valve shaft. Careful consideration of the design of these valves and sourcing from reliable manufacturers is imperative in these cases. Given the highly specialized nature of these valves, the delays experienced in the startups ranged from 3 days for instrument side component failure which involved local fabrication, to a week for removal of the valve with broken shaft and be replaced with a spool. Additional HAZOPs and safety engineering reviews were required. The valves itself had to be flown to the vendor for an upgrade.

One of the major upsets causing more than a weeks' delay during a startup involved the breaking away off SHP warm up line upstream of turbine TTV. It involved blow-off through a section of the roof of the compressor platform, fortunately with no other consequence. This happened as the steam line was kept under continuous warm up while the plant had constraints in feed forwarding. This led to failure of the line and the possible phenomena identified was acoustic induced / flow induced vibrations. Careful consideration in design is needed even for the smallest of piping.

DCS design/sizing has been one of the most common and biggest non process related challenges in many startups. Often DCS load is determined at a very early stage and as with all project items, it is designed with basic margins to minimize costs. However given that

HAZOPS, enhanced SIL requirements, ever evolving process safety practices not mentioning the 'good to have' from operations keep adding up quickly, the DCS loading increases exponentially. It has been repeatedly observed that DCS has posed major constraint for a smoother startup experience in many sites.

## Construction/Erection

One of the major challenges specific to mega crackers is the time that it needs to align and box-up the large piping in cold section, especially cold box connections. Longer the time it takes to get these large piping/flanges boxed up, longer the exposure to elements. At a mega cracker startup, chilling train had to be warmed up and purged due to hydrate formation and one of the possible causes identified was weather patterns during the cold section box-up. One of the ways to address this is to ensure completion of the fitment before end of the day or provide adequate covering assuming the worst of conditions.

Availability of large process compressors has become a necessity for effective precommissioning activities. Portable systems used for smaller /medium sized units are no more effective. This needs to be a consideration right from the project stage through procurement and erection stage. Availability of utilities like IA, steam, CW etc. needs to be ensured to allow these machines to be operated. A recent evaluation at a site to replace CGC with portable compressors for air blowing required a battery of more than 20 units which is impractical. In addition to the availability, flexibility of these machines to operate on air/nitrogen needs to be assessed with vendors.

Construction & Erection scheduling needs to consider commissioning and startup requirements as a critical factor

Longer construction schedules (given the larger plant size) also imply need for robust preservation practices.

## Precommissioning & Preservation

Precommissioning had historically been the most rushed activity as it is difficult to fathom not producing with the facility already built. From experience it is evident that inadequate precommissioning is one of the largest contributor to poor commissioning and startup of facilities. While the importance is well recognized, it lacks in execution and the same lessons are being learned over and over again from site to site. A mega cracker due to its large piping and equipment pose additional challenges.

Inadequate cleaning of the *interconnecting piping* was one of the biggest concerns during an actual large cracker startup. The delay to startup and feed ramp up was actually more than 6 weeks (conservative estimate). As it is an integrated complex with a refinery, cracker and other units, multiple parties handled the precommissioning activities. While the upstream and downstream (cracker) units were handled by very experienced engineering companies, the interconnecting piping, which ran approx. 3~5 KM each, was handled by the client themselves using a local contractor. While it is not unusual for a client to handle precommissioning activities, in this particular case, both the personnel and the contractor involved lacked experience in precommissioning and the fact that these are large diameter pipes made it worse.

Some of the many challenges that were faced at various sites due to poor precommissioning of interconnecting piping are noted below. This also holds true for plants with supply streams coming in from far away.

*Feed Piping*: The facility had a direct feed coming in from the upstream unit and fine micron filters were installed to prevent metal contamination in the feed stream to cracking heaters. The filters kept chocking at such a high frequency that it wasn't possible to get the spare available in time. Feed was flared to reach a point where filters can sustain enough to get the spare ready. As liquid feeds typically do not have flare connections, heaters had to be lined up just to ensure that there is flow through the piping and flared from the quench tower. However since enough heaters cannot be lined up just to be flared, the velocity through the piping was never enough and the filters kept chocking. Needless to say the economic loss and environmental impact was immense.

It is imperative that such streams that are installed with fine mesh need special attention during precommissioning. Based on experience, irrespective of the best efforts, for such high impact aspects, it is necessary to have a plan B and C, just in case a similar situation arises. A standalone temporary skid could be one of those.

*Large pump seals*: Seals on a few very large pumps (>1000m3/hr.) had to be repaired/replaced due to the poor cleanliness of the pipes. Given that they were massive and the nature of the damage was severe and given the lack of expertise / tools within the site, they had to be shipped out causing major constraints in startup.

*Large cage type CVs*: Similar to the feed piping as above, a lot of rust was brought in when the steam backup to dilution steam was lined up initially. The letdown valves, which were of cage type, chocked badly. The header had to be isolated and startup delayed. This repeated three times and when opened damage to the seat was observed. Impact of dilution steam loss back up during a cracker startup needs no mention.

*Critical Piping*: Similar challenges were encountered in several other systems.

- Hydrogen CV to the backend Acetylene converters. Not only did the CVs choke, the flow transmitters were affected badly posing process safety concerns.
- Foreign material was found in control valves on FG import lines to boilers, forcing the boilers to be shut down and holding the cracker startup.
- Foam was found in the refrigeration quench mixer (min flow and liquid quench) resulting in the unit shutdown and decommissioning parts of the refrigeration system resulting in major delay in startup.
- Most noteworthy (unfortunately so) is the experience where the strainers on the refrigeration system had to be cleaned more than 50 times (carbon steel piping going to cold box exchanger). During the initial days the strainers wouldn't sustain even until the spare is made available resulting in multiple upsets.
- In a particular startup the ethylene export line was used initially for inventorying the plant. However during a subsequent startup, there was no flow through the line. After several troubleshooting efforts the line was open in sections and part of the line was found completely iced-up (across the full diameter) for a length of no less than 50M. It was noted that the piping made a loop at the pipe rack and there was no drain at the low point. Possible hydro test water. This also raised several questions on the dew point measurements which seem to have been lost in the communication between plant operations personnel and off-sites.

*Size Matters*: The biggest challenge in mega crackers is the size of the pipes. Conventional methods used for small or medium sized crackers are no more effective. Piping sizes of up to 84" are seen in mega compressors. It is not feasible to achieve required velocities for effective blowing using portable compressors. This is true for pipe cleaning and also for dry out of systems. Availability of plant compressors and startup/temporary piping to ensure all large loops are covered is a must.

*Preservation*: It's widely recognized that preservation is one of the most critical parts of the plant pre-startup activities. Nevertheless, failures related to preservation have been a fairly common observation in many startups. Pump seals have been one of the most common observations. On a couple of auxiliary units the loaded activated carbon had to be completely replaced. Of course rust from poorly preserved piping is observed in most of the startups.

## Commissioning & Startup

Flare minimization during commissioning and startup has become a primary focus for the industry. While zero flaring is the most desirable outcome, a certain flame size/duration of flaring is reluctantly accepted. For small or medium sized crackers, this could mean that the plant is reasonably loaded and it is possible to stabilize them. Needless to say, allowable flare sizes do not increase with plant sizes, which means the available rates are much below the turndown rates for mega crackers, which implies it is extremely challenging to stabilize plants.

During a few recent mega cracker startups, the startup loading was much lower than 30% and it takes a long time to stabilize the unit. It also poses challenges in managing minimum loads through reactors etc. A topic that is repeatedly discussed at sites is whether the plant should start with adequate load and stabilize quickly or start slow even though it might take a while to stabilize. Community and regulatory requirements seem to be additional non-technical factors to be considered in this respect.

Having said that, it is imperative that plants will grow bigger. Policies, regulations and practices become more robust. It is also true that crackers, especially cold section stabilize with better loads. It has been an established practice to install recycle loops to achieve the same. With mega crackers this becomes very significant and needs additional impetus on the similar design considerations

Another challenge that a mega cracker throws in is the difference between startup and normal make-up inventories and corresponding tankage. While this difference always existed and companies were managing the same by staggered mobilization, the same gets more challenging with large crackers. Early planning and mobilization needs to be considered to avoid logistical challenges. On the flip side, early mobilization also means better preservation and monitoring.

One of the most undesirable practices seen repeatedly at many sites is utilization of heaters as boilers. The metallurgical impact on the heater coils due to prolonged HSSB operation and the long term impact on the convection section and an occasional impact on refractory is often neglected. Appreciation of the impact and robust feed planning should help minimize these unacceptable operations.

During any commissioning and startup, a few failures are always anticipated and facilities maintain adequate personnel on site to address the same. However with large crackers, the equipment becomes bigger and more complex and it is observed that the site expertise is not always adequate to address the issues. Delays were observed whenever such expertise is not available. It is important to recognize the criticality of such aspects and appropriate expertise be maintained at site during the startups.

It has been a best practice to provide recycle loops and external methane streams to pre-chill the chilling train so as to minimize flaring and quicken the startup. A few cold box failures have been noticed due to poor temperature gradient control and impatience shown during startup. This is especially true for large crackers as it is difficult to maintain large flows required for such operation. Careful evaluation and adequate provisions should be made during the design phase to prevent such failures.

At a large cracker, similar provisions were made to pre-chill the system. As the source gas was from a cold box it was directly introduced into the cold box and a mixed refrigeration compressor, bypassing the dryer. However the source was more than 100 miles away, and although the particular pipe was seeing flow (lined up to other units a few months earlier) with the increase in flow, the resultant velocity increase picked up leftover moisture. Unfortunately both the chilling train and the mixed refrigeration loops had to be shut down and dried up. The loss to schedule was more than 2 weeks. The basic principle of 'never assume' and double check everything holds good for a startup.

In a couple of large units, minimum flow/anti-surge control valves were not opening due to their size with default instrument connections. Tubing and positioners had be modified so that these valve perform as per machine integrity requirement. Similar issues were faced in other very large control valves too.

Lab and analyzers dependability has been one of the biggest challenges during recent startups. While the companies are trying to bank on internal expertise and optimize costs by minimizing vendor specialist's presence, it needs to be noted that the equipment and methods continually evolve and it is not always necessary that personnel with previous experience are capable of handling the latest. This has been a major factor in stabilizing units during startups.

## Human Factor

While precommissioning is the most rushed activity, human factor seems to be one of the most underrated aspects when it comes to a project. Earlier crackers were mostly standalone (with a few downstream units), and the project was typically populated with relevant experience. However, as most mega crackers tend to be multi-feed, possibly part of an integrated complex with a refinery, a lot of project personnel tend to overlap, implying cracker projects could be handled by personnel with mostly refinery background. While this is not necessarily an issue, opportunity to anticipate commissioning and startup issues, raise pertinent concerns and seek appropriate technology solutions at the very beginning of the project, could be lost. Each technology has its own intricacies and the least that can be done is to start well.

To keep the overall costs down projects are broken into parts and multiple technology licensors and EPCC companies are engaged. While some companies have been extremely successful at it, some struggled to cope with challenges associated with an integrated complex. Experiences clearly indicated missed opportunities and poor evaluation of potential scenarios among different units resulting in major delays during startups. While many of them have excellent operations and technical personnel, it needs special skills to integrate a complex optimally. It is important to engage the right support.

Operating companies are increasingly looking inward for pre-commissioning on cost considerations. The difference between a turnaround preparation and precommissioning is often not adequately appreciated. Expertise is on the wane while plants are becoming large and complex.

As with any industry, plants are becoming more complex while skill levels are not able to keep pace. As the plants are getting bigger the schedules are getting shorter, to keep up with the schedules every company is manning aggressively at every stage of the project. Unfortunately though, since skills don't grow overnight, compromise in quality of manpower is inevitable.

Lack of intensive training for plant operations and maintenance is another aspect that has been observed repeatedly.

While most producers have excellent systems in place for each aspect of the project, sometimes the competency is limited to the system itself and not core technical expertise and experience that is needed to analyze the risks by using the system.

## Conclusion

There were numerous challenges other than those detailed in the paper, which had to be addressed during multiple startups and the list grows and gets more complex as the plant size increases. Unfortunately many lessons are learned repeatedly too.

Having said that each of these have been addressed successfully and all the facilities are today operating at higher than nameplate capacities with better than design efficiencies. In the end, each of them have been very successful projects. The question is, how could the experience be better?

Having said that with the hindsight of a few mega cracker startups, it can be confidently stated that with the right experience and expertise and a genuine evaluation of the potential risks, commissioning and startup of mega crackers would be no different from any other normal startup. The need to anticipate potential challenges during early phases of the project, the ability required to integrate large complexes and capability to execute precommissioning comprehensively cannot be overstated for a successful commissioning and startup experience.

# 619165 Case Study in M+a – Communication and Understanding Hazards and Safety Concepts

**Abstract:**

A multinational corporation acquired a local consumer products manufacturing facility. Hazard awareness and process safety protocols were lacking at the local company, especially as measured against the parent company's corporate standards. The primary root cause was lack of a strong management of change procedures, as the processes in the facility were originally designed for non-flammable products but over time the facility transitioned to handling flammable liquids and combustible dusts.

Hundreds of unique chemicals/products stored and handled at the facility were analyzed to determine compliance with local fire codes and serve as a design basis for safety program improvements. Specific challenges around flammable liquids and combustible dusts were deemed to be highest risk and were addressed immediately.

This case study presents high level findings and both capital and operational budget related recommendations applicable to companies in similar situations. By applying a combination fire/building code evaluation and hazardous materials / process safety gap analysis, the company was able to prioritize risk reduction actions in alignment with their strategic objectives. This work presents a basis for hazardous material safety in facilities that store and handle a wide range of hazardous materials and aligning risk remediation activities with corporate objectives.

# 618542 Is Your Plant at Risk from Climate Change?

**Abstract:**

Climate change is occurring and some of the changes will impact the safety of process plants. This paper summarizes potential adverse impacts on process plants and then outlines how to conduct a Climate Risk Vulnerability Assessment (CRVA) for a process plant. One product of the CRVA is a plan of action to eliminate or reduce adverse climate effects on the process plant.

# 619167 Is Emergency Response for Facilities Impacted By Climate Change?

**Abstract:**

Climate change essentially bring about alteration to our natural environment. This can be seen by way of changes in the weather pattern. We see this here in the Gulf region of the US, almost every year during our most infamous hurricane/storm season.

Emergency response plans and preparedness during this season is always critical. It is based on science, best practices and lessons learned. At these times, not only is the shutdown process very critical , however the start up and recovery measures are also critical..

This paper will discuss thoughts and views of the author of changes required to emergency response and preparedness plans that facilities have to account for, due to changes in the weather pattern. The author will also discuss how the risks to people and process during the recovery measures can best be minimized. Irrespective of the conditions and planning, the human lives must be preserved.

# 619069 Transitioning from One Explosive Fuel to the Next – Safety Considerations in the Emerging Hydrogen Economy

**Abstract:**

Since the turn of the century, there has been a focus on the development of the hydrogen economy, driven by the desire to reduce greenhouse gas emissions through utilization of alternative energies and cleaner fuels. With both the current and future interest in utilizing hydrogen, it is crucial to evaluate and mitigate the fire and explosion risks posed by hydrogen during storage and transfer. While many other fuels such as natural gas, propane, LNG, and petroleum also present flammability and explosive hazards, hydrogen has several unique characteristics. These include, but are not necessarily limited to, a relatively low ignition energy, low density, and wide flammability limits. These properties present unique challenges for risk management compared to other flammable liquids and gases that will need to be addressed as the hydrogen economy further expands into its emerging market.

In this paper, the properties of hydrogen and how they impact the fire and explosion hazard compared to other fuels will be analyzed. Then, best practices and risk management will be discussed for storage and transferring fuels, with a focus on highlighting differences that are important considerations for hydrogen. Finally, a case study will be presented that reviews an incident involving a hydrogen leak and ensuing jet fire that ignited during the removal of a portable tube trailer. The events that led to the unintended release and ignition of pressurized hydrogen, and the lessons learned from the incident will be discussed. The case study will highlight several aspects related to the safe storage and transfer of hydrogen in commercial applications.

# 619275 Application of Causal Analysis Based on System Theory (CAST) to the T2 Laboratories Incident

**Abstract:**

## Application of Causal Analysis based on System Theory (CAST) to the T2 Laboratories incident

Joshiba Ariamuthu Venkidasalapathy[a,b,*]

[a]*Artie McFerrin Department of Chemical Engineering, Texas A&M University, College Station, TX 77840, USA*

[b]*Mary Kay O Connor Process Safety Center, Texas A&M University, College Station, TX 77840, USA*

[*]*Presenter email:* joshvenkat03@tamu.edu

### Abstract

STAMP (Systems-Theoretic Accident Model and Processes) is an accident causalty model that is based on systems thinking and systems theory. STAMP can be applied for incident investigations through the methodology called causal analysis based on system theory (CAST)[1]. It is a socio-techinial causal analysis method that systematically helps to identify causes that led to the accident. Unlike traditional accident models that rely on linear causation effects such as swiss-cheese model and domino effect models, CAST adopts a control-centric approach that helps account for causes that arise due to nonlinear interactions. Some unique features of CAST include treating safety as an emergent property, and providing the scope to account for software interations and temporal aspects of control actions. However, it is seen that applications of CAST to chemical engineering systems is scarce so far[2-4]. Considering the increasing complexity of chemical processes, accident models based on simple linear causation effects may not be sufficient in identifying the root cause of an incident.

CAST methodology is implemented in five steps: assembling the basic information, modeling the safety control structure, analysing the control structure to identify failed control elements, analysing the systemic factors and control structure flaws and providing recommendations to improve process safety. The second step, which is to model the safety control structure, is the essence of the CAST methodology. The control structure is heirarchical and requires every control block to have well defined functional objectives and control elements to achieve its goals.

In this paper, the objective is to apply CAST to a chemical incident to identify the social and technical causes that led to a chemical indident. The T2 laboratories explosion that occurred in Florida, 2007 is chosen to illustrate the CAST methodology[5]. The explosion led to four fatalities and several injuries. The CAST methodology along with extensive simulation based on MATLAB environment led to identifying both social and technical proximate causes. Most importantly, the results showed that temporal aspects of control actions, such as, how long the action was applied or how early/late it was applied also played an important role in the T2 Laboratories incident.

References

1. Thomas, J. and C. Malmquist, *Learning from Accidents that are a Consequence of Complex Systems.* Presented at the 48th Annual International Society of Air Safety Investigators, 2017.
2. Elliott, J., *CAST analysis of the Buncefield incident.* Procedia Engineering, 2017. **179**: p. 23-33.
3. Leveson, N.G., *CAST Analysis of the Shell Moerdijk Accident.* 2016, MIT. Cambridge: EU Major Accident Hazards Bureau.
4. Muñoz, M.F.T., *The Underestimated Value of Safety in Achieving Organizational Goals: Cast Analysis Off the Macondo Accident.* 2017, Massachusetts Institute of Technology, System Design and Management Program.
5. CSB, *T2 Laboratories Inc. Reactive Chemical Explosion.* 2009.

SPRING21 +17TH GCPS
A Joint AIChE and CCPS Meeting

# Process Safety Manuscript Archives Available to the Practitioner

**Raymond "Randy" Freeman**
**S&PP consulting**
**12303 Lake Shore Ridge**
**Houston, TX 77041**
**Rafree@Yahoo.com (for correspondence)**

**Ronald J. Willey, PE, PhD**
**Professor and Associate Chair Chemical Engineering**
**Northeastern University**
**360 Huntington Ave. 201 CN**
**Boston, MA 02115**
**R.Willey@northeastern.edu (for correspondence).**

Prepared for Presentation at
American Institute of Chemical Engineers
2020 Spring Meeting and 17th Global Congress on Process Safety
Virtual
April 18 - 22, 2021

# Process Safety Manuscript Archives Available to the Practitioner

**Raymond "Randy" Freeman**
**S&PP consulting**
**12303 Lake Shore Ridge**
**Houston, TX 77041**
**Rafree@Yahoo.com (for correspondence)**

**Ronald J. Willey, PE, PhD**
**Professor and Associate Chair Chemical Engineering**
**Northeastern University**
**360 Huntington Ave. 201 CN**
**Boston, MA 02115**
**R.Willey@northeastern.edu (for correspondence).**

## ABSTRACT

Since 1967, the American Institute of Chemical Engineers (AIChE) has sponsored a series of safety symposia. The goal of these symposia is to reduce the frequency and severity of accidents in the chemical process industries (CPI). The Safety and Health Division of AIChE was established to serve as a central hub for collection and dissemination of this process safety information. Over the years, this dissemination expanded to several process safety symposia tracks at the Global Congress on Process Safety. Three symposia manuscripts have been assembled and archived into a data base accessible on the Internet through the AIChE Academy.

- Loss Prevention Symposium (LPS)
- Process Plant Safety Symposium (PPSS)
- Center for Chemical Process Safety (CCPS) Symposium

The technical papers presented at these symposia cover a wide range of topics such as fire protection, risk reduction, explosion analysis, plant maintenance practices, process safety management and many other topics.

A total of 2,524 papers were scanned and organized into a form which can be used to search for information on process safety topics. A spreadsheet of the contents of the archives was created. This paper describes the organization of the Process Safety Archives and presents an example of how the archives can be used to find information.

## 1. BACKGROUND

As the English philosopher George Santayana[1] (1905) eloquently stated:

"*Those who cannot remember the past are condemned to repeat it*"

Trevor Kletz[2] (1993) restated the importance of the failure to remember the past in his book entitled: *Lessons from Disaster: How Organizations have no memory and accidents recur.* Only people have memory. Unfortunately, our memories are very poor. Therefore, to preserve knowledge two conditions must be met:

- It must be written down
- It must be findable.

The ability to find information is very important. It does very little good to state that the needle you must rapidly find is located in one of many haystacks in the back of 40 acres of a field. Likewise stating that a document of interest may be in the Library of Congress or the National Archives achieves little in actually finding the document of interest. For information to be useful, it must be timely and rapidly findable.

Previously, the Safety and Health Division sponsored projects to scan and digitize technical papers presented at various meetings. The papers were loaded into a Folio database and made available to the Safety and Health Division membership. Several items are of interest:

- The last update of the Folio database was in 1995.
- The process safety databases (versions 1, 2, 3) use a proprietary FOLIO format and did not include PPSS.
- Better scan technology today than was available 25 years ago.

Why this is important and why is it useful? Many safety research programs were only presented at Loss Prevention Symposia and were never formally published. These include accident investigations, chemical stability tests, hazardous material property test results and other information. Such research reports may only be available in the printed conference proceedings. Rapid access to these reports and papers may improve designs and prevent future process safety incidents.

## 2. LOSS PREVENTION SYMPOSIUM (LPS)

### 2.1 History of LPS

The first LPS was organized and held in Houston, TX in February of 1967. A total of 23 technical papers were given on topics such as:

- Building Design Criteria
- Equipment Design
- Reactive Chemicals
- Process Safety Management

The final session of the first LPS was devoted to a review of six process incidents.

Since the first 1967 symposium, the Safety and Health Division has sponsored an annual LPS. More details on the history of the LPS may be found in papers by Davenport[3] (2006) and Freeman[4] (2016).

### 2.2 Number of LPS papers

In the time period from 1967 to 2005, a total of 1,324 technical papers have been presented and are included in the Process Safety Archives. The papers include many summaries of process accidents, as well as discussions of such topics as fire protection, safety instrumented system analysis, dispersion analysis, and nitrogen inerting to name but a few.

## 3. CENTER FOR CHEMICAL PROCESS SAFETY (CCPS)

### 3.1 History of CCPS

In the second half of 1984, several incidents occurred which had international impact:

- In October 1984 a lubricating oil tanker caught fire and exploded immediately after leaving San Francisco. One sailor died and a large oil spill resulted.
- In November 1984 a large fire and explosion occurred in a PEMEX storage and distribution center in Mexico City. Between 500 and 600 people were killed and houses were damaged to a distance of over 5 miles.
- In Early December 1984, water entered a storage tank of methyl isocynate (MIC) at a Union Carbide plant site in the Indian town of Bpohal. The resulting reaction of MIC and water caused in a runaway exothermic reaction. A large cloud of toxic material (MIC and others) was released killing at least 2,000 people and injuring 10s of thousands of people.

As a result of these disasters, 16 U.S. chemical companies approached AIChE with a request to help improve the process safety practices in the chemical industry. This request was the start of CCPS. Details on the early history of CCPS may be found in the paper by Berger[5] (2009).

### 3.2 Topics and intent of CCPS Symposia

The first CCPS conference was held in 1987. The conference was focused on the following areas:

- Hazard Identification
- Storage and Transfer of Hazardous Chemicals
- Reactive Chemicals
- Post-Release Mitigation
- Human Factors
- Vapor Cloud Modeling

All of these topics are of direct relevance to the Bhopal Incident.

### 3.3 Number of papers

In the time period of 1987 to 2006, a total of 736 technical papers were presented at the annual CCPS conference. These technical papers are focused on specific problems and included detailed reviews of lessons learned from accidents.

## 4. PROCESS PLANT SAFETY SYMPOSIUM (PPSS)

### 4.1 History of PPSS

In 1989, there were two major process incidents in the Houston, TX area.

- A fire/explosion at the Phillips Pasadena plant killed 23 workers
- A fire/explosion at the Arco Channelview plant killed 17 workers

In 1992, the South Texas Section of AIChE organized a technical symposium on process safety. The symposium was held in Houston, TX to minimize the travel to attend the meeting and to encourage engineers and managers from the Houston area chemical and process industries to attend. The first PPSS was a great success with attendance by over 700 people.

### 4.2 Topics and intent of PPSS Symposia

The focus of the PPSS is on plant operations. PPSS papers are focused on practical application of process safety programs. More detailed technical material is found in LPS and CCPS symposia series. Papers on the management and control of process safety programs are a popular topic. Details on the history of the PPSS may be found in the Congress paper by Edwards[6] (2012). Edwards summarizes the focus of the PPSS as:

"*Practical Application of Process Safety Management at the Plant Level: Ensuring the Protection of People, the Environment, Capital and Business.*"

### 4.3 Number of papers

In the time period of 1992 to 2005, a total of 515 technical papers were presented at PPSS conferences. These technical papers are focused on specific problems and included detailed reviews of lessons learned from accidents.

## 5. CONTENTS OF PROCESS SAFETY ARCHIVES

Contents of the process safety archive include the following collections:

- Loss Prevention Symposium manuscripts from 1967 to 1980, originally Published annually as "Loss Prevention Symposium, Vol XX" by AIChE, New York, New

York, where XX went from 1 through 14. 326 titles across 14 volumes are in the archives.
In 1981, Loss Prevention Symposium manuscript numbering changed over to a AIChE session number, alpha character (e.g. 60b), mode of numbering and the AIChE discontinued the annual compilation of manuscripts into a single issue. Fortunately, John Davenport and others collected these papers and when this archive was announced, shared their personal collections with one of the authors where these were scanned in. In 1982, selected papers from the 1981 session forward began to appear in a new AIChE publication, initially called *Plant/Operations Progress*. In 1993 *Plant/Operations Progress* was renamed to *Process Safety Progress*. This collection of LPS manuscripts may also include both the *Plant/Operations Progress* version and the original manuscript delivered at the conference. Approximately 397 manuscripts and publications from 1981 to 1989 have been archived.

In 1990, AIChE moved to posting manuscripts onto digital storage media, and all manuscripts have been extracted from various digital media that existed at the time for the years of 1990 to 2005. Approximately 601 manuscripts exist in this part of the collection. After 2005, AIChE retained manuscripts from various Global of Process Safety on its web site AIChE Academy, and these can be accessed by going to the appropriate year and congress.

- The Center for Chemical Process Safety began congresses in 1987 and all manuscripts from 1987 to 2006 have been found, scanned, and archived in this database. Approximately, 736 manuscripts make up this part of the archive. Manuscripts after 2006 are available via the AIChE Academy.
- The final collection of manuscripts continues with Process Plant Safety Symposium, beginning 1992 and continued to 2005. 515 manuscripts make up this part of the archive.

Manuscripts published after 2005 are available on at www.aiche.org under this web link by year: https://www.aiche.org/conferences/aiche-spring-meeting-and-global-congress-on-process-safety

# 6. INTERNET ACCESS OF PROCESS SAFETY ARCHIVES

## 6.1 Overview of Access Protocol

Accessing the Process Safety Archives involves two major steps:

- Determine what you want to look at. This is accomplished by using the Master Index to the Process Safety Archives. The Master Index (Revision 0) is presented as supplementary material to this paper in the form of a Microsoft Excel spreadsheet. Details on how to use the Master Index are presented in section 7 of this paper,
- Once you know which volume of the LPS, PPSS or CCPS that the paper of interest is located, you must purchase the volume. The purchase price for a member of the Safety and Health Division is $0.00 (FREE). Purchase of the volume is via the AIChE transaction portal. The following section of this paper demonstrates the process for obtaining a paper from the Process Safety Archives.

## 6.2 Detailed Description of Method to Access Process Safety Archives

Important note: this example contains working links that if you are a member of the safety and health division, should take you to the images shown.

The process safety archives can be accessed at AIChE thru this web link: https://www.aiche.org/ccps/resources/safety-health-proceedings.

Once at the web site clicks "sign in" if you are a member of safety and health division (see Figure 1). The Home Page of CCPS will appear.

Re-click this web link: https://www.aiche.org/ccps/resources/safety-health-proceedings. to return to the archives.

This part of the archive is organized sequentially by year from 1976 to 2005. There are 7 subpages. Within each year is a link to the AIChE academy where the papers are actually stored. This approach requires the user to know what year and what conference the paper of interest exists. The easiest way to find out what to download is to use the Master Index as presented in the Supplementary Materials. For example, clicking on page 6 (bottom). Then scrolling, you will find two conferences as shown in Figure 3.

Click "READ MORE" for CCPS 17$^{th}$ conference. You reach this page (Figure 4).

Then you click TECHNICAL PAPERS and Figure 5 will appear.

Click the first title, Figure 6 will appear.

Click Checkout (this link is unique to the user and thus is not included).

This page appears, Figure 7.

Click Process my order. Figure 8 Will appear.

Next click CONTINUE and the following page appears (Figure 9).

You need to re-navigate after paying or receiving permission to access the conference papers. Click TECHNICAL PAPERS, and then click the first title again; this time a pdf file appears in the window (Figure 10).

This page now shows the pdf file of the original paper. Click on the paper abstract and a pdf display of the first page of the paper will appear (Figure 11).

Figure 11 shows the pdf file that can be read in a web reader or downloaded.

Another link, shown back in Figure 9 and highlighted in Figure 12, on the lower right hand side is a zip file download that will contain all papers presented in that conference, as well as the table of contents.

You can also download a zip file with all papers presented in that conference. Simply click on the Downloadable version and put the zip file onto your computer.

## 7. HOW TO USE THE PROCESS SAFETY MASTER INDEX -EXCEL FILE AN EASY WAY TO FIND A PAPER ON THE WEB SITE

### 7.1 The Process Safety Master Index

As the project progressed, an Excel File was assembled of year, symposium name, titles, authors, paper numbers. This file, to be available on the process safety web site, brought in two other data bases, the Process *Beacon* collection titles from 2001 to 2019 and the Hyperlinks to all *Plant/Operation Progress* and *Process Safety Progress* Papers published from 1982 to 2020, Issue 2. To the process safety professional, in need to seek past papers on the topic, this spreadsheet can proven to be very helpful. Two examples follow.

### 7.2 Ammonia Nitrate Example

An ammonia nitrate accident occurs (example Beirut, Lebanon, August 2020). The process safety expert desires to know all titles that have "ammonium nitrate" that are available in the archive:

Following these steps,

- Open the Process Safety Master Index in Microsoft Excel
- Initiate a Find and Replace dialog Box ("Ctrl F") – Figure 13
- Enter the search term "ammonium nitrate", and select the whole Workbook under option button to see all titles: - Figure 14
- Click Find All

The following list appears (in this case two manuscripts related to ammonium nitrate appear as the key word appears twice for each manuscript).

### 7.3 Ethylene Oxide Example

Let's try one more example, one that will have more results by searching on the combination "ethylene oxide". The following Figure 15 shows 19 titles appearing 7 titles from LPS, 1 title from CCPS (a duplicate appears), and 11 titles from *Process Safety Progress*.

Let's pick paper 3 in the list Ethylene Oxide Explosion at Doe Run. We see in Figure 16 that the paper was presented in the Loss Prevention Symposium in 1968.

Next arrow over to column M and click on the hyperlink provided:

Then, arrow over to column S and click on the hyperlink provided there (See Figure 18):

After clicking the hyperlink in Column S, Figure 19 should appear.

**Figure 19. Paper Title Page**

The Checkout link exists to purchase this paper. See the previous section for directions on how to purchase. If you have already purchased this volume, the pdf file link should appear below the authors (not shown in this figure).

## 8. DISPOSITION OF PAPER COPY OF THE ARCHIVES

The paper copies of the Process Safety Archives were scanned by InterActive of Houston, Texas. After scanning, the paper copies of the archives were deposited in the Linda Hall Library of Kansas City, MO for safe keeping. This step is still in progress as of March 2021.

## 9. ACKNOWLEDGMENTS

The creation of the Process Safety Archives was sponsored by the Safety and Health Division of AIChE. Creation of the Process Safety Archives using historical documents was a major task.

The Process Safety Archives includes over 2,000 technical papers, reports and documents. The Safety and Health Division Committee members were Randy Freeman (S&PP Consulting), Ron Willey (Northeastern University), Tracy Whipple (BP), Philip Myers (Advantage Risk Solutions), and Dan Crowl (Professor Emeritus, Michigan Technological University). Special thanks go to Jing Chen and Gennady Shafirovich from the New York AIChE staff. Scanning, initial file organization, and meta- data creation was completed by Maricela Manteca and her team at InterActive of Houston, TX.

The Proof Checking, Quality Assurance Team was organized by Ron Willey at Northeastern University. This team was instrumental in finding scanning issues, and cross checking the scanned material against the original conference programs to find missing papers. In several cases the QA/QC team identified papers that were missing and were retrieved by contacting the original paper author. This team included Zachary Kylar, Isabella Cardona Barber, Saule Bussing, Michael Nguy, Naa Monoh Odateifio, Joshua Garvey, and Eric Weinstein, as well as others.

### 9.1. Special Acknowledge for John Davenport

During the period of 1981 through 1989, the papers presented at the Loss Prevention Symposia were not collected by AIChE and published in a book as had been done in the earlier years. The years 1981 through 1989 are called the "Lost Years" by the officers of the Safety and Health Division. During this period, John Davenport of IRI collected the papers from each of the "Lost Years" as well as any associated published technical papers. Thanks to John Davenport, the collection of the "Lost Years" Loss Prevention Symposia papers have been digitized and are included in the Process Safety Archives. In many cases the collection of material by John Davenport contains both the technical paper presented at the LPS symposium and the subsequent paper as published in a journal by AIChE. Thank you John for a job well done!

## 10. REFERENCES

1. George Santayana, *The Life of Reason: The Phases of Human Progress, Volume 1 – The Life of Reason*, Dover Publications, New York, 1905. (Available on line at the Gutenburg Project, Ebook number 15000.)
2. Kletz, T., *Lessons from Disaster: How Organizations have no and memory and accidents recur*, Institute of Chemical Engineers/Gulf Publishing Co, Houston, TX, 1993.
3. Davenport, J., History of Loss prevention symposia: Forty years, 1967 -2006, *Process Safety Progress,* Volume 25, 2006, pp 268-273.
4. Freeman, R. "History of the Loss prevention symposium: The first 50 years – There are no secrets in safety," *Process Safety Progress*, Volume 35(1), pp 32-38, 2016.
5. Berger, S. *CCPS – 20 Years and more, Process Safety Progress,* Volume 25(1), 2006, pp 83-85.
6. Edwards, et al., *History of Process Plant Safety Symposium, Process Safety Progress,* Volume 31(3), 2012, pp 200–202.

## SUPPLEMENTARY MATERIALS

1. Master Index – Excel Spreadsheet

   Contents – LPS, CCPS, PPSS, Beacon List, Index to PSP papers

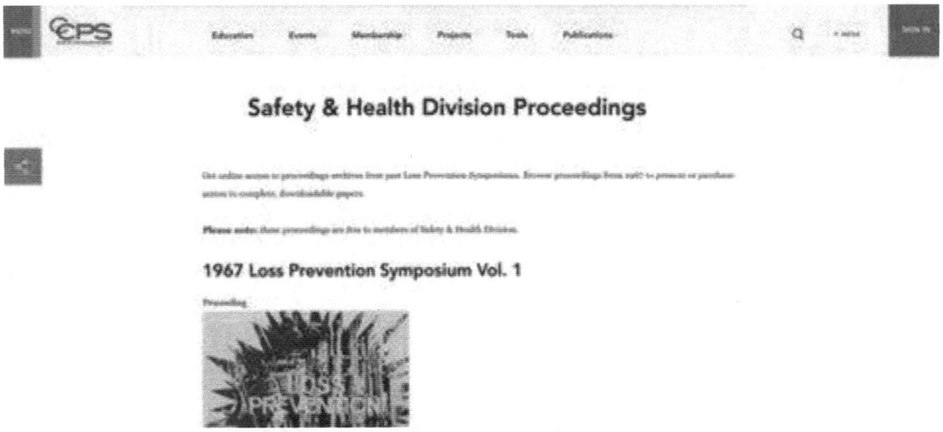


Figure 1. Initial entry page to the Safety and Health Proceedings.

Figure 2. CCPS home page.

**Figure 3. Example index page that take one to the AIChE Academy Web Page.**

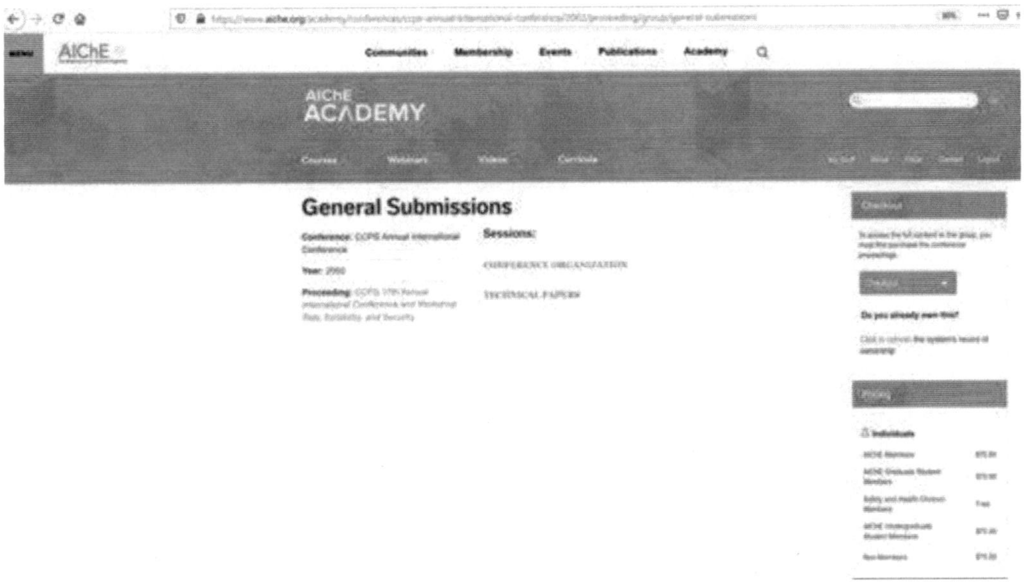


**Figure 4. How to reach technical papers.**

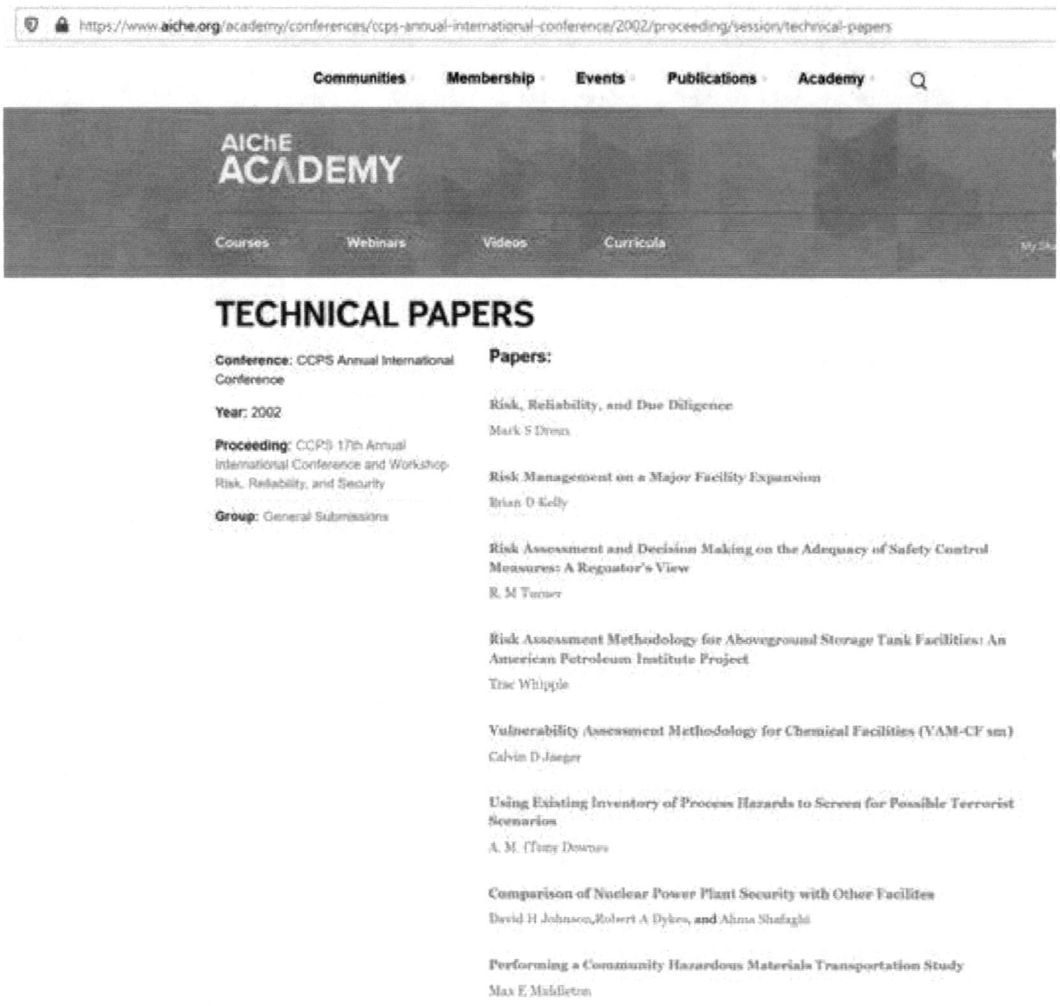


**Figure 5 List of papers available for the CCPS Annual International Conference, 2002.**

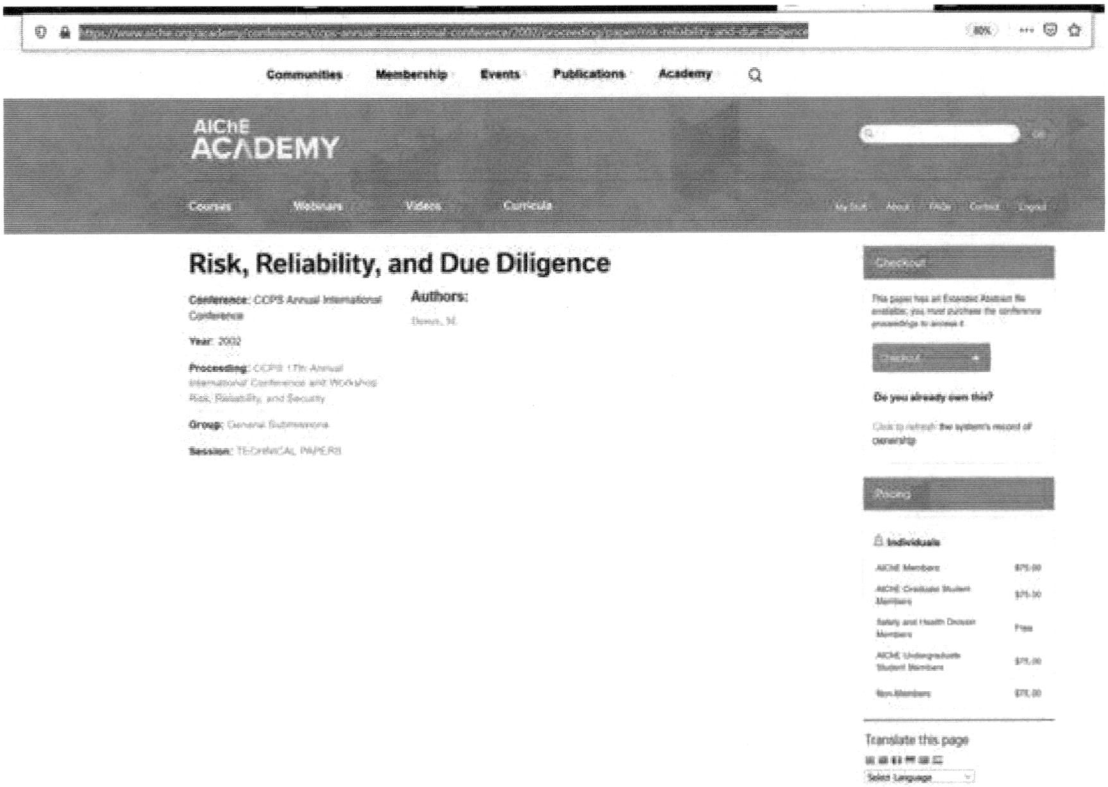


**Figure 6 AIChE Academy representation of Risk, Reliability, and Due Diligence, by M. Dreux.**

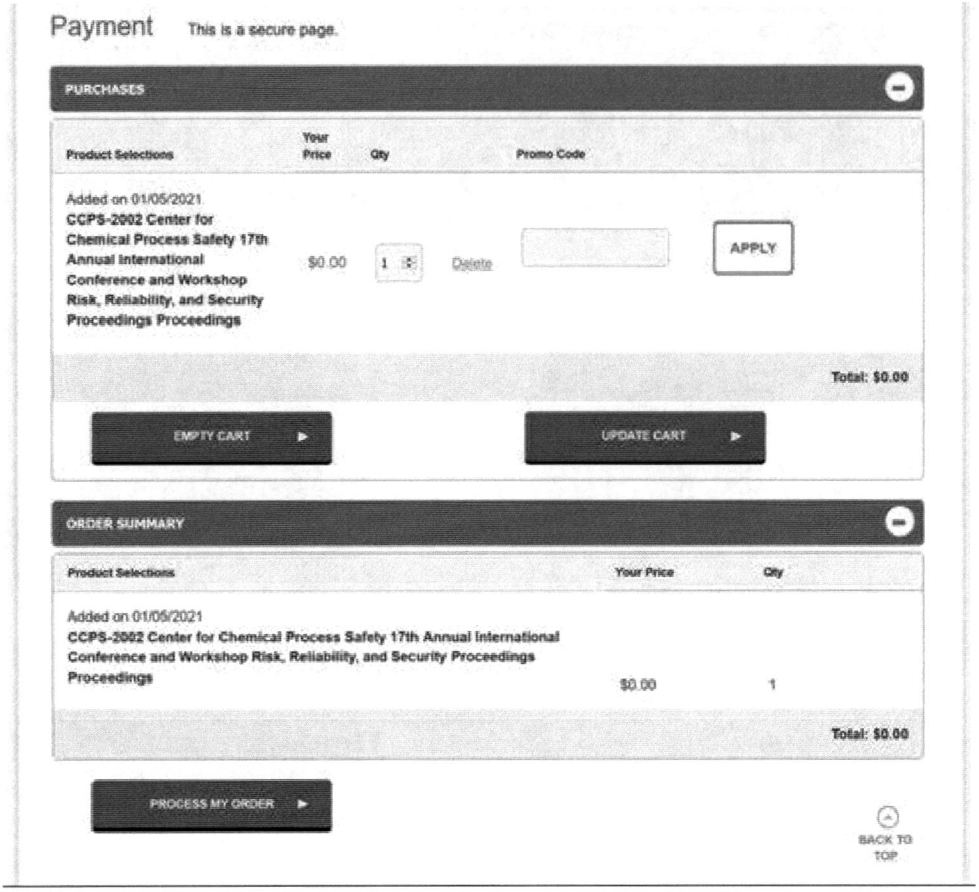


**Figure 7. Purchase confirmation page.**

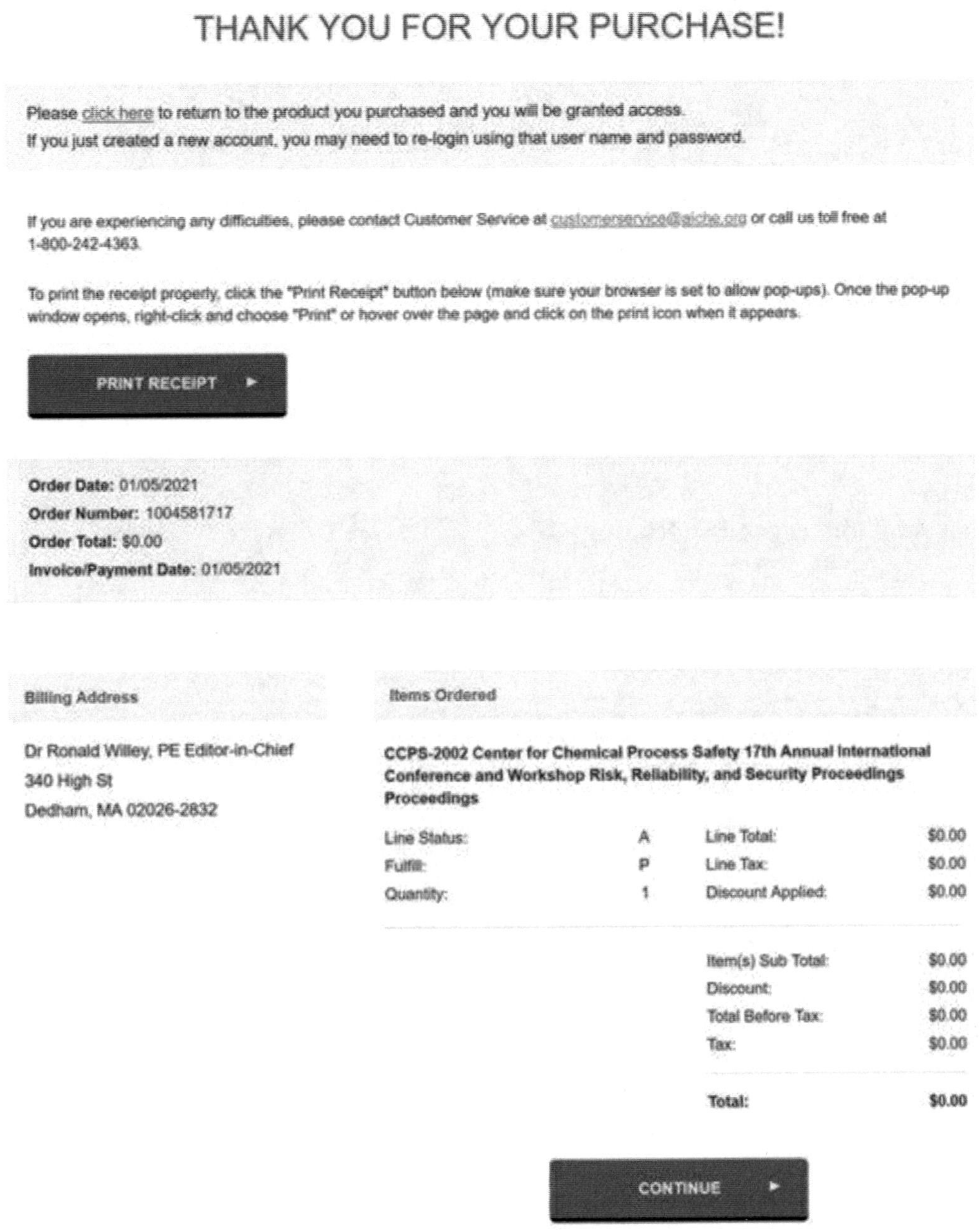

# THANK YOU FOR YOUR PURCHASE!

Please click here to return to the product you purchased and you will be granted access.
If you just created a new account, you may need to re-login using that user name and password.

If you are experiencing any difficulties, please contact Customer Service at customerservice@aiche.org or call us toll free at 1-800-242-4363.

To print the receipt properly, click the "Print Receipt" button below (make sure your browser is set to allow pop-ups). Once the pop-up window opens, right-click and choose "Print" or hover over the page and click on the print icon when it appears.

PRINT RECEIPT ►

**Order Date:** 01/05/2021
**Order Number:** 1004581717
**Order Total:** $0.00
**Invoice/Payment Date:** 01/05/2021

**Billing Address**

Dr Ronald Willey, PE Editor-in-Chief
340 High St
Dedham, MA 02026-2832

**Items Ordered**

**CCPS-2002 Center for Chemical Process Safety 17th Annual International Conference and Workshop Risk, Reliability, and Security Proceedings Proceedings**

| | | | |
|---|---|---|---|
| Line Status: | A | Line Total: | $0.00 |
| Fulfill: | P | Line Tax: | $0.00 |
| Quantity: | 1 | Discount Applied: | $0.00 |

| | |
|---|---|
| Item(s) Sub Total: | $0.00 |
| Discount: | $0.00 |
| Total Before Tax: | $0.00 |
| Tax: | $0.00 |
| **Total:** | **$0.00** |

CONTINUE ►

**Figure 8. Receipt page**

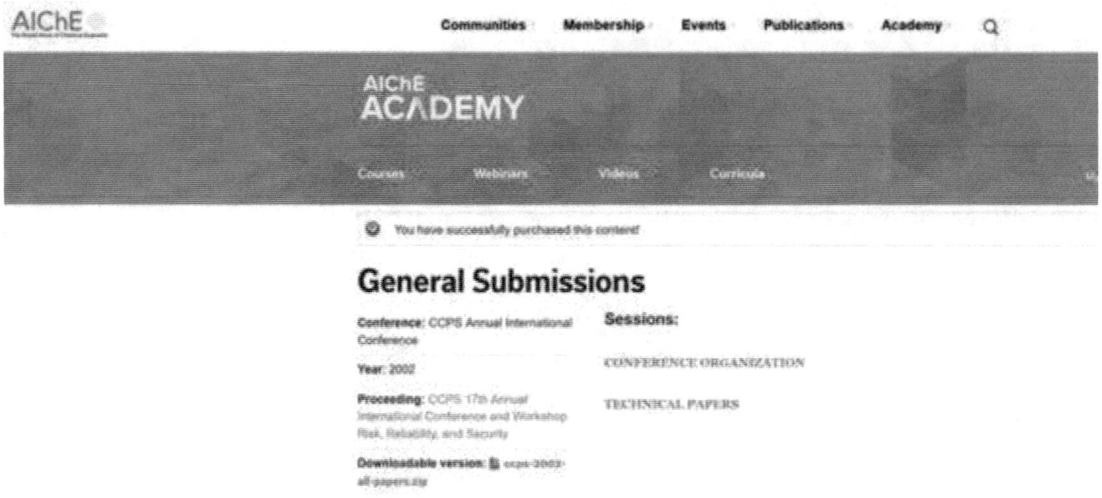


**Figure 9. New View of Conference Proceedings**

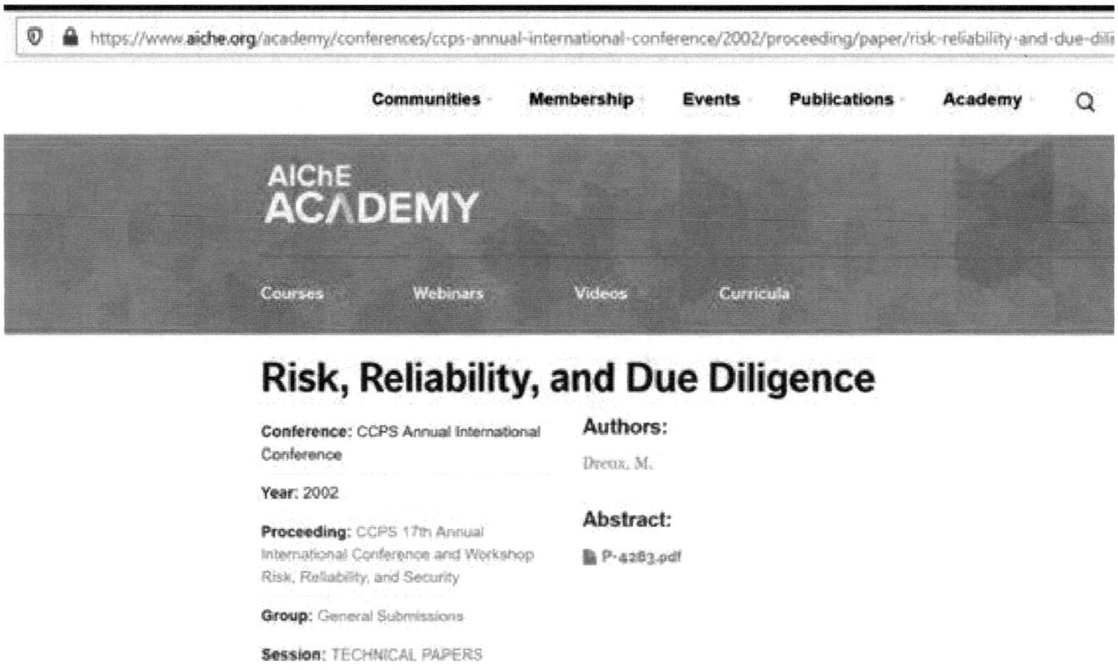


**Figure 10. Paper Title Page.**

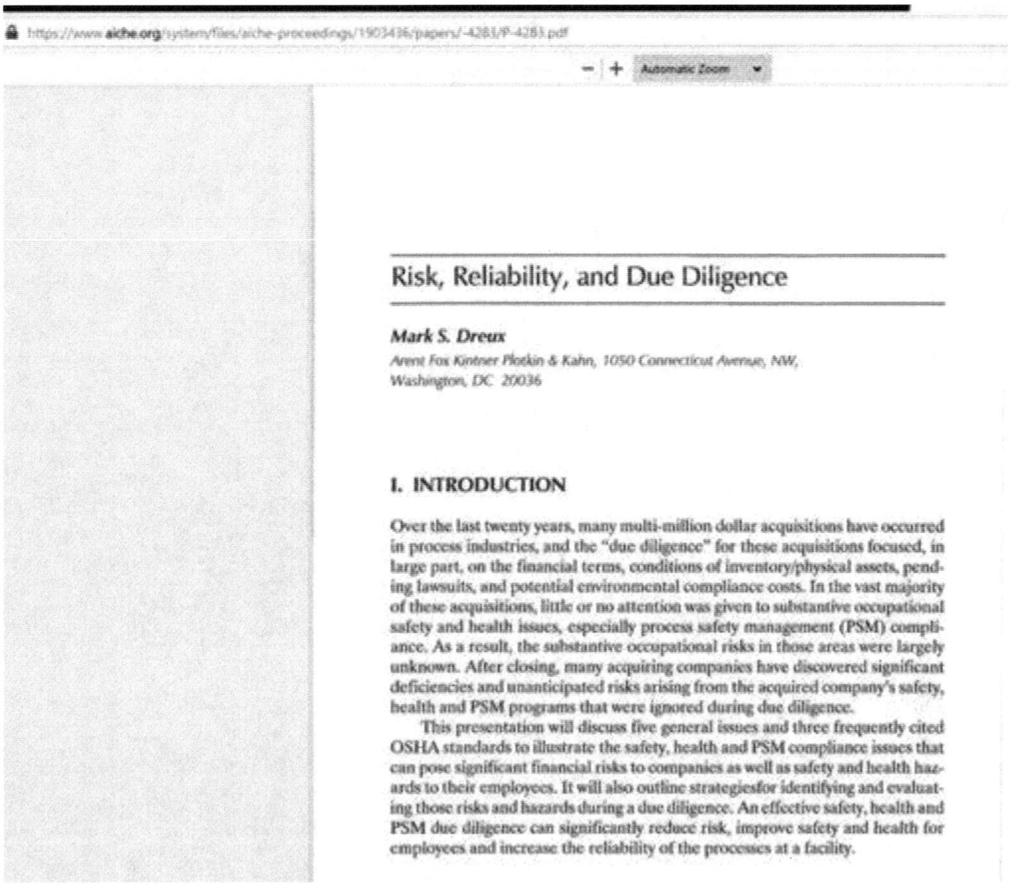

https://www.aiche.org/system/files/aiche-proceedings/1903436/papers/-4283/P-4283.pdf

Automatic Zoom

Risk, Reliability, and Due Diligence

Mark S. Dreux
Arent Fox Kintner Plotkin & Kahn, 1050 Connecticut Avenue, NW,
Washington, DC 20036

I. INTRODUCTION

Over the last twenty years, many multi-million dollar acquisitions have occurred in process industries, and the "due diligence" for these acquisitions focused, in large part, on the financial terms, conditions of inventory/physical assets, pending lawsuits, and potential environmental compliance costs. In the vast majority of these acquisitions, little or no attention was given to substantive occupational safety and health issues, especially process safety management (PSM) compliance. As a result, the substantive occupational risks in those areas were largely unknown. After closing, many acquiring companies have discovered significant deficiencies and unanticipated risks arising from the acquired company's safety, health and PSM programs that were ignored during due diligence.

This presentation will discuss five general issues and three frequently cited OSHA standards to illustrate the safety, health and PSM compliance issues that can pose significant financial risks to companies as well as safety and health hazards to their employees. It will also outline strategiesfor identifying and evaluating those risks and hazards during a due diligence. An effective safety, health and PSM due diligence can significantly reduce risk, improve safety and health for employees and increase the reliability of the processes at a facility.

**Figure 11. Paper finally appears.**

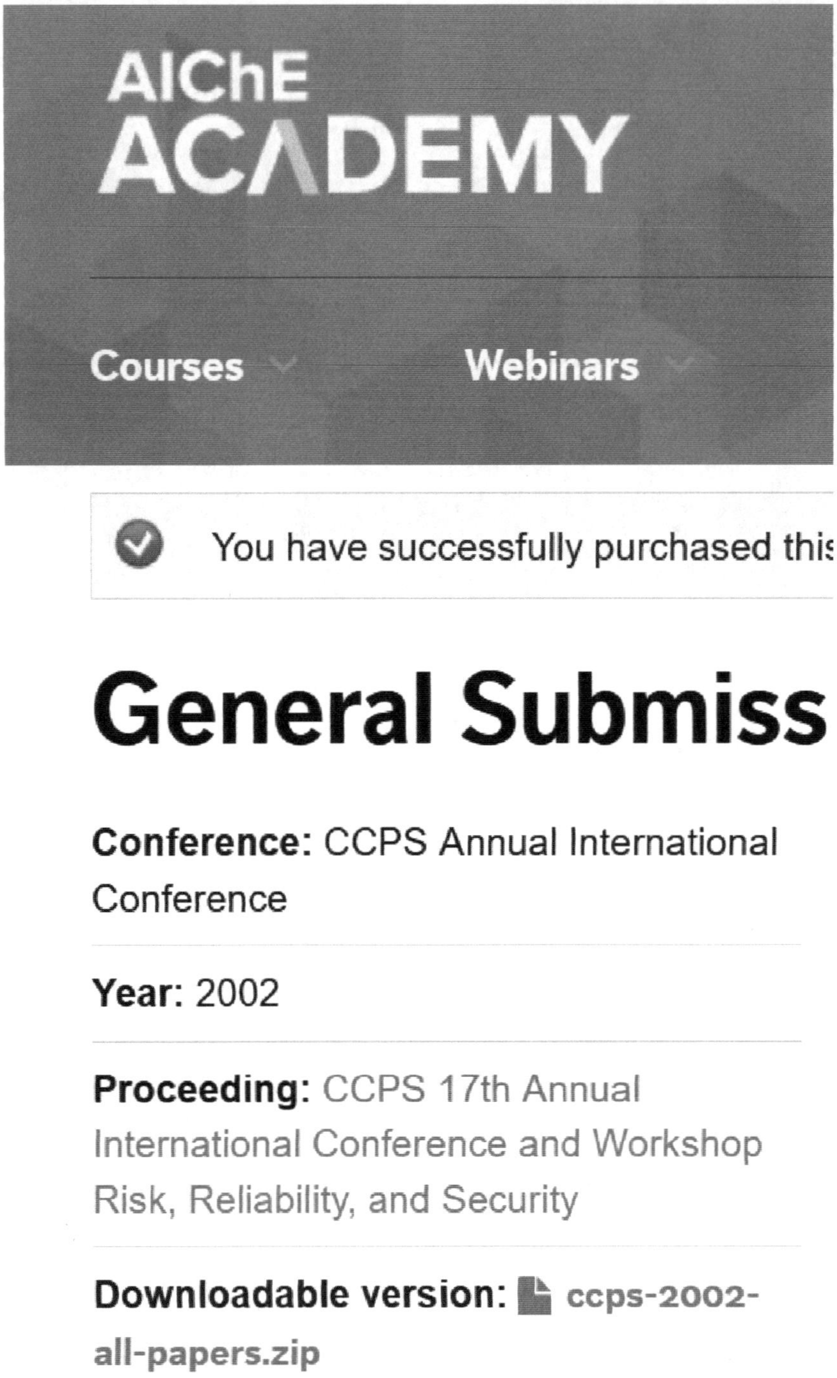


**Figure 12. Conference Download Option.**

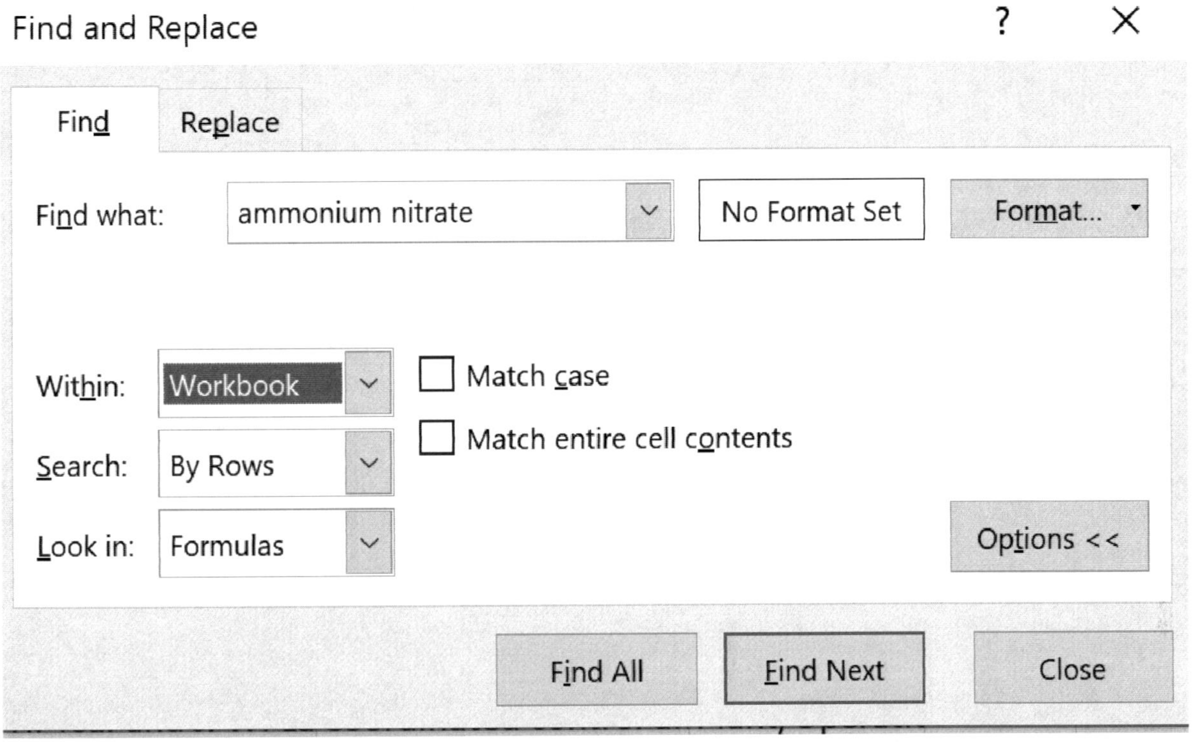


**Figure 13. Setting up an Excel Search for the key word combination "ammonium nitrate".**

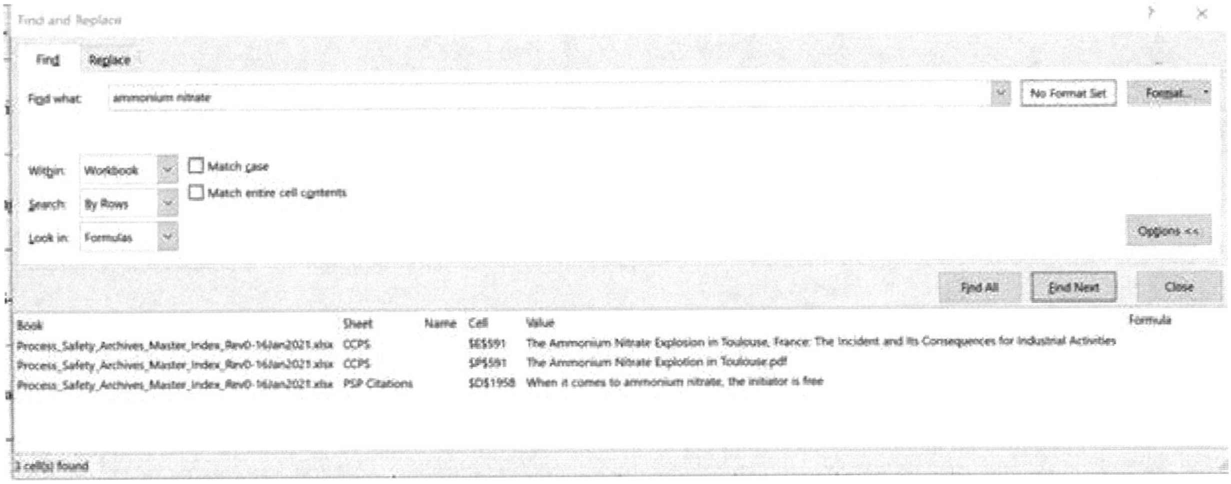


**Figure 14. Search results using the term "ammonium nitrate" in the Excel Spreadsheet.**

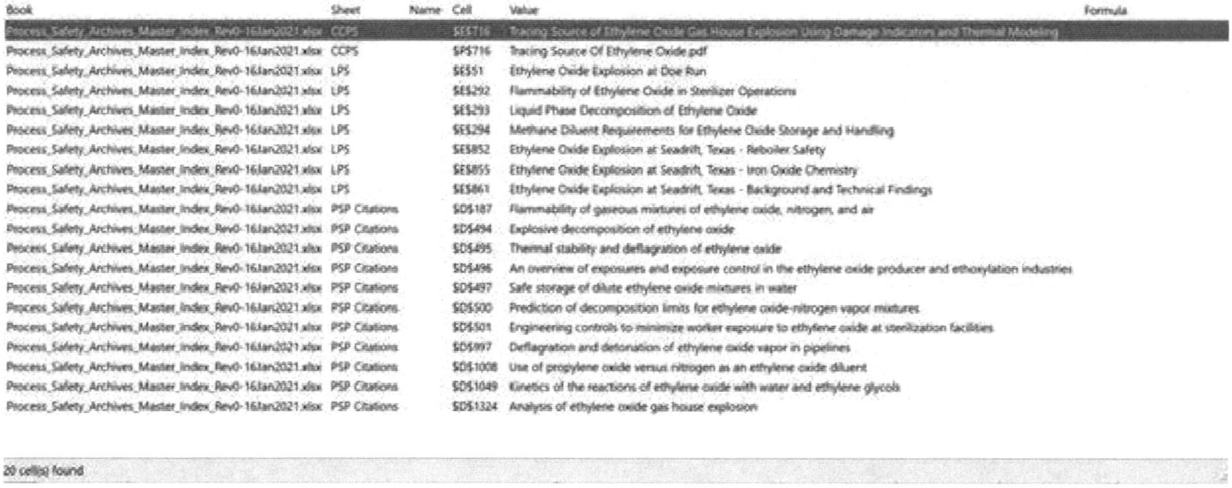

| Book | Sheet | Name | Cell | Value | Formula |
|---|---|---|---|---|---|
| Process_Safety_Archives_Master_Index_Rev0-16Jan2021.xlsx | CCPS | | $E$716 | Tracing Source of Ethylene Oxide Gas House Explosion Using Damage Indicators and Thermal Modeling | |
| Process_Safety_Archives_Master_Index_Rev0-16Jan2021.xlsx | CCPS | | $P$716 | Tracing Source Of Ethylene Oxide.pdf | |
| Process_Safety_Archives_Master_Index_Rev0-16Jan2021.xlsx | LPS | | $E$51 | Ethylene Oxide Explosion at Doe Run | |
| Process_Safety_Archives_Master_Index_Rev0-16Jan2021.xlsx | LPS | | $E$292 | Flammability of Ethylene Oxide in Sterilizer Operations | |
| Process_Safety_Archives_Master_Index_Rev0-16Jan2021.xlsx | LPS | | $E$293 | Liquid Phase Decomposition of Ethylene Oxide | |
| Process_Safety_Archives_Master_Index_Rev0-16Jan2021.xlsx | LPS | | $E$294 | Methane Diluent Requirements for Ethylene Oxide Storage and Handling | |
| Process_Safety_Archives_Master_Index_Rev0-16Jan2021.xlsx | LPS | | $E$852 | Ethylene Oxide Explosion at Seadrift, Texas - Reboiler Safety | |
| Process_Safety_Archives_Master_Index_Rev0-16Jan2021.xlsx | LPS | | $E$855 | Ethylene Oxide Explosion at Seadrift, Texas - Iron Oxide Chemistry | |
| Process_Safety_Archives_Master_Index_Rev0-16Jan2021.xlsx | LPS | | $E$861 | Ethylene Oxide Explosion at Seadrift, Texas - Background and Technical Findings | |
| Process_Safety_Archives_Master_Index_Rev0-16Jan2021.xlsx | PSP Citations | | $D$187 | Flammability of gaseous mixtures of ethylene oxide, nitrogen, and air | |
| Process_Safety_Archives_Master_Index_Rev0-16Jan2021.xlsx | PSP Citations | | $D$494 | Explosive decomposition of ethylene oxide | |
| Process_Safety_Archives_Master_Index_Rev0-16Jan2021.xlsx | PSP Citations | | $D$495 | Thermal stability and deflagration of ethylene oxide | |
| Process_Safety_Archives_Master_Index_Rev0-16Jan2021.xlsx | PSP Citations | | $D$496 | An overview of exposures and exposure control in the ethylene oxide producer and ethoxylation industries | |
| Process_Safety_Archives_Master_Index_Rev0-16Jan2021.xlsx | PSP Citations | | $D$497 | Safe storage of dilute ethylene oxide mixtures in water | |
| Process_Safety_Archives_Master_Index_Rev0-16Jan2021.xlsx | PSP Citations | | $D$500 | Prediction of decomposition limits for ethylene oxide-nitrogen vapor mixtures | |
| Process_Safety_Archives_Master_Index_Rev0-16Jan2021.xlsx | PSP Citations | | $D$501 | Engineering controls to minimize worker exposure to ethylene oxide at sterilization facilities | |
| Process_Safety_Archives_Master_Index_Rev0-16Jan2021.xlsx | PSP Citations | | $D$997 | Deflagration and detonation of ethylene oxide vapor in pipelines | |
| Process_Safety_Archives_Master_Index_Rev0-16Jan2021.xlsx | PSP Citations | | $D$1008 | Use of propylene oxide versus nitrogen as an ethylene oxide diluent | |
| Process_Safety_Archives_Master_Index_Rev0-16Jan2021.xlsx | PSP Citations | | $D$1049 | Kinetics of the reactions of ethylene oxide with water and ethylene glycols | |
| Process_Safety_Archives_Master_Index_Rev0-16Jan2021.xlsx | PSP Citations | | $D$1324 | Analysis of ethylene oxide gas house explosion | |

20 cell(s) found

**Figure 15. Results using "ethylene oxide" as a search term in the Excel Spread Sheet.**

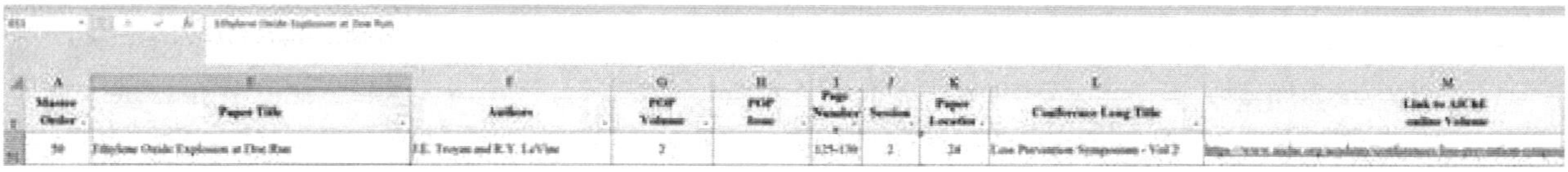

| Master Order | Paper Title | Authors | POP Volume | POP Issue | Page Number | Session | Paper Location | Conference Long Title | Link to AIChE online Volume |
|---|---|---|---|---|---|---|---|---|---|
| 50 | Ethylene Oxide Explosion at Doe Run | J.E. Troyan and R.Y. LeVine | 2 | | 125-130 | 2 | 24 | Loss Prevention Symposium - Vol 2 | [illegible] |

**Figure 16. Clicking on the third entry in Figure 15 search takes one to Cell E51 of the LPS tab.**

| Master Order | POP Issue | Page Number | Session | Paper Location | Conference Long Title | Link to AIChE online Volume | Citation |
|---|---|---|---|---|---|---|---|
| 50 | | 125-130 | 2 | 24 | Loss Prevention Symposium - Vol 2 | [illegible] | J.E. Troyan and R.Y. LeVine, "Ethylene Oxide Explosion at Doe Run," Loss Prevention Symposium - Vol 2, AIChE, 1968 pp. 125-130. |

**Figure 17. Column M contains a direct link to all proceedings presented in that conference.**

| Master Order | Citation | Link to AIChE online Paper |
|---|---|---|
| 50 | J.E. Troyan and R.Y. LeVine, "Ethylene Oxide Explosion at Doe Run," Loss Prevention Symposium - Vol 2, AIChE, 1968 pp. 125-130. | https://www.aiche.org/node/519271/group/8981/session/116896/paper/812846 |

**Figure 18. Column S contains a direct link to the actual paper.**

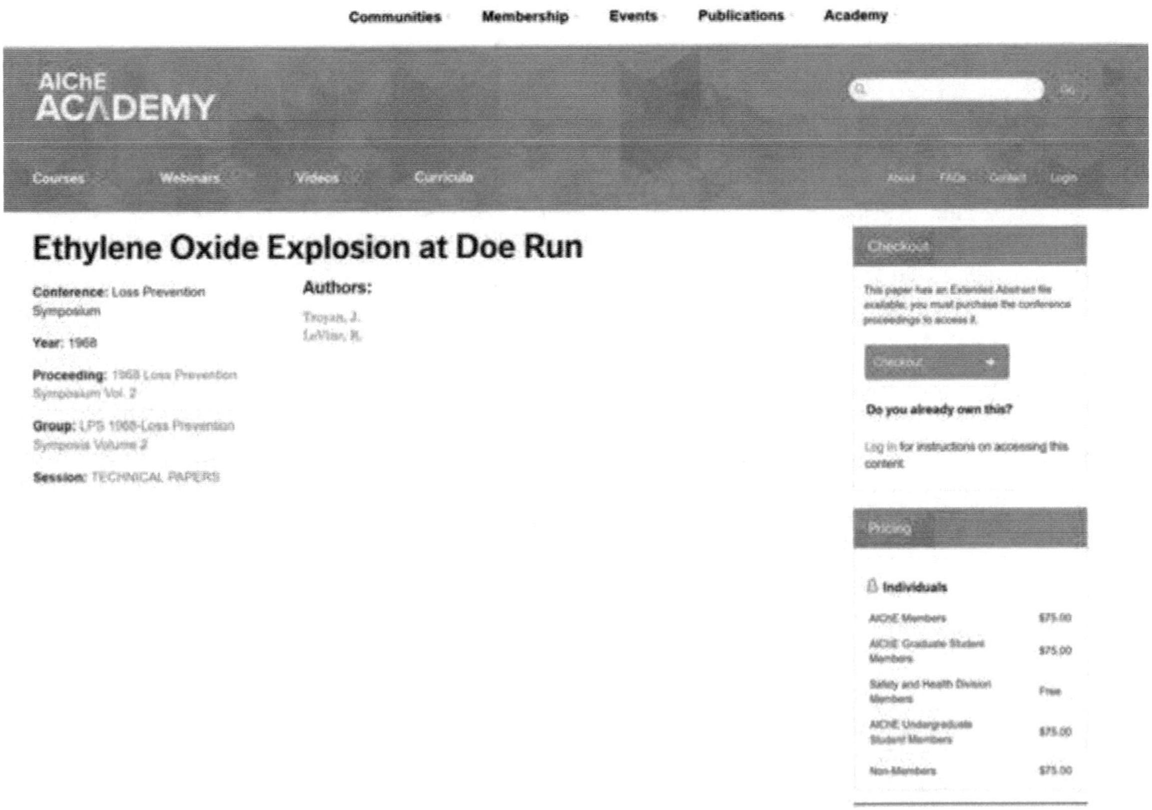


**Figure 19. Paper Title Page**

# 619027 Millennial Strategies to Capture Baby-Boomer Knowledge

**Abstract:**

Risk data is some of the most difficult data in the world to read, digest and understand. Focusing on Process Hazards Analysis (PHA) reports specifically, although the framework and guidewords are followed, for the most part, universally it is still difficult for someone with fresh eyes to pick up a report, read it and know what is important. Never mind having the time to wade through 500-5,000 pages of risk assessment information or multiple risk assessments looking for the specific piece of equipment that they need to know about.

Wouldn't it be great if a new operator or engineer, starts on Monday morning and by noon hour knew the most important risk information relevant to the facility or project they're working on? Why does this matter? Thinking about critical defeats processes, understanding critical procedures and which alarms matter most and need to be prioritized there are many times and processes where this information will be important.

A solution will be presented to make your risk data user friendly for millennials and all generations who are familiar with basic search engine functionality. Being able to search risk data for specific equipment tags and seeing that data in easy to understand, interactive visualizations puts it in context for what they need when they need it. Understanding the relationships between initiating events, safeguards/ IPLs (mitigative/ preventative) and consequences using bowties to tie it all together.

In addition, gamification of risk data will be presented with online, modern and interactive puzzles that can help people understand in minutes the strategies used to safeguard facilities and process units. It also adds a fun factor, competition and other nuances to engage and excite people about safety.

The future workforce needs better solutions to help enable them to leverage the knowledge of the baby boomers and generations that went before. Tens of thousands of experienced workers, who are often the SME in PHA sessions, are retiring on a weekly basis each with decades of knowledge. Although significant knowledge is being lost, the information captured in documents like PHA reports can help the next generations to learn, in second and minutes, what others took a lifetime to build. This knowledge transfer is only possible when presented in an accessible and consumable format.

# 619104 Lessons Learned for Nextgen PHA Facilitators

**Abstract:**

There are many training sessions available to train the next generation of PHA facilitators on how to execute a PHA; however, as seasoned facilitators look back on their experiences, they realize there are many details that did not get transferred in their classroom sessions. They learned these details on the job through years of valuable experience and through face-to-face and virtually-led workshops. These learnings can improve the quality and efficiency of the PHA workshop, the documentation of the workshop and the PHA experience. In 2020, this knowledge transfer became even more necessary during the COVID-19 pandemic. Organizations had to move toward conducting workshops virtually while still striving to achieve the same standards needed to ensure a safe facility and satisfy OSHA PSM requirements.

In this paper, seasoned facilitators capture both the small and the significant findings they have learned in their years of PHA facilitation from the very beginning of the PHA (at the preparation phase) to the reporting and follow-up phase. These learnings touch on setting expectations in the terms of reference, maintaining consistent documentation throughout the workshop to aid in effective reporting, maintaining consistency when the team members change (which happens in longer-term workshops), increasing engagement, especially in virtual workshop settings, and increasing use of tools (such as PHA Pro) to improve consistency and efficiency throughout the documentation process.

It is our desire to pass these lessons learned to the next generation of PHA leaders to supplement their classroom learnings and help them succeed in different settings and environments.

# 618410 "Begin at the Beginning and Go on till You Come to the End: Then Stop" Lifecycle of a Safety Instrumented System

**Abstract:**

Safety Instrumented Systems have specific steps in their lifespan. Many engineers are responsible for managing these safety systems at a specific point in the lifecycle. This seminar will describe all of the phases of the SIF lifecycle, beginning at the beginning and going through to the end.

# 617758 Adopting an Integrated Process Maturity Risk Model Approach to Process Safety

**Abstract:**

Adopting an Integrated Risk Maturity Model for Process Safety

Process Safety can mean different things to different people but one thing is for sure, that when it comes to automation we need to address the three disciplines of: alarm management cybersecurity and functional safety. The reason is that there are consequences of not addressing these from a risk point of view, such as:

- the consequence of inaction due to operators being flooded with alarms;
- the consequence of compromise due to malware or viruses that disable and/or take over the Safety Instrumented System (SIS) or Basic Process Control System;
- the consequence of failing on demand due to the SIS not being maintained properly.

The key point here is that all three disciplines are managed via performance-based standards that have a lifecycle approach with the same three phases: Analysis, Design/Implementation and Operation and Maintenance. By adopting an integrated risk maturity model, companies can assess how well they are doing at implementing these three disciplines and managing the risks associated with them, to meet tolerable levels. This paper outlines an approach and methodology to help companies establish where they are from a maturity point of view, help identify where they want to get to and how to address any gaps. The paper will also present some examples of where companies have benefited from this type of approach and that it is not just large companies that can benefit but companies of all sizes within the process industries.

# Development of a Deep Learning Model for Predicting the Drag Coefficients of Spherical and Non-spherical Particles

*Maria E. Presa-Reyes*[*], *Pratik Mahyawansi*[†], *Beichao Hu*[†], *Cheng-Xian Lin*[†], *Shu-Ching Chen*[*]
[*]*Knight Foundation School of Computing and Information Sciences*
[†]*Department of Mechanical and Materials Engineering*
*Florida International University, Miami*

## Abstract

There is yet to be a well-established drag model for non-spherical particles required in a particle-laden flow that could cover a wide range of sphericities. This talk will explore the development of a general drag model for non-spherical particles by applying deep learning using available experimental data available in the literature. The integration of several raw experimental measurements from different sources and research directions allows the training of robust Artificial Intelligence and Machine Learning (ML) models. Neural networks are an ML approach inspired by the inner biological workings of the brain. This work aims to develop a Deep Neural Network (DNN) that predicts drag coefficient values with the ability to adapt appropriately to unseen data. Given the limited number of data points available and the variance found within the data collected from various sources, challenges may arise when looking to train the model. Our study tests and implements various model regularization techniques and assesses different loss and activation functions for the proposed DNN. The proposed model considers a broader range of features other than sphericity and Reynold number. These features include density ratio, solid volume fraction, lengthwise and crosswise sphericity, and more. Furthermore, we present the features that play a significant role in predicting different drag coefficients through feature importance.

Within the investigated parameter ranges in this study, the following conclusions can be achieved and summarized below:

- An improved drag coefficient model can be developed by considering more features such as, aspect ratio, lengthwise sphericity, crosswise sphericity, and density ratio.
- DNN model can predict better results compared to traditional methods using MAE metric.
- The proposed model addresses data challenges such as limited data and extreme data points through expanded feature-set and regularization.
- Three major features that mostly affect the drag coefficient were identified from a feature importance analysis.

# 617249 Reimaging Management of Change

**Abstract:**

For many companies, Management of Change opportunities ranks among the higher number of findings and observations of all OE processes. Additional experience has found that MOC is one of the more difficult process safety management elements to implement and to get and keep it right. But how can we improve MOC, improve leadership appreciation for what the MOC process is trying to prevent and help eliminate the "get 'er done" culture to make the change and document when we have time? Reimagine an improved overall MOC system, including people, tools, competency and efficiency.

Reimaging a MOC system would include:

- MOC system(s) are more fault tolerant and resistant to circumvention or human error
- Competent, knowledgeable personnel are involved upfront to decide what changes need to be put into the MOC process
- Qualified personnel, normally independent of the MOC originator, review the request to identify any potentially adverse impacts
- Achieve better results with fewer, qualified resources
- Monitor MOC performance and efficiency in a practical way

# 618537 Using Data Analytics to Generate PHA Checklists – Saving SME Time and Supporting SME Knowledge

**Abstract:**

The majority of those that have been exposed to Process Safety are aware of Hazard and Operability Studies (HAZOPs); a Process Hazard Analysis (PHA) methodology that is regularly used in industry because of its scalability and relative ease of execution when analyzing the risks of a process unit. Those who have been exposed to HAZOPs are also aware that these studies are qualitative, subjective and incredibly dependent on the people involved in the analysis session. Everyone who participates in a HAZOP does the best they can to do a complete and comprehensive analysis, however their contributions can be limited and biased based on their human knowledge and personal experience. This can lead to inconsistencies in identifying and capturing risks in comparable facilities/units, which can result in increased risks at site.

In order to strive towards higher quality, more comprehensive and more consistent HAZOPs, operation companies are starting to create HAZOP/PHA Checklists to help their teams better prepare for their sessions and be aware of potential high risk scenarios.

Unfortunately, HAZOP/PHA Checklists often take a large amount of time and effort for a Subject Matter Expert (SME) to create, and this amount of SME time is not always available in smaller operating companies.

So, what to do? There is obviously a benefit in having HAZOP/PHA Checklists, so how can the time of an SME required to generate one of these checklists be reduced while still creating a high quality reference tool for HAZOP teams? Also, if the insights of a few SMEs are so beneficial, what about the knowledge of ten or a hundred SMEs?

From these wants, a new method has been found to accomplish this task of generating HAZOP/PHA Checklists by leaning on the power of automatic data processing and data analysis of already completed PHAs. By combining the SME knowledge of tens to hundreds of HAZOPs of the same technology process together, a complete HAZOP/PHA Checklist can be generated with minimum manual SME input.

This paper shows a case study comparison between an SME generated HAZOP Checklist and a Big Data generated HAZOP Checklist, and highlights the high similarity/accuracy when compared, as well as the time savings of the a Big Data generated HAZOP Checklist methodology.

# 617639 How Effective Is Your Fire and Gas System?

**Abstract:**

A well-conceived fire and gas system enables the early detection of loss of containment incidents, and the resultant actions can significantly reduce the magnitude and duration of potential consequences. With a wide range of variables affecting fire and gas hazard behavior, and an unlimited number of potential fire and gas detector configurations, how does one ensure that the system is effective and fit-for-purpose?

This presentation will outline an adaptable and consistent approach to assist with the strategic development of a fire and gas detection philosophy that is applicable to both onshore and offshore environments. Emphasis is placed on the importance of hazard assessment to gain an understanding of the behavior of the specific hazards and how this can be used to inform the detection strategy, in conjunction with relevant standards and codes. The presentation goes beyond what may be included in a typical fire and gas philosophy document and covers detector layout strategies, describes modern verification methods to assess detector coverage (e.g. 3D detector mapping), and provides approaches for the management of ongoing inspection and testing of the fire and gas detection system.

# 618809 Factors Contributing to US Chemical Plant Process Safety Incidents from 2010 – 2020

**Abstract:**

Process safety incidents occur more frequently than desired – they can result in injuries, fatalities, environmental impacts, facility damage, downtime & lost production, as well as impacts on a company's and industry's reputation. While such incidents occur across many industries, this study is focused on an analysis of the most commonly reported contributing factors to process safety incidents in the US chemical industry.

The database for the study contained 79 incidents from 2010-2019, investigated by the Chemical Safety Board (CSB) and/or by companies. To be included in the database, the incident had to occur at a company classified as 325 in the North American Industry Classification System (NAICS), assigned to businesses who participate in chemical manufacturing. Unlike other summary studies of process safety incidents, the subject analysis is specific to the US chemical industry and based on contributing factors to the incidents reported post investigation.

All of the incidents included in the study were classified as Tier 1 incidents according to the American Petroleum Institute (API) RP 754 definition. Using this methodology for determining incident severity, each incident in the database included a severity score, and was placed into one of four severity categories, identified as low, medium, high and very high severity. All incidents investigated by the CSB were deemed very high severity. The list of contributing factors included in this analysis also provided more granularity and specificity than typically in other studies, and included factors such as preventive maintenance; safeguards, controls & layers of protection; and hazard awareness and identification.

Reported contributing factors were catalogued for each incident by severity category, and then a prioritized (weighted) list of contributing factors was developed for the database of incidents. The top three weighted contributing factors for the chemical industry were: design; preventive maintenance; and safeguards, controls & layers of protection. The finding that the top contributing factors to process safety incidents included factors such as operating procedures, preventative maintenance, and emergency preparedness and response, areas that are traditionally covered under the occupational health and safety administration (OSHA) regulations, prompted a review of the federal OSHA citations database. The investigation and citation history for NAICS 325 companies was analyzed to assess whether there was any overlap between the top reported contributing factors to process safety events and the top OSHA citations recorded for the industry.

A database consisting of the inspection and citation history for the chemical industry identified by NAICS code 325 was assembled for inspections occurring between 2010-2020 (August). Analysis of the thousands of entries provided an understanding of which regulations were most commonly cited, and which were most commonly cited following fatalities. Findings were presented for NAICS code 325, as well as for each sub grouping, including NAICS 3251, 3252, 3253, 3254, 3255, 3256, 3259.

The analysis of the citation history for the chemical industry specifically, identified that the key factors reported as contributing to incidents overlapped with the most common OSHA violations. The overlap of contributing factors for process safety incidents with the most common OSHA violations suggests that addressing regulatory issues may also pay dividends in reducing process safety incidents. These findings will likely be most beneficial if shared among those with the ability to create influence within the chemical industry, assess practices in these key areas, and identify improvement opportunities seeking industry best practices.

# AIChE Paper 113b

# Distillation Tower Startup Considerations in Ethylene Plants

*André Bernard*
*Process Design Engineer*
*NOVA Chemicals Corporation*

Prepared for Presentation at the 2021 Spring National Meeting
Virtual, April $18^{th}$ – April $22^{nd}$, 2021

# AIChE Paper 113b

# Distillation Tower Startup Considerations in Ethylene Plants

*André Bernard*
*Process Design Engineer*
*NOVA Chemicals Corporation*

Abstract

Ethylene plant startup involves load and composition transition on distillation towers, that if not properly handled could result in un-expected constraints as rates are being increased to production target. Beside affecting unit capacity, startup malpractices may prolong unit downtime along with loss of valuable chemicals to flare. This paper discusses, through a case study focused on Demethanizer, Deethanizer and C2 Splitter, potential startup problems and describes practices enabling smooth transition of towers to steady state conditions. The discussion covers individual tower startup parameters but also includes interactions through the distillation train. Potential for cold temperature and over pressure excursions is also discussed.

## Introduction

Ethylene plants have been striving for longer plants run lengths over the years with typical turnaround frequencies that could extend beyond 5 years. Process upsets in some case may require a unit re-start, but those are also infrequent. Therefore, Operating crews have limited exposure to startup activities. It is not unusual for unit process engineers to rotate position at rates exceeding turnaround frequency, which in turn affects technical support to operating crews during plant startup. Given the extended time between plant turnarounds, it is not unusual to implement retrofit projects during those schedule outages, resulting in potential changes on process unit characteristics. This compounded with the lack of startup exposure emphasize the need to review startup procedures and updates process targets in alignments with any changes in equipment characteristics.

While reviewing startup practices, distillation tower evaluation should extend beyond a process box achieving separation. Tower transition from a stagnant vapor filled vessel at low pressure to a functional vapor/ liquid fractionator requires a sequence of events and process conditions in alignment with the equipment characteristics. Understanding startup expected transient conditions, equipment limitations and potential pitfalls ahead of startup; will help adjusting startup procedures, speedup problems diagnostics and minimize improvisation during startup activities.

## Understanding the Process

Distillation is a vapor/ liquid separation process which leverage on differences in volatility among the components involve. The more volatile concentrating in the vapor phase and the less volatile in the liquid phase. Tower internals are designed to enable vapor/ liquid contact for promoting mass transfer between the two phases. Vapor/ liquid circulation must therefore be present for fractionation to occur. The challenge of tower startups involves transient conditions that typically begins with a stagnant vapor filled vessel. Figure 1 illustrates a typical distillation tower with a reboiler and condenser. The tower has a single feed point, distillate, and bottom products. Heat to the reboiler drives the vapor circulation while the condenser enables liquid traffic through reflux. In a simplistic way, starting such tower from a nitrogen filled vessel would involve initiating feed to the tower. The feed conditions, vapor or liquid will dictate the initial traffic inside the tower and drive the startup sequence. For example, a liquid feed will accumulate in the tower sump and allow starting heat to the reboiler. This will initiate vapor traffic throughout the tower. Given the liquid nature of the feed, vapor/ liquid circulation becomes enable in the stripping section of the tower at that point. Vapor rising through the rectifying section would reach the condenser, with condensed liquid accumulating in the reflux drum and nitrogen purged out through a vent. A reflux pump would be started once sufficient inventory has accumulated in the reflux drum. Vapor/ liquid traffic is then established throughout the tower once introducing reflux.

If the tower is fed from a vapor stream, the sequence is reversed, where vapor would flow through the rectifying section and condense in the condenser. Until the reflux pump gets started, traffic in the tower is limited to vapor flow in the rectifying section. Once reflux is introduced in the tower, vapor/ liquid traffic is established in the rectifying section and liquid traffic in the stripping section. When sufficient liquid inventory fills the sump, heat is provided to the reboiler and vapor/ liquid circulation is then initiated throughout the tower.

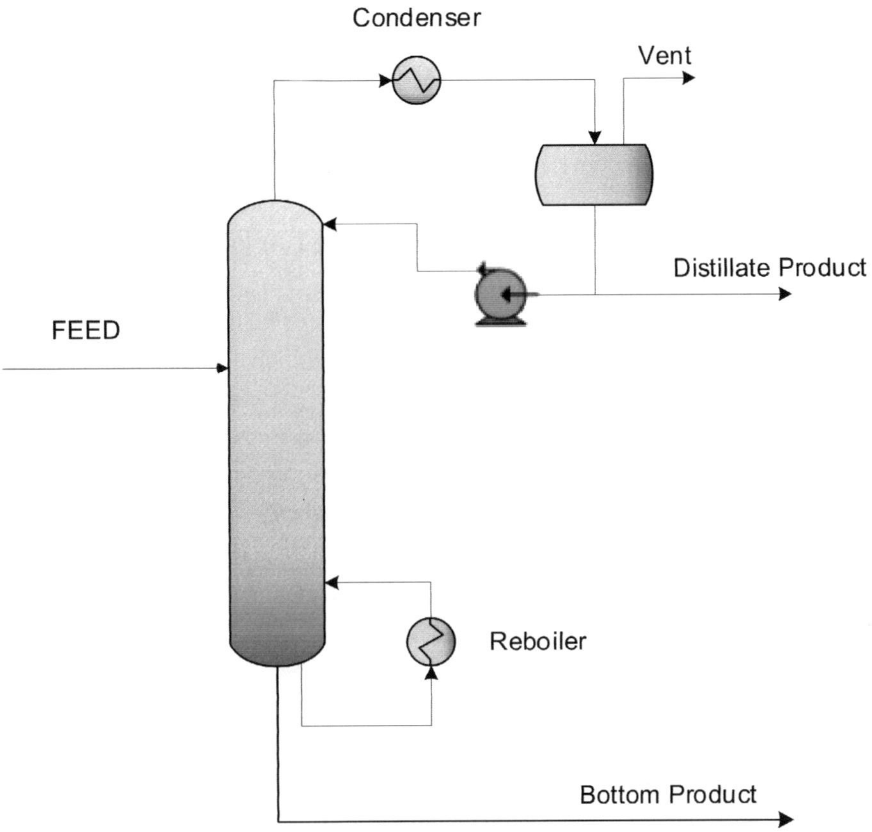


**Figure 1: Typical Distillation Tower**

The progression to effective liquid/ vapor circulation in a tower relies on consistent thermodynamic and hydraulic transient conditions. The difference between the initial and normal operating pressure of a tower defines the magnitude of potential thermodynamic inconsistency that could arise throughout the transition. For instance, a tower operating at or near atmospheric pressure is already at its expected steady state pressure in the initial stage of startup. As such, the boiling points/ compositions relationship is not transitioning as circulation is initiated in the tower per the above description. However, a transition from low to high pressure is associated with variation in boiling points/ composition that influence the ability of the reboiler and condenser in achieving their functions.

Thermodynamic consistency and initiation of vapor/ liquid circulation is not sufficient for achieving fractionation in a tower. The liquid and vapor must follow their intended flow path, which is driven by hydraulics through tower internals. Operation outside the design range of those internals may result in little to no fractionation. Development or evaluation of startup strategies should involve:

- Understanding of thermodynamic consistencies
- Understanding of equipment capabilities
- Potential environmental impacts
- Potential process safety risk

## Thermodynamic Consistency

Although basic and fundamental, it is not unusual to overlook the thermodynamic consistency of a startup plan. Figure 2 illustrates a Txy diagram for a binary mixture of component A and B at the initial startup and steady state pressures. A bottom product at a mole fraction purity near 1 for component B and distillate mole fraction purity near 1 for component A at steady state conditions, implies that the heating medium supplying the reboiler be warmer than the boiling point of B and that the cooling medium in the condenser be colder than the boiling point of A. Assuming a design temperature approach of 10 °F for the reboiler and condenser, infers a heating and cooling medium of 165 °F and 25 °F. The boiling point of A at the initial pressure (-39 °F) being significantly colder than the cooling medium reveals that the tower would not operate at the initial pressure. However, it is thermodynamically feasible to boil component A in the reboiler if it supplies the tower in the liquid phase at the initial pressure. Similarly, if component B enters the tower in the liquid phase, it can also be boiled in the reboiler. However, it is more likely when looking at the initial pressure boiling points that component B could enter the tower in the liquid phase and A in the vapor phase if the tower was to be initially fed with pure components. Feeding a mixture of A and B at a temperature equivalent to the initial pressure boiling point of B suggest that the mixture would enter the tower as a vapor. Allowing the vapor flow to pressure up the tower will result in partial condensation as the pressure transition to its steady state value. The extent of condensation depends on the mixture composition. Condensation in the overhead condenser will begin once the dew point of the vapor mixture rises above the cooling medium. The above discussion suggests 3 approaches for starting the tower.

1. Feeding component B in the liquid phase:
    a. Accumulate B in the sump
    b. Start heat to the reboiler
    c. Use the vapor generated by the reboiler for pressuring the tower
    d. The tower can be operated on total reflux followed by feed transition from B to mixture A/B

2. Feeding component A in the vapor phase:
    a. Use the rising vapor for bringing the tower up to target pressure
    b. Condensing will start once the dew point rise above the cooling medium temperature
    c. Introduce reflux
    d. Liquid will accumulate in the sump
    e. Start the reboiler

3. Feeding a mixture of A and B (A mole fraction of 0.7 assumed on Figure 2).
    a. Use the rising vapor for bringing the tower up to target pressure
    b. As pressure increases, condensed liquid will start filling the sump.
    c. The liquid flow will increase as pressure increases and liquid reflux gets introduced to the tower
    d. The reboiler can be started once sufficient liquid accumulates in the sump

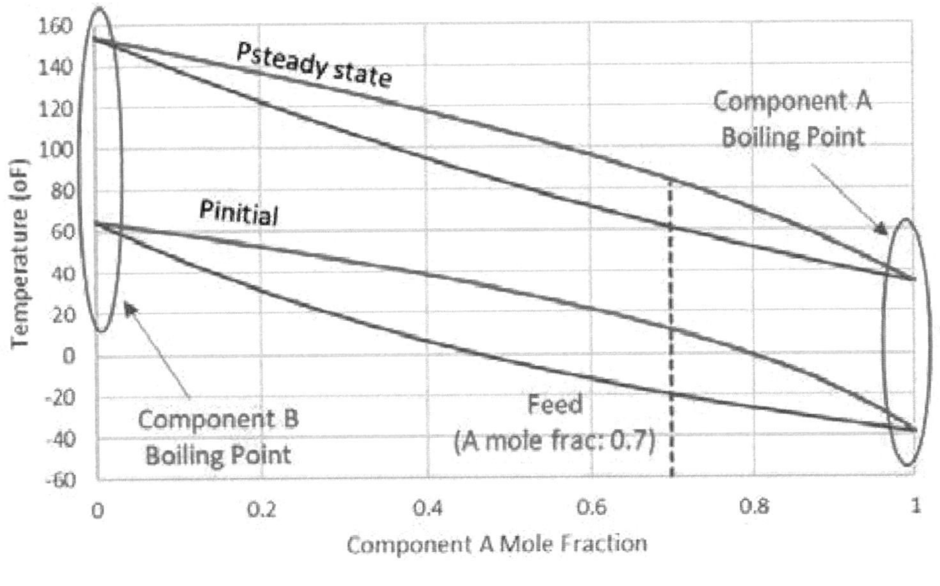


**Figure 2: Binary Mixture Txy Diagram**

Other than the sequence in which vapor/ liquid circulation is established, these 3 cases differ from each other in terms of the initial composition of the pool of liquid filling the sump and the reboiler. As shown on Table 1, this translates in different temperature differential between process and the heating medium.

Although still consistent from a thermodynamic standpoint, large temperature differences between the initial boiling point of the liquid mixture accumulating in the reboiler and the heating medium may result in film boiling, which is discussed later.

**Table 1: Reboiler Mixture Initial Boiling Point**

| Case | Reboiler Heating Medium Temperature (°F) | Reboiler Mixture Initial Boiling Point (°F) | Temperature Difference (°F) |
|---|---|---|---|
| 1 | 165 | 64.5 | 100.5 |
| 2 | 165 | 35 | 130 |
| 3 | 165 | 60 – 82[1] | 105 - 83 |

***1: Depending on the extent of condensation***

A Txy diagram does not apply to multi-components systems. However as shown on Figure 3, boiling curves for the pressure range expected from startup to steady state conditions can provide valuable information relative to thermodynamic consistency. The system illustrated on Figure 3 is based on a multi-component mixture feeding a tower at a temperature of -35 °F. The boiling curve of the feed at the initial pressure indicates that 96 % of the mixture is in the vapor phase. Feeding the tower under those conditions suggest that the heavy tail representing 4% of the stream will initially build a liquid pool in the sump and reboiler. The initial boiling point of that liquid mixture is -35 °F but would rise to 162 °F when increasing the pressure to the steady state target. Supplying heat to the reboiler once the tower reaches its targeted pressure would likely result in a temperature pinch due to thermodynamic inconsistency when assuming a heating medium temperature of 47 °F.

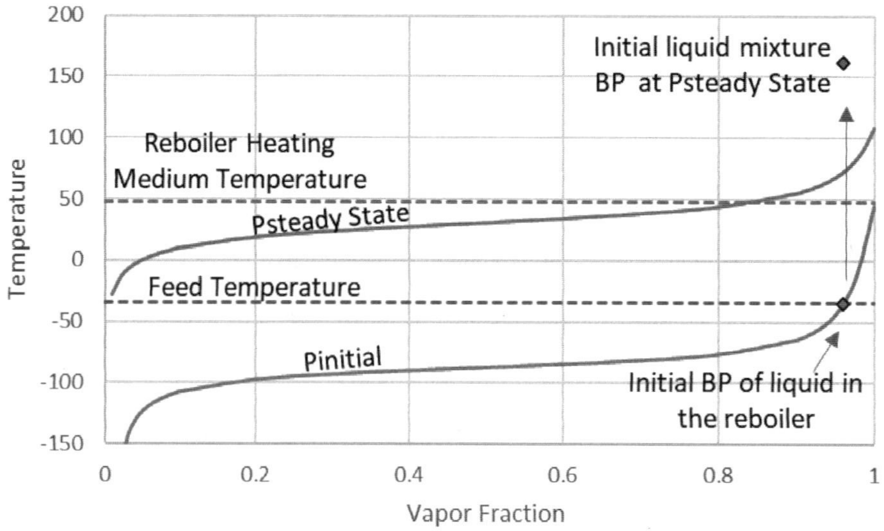


**Figure 3: Multicomponent Mixture Boiling Curve**

Starting the reboiler at the initial pressure looks thermodynamically feasible. However, the initial pool of liquid contained in the sump and reboiler will get heavier as boiling starts. This can be demonstrated through a material balance around the reboiler. It is understood that prior to introducing heat to the reboiler, the boiling point of the pool of liquid contained in the reboiler is the same as the liquid mixture from the feed. Once heat is introduced to the reboiler, a certain degree of vaporization would occur. The liquid supply to the reboiler at that point is no longer equivalent to the liquid mixture coming from the feed but is the combination of the liquid return from the reboiler and the liquid mixture from the feed. The liquid return from the reboiler being heavier due to evaporation of its light fraction causes the combined stream feeding the reboiler to be heavier than the liquid mixture from the feed. Figure 4 illustrates this transition through a dynamic simulation.

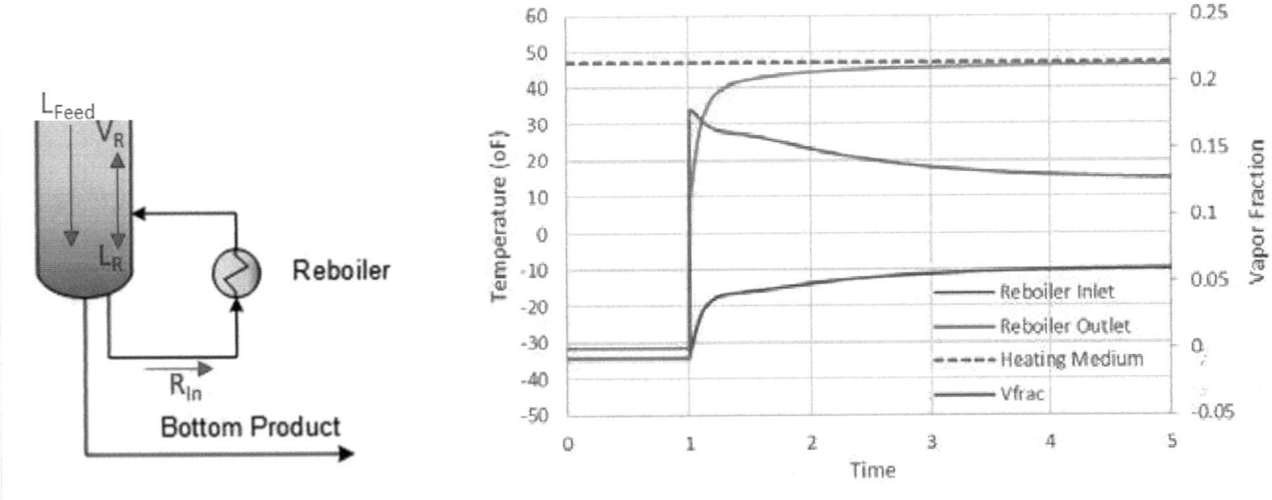


**Figure 4: Dynamic Evaluation at Initial Pressure**

Increasing the tower to its targeted steady state pressure and draining the initial pool of liquid building up in the reboiler would provide a liquid mixture consistent with the boiling curve of the feed stream at its targeted steady state pressure. The initial boiling point for the volume of liquid contained in the reboiler would then be -28 °F, which is colder than the heating medium at 47 °F; making this approach thermodynamically consistent. As described above, the initial boiling point of the mixture entering the reboiler will increase as the thermosyphon circulation gets established around the reboiler. However, the inlet and outlet temperature of the reboiler are more likely to remain colder than the heating medium with those conditions then the previous example. Simulation suggests that with 20% vapor fraction in the reboiler outlet and 70% recirculation rate from the thermosyphon that the inlet and outlet temperature of the reboiler would be 19.6 and 30 °F.

## Equipment Capabilities

The effectiveness of fractionation relies on proper contact between vapor and liquid. Tower internals are designed to promote this contact but have a limited operating range. It is not unusual to feed towers at low rates during startup operation to minimize product losses and environmental impacts due to flaring. It is therefore paramount to understand the turndown limitations of the equipment when establishing a startup plan.

### *Reboiler*

Reboiler operation is critical during startup and normal operation as it supplies the driving force for initiating and sustaining vapor traffic in the tower. Thermosyphon reboilers are commonly used in ethylene plants and their operation and design have been widely discussed in literature. They are known to be problematic on a few services during startup operation. Those problems are often related to difficulties in establishing the thermosyphon, which in many cases is not caused by the reboiler system but as discussed previously to a startup path that leads to thermodynamic inconsistency. Multicomponent systems involving wide boiling range are more at risk to encounter those problems.

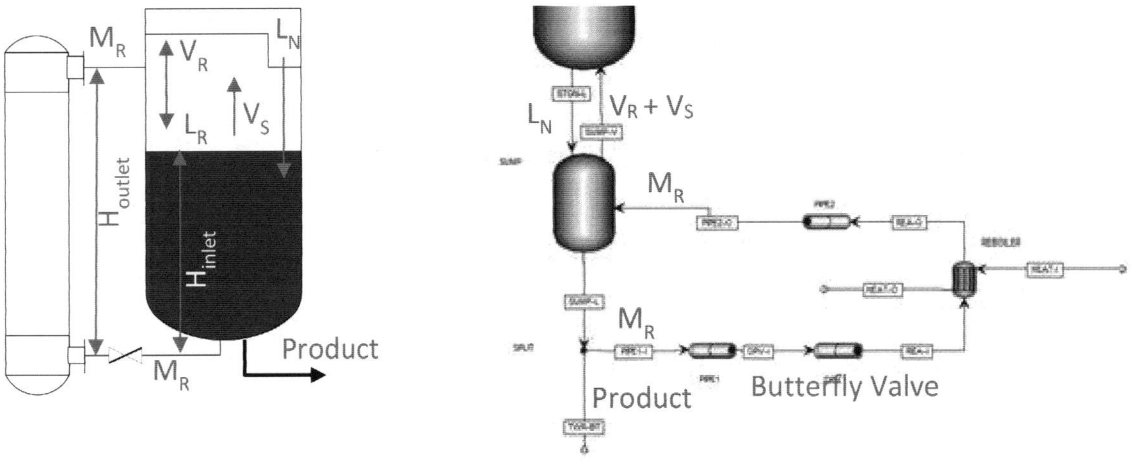


**Figure 5: Thermosyphon Reboiler**

For discussion, the operating rangeability of Thermosyphon reboiler is evaluated through an example involving a recirculating thermosyphon system without baffle as illustrated on the left side of Figure 5. The circulation rate of the thermosyphon depends on the pressure balance between the inlet and outlet nozzles, which is defined by the difference in static head between the sump and reboiler along with friction losses. Fluid properties, heat input, exchanger geometry and piping configuration are all components involved in the pressure balance. Therefore, the Thermosyphon analyses cannot be limited to the exchanger but must include the tower and piping configuration. The right side of Figure 5 illustrates a simulation flowsheet used for evaluating the system. As shown the reboiler is modelled outside the tower block with a rigorous exchanger. A flash block is used for modeling the sump. Liquid from the bottom tray ($L_N$) mixes with the reboiler return stream ($M_R$) in the flash block, which assumes a perfectly mixed volume resulting in some of the cold liquid from the tower to flash ($V_s$) as it mixes with the warmer reboiler return stream ($M_R$). The combined vapor from the flash block ($V_R$ and $V_S$)

are moving up to the tower. The combined liquid stream exits the flash block and is split between the reboiler inlet and the tower bottom product. The split ratio is calculated such that the reboiler outlet stream pressure feeding the flash block is identical to the tower liquid stream entering the flash block. Pipe blocks are used for modeling the reboiler inlet and outlet piping. A pipe block is used for modeling a reboiler inlet butterfly valve. The use of a pipe instead of a valve block allows specification of a valve fixed flow resistance express in terms of pipe lengths rather than specifying a fix pressure drop. It is understood that, in reality, the tower bottom liquid ($L_N$) would not mix perfectly with the reboiler return stream ($M_R$). Non-ideal mixing can be model by bypassing a fraction of the tower bottom liquid ($L_N$) from the flash block. Adjusting the extent of bypass would require tuning with plant data. In the absence of data, perfect mixing can be used as an initial assumption.

A multicomponent mixture with a similar boiling curve as described in Figure 3 was used for simulating the tower and reboiler system. The flow resistance or equivalent length of the pipe block emulating the butterfly valve was set such that the system would reach a pressure balance with a reboiler outlet vapor fraction of 25% mole at the 100% reference rate. The tower feed rate was then reduced by increments down to 15% of the reference rate. The flow resistance of the butterfly valve was kept constant. The heating medium rate was set such that the tower bottom stream meets its intended specification. Figure 6 shows the impact of rate reduction on the heat transfer coefficient, the reboiler outlet vapor fraction and the thermosyphon recirculation rate as a fraction of the total liquid flow exiting the tower sump (product plus reboiler inlet).

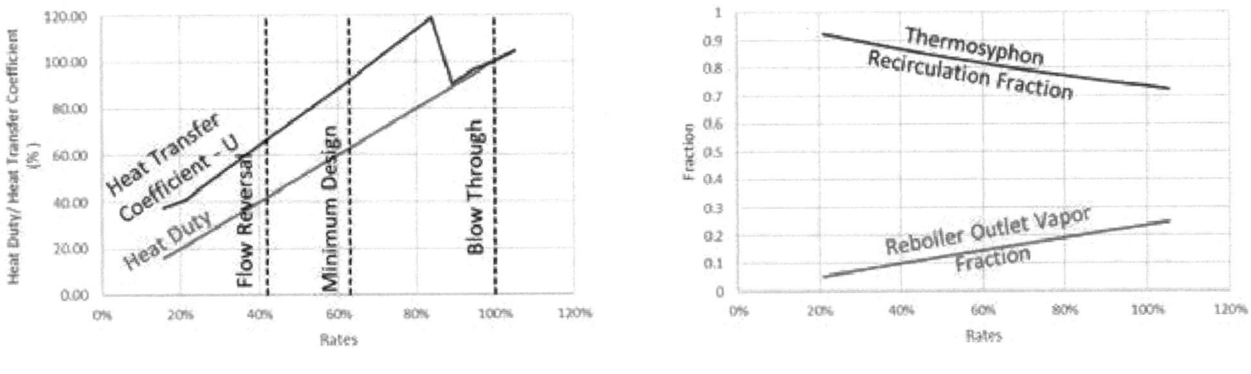


**Figure 6: Thermosyphon Reboiler Operation vs Rate**

As illustrated on Figure 6, the maximum tower rate sustainable by the reboiler is limited to the 100% rate case. Incomplete condensation resulting in vapor blow through of the hot stream exiting the exchanger would be experienced beyond 100% rate. Except for a reduction in the overall heat transfer coefficient from 84 to 89% rate, a linear decrease in alignment with the reduction in heat duty is observed. The film coefficient on the tube and shell side is a function of the Prandtl (Pr) and Reynolds (Re) numbers. The Prandtl number, being based on thermal properties and viscosity is barely affected by the change in rates. Although the Thermosyphon recirculation fraction varies between 0.7 and 0.9, the absolute flowrate in the Thermosyphon circuit only changes by 6% between the minimum and maximum cases evaluated. The heating medium flow reduction, which is directly proportional to the tower feed rate, is primarily responsible for the reduction in heat transfer coefficient as driven by the reduction in Reynolds number on the hot side of the exchanger.

The jump in the overall heat transfer coefficient between 89 and 84% rate is believed to be caused by the extent of sub-cooling of the hot stream. The hot stream outlet temperature pinches with the process at rates lower than 84%.

The lower operability limit of the Thermosyphon reboiler, from an equipment limitation perspective, is assumed defined by flow reversal in the exchanger tube. Flow reversal is predicted by the Wallis criteria[1,2], which is based on a dimensionless vapor velocity per the following equation.

$$u_g^* = \frac{u_g \rho_g^{0.5}}{\left[g_n d_i (\rho_l - \rho_g)\right]^{0.5}} \tag{1}$$

Where $u_g$ is the superficial vapor velocity inside the tubes as calculated by:

$$u_g = \frac{x_g \dot{m}}{\rho_g} \tag{2}$$

$\rho_g$ and $\rho_l$ being the gas and liquid densities, $\dot{m}$ the mass flux, $x_g$ the mass vapor fraction, $d_i$ the tube inside diameter and $g_n$ the gravitational acceleration.

Heat exchanger design and rating software suggest a value of $u_g^*$ higher than 0.5 to prevent flow reversal. Flow reversal is a condition where the vapor velocity becomes insufficient for dragging the liquid upward in the tube. This condition result in Thermosyphon instability which can ultimately stall the reboiler due to the liquid becoming heavier. The vapor rate is the parameter that has the most influence relative to the Wallis criteria, since the exchanger geometry is fixed and changes in liquid and vapor densities are not expected to vary significantly with circulation rate. As such, manipulating the circulation rate through the butterfly valve does not provide much sensitivity relative to expanding the turndown capability of the reboiler. The circulation rate will cause variation in vapor fraction but the overall $x_g \dot{m}$ value will remain the same. As for trays, false loading the tower by operating at higher reboil rate would provide the necessary stability for low-rate operation.

Based on the above criteria, the Thermosyphon reboiler turndown evaluated would be limited to rates above 62%. This limit is defined as a minimum design conditions on Figure 6. It is believed that depending on the mixture boiling curve, that turndown be expanded to value of $u_g^*$ slightly above tube flooding velocities. Flooding is the opposite of flow reversal and is define by the velocity at which some of the liquid starts moving upward with the vapor. The transition between flow reversal and flooding is characterize by liquid flowing in both directions. The reboiler should still operate providing that the fraction of liquid moving out is sufficient to prevent a temperature pinch. Therefore, the turndown of the above example can be extended to rates above 42% based on vapor velocity slightly higher than the flooding velocity. Flooding and flow reversal are complex phenomenon where variability between correlations and actual data may differ significantly. Complementing theoretical work with plant data will help in predicting more accurate reboiler turndown stable operation.

Thermosyphon reboilers are normally designed to operate in nucleate boiling. Film boiling conditions could however developed at high heat flux and high temperature differential between the process and heating medium. The heat flux is normally at the low end under startup conditions. However, depending on the startup sequence, the temperature differential could be significantly higher with the initial liquid inventory accumulating in the reboiler. As indicated in the example presented on Table 1, the initial temperature differential could be as high 100 – 130 °F on that system which assumed a 10 °F temperature approach. It is not clear once getting into film boiling if a system will revert to nucleate boiling as normal operating conditions are re-established. It is speculated that the thickness of the film layer decreases if the wall temperature falls. Since the heating medium temperature remain constant and the process side temperature rises on the transition to steady state, it is very likely that film boiling conditions will remain. The negative impact on the heat transfer coefficient will cause a premature shortfall on the reboiler capacity. Recovering from this condition may require slumping the tower by pulling heat from the reboiler to re-wet the tubes and then gently transitioning back to normal conditions.

### *Condenser & Reflux System*

The overhead condenser is typically critical relative to the tower pressure control. Exceeding the condenser capacity is normally associated with increase in tower pressure. Accumulation of non-condensables can stall the condenser and result in high pressure. The presence of non-condensables must be recognized and venting capabilities provided.

Low rate or turndown operation is not necessarily limited by the condenser itself but by the rangeability of the reflux system. The initial introduction of reflux to the tower should be conducted through a gentle rise in flowrate such that it does not starve the overhead condenser. Aggressive introduction of reflux could result in a rapid drop of the reflux drum inventory resulting in stop and go reflux sequence. It could also result in upset on refrigeration systems for tower using refrigerant in the condenser.

### *Trays*

Analyses and discussion around tray operation at low rates is best describe through a tray operating window[3,4] or stability diagram[5]. Figure 7, illustrates such a diagram, which provides the tray operating limit as a function of vapor and liquid loads. The vapor load is typically expressed as VLOAD or $C_{factor}$ where:

$$VLOAD = V_{rate}\sqrt{\frac{\rho_V}{(\rho_L - \rho_V)}} \quad (3)$$

$$C_S = \frac{VLOAD}{A} = \frac{V_{rate}}{A}\sqrt{\frac{\rho_V}{(\rho_L - \rho_V)}} \quad (4)$$

A plot of the operating line for a given separation provides the operating range expected by the tray. Startup involves transient conditions starting well below the minimum limit of the tray, which then progress towards the stability region of the operating window.

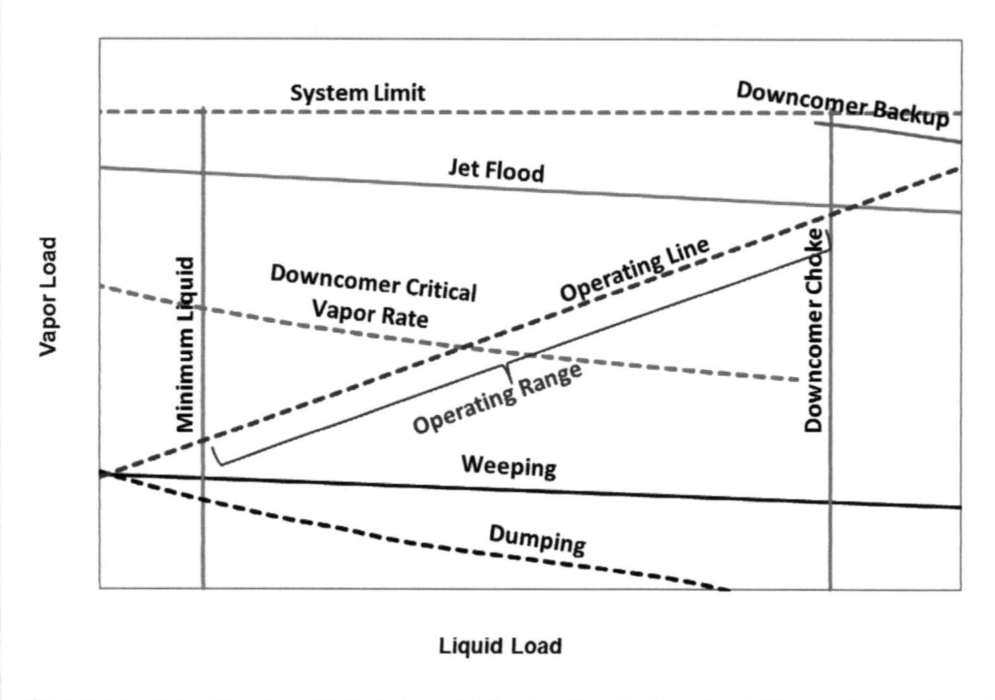


**Figure 7: Tray Operating Window[4]**

The following operating limit defines the low end of the operating window:

- The Minimum Liquid load depends on the tray design and is defined relative to weir loadings or downcomer seal limit. Low weir loadings can cause maldistribution impacting tray efficiency. In low liquid load applications, low weir loads may cause a transition to the spray regime which is associated with higher liquid entrainment. For downcomers designed with dynamic seal, the minimum liquid load can be defined relative to downcomer sealing. Loss of downcomer seal will cause the vapor to flow across the downcomers and bypass the active area of the tray resulting in poor fractionation.

- Weeping is defined by liquids falling through the tray deck perforation rather than following the intended path through the downcomers. It is cause by insufficient vapor rate, which does not provide sufficient dry tray pressure drop for holding the liquid on the tray deck. Fractionation will still occur but depending on the extent of weeping efficiency will be affected.

- Dumping is an extension of weeping where most or all the liquid is falling through the tray deck perforation. The downcomers could likely become unseal causing little vapor/ liquid contact. Sever loss of fractionation would be observed under those conditions.

- The Downcomer Critical Vapor Rate is a tray limitation that only applies until the downcomers becomes liquid sealed. This limitation is defined by a critical vapor velocity in the downcomer that causes liquid entrainment preventing the downcomer from holding a liquid inventory. Little to no fractionation is expected under those conditions as both vapor and liquid are not following their intended flow path resulting in poor vapor/ liquid contact. This limitation disappears once the downcomer gets a liquid seal, which then enables the full operating range of the tray operating window.

The initial traffic in the tower is driven by the feed conditions. A vapor feed will see the vapor flowing upward through the downcomers and the perforation across the tray deck. A liquid feed will be associated with liquid raining down the tray deck perforation and may or may not flow through the downcomers depending on rates and weir heights. A mix feed of vapor and liquid will follow the same path as above with disengagement of vapor and liquid at the feed point. Until the reboiler and/or reflux is commissioned no vapor/ liquid contact exists. It is imperative that the initial vapor rate to the tower does not exceed the downcomer critical vapor rate such that when reflux is initiated, the downward liquid be capable of sealing the downcomers. Similarly, the vapor load created by the reboiler when commissioning heat to the reboiler must remain below the downcomer critical vapor rate until circulation is established throughout the tower and tray loads falls within the stable portion of the operating window while still below the downcomer critical vapor rate. The concept is to transition load following the operating line up to the low end of the operating window stable region. Weeping and poor liquid distribution due to low weir load is expected during the transition but none of those conditions will results in material accumulation and should self-correct as load transitions toward stable conditions.

In many cases the startup target rates are below the operating range of the trays. In this case the tower can be false loaded by operating at higher reflux and reboil rates such that the vapor/ liquid traffic meets the turndown capability of the trays. The false load conditions should be initiated after commissioning both the reflux and reboiler. Those conditions can be established based on process simulation and tray load calculations and communicated to operations in term of heating medium flow to the reboiler and reflux rate in relation to feed rates. The use of a tray operating window—besides being useful in avoiding numerous tray load calculations during pre-startup evaluation—provides an excellent tool for monitoring load transition during startup. It is important to re-emphasize the need for smooth transition in reboiler duty to prevent exceeding the downcomers critical vapor rate as they may not be sealed at the beginning of this transition. Due to the lag between the reboiler and reflux drum, the sequence for false loading a tower should be first initiated by small increment of reboil heat with reflux increment once the reboiler action is confirmed by additional liquid accumulation in the reflux drum. False loading the tower by increasing reflux first can be constraint by the reflux drum inventory. Furthermore, the outcome of the reflux increase is more difficult to confirm relative to the reboiler operation. Due to the false load operation, the tower bottom and overhead product quality should exceed the targeted specification once the tray loads fall within the stable area of the tray operating window. The tower is then ready for a feed increase, which will require reboil and reflux adjustments to ensure that as rates approach the targeted steady state conditions, tray loads remain inside the stability area of the tray operating window and not exceed flood limits.

### *Packing*

Packing is far less complex then trays relative to turndown operation. The turndown criteria in Literature[6] is based on a minimum wetting rate of 0.5 to 2 gpm/ft$^2$ for random packing and 0.1 to 0.2 gpm/ft$^2$ for structure packing. Operation below turndown is associated with a reduction in efficiency due to some of the packing surface being dry. Liquid distribution in packed towers is critical and is achieved primarily through feed distributors, which in most case will dictate the tower turndown. Extended operation below the distributor turndown, especially in large diameter tower; will cause maldistribution and potential channeling across the packed bed, which may not recover as rates are increased. The initial liquid feed to the packed bed must therefore be ramped to turndown conditions relatively quickly to prevent channeling.

### *Feed Distributor*

Turndown sensitivity of feed distributors varies with design. The criticality of operating the distributor within its turndown criteria relative to a tower startup depends on the sensitivity of the internals relative to mal-distribution and the ability for the distribution to self-correct itself as rates are increased. The creation of segregated active areas on multi-pass trays makes them more sensitive to maldistribution than tray designs where both vapor and liquid are free to travel across the entire tray deck. As indicated above, the effectiveness of packed towers is closely related to effective distribution. Trough distributors are probably the most sensitive to turndown conditions as a minimum liquid head of ½ to 1 in[7] is required in the trough to ensure even distribution across the holes or drip tubes associated with a given trough.

### *Feed Piping*

Other than process safety consideration discussed later, feed piping is typically not a problem relative to startup transient conditions. However, tower feed preceded by feed pre-heat resulting in two phase flow could potentially result in a flow regime transition when operated at low rates. Transition to slug flow could potentially cause damage to tower internals and should be avoided.

## Environmental Impacts

Tower startup involves product streams outside specifications until establishing stable fractionation and the desired composition profile. Off specification storage if available can be used for minimizing losses. In some case, a recycle line between off specification storage and the process can be utilized for false loading the unit. However, often a number of streams cannot be recovered and must be released to flare. Flare releases and smoky flares are normally regulated relative to duration and intensity. Beside the environmental impact, flaring is also a loss of valued chemicals. Startup procedures are therefore developed to minimize flare release, hence explaining why towers are often started at rates below turndown. It also emphasizes the need to quickly bring the tower product streams within specifications. Figure 8 illustrates typical cracking yields for different feedstocks typically processed in ethylene plants. Some plants are specific to ethane cracking while others have feed flexibility. This plot suggests that focusing on getting the offgas and C2's streams within specification when cracking ethane would take care of about 95% of the hydrocarbon released to flare. This number drops down when cracking naphtha, propane and butane but remains in the low 70%'s when assuming that liquid streams such as fuel oil and RPG can be stored. The C3's, C4's and RPG from debutanizer bottom despite representing a significant portion for naphtha, propane and butane cracking, require the front end of the plant to meet specification. Therefore, meeting the offgas and C2 streams specification should be a priority when developing an ethylene plant startup strategy. As shown on the plot, startup feedstock selection may have a significant impact on flare losses but shall be consistent with unit turndown capabilities. Many plants with feed flexibility have expanded their ethane cracking capabilities over the last decade. However, their liquid cracking capabilities may constraint their turndown abilities relative to starting up on ethane feed.

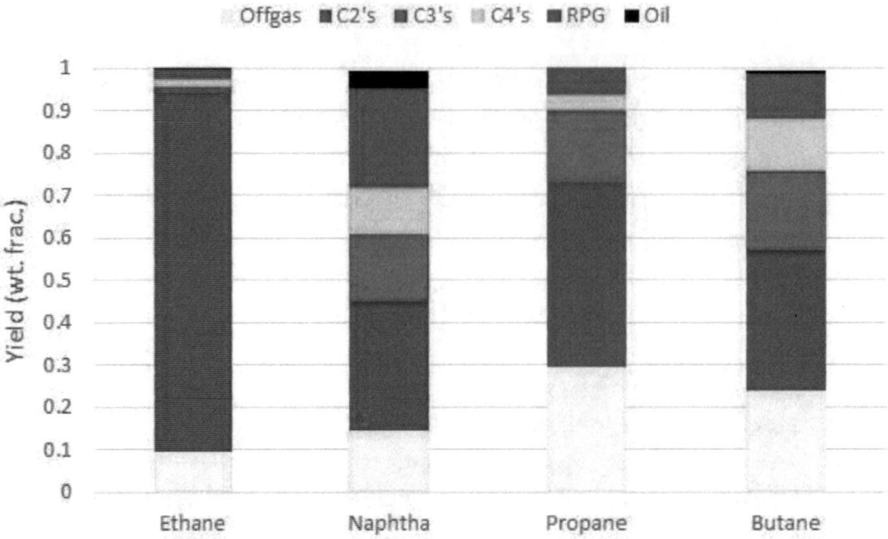


**Figure 8: Ethylene Furnace Typical Cracking Yields**

## Process Safety Considerations

The main process safety risk related to tower startup in ethylene plants are related to potential overpressure and risk of brittle fracture in carbon steel vessels. Other risks are related to potential equipment damage caused by improper operation or faulty instrumentations. Tower sump levels operated above the reboiler return nozzle, either due to faulty level indication or improper operating procedure is an example that could result in tray damage requiring an outage.

### *Risk of Over Pressure*

Pressure in distillation towers involved within the separation train of an ethylene plant is the combination of the crack gas compressor that provides the driving force for material to travel through the unit, heat/cooling inputs in reboilers/ condensers and control valves directing the flow path of material. Startup operation normally occurs after plant turnaround, during which several isolation valves are in the opposite position as found during normal operation. A commissioning check list would typically be used to ensure that those valves are set in their appropriate position. It is however a good practice to validate the flow path of streams that can lead to over pressure before the tower reach its intended operating pressure. In many cases, this would involve validating control valves releasing material to flare and downstream unit through the normal flow path. Although most of the towers are protected by pressure relief valves (PSV), lifting of any of those valves should be treated as a process safety incident.

### ***Potential for Cold Brittle Fracture***

Cold brittle fracture of a vessel is the result of a low temperature excursion below a level where the vessel material loses its ductility and becomes fragile. The minimum acceptable temperature (MAT) of vessels can be expressed in terms of single temperature (level 1) or pressure/ temperature relationship (level 2). Normal design operating conditions must be above the MAT of the vessel. The potential for cold brittle fracture is the result of process temperature excursions resulting from abnormal conditions such as loss of heating, sudden loss of pressure and others. Ethylene plant startups involve transient conditions associated with auto-refrigeration of streams expanded from high to low pressure. The presence of nitrogen causes further drop in temperature if liquid hydrocarbons are introduced in process vessels, due to imbalance in partial pressure. Figure 9 illustrates this behavior and the implication relative to the vessel MAT on front end bi-metallic Demethanizers.

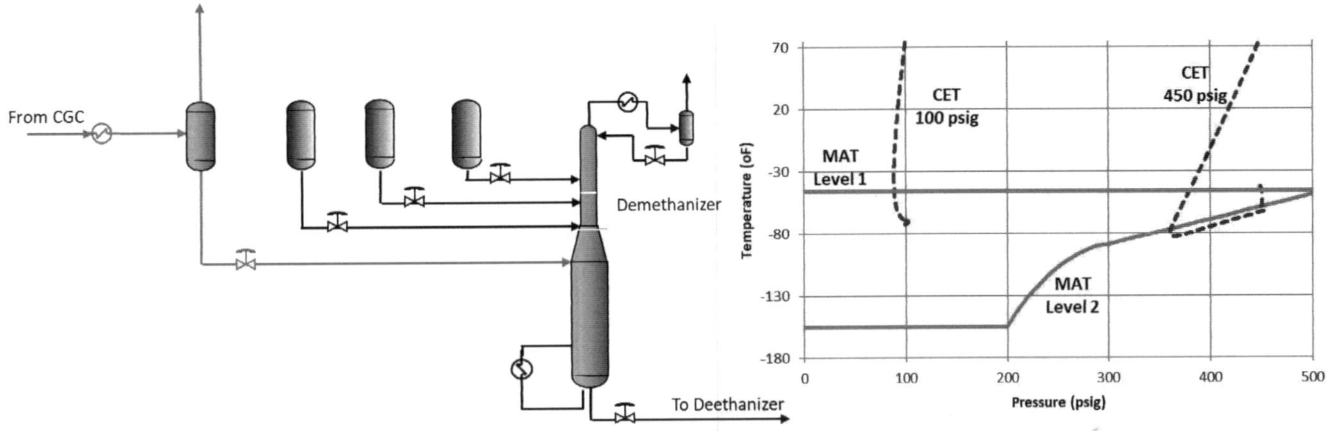


**Figure 9: Critical Exposure Temperature (CET) vs Vessel Minimum Allowable Temperature (MAT) – Demethanizer Startup**

This example assumed a bi-metallic Demethanizer with the bottom section MAT characterized in red on Figure 9. 2 MAT curves are provided: a level 1 MAT independent of pressure and a level 2 MAT curve providing the pressure dependency. The temperature/ pressure transition of the tower bottom from a hydrocarbon feed is illustrated in blue for a vessel initially filled with nitrogen at 100 and 450 psig and a temperature of 70 °F. The hydrocarbon feed stream is assumed supplied to the first feed knock out drum at a temperature of -40 °F and pressure 50 psi higher than the tower. The expansion across the control valve between the vessel and the tower results in a temperature drop of about 1 °F. The temperature at the bottom of the tower, transition from its initial temperature of 70 °F to a final temperature equivalent to the hydrocarbon stream feeding the tower, if sufficient time is provided for reaching a steady state condition. However, the minimum process temperature between the initial hydrocarbon feed and steady state would reach -73 and -82 °F for a tower pressure of 100 and 450 psig. As shown on Figure 9, the temperature excursion with the tower pressure set at 450 psig would transition below the MAT of the vessel putting the tower at risk for brittle fracture. The temperature with the tower at 100 psig remains well above the MAT (level 2 curve) due to the lower pressure. However, the tower would be at risk of brittle fracture if the pressure was increased to 450 psig while the temperature remains at -73 $^{o}$F.

The above example illustrates the importance of assessing the risk for potential brittle fracture in the development of a startup sequence.

### *Instrumentations*

Level instruments issues on tower sump are not uncommon during startup. Improper level indication may result in tower flooding and/ or difficulties in starting Thermosyphon reboilers. Liquid accumulation over the reboiler return nozzle may cause internal damage and should be avoided.

Startup transitions from initial to steady state conditions cover a range for process variables that is not necessarily in alignment with the instruments range calibration. Failing to recognize this limitation may result in startup procedural step change based on improper process reading. A near miss incident reported in literature[8] involved introduction of a liquid hydrocarbon mixture into a vessel above a specified pressure. The pressure indication used as reference was calibrated with a minimum pressure value above the target procedure value. This resulted in a cold temperature excursion cause by a lower-than-expected vessel pressure.

## Case Study

The use of the above concepts is demonstrated through a case study involving startup of the C1/ C2's distillation train on a front end Demethanizer ethylene plant as illustrated on Figure 10. The process scheme involves a Pre-Stripper operating with a Demethanizer tower. Propylene vapor streams from the propylene refrigeration system (C3R) provides heat to the reboilers. The Demethanizer overhead condenser use 1$^{st}$ level ethylene refrigeration (C2R) for cooling. The Pre-Stripper and Demethanizer bottom streams are feeding the Deethanizer at different locations. C3R ensures cooling in the Deethanizer overhead condenser while heat to the reboiler is supplied by low pressure (LP) steam. The Deethanizer uses a partial condenser where the vapor distillate feeds the Ethylene Fractionator after hydrogenation of acetylenes in acetylene reactors. Cooling and heating in the Ethylene Fractionator condenser and reboiler is ensured by C3R.

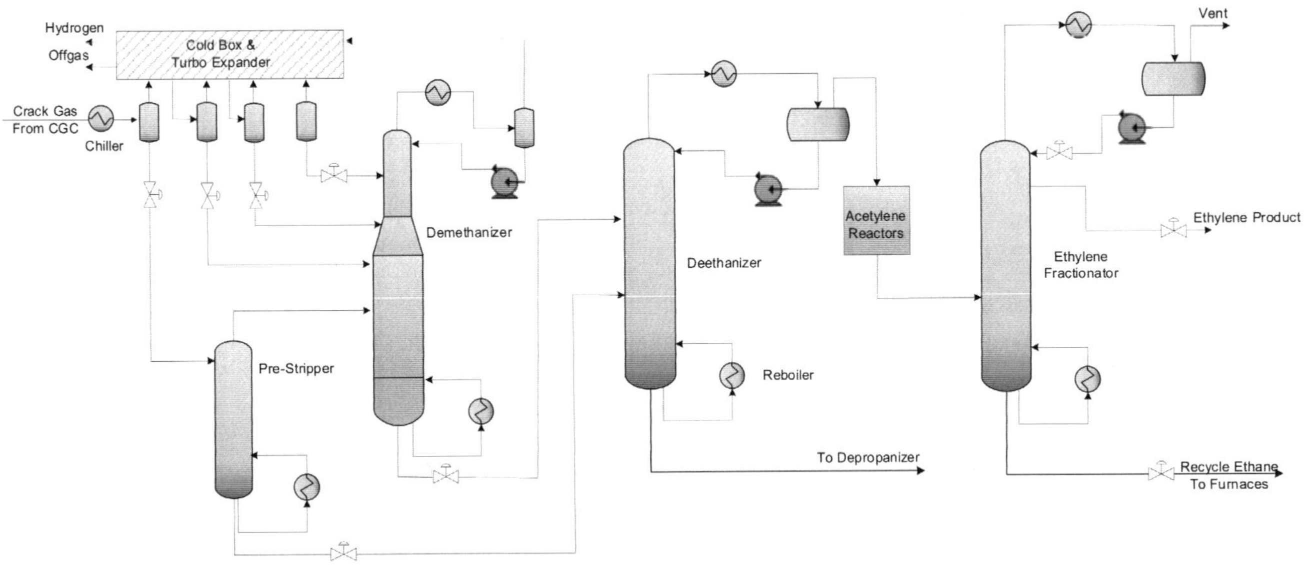


**Figure 10: C1/C2's Distillation Train**

The evaluation assumed random packing in the Pre-Stripper and tray towers for the Demethanizer, Deethanizer and Ethylene Fractionator. The system initial conditions are assumed to be under low nitrogen pressure and ambient temperature. The startup sequence segregates the Ethylene Fractionator from the front-end tower as it can be started on total reflux prior to hydrocarbon feed to the furnaces. The Pre-Stripper, Demethanizer and Deethanizer relies on forward feed for transitioning to normal operating conditions.

### *Ethylene Fractionator*

It is not uncommon to start the Ethylene Fractionator on total reflux before feeding hydrocarbon to the cracking furnaces. This practice provides the advantage of having the tower ready to receive feed once the front-end towers meet their specifications and expedites the time required to meet ethylene product specification. It should not have any impacts on the startup energy balance as the propylene refrigeration loads remain within the minimum power operation of a false loaded propylene refrigeration compressor (PRC).

Ethylene from high pressure storage or pipelines is first used for purging nitrogen off the tower through a series of pressure purges. Due to the presence of nitrogen, ethylene must enter the tower in the vapor state and temperature that would not cause a low temperature excursion below the MAT of the vessel at the expected operating pressure. Ethylene from storage and pipeline being under a supercritical state, requires preheating as it expands to lower pressure for avoiding a low temperature excursion.

Once nitrogen free, the tower is brought up to pressure by importing ethylene from storage or pipeline. As the pressure climbs, a point is reached where condensation of ethylene begins in the overhead condenser and liquid accumulates in the reflux drum. Once sufficient liquid inventory becomes available, reflux gets commissioned to the tower. Introduction of liquid reflux to the tower will eventually be associated with liquid accumulation in the tower sump, which enables startup of the reboiler. The vapor load in the tower is limited to the external ethylene supply until vapor generation from the reboiler is established. That transition is associated with a significant increase in vapor traffic and condenser duty. It is also followed by an increase in reflux flow.

Operational problems occurring in the C3R system have been reported relative to the total reflux transition conditions. Those problems are associated with the heat pump circuit created between the Ethylene Fractionator condenser and reboiler. Figure 11 illustrates a typical propylene refrigeration system using a 4 stage compressor. Note that the 3$^{rd}$ and 4$^{th}$ stage suctions are not bidirectional streams. They have been illustrated as such to reflect that in some systems they are extracting stages and on others, side load to the compressor. Prior to transition the Ethylene Fractionator to total reflux operation, the C3R system runs without any process loads. Surge margins are met through minimum recycle flows (not shown on Figure 11). Operating the Ethylene Fractionator (C2S) on total reflux results in propylene circulation between the Ethylene Fractionator reboiler (C2S Reboilers) and condenser (1$^{st}$ Level User) as illustrated in red and blue on Figure 11. Initiating this heat pump circuit provides loads on the compressor that are inconsequential on power as the anti-surge control compensates the load increase by reducing the minimum flow required for keeping the compressor operation within its surge margin.

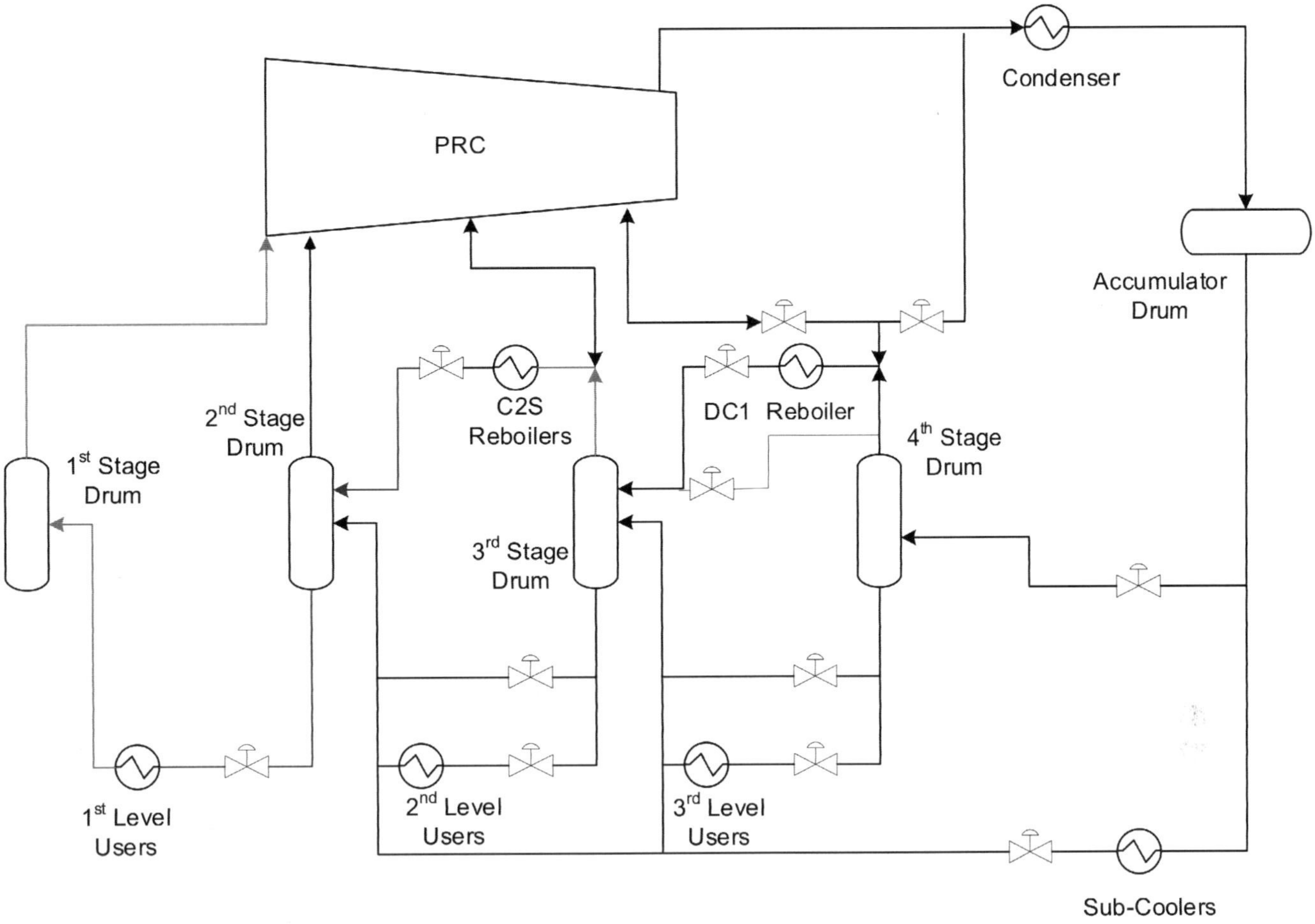


**Figure 11: Typical Propylene Refrigeration System**

The tower ethylene inventory, pressurization, reflux and reboiler transition has been evaluated through a dynamic model to gain understanding relative to the operational problems reported. The results were compared to plant data for similar transition as illustrated on Figure 12. The model results are presented on the left side of Figure 12 and shows that as pressure increases, ethylene will start condensing and accumulating in the reflux drum. Once the reflux drum level reaches its set point (50% in the model), the reflux flow ramps up until it becomes balanced with the incoming flow, which at this time is limited to the external ethylene supply. Liquid ethylene from the reflux drum flows down the tower and accumulates in the sump. Heat is then supplied to the reboiler once sufficient inventory builds up. The model shows a significant increase in reflux and reboiler duty following the initial kick on the reboiler. The reflux and reboiler duty plotted by a dash line on the bottom left plot of Figure 12 are scaled up by a factor of 100 when compared to their associated values plotted as a solid line. This scaling was done to illustrates the difference in magnitude when the reflux is only driven by the external supply of ethylene. The tower circulation becomes significantly higher once the reboiler get started.

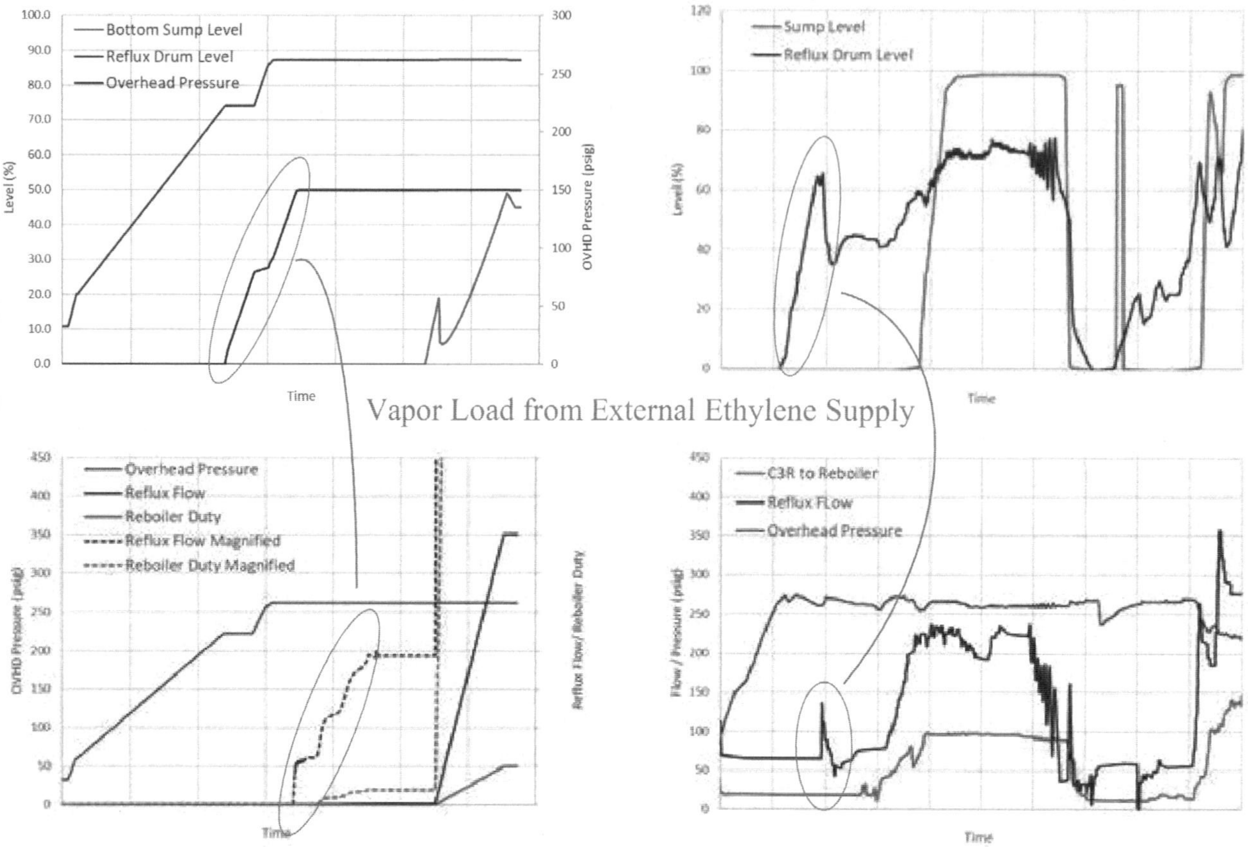


Simulation　　　　Plant Data

**Figure 12: Ethylene Fractionator Total Reflux Transition**

The first observation when comparing the model results to actual plant data reside in the reflux drum level transient after the initial ramp of the reflux flow. While the model suggests an upward ramp followed by a constant reflux flow until driven higher by the rise in reboiler duty; the plant is characterized by a sharp rise in reflux flow followed by a relatively quick decline. The reflux drum level, which regulates the reflux flow through a level controller; follows the same path. Unlike the plant, the model is not limited by equipment turndown and valve rangeability. As such, the initial reflux flow is significantly smaller than what is experienced in the plant. The larger reflux flow experienced by the plant causes a rapid decline in reflux drum level due to imbalance with the ethylene supply but also due to potential condensation of the rising vapors as they become in contact with cold reflux.

The plant data shows a more serious problem following start of the reboiler where for a limited period the reflux flow, level and reboiler duty were relatively constant; followed by a quick drop in sump and reflux drum level. Those instabilities are detrimental to the stability of the C3R system. The worst case occurs when excessive reflux starves the overhead condenser. This creates a transient condition where the C3R demand to the condenser can stop, causing C3R liquid accumulating in the C3R $2^{nd}$ stage drum as condensed C3R from the reboiler is still supplying the drum. This situation could likely result in trip of the PRC caused by high inter-stage level. The oscillation in reflux drum and sump level shown on Figure 12 could also be associated with insufficient ethylene inventory in the tower for enabling a stable total reflux operation. Those oscillations could also result from out of synchronization reboiler/ condenser duty resulting in high level in the C3R 2nd stage drum.

Transition to stable total reflux operation should be driven by the reboiler. However, sufficient liquid ethylene inventory is required and must initially be supplied from the reflux drum. Given that the overhead condenser duty is relatively small before introducing heat to the reboiler while supplying external ethylene to the tower, makes the C3R system less vulnerable to upset at that stage. Once initiating heat to the reboiler, the stability of the C3R circulation through the reboiler/ condenser path must be maintained. The reflux flow should be set to keep the reflux drum in material balance (constant level). Operation at low reflux rate beyond the capability of the system may be required initially. This can be achieved by regulating the reflux control valve on manual and meeting the reflux pump minimum flow through a spill back line. The reflux drum level should provide margin for increasing the reflux rate or allowing temporary accumulation. Heat to the reboiler should not be initiated until sufficient liquid inventory is established in the sump. Heat should be ramped up gently on the reboiler and sump level should be monitored. Due to the lag between vapor generation from the reboiler and accumulation in the reflux drum, adjustment in reflux rate should be done only once noticing accumulation in the reflux drum. Those adjustments must be done gently to prevent starving the condenser. The sump level must also be monitored to prevent the reboiler from starving. Heat to the reboiler is to be increased by increments with each increment implemented after noticing the response of the previous increment on the reflux drum and appropriate reflux flow adjustments. As the circulation increases, liquid inventory will build up on the trays and can be associated with a reduction in level in the sump and reflux drum. This situation can be rectified by importing more external ethylene. It could also be a sign that the ramp in reboiler duty is too fast relative to the external ethylene import rate if still ongoing. A steady and stable total reflux operation is reached once the reflux drum and sump level remain constant and external ethylene supply is no longer required. The internal circulation rate at that point should allow automatic reflux drum level control through regulation of the reflux flowrate.

The total reflux circulation rates should be set such that it meets the tray hydraulic operating window. In some case this may get constrained by the C3R system configuration. As shown on Figure 11, in some systems, a significant fraction of the C3R vapor supplied to the reboiler is generated by 3rd level users such as the Deethanizer condenser. The total reflux operation, being established prior to any other process loads on the C3R system, can become constrained in supplying the Ethylene Fractionator reboiler. At that point, pushing the total reflux circulation may drive the C3R 3rd stage drum pressure low and cause a temperature pinch. The circulation should then be set within the capability of the C3R system. As the Deethanizer is brought on stream, the circulation rate could then be ramped up as the C3R system constraints disappear.

It is important to understand the incremental load expected on the tower once feed from the upstream unit becomes available. The system evaluated in this case study is based on back end acetylene reactors. In such system, the outlet of the reactors would be routed to flare until meeting an acetylene composition specification. At this point, the flow path transitions from the flare to the tower. This transition must be done through a gentle ramp as it will impact the C3R system through its connectivity with the condenser and reboiler. The feed condition impacts the load distribution where a vapor feed would increase the vapor load in the rectifying section associated with an increase in condenser duty and reflux rate. On the opposite end, a liquid feed would increase the liquid load in the stripping section and may require increasing heat to the reboiler. Under those conditions, the tower will be fractionating as soon as it receives feed and ready for on specification ethylene product withdrawal.

The transition between feed-in and steady state operation can be challenging relative to the initial ethylene volume accumulated in the sump. Under normal operation, ethane accumulating in the tower bottom is recycled to the furnaces for cracking. As feed is introduced with the tower on total reflux, the initial volume of ethylene contained in the sump cannot be released to the furnaces as it would upset their operation and increase coking. This transition can cause high level in the sump, which could easily exceed the reboiler return nozzle. Understanding this problem requires a review of the transient heat and material balance through the feed ramp period. It is understood that ethylene contained in the sump cannot be released to the furnaces. Therefore, as feed is introduced, the only exit points to prevent accumulation in the tower is the ethylene product draw. Product must then be withdrawn at the same rate as the incoming feed. The ethylene reflux will cause ethane from the feed to initially move downward in the tower. The ethane volume will start displacing ethylene contained in the sump by stripping through rising vapor generated in the reboiler. The ethane composition profile will get established over time throughout the tower. A transition in material balance must take place as the ethane profile rises in the tower and put the ethylene product at risk of exceeding specifications. This transition would require withdrawing ethane from the tower bottom and reduce the product draw in alignment with the tower material balance.

This strategy has been evaluated with a dynamic simulation. The Ethylene Fractionator system modelled is illustrated on Figure 13. A pasteurization section is included based on the assumption of a front end Demethanizer Ethylene plant with back end acetylene reactors. The model initial conditions assume a tower operating on total reflux with internal circulation rates falling within the turndown capabilities of the trays. The conditions being stable allows for a constant heat duty on the reboiler and condenser prior to feed in. The Condenser although constant before feed-in is regulated by the tower pressure controller. Similarly, the reflux flow is on a level to flow control cascade loop. On-line ethylene/ ethane analyzers are assumed available on the bottom tray and product draw location.

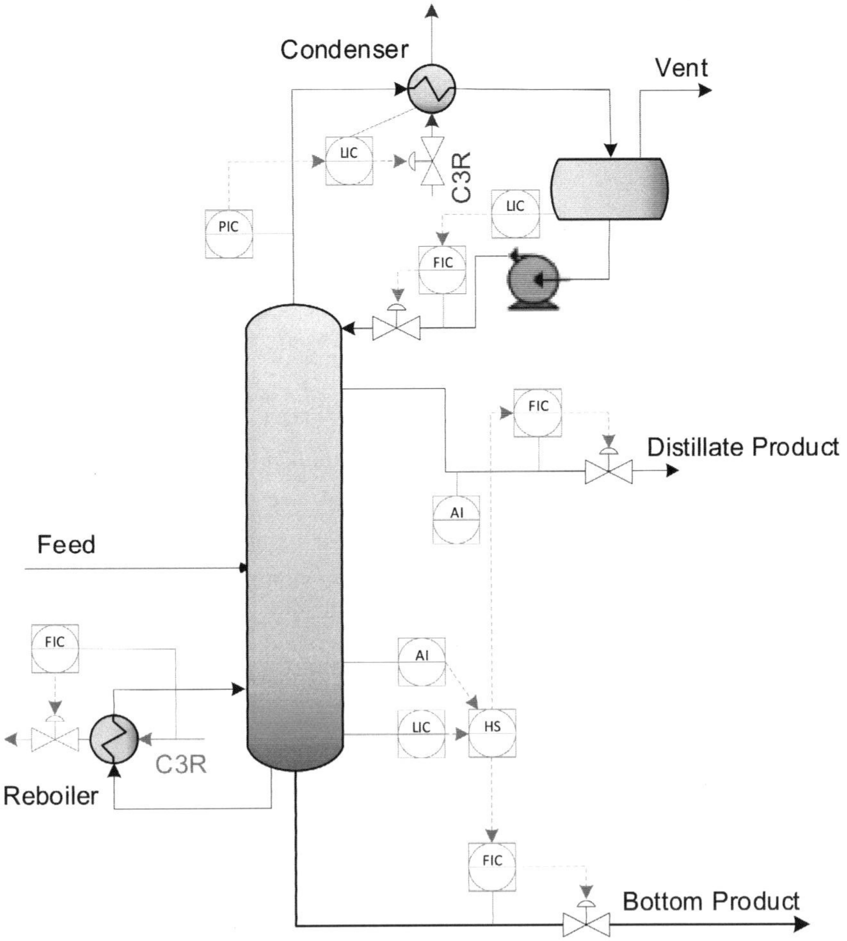


**Figure 13: Ethylene Fractionator**

An automatic sequence was configured in the model for ramping the feed over a period of 20 minutes. The tower sump level control is initially set for regulating the ethylene product draw rate from the tower. The Level controller output transitions from the product draw to the tower bottom once the ethane composition on the bottom tray reaches or exceeds 92% mole. The product draw rate at that point is re-set to align with the tower ethylene balance. A small vent rate is set on flow control when feed is introduced to the tower to prevent accumulation of non-condensables. The ethane and methane content of the product draw is monitored through the product stream analyzer. The results from the simulation run are presented on Figure 14 where the feed-in ramp starts at time 12 minutes over a period of 20 minutes. The increase in condenser duty follows the same rate as the feed increase due to the vapor state of the feed. While the incremental condenser duty is associated with an increased reflux flow, the reboiler duty remains constant. The tower bottom tray analyses shows that the ethane composition rises above 92% mole 7 minutes after initiating the feed ramp. The sump level controller output then transitions from the ethylene product draw to the tower bottom stream. The transition is associated with small decrease in ethane composition in the bottom tray. The bottom right plot on Figure 14 shows that the ethane and methane content of the ethylene product remained well below the stream specification over the transition.

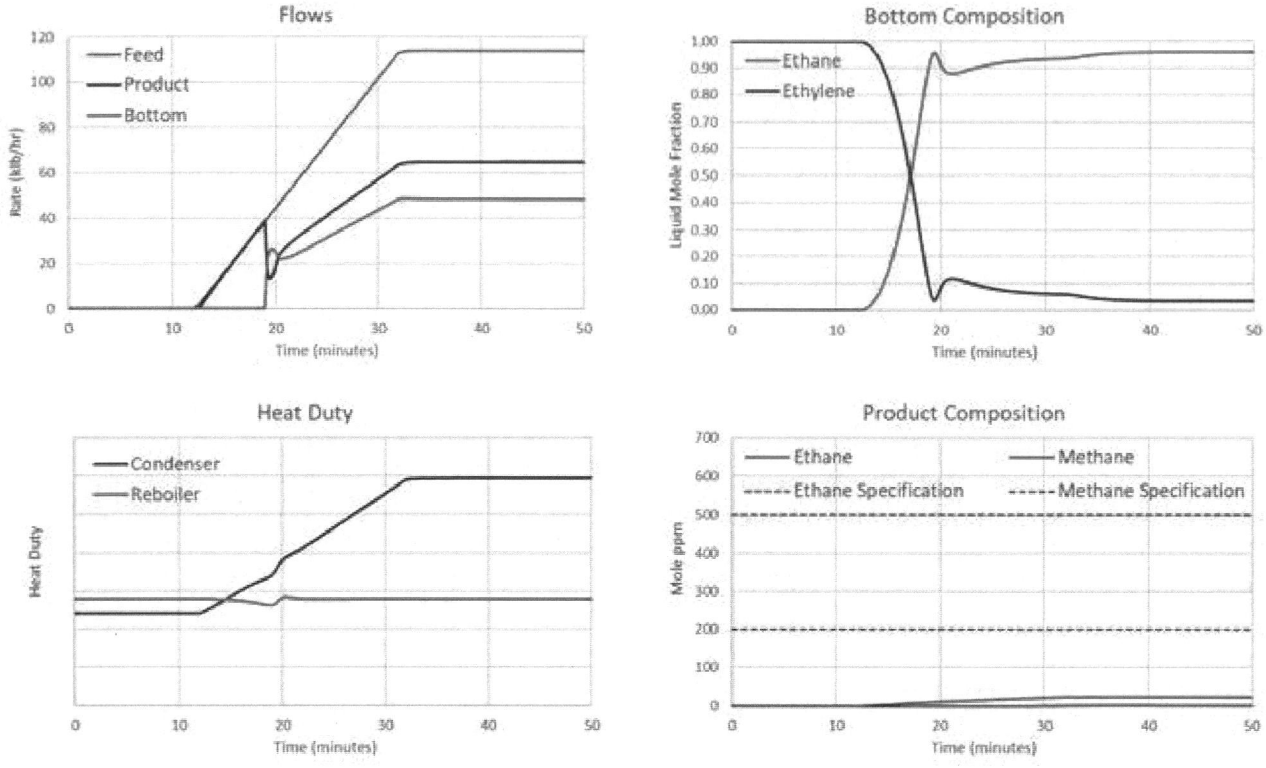


**Figure 14: Ethylene Fractionator Feed-In Transition – Dynamic Simulation Results**

This example demonstrates that the initial ethylene volume contained in the tower sump can be recovered as product by building the composition profile in the tower prior to withdraw material from the sump. This prevents flooding of the bottom and eliminates product losses through unnecessary drainage of material contained in the sump. The procedure described can be automated or accomplished manually by a board operator. The material balance transition period is of relatively short duration in the above example and depends on the tower characteristics and the magnitude of the feed ramp time and final flowrate.

## ***Pre-Stripper & Demethanizer***

Unlike the Ethylene Fractionator, startup of the Pre-Stripper and Demethanizer requires forward feed to the unit. As illustrated on Figure 15, crack gas from the furnaces supplies feed to the Pre-Stripper and Demethanizer after compression by the Crack Gas Compressor (CGC). The feed distribution is closely linked with the Cold Box operation through feed separator drums. Front end chilling of crack gas typically feeds the 1st separator drum at a temperature of about -35 $^{o}$F. The normal operating pressure is set by the Demethanizer overhead and regulated by manipulating the vapor distillate flow to the turbo-expander/ cold box section. A positive pressure differential is maintained between the separator drums and the Pre-Stripper/

Demethanizer by regulating the flow rate of the hydrogen stream exiting the cold box to ensure forward liquid flow from the drums to the towers. The crack gas temperature is decreasing as it flows across the cold box through heat exchange with different levels of ethylene refrigeration and cold process streams. As such the temperature and composition of the liquid collected in the separator drums, which feed the Demethanizer, get colder and lighter in the drums forward sequence.

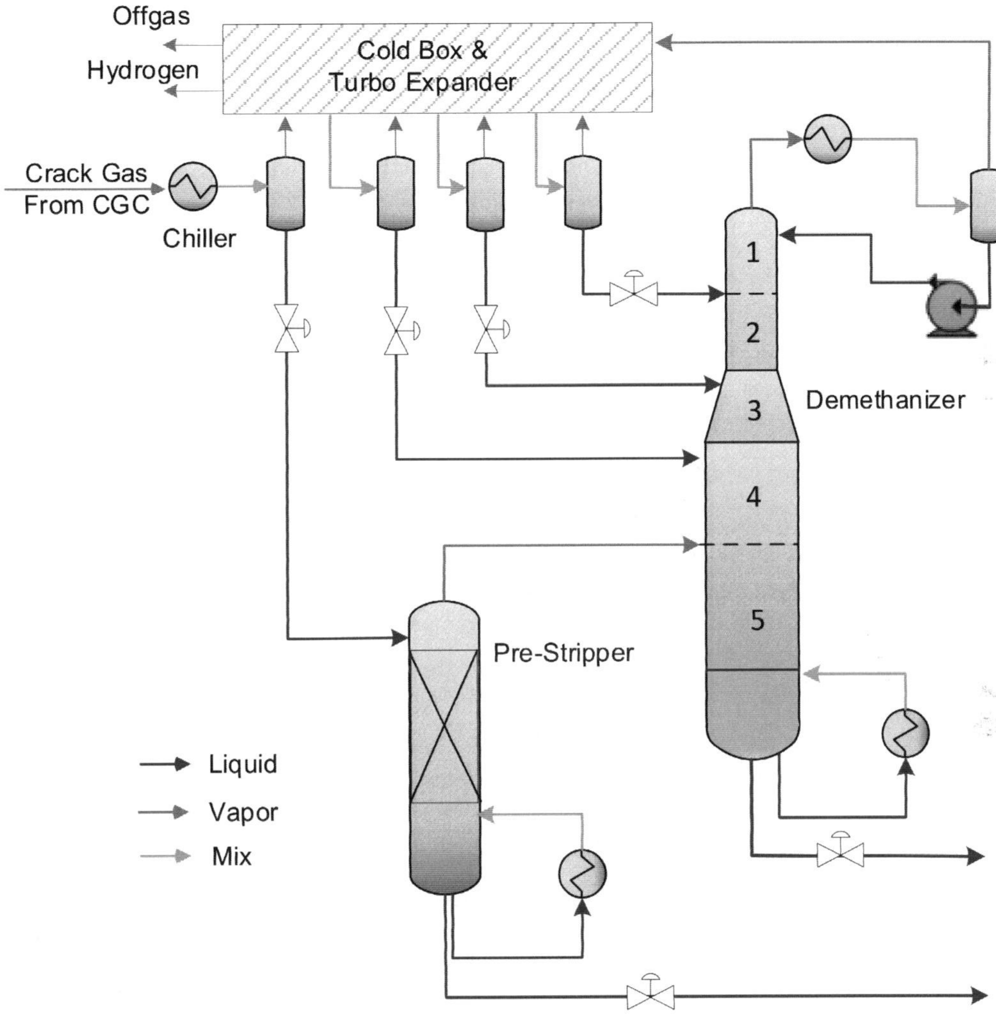


**Figure 15: Pre-Stripper/ Demethanizer**

Hydrocarbon feed to those towers becomes available after startup of the CGC. The discharge of the compressor would normally be routed to flare initially and then transitioned to the Pre-Stripper/ Demethanizer/ Cold Box system (DC1). The initial hydrocarbon introduction to the system has been evaluated with a dynamic model that includes the crack gas inlet stream, the 1st Feed Separator Drum and the Pre-Stripper. The crack gas inlet stream is assumed supplied from the Crack Gas Dryers at a pressure of 515 psig and 65 °F. The flow is regulated through a control valve, which mimics the transition from flare to the unit. The 1st Feed Separator Drum and the Pre-Stripper are assumed filled with nitrogen at pressure of 100 and 50 psig.

Two scenarios have been evaluated. The first scenario, referred as Case 1, assumes a startup transition following the intended flow path of the system. Forward flow of crack gas is ramped up to 20% of the startup plan flowrate over a period of 10 minutes. Heat is introduced to the reboiler once sufficient liquid inventory fills the tower sump. The tower pressure is raised once the system is nitrogen free and temperature have stabilized.

A second scenario, Case 2, Follows a similar ramp rate as Case 1 relative to the hydrocarbon forward feed. However, no cooling is provided on the front-end chiller. The 1st Separator Feed Drum is then supplied by a warm vapor stream. The level control valve of the drum is assumed fully open so that vapor can blow through and supply the Pre-Stripper and Demethanizer. The level control valve in the bottom of the Pre-Stripper and Demethanizer are assumed fully open, which ensure vapor flow reversal in the tower. This flow reversal prevents the presence of a stagnant volume of nitrogen locked in the bottom of the towers and also supplies warm crack gas to the Deethanizer. Drain valve on the Deethanizer would be assumed open such that nitrogen contained in the tower below the feed gets purged out as well. The warm crack gas stream is used to pressure up the Pre-Stripper, Demethanizer and Deethanizer once the nitrogen content in the towers becomes negligeable. Chilling is re-introduced in the front-end chiller once the tower approaches the normal operating pressure. The 1st Separator Feed Drum level control valve is then pinched back to allow liquid accumulation in the drum. The drum level controller is then operated on automatic once liquid accumulation reached its intended operating target. The Pre-Stripper reboiler is started once sufficient liquid accumulate in the sump.

The reboil heat for Case 1 and 2 is emulated in the model by ramping the UA value of the reboiler to a fixed number, identical for both cases over a period of 30 minutes. Hydrocarbon/ nitrogen contacting, and pressurization sequence differentiate those two different startup approaches. Figure 16 compares the main differences between those two cases. As previously discussed, introducing liquid hydrocarbons into a nitrogen environment is associated with a significant drop in temperature. As shown on Figure 16, the Pre-Stripper bottom experience a sharp drop in temperature from 55 to -133 °F in about 4 minutes for Case 1. The temperature then rises and eventually approaches the tower feed temperature with depletion of the nitrogen content over the following 25 minutes. Case 2 in contrast uses warm crack gas vapor for displacing nitrogen from the system. The nitrogen content inside the Pre-Stripper drops to negligeable level within 2 to 3 minutes after initiating forward feed. Crack gas is then used for pressurizing the Feed Separator Drums and the Towers. The pressurization is associated with a temperature rise in the tower from 55 to 65 °F. Once the system approaches its targeted pressure, chilling is re-introduced in the front-end exchanger, which is observed by a sharp reduction in temperature from 65 to -35 °F over less than 2 minutes.

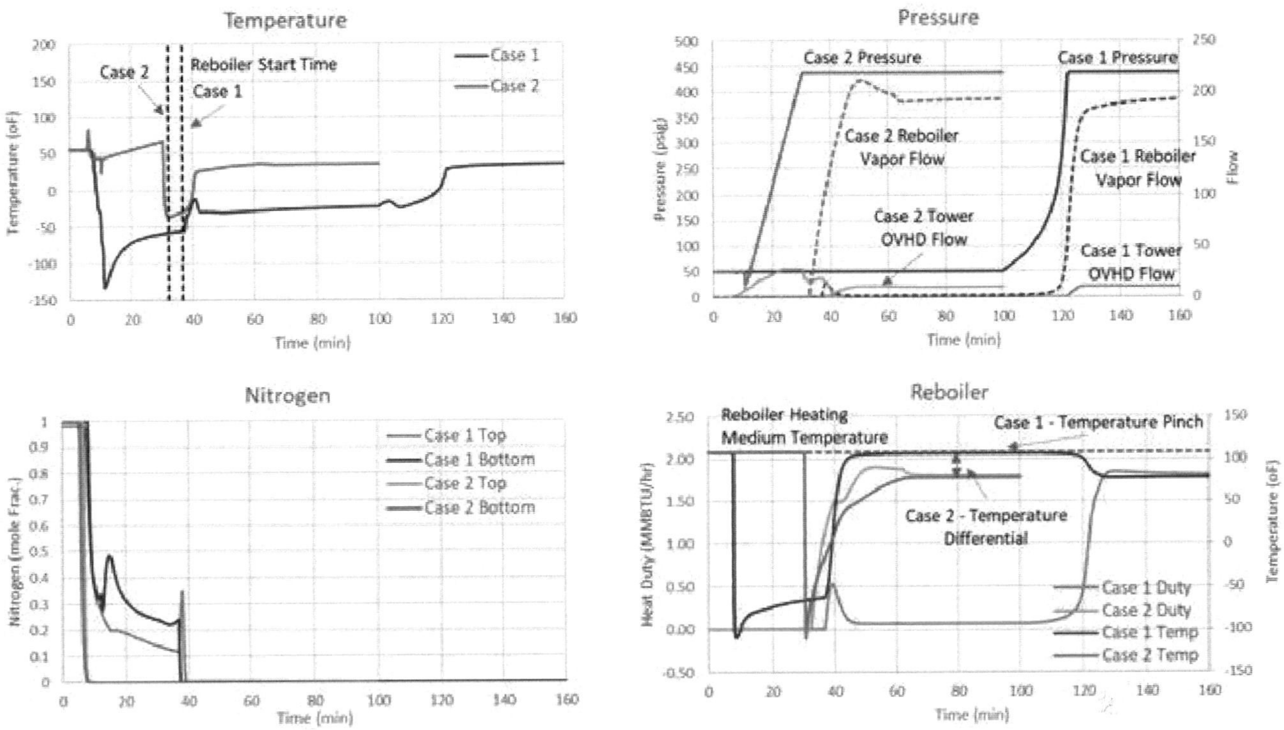


**Figure 16: Pre-Stripper Dynamic Simulation Results**

Pressurization of the tower is done after introducing heat to the reboiler in Case 1, while in Case 2, pressurization precedes the reboiler heat ramp. The tower pressure transient illustrated on Figure 16 demonstrate different behavior between Case 1 and 2. Note that although the reboiler was started at time 37 minutes for Case 1, the rise in pressure was not initiated until time 100 minutes. The pressure rise timeline is relatively similar for both cases at about 20 minutes. However, While Case 2 exhibit a linear increase, Case 1 aligns better with an exponential rise. The linear increase in Case 2 is easily explained by a consistent forward vapor flow in conjunction with a linear ramp of the pressure controller set point. Case 1 differs from Case 2 as the pressure rise is driven by the reboiler. As shown on Figure 16, the reboiler in case 1 is at a temperature pinch until the pressure reaches 280 psig, at which point the model predicts a sharp rise in vapor generation from the reboiler. Until that point, the model predicts a relatively small vapor generation from the reboiler, which explain the slower rise in pressure initially.

The low temperature excursion in Case 1 raises concern relative to potential for brittle fracture on killed carbon steel (KCS) vessels. Evaluating the risk of cold brittle fracture can be done through a level 1 or level 2 evaluation. A level 1 evaluation compares the coldest temperature to the vessel coldest allowable temperature at the maximum operating pressure. A level 2 evaluation accounts for the pressure dependency of the vessel minimum allowable temperature (MAT). The Pre-Stripper temperature/ pressure excursions for Case 1 and 2 are plotted against the MAT of a carbon steel vessel on Figure 17. Note that the MAT curve displayed on Figure 17 is used for example purposes and shall not be used as reference for other carbon steel vessel as it is specific to the vessel construction characteristics.

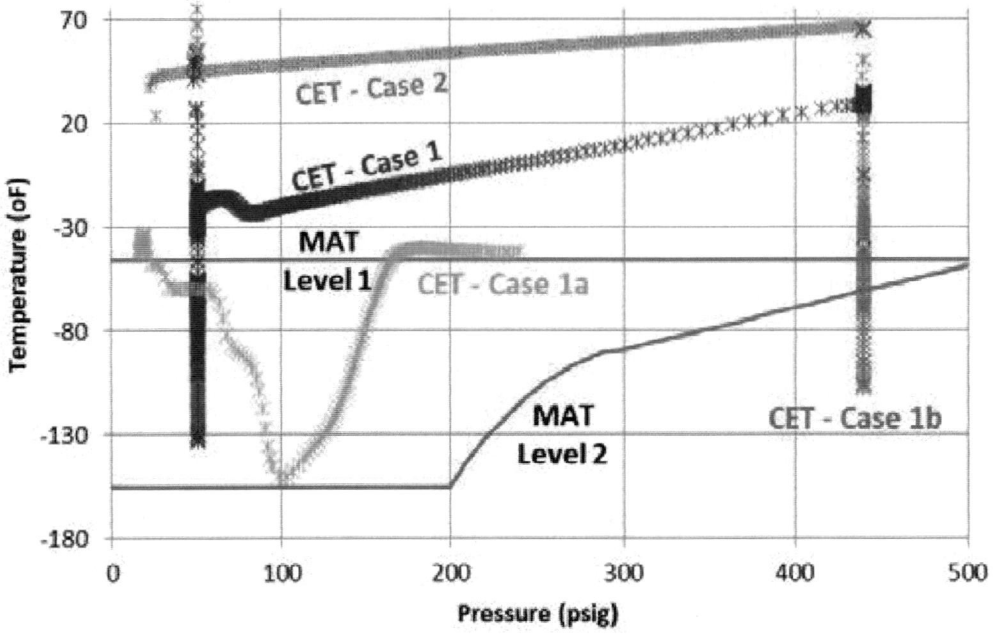


**Figure 17: Pre-Stripper Startup Critical Expose Temperature (CET) vs Vessel Minimum Allowable Temperature (MAT)**

The level 1 evaluation is an initial assessment, that if passed does not require further evaluation. Therefore Case 2 can be limited to a level 1 evaluation. Case 1 fails the level 1 evaluation but passes the level 2 due to the low pressure under which the low temperature excursion occurs. Two extension of Case 1 were evaluated relative to the risk of cold brittle fracture. Case 1a assumes that the tower pressure rises as forward feed is initiated. Without heat to the reboiler, the system becomes constrained relative to pressure increase. The cold temperature excursion, although more severe than Case 1, remains within the limitation of the MAT level 2 curve. Case 1b assumes that the Pre-Stripper has been pressurized with nitrogen prior to hydrocarbon feed. This case is becoming well known as a "not to do" practice in the industry and has been discussed in literature. As shown on Figure 17, the CET of Case 1b would severely cross the MAT of the vessel.

Although Figure 17 shows that Case 1 and Case 2 do not represent a risk for cold brittle fracture, the sharp drop in temperature experienced from both cases can cause thermal stress that could result in failures and/ or leaks. As stated above, Case 1 exhibits a 188 °F temperature drop over 4 minutes while Case 2 a drop of 100 °F over 2 minutes. While Case 1 temperature excursion is the thermodynamic outcome of introducing liquid hydrocarbon into a nitrogen environment; Case 2 can be mitigated by a more appropriate operating sequence. The rapid temperature reduction in case 2 is driven by the simulation model, which stops the crack gas vapor flow to the Pre-Stripper while ramping the chilling duty. The flow to the Pre-Stripper transitions from hot vapor to cold liquid as a result. This sharp decline in temperature can be prevented by pinching back on the 1st Feed Separator drum level control valve rather than fully closing it. The opening of the valve should be such that a vapor/liquid stream is free to flow to the tower while chilling is gently re-introduced in the front-end chiller.

The different pressurization path between Case 1 and 2 highlights differences in reboiler operation. As shown on Figure 16, the reboiler operates at a temperature pinch on Case 1 until the pressure rises. While the simulation model seems forgiving, it is believed that the reboiler would stall under those conditions. Boiling curves on the liquid supplying the reboiler have been calculated based on the initial composition prior to input heat on the reboiler and once a temperature pinch or stable operation were obtained. The boiling curves for Case 1 and 2 are illustrated on Figure 18.

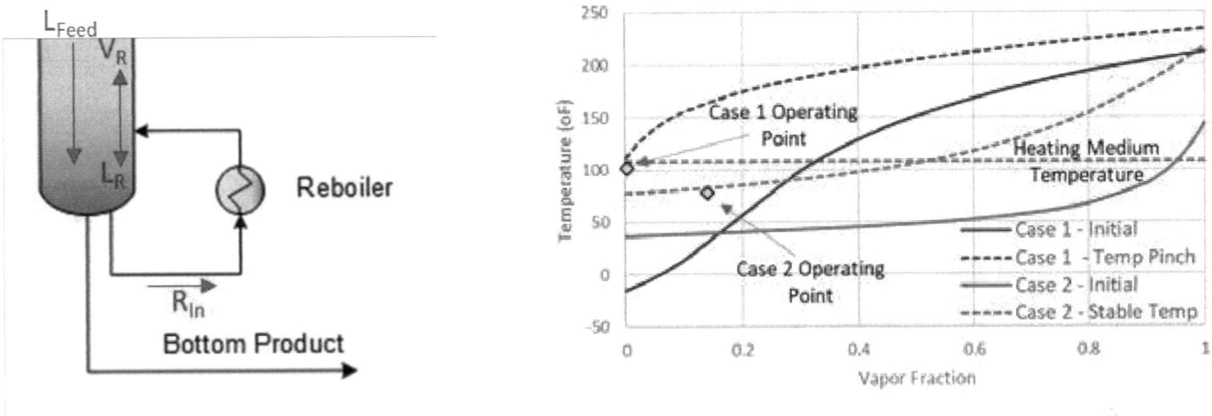


**Figure 18: Reboiler Liquid Supply Boiling Curve**

As previously explained, the initial pool of liquid filling the reboiler volume is equivalent to the liquid composition feeding the tower. As heat is introduced to the reboiler, a fraction of the material is evaporated and provides rising vapor to the tower. The remaining liquid mixes with liquid from the bottom tray making the combined stream heavier than the liquid falling from the bottom tray. The boiling curve of the liquid supplying the reboiler is therefore expected to be heavier than the initial liquid supply. The initial boiling curve for Case 1 is based on liquid collected in the sump at low pressure. Although its initial boiling point is significantly less than the initial boiling point of Case 2, 85% of its content is heavier or has a higher boiling point than the Case 2 initial curve. As a result, the liquid supplying the reboiler quickly reaches a temperature pinch with the heating medium. The boiling curve of the mixture supplying the reboiler at the pinch point has an initial boiling point equivalent to the heating medium, which suggest a stalled reboiler. The initial boiling curve for Case 2 would pinch with the heating medium at a vapor fraction of 0.92, providing significantly more latitude than for Case 1. As a result, the boiling curve for Case 2, once reaching stable operation, still has a temperature margin from the heating medium. This evaluation demonstrates that the conditions provided by Case 1 would result in thermodynamic inconsistencies relative to starting the reboiler. This would require many starts and re-starts in the field associated with sump drainage. The methodology provided by Case 2 sounds more robust and should ensure thermodynamic consistency relative to starting the reboiler. Case 2 should also reach stable operating conditions sooner than the methodology used for Case 1.

The above discussion although focused on the Pre-Stripper would also apply to the Demethanizer and the Deethanizer. The sequence proposed by Case 2 at this point would continue by transitioning the crack gas vapor path from the 1st to the 2nd separator drum through the level control valve. Once a stable liquid flow gets established from the 1st separator drum, the vapor path to the tower should transition to the 3rd separator drum through its level control valve while closing the 2nd feed separator level control valve to establish a liquid inventory. Transition to the 4th separator drum would occur in a similar fashion. Heat to the Demethanizer reboiler may get started in a similar fashion. Forward flow from the Pre-Stripper and Demethanizer bottom to the Deethanizer will be initiated once sufficient liquid inventory accumulates in their respective sumps. Early pressurization and nitrogen purged of the Deethanizer would prevent low temperature excursion and expedite reboiler startup.

Ethylene refrigeration can be commissioned to the overhead condenser once the Demethanizer temperature profile stabilizes, after establishing a stable liquid flow from the 4th separator drum. It is assumed at this point, that although the cold box might not have reached its steady state profile, streams from feed separator drums have pinched on temperature with their associated refrigeration level.

Unlike the Ethylene Fractionator and Deethanizer, the Pre-Stripper cannot be false loaded. The tower load is driven by feed rates. A packed tower using random packing is assumed for this example. As discussed earlier, the minimum operating limit of the packed bed is defined by a minimum wetting rate of 0.5 to 2 gpm/ft$^2$ and a minimum liquid head of ½ to 1 inch in the trough of the liquid distributor. While process loads predicted by the simulation satisfies the turndown requirement for the packing and distributor, they fall short for the reboiler/ Thermosyphon circuit. A reboiler/ thermosyphon design having an operating range between 42 and 100% of full rate as illustrated on Figure 6 is assumed for the example. The startup simulation suggests a reboiler operating at 30% of full rate. However, since the simulation was set for meeting a bottom stream specification, the duty can be increased for meeting the reboiler turndown criteria. Increasing the reboiler duty cannot always allow meeting turndown limitations as it may violate the integrity of the heat and material balance. In such case, the use of lift gas is a practice that has been done for increasing the vapor velocity and ensure operation above flow reversal. Note as discussed in more detail below, that higher reboiler duty and/ or the use of lift gas has implications on the Demethanizer operation.

As discussed above, the Pre-Stripper has limited options relative to false load and can in some case define the minimum startup operating rate. This would likely be the case with a tray tower, which has significantly less turndown capabilities than random packing.

The Demethanizer, although equipped with an overhead condenser, has limited false load capabilities with liquid loads mostly driven by feed rates. The liquid load from the overhead condenser is effective on the upper section but has a reduced impact on the lower section. The extent of false load could become limited by the overhead condenser capacity and/ or temperature pinch. From a startup standpoint, minimum feed rates shall not be limited by fully efficient Pre-Stripper/ Demethanizer operation. The towers need to be capable of stripping methane from the C2+ streams feeding the Deethanizer such that on specification ethylene product be withdrawn from the Ethylene Fractionator. Figure 15 breaks the Demethanizer in 5 different sections, which are delimited by feed points. External loads from the feeds causes each of those section to have different vapor/ liquid internal circulation. Due to the difference in process loads, each section is likely characterized by different tray geometry. The relevance of those sections relative to methane stripping is illustrated on Figure 19. The tower has been simulated per typical product specification, in a false load operating mode and in a reduced efficiency (half the normal stage count). All cases exhibit a sharp decrease in section 1 of the tower. The normal simulation based on stream specifications follows a linear reduction starting on section 2 that extend to section 5. The low efficiency case follows a significant reduction in section 2 and aligns with the false load case for section 3, 4 and 5. This plot shows that regardless of efficiencies and/ or false load, section 5 is critical for meeting the bottom stream specification. Lower efficiencies cause by weeping or low weir loadings can be tolerated. However, absence of fractionation due to unsealed downcomers or dumping will result in methane content above specification.

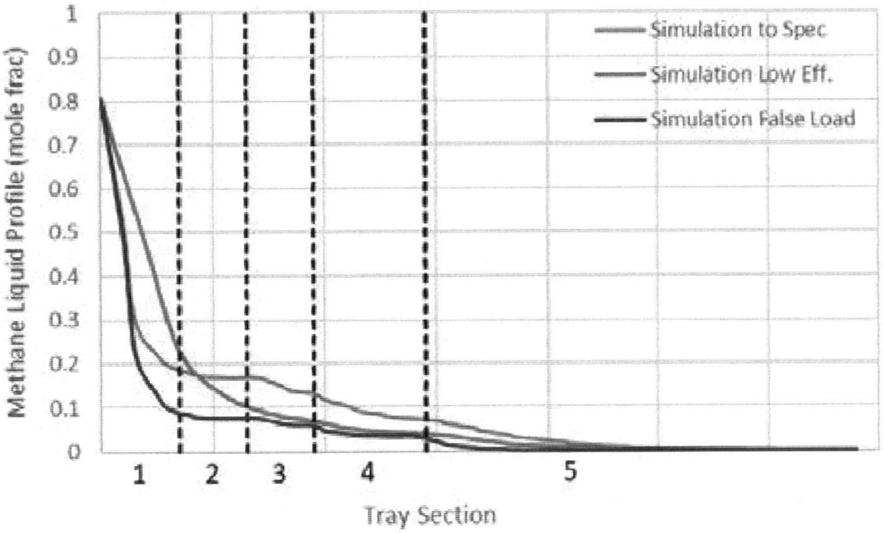


**Figure 19: Demethanizer Methane Composition Profile**

In the above startup sequence, the Pre-Stripper will supply vapor to the Demethanizer before any liquid feeds the tower sections 2, 3 and 4 from the Feed Separator Drums. This makes section 4 of the Demethanizer vulnerable to high vapor velocity in the downcomer, which may prevent it from liquid sealing itself. It is therefore critical that the vapor rate from the Pre-Stripper to the Demethanizer remains below the downcomer critical vapor rate until liquids from the Feed Separator Drums seal the downcomers. Although expected to be sealed at that point, caution must be exercise while ramping heat to the Demethanizer reboiler.

As stated earlier, for meeting turndown the Pre-Stripper reboiler duty in this example must be operated above the level predicted for strictly meeting the product stream specification. The increase in reboiler duty is associated with an increase in vapor traffic in section 4 and above on the Demethanizer and provides tray load profile similar to a false loaded Demethanizer. As shown on Figure 20, except for one stage, this moved the operating points within the stable region of the tray operating window. The points outside the window in section 4 are below weeping but believed to be above dumping and right at the minimum liquid limit. Caution should be taken if the minimum liquid limit of the tray is based on downcomer sealing. However, in this case the vapor load is most likely below the downcomer critical vapor rate. The downcomer will likely seal as feed rates increase providing that the L/V ratio is maintained.

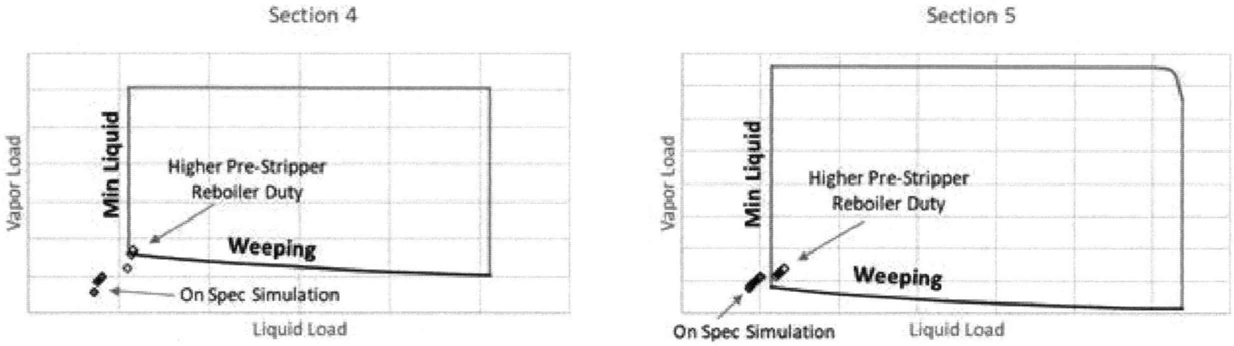


**Figure 20: Demethanizer Section 4 & 5 Tray Operating Window**

### *Deethanizer*

The Case 1 startup strategy would involve feeding liquid hydrocarbon into a nitrogen filled vessel associated with similar problems as previously discussed for the Pre-Stripper evaluation. The startup strategy defined by Case 2 provides a Deethanizer free of nitrogen and near or at normal operating pressure. As such, the initial liquid mixture filling the sump and reboiler should allow for smooth startup. Meeting the tray hydraulics will likely require false loading the tower, which can be accomplish by ramping heat to the reboiler in synchronization with the reflux rate. As for other towers, the extent of reflux can be pre-determined by process simulation to ensure process loads above minimum liquid and the weep point. Typically meeting the tray hydraulics should allow operation above the reboiler turndown point. However, if historically identified as a problem, the reboiler Thermosyphon circuit should be reviewed.

Once meeting the overhead stream specification, the Deethanizer vapor distillate gets routed to the Acetylene Reactors. Forward feed can be transitioned to the Ethylene Fractionator once the Reactors meet their acetylene specification. As previously described, the Ethylene Fractionator total reflux operation allows the C2 stream upstream of the tower to transition from flare to on specification ethylene product. This transition is associated with a significant reduction in flaring and defines a point where feed rates can ramp up to target production without significant flare increases.

## Conclusion

Ethylene Plant startup involves rate transitions on many process equipment. Startup rates can be limited by energy balance or turndown constraints on some equipment. Distillation towers are a component of the startup sequence. Like other pieces of equipment, their functionality is limited by a number of factors. Failure to understand their limitations may prolong startup time and flaring due to unnecessary trouble shooting of event that could have been predicted ahead of time. The intent of this paper was to describe how fundamentals concepts can be used for evaluating the startup sequence of distillation towers. Although developed for ethylene plants, the concepts are believed applicable to other chemical plants.

The methodology discussed relies on thermodynamic consistency and equipment capability. The thermodynamic consistency refers to heat & material balance and process load calculations from process simulation that are independent from the equipment characteristics. There is no point in evaluating the capability of the equipment until after development of thermodynamically consistent heat & material balance and process loads. The process loads are then used for rating the equipment. The deviations or gaps from the equipment capability will drive heat & material balance and process load adjustments. This process may take a few iterations but once completed provides clear process targets.

## Nomenclature

| | |
|---|---|
| A | Tower Cross Section |
| C2R | Ethylene Refrigeration |
| C3R | Propylene Refrigeration |
| CET | Critical Exposure Temperature |
| CGC | Crack Gas Compressor |
| KCS | Killed Carbon Steel |
| MAT | Minimum Allowable Temperature |
| PRC | Propylene Refrigeration Compressor |
| $P_r$ | Prandtl Number |
| Re | Reynolds Number |

| | |
|---|---|
| Txy | Temperature, Liquid/ Vapor mole fraction of the more volatile component in a binary mixture |
| VLE | Vapor – Liquid Equilibrium |
| VLOAD | Vapor load for trays |
| $V_{Rate}$ | Vapor Volumetric Rate |
| $d_i$ | Tube inside diameter |
| $g_n$ | Gravitational acceleration |
| $\dot{m}$ | Mass Flux |
| $u_g$ | Superficial Vapor Velocity |
| $x_g$ | Mass Vapor Fraction |
| $\rho_g$ | Vapor Density |
| $\rho_l$ | Liquid Density |

## References


1. P.B. Whalley, "Vertical Thermosyphon Reboilers. Part 2: Analysis of the System", HTFS DR1 Rev 2, January 1984
2. Aspentech, "What does the flow reversal and flooding criterion mean?", Article ID: 000084016, November 3rd, 2020
3. D. Summers, "Performance Diagrams All Your Tray Hydraulics In One Place", 2004 AIChE Annual Meeting, Austin, TX
4. A. Bernard, "Tray Operating Window – Description and Applications", 2017 AIChE Spring Meeting , San Antonio, TX
5. H. Z. Kister, "When Tower Startup Has Problems", Hydrocarbon Processing, February 1979
6. H. Z. Kister, "Distillation Design", McGraw-Hill.
7. H. Z. Kister, "Distillation Operation", McGraw-Hill.
8. Michael J. Kennedy, Autorefrigeration Near-Miss Incident at Equistar Chemicals, a Lyondell Company, Ethylene Producer Conference, 2005 AIChE Spring Meeting, Atlanta, GA.
9. H.Z. Kister, Distillation Troubleshooting, Wiley-Interscience, 2006
10. A. Bernard, "Evaluating the Potential for Cold Brittle Fracture in Ethylene Plant Through Dynamic Simulation", 2015 AIChE Spring Meeting, Austin, TX
11. A.W. Sloley, "Properly Design Thermosyphon Reboilers", Chemical Engineering Progress, March 1997
12. G.R. Martin, A.W. Sloley, "Part 1 - Effectively design and simulate thermosyphon reboiler system", Hydrocarbon Processing, June 1995
13. G.R. Martin, A.W. Sloley, "Part 2 - Effectively design and simulate thermosyphon reboiler system", Hydrocarbon Processing, July 1995
14. M.D. Hagan, V.N. Kruglov, "Understand Heat Flux Limitations on Reboiler Design", Chemical Engineering Progress (CEP), November 2010
15. O.C. Iloeje, D.N. Plummer, W.M. Rohsenow, "Transition from film boiling to nucleate boiling in forced convectionvertical flow", Report No. DSR 72718-78, Massachusetts Institute of Technology, March 1972

# 618590 Optimal Machine Learning Models for Liquid Aerosolization Safety Contributors

**Abstract:**

To date, many attentions have been paid to the new types of marine fuels to meet the requirement of the long-term sustainable strategy issued by the International Maritime Organization. Speaking to marine fuel selection, the safety factors should always be top prioritized. Flammable and explosive hazards, which have been well studied, are the major concerns of safety aspects of fuel application. However, the liquid aerosolization issue, making bulk liquids more hazardous on combustion and explosion, has not widely recognized in industry or academia.

This study is aimed at identifying the contributors of liquid aerosol formulation and solving their data deficiencies by developing the corresponding quantitative structure–property relationship models. 1215 liquid chemical substances and 14 predictors have been input to train the machine learning models via k-fold cross validation with the consideration of principal component analysis. Three rounds of comparisons were executed to find the final optimal models for liquid dynamic viscosity (LDV), surface tension (ST) and liquid vapor pressure (LVP). The most persuasive model for LDV is obtained by exponential Gaussian process regression (GPR) approach with seven principal components while the Matern 5/2 GPR algorithm is the most robust one for ST and LVP to formulate the optimal models. Since the reasonably good interpretation and prediction performance provided by the developed models, they can be served as effective tools to expand the database for the liquid aerosol formulation properties of organic compounds.

# 619491 Lessons-Learned from Remote HAZOP/LOPA Applications in 2020

**Abstract:**

Going into Year 2020, basic brainstorming team session approaches for the performance of a Hazard and Operability (HAZOP) Study or a Layer of Protection Analysis (LOPA) had not changed significantly over the years; however, the infection prevention measures necessary to address COVID-19 resulted in a rapid transition to widespread application of telecollaboration tools that previously had only been applied sporadically. Prior to 2020, telecollaboration tools for Process Hazard Analysis (PHA) had been used to short-circuit the need for pulling together international teams, for very short PHAs or MOCs, and other select applications. As 2020 drew to a close, numerous applications of HAZOP/LOPA Team Exercises using telecollaboration tools resulted in valuable insights into tools and techniques that were productive for different PHA applications. Properly applied, applications ranging from shorter Process Hazard Analyses (PHAs) and Management-of-Change (MOC) evaluations to longer studies addressing Five-Year PHA Updates and Capital Projects can be implemented productively using telecollaboration approaches.

Post COVID-19, many PHA applications will still be best addressed with the HAZOP/LOPA Team being co-located, but the savings in travel costs, the logistics associated with personnel travel, and the ability to quickly convene and implement a HAZOP/LOPA review can allow remote HAZOP/LOPA, through the use of contemporary technologies, to be an attractive option that can increase overall efficiency, without compromising quality. Experiences in the successful application telecollaboration tools in 2020 mean that Remote PHAs are here to stay.

A core objective of this paper is to provide insights into good applications and tips that can directly be used by the HAZOP/LOPA Facilitator to optimize the logistics to get the most out of a HAZOP/LOPA, where some or all team members are geographically distant.

SPRING21
+17TH GCPS
A Joint AIChE and CCPS Meeting

# Research on Human Resource Development Exercises for Resilience to Cyber Incidents in Critical Infrastructure

**Yuitaka Ota**
**Nagoya Institute of Technology,**
**System Management and Engineering**
**Showaku, Gokiso, 4668555 Aichi, Japan**
**cjr17009@nitech.jp**

**Erika Mizuno, Asai Haruna, Tomomi Aoyama, Yoshihiro Hashimoto, Ichiro Koshijima**
**Nagoya Institute of Technology,**
**System Management and Engineering**
**Showaku, Gokiso, 4668555 Aichi, Japan**
**ckv17032@nitech.jp, h.asai.852@nitech.jp, aoyama.tomomi@nitech.ac.jp,**
**hashimoto@nitech.ac.jp, koshijima.ichiro@nitech.ac.jp**



Prepared for Presentation at
American Institute of Chemical Engineers
2020 Spring Meeting and 17th Global Congress on Process Safety
Virtual
April 18 - 22, 2021

# Research on Human Resource Development Exercises for Resilience to Cyber Incidents in Critical Infrastructure

**Yuitaka Ota**
**Nagoya Institute of Technology,**
**System Management and Engineering**
**Showaku, Gokiso, 4668555 Aichi, Japan**
**cjr17009@nitech.jp**

**Erika Mizuno, Haruna Asai, Tomomi Aoyama, Yoshihiro Hashimoto, Ichiro Koshijima**
**Nagoya Institute of Technology,**

**Keywords:** OT-Security, cyber incident exercise, business continuity management

## Abstract

Current cyber-attacks target IT systems and OT (Operational Technology) systems that are indispensable for the automatic operation of critical infrastructure industries. There have been various reports of cyber incidents targeting the OT system. In 2019, Norsk Hydor was infected with particular ransomware worldwide through its corporate network, forcing the manufacturing site to operate manually. In May 2020, Taiwan's CPU, responsible for Taiwan's oil supply, was targeted by a cyber-attack.

COTS (Commercial Off The Shelf, such as Windows OS, Intel PC, and open source applications) devices have been proposed to reduce the deploying OT system budget. Also, DX efforts are being made to shift from physical operation to virtual operation by using virtualization with IoT, AI, and Cloud. Due to this technological trend, the expanded attack surface is forced to manage the OT system as a "Zero Trust System" that assumes cyber-attacks. In IT systems, countermeasures such as whitelisting and firewalls prevent attackers from outside the system. For OT system, Network Security Zoning described in IEC62443 is recommended to mitigate the effects of cyber-attacks.

However, attackers exist all over the world, and they are continually developing new attack methods. For this reason, the authors believe that it is impossible to prevent all attacks, no matter what measures the defenders take. Therefore, when responding to cyber-attacks against the OT system, it is essential to consider countermeasures by focusing on plant accidents caused by control system failures, most of which are assumed as safety measures.

To ensure the plant's safety and continue the business operation even if it has a cyber-attack on the OT system, it is indispensable to develop human resources to respond to the incident flexibly. For this reason, we had been developing cybersecurity exercises to acquire the meta-knowledge necessary to respond to cyber-attacks targeting the OT system.

In this paper, the authors present the design framework, the implementation method, and OT expert's evaluation results of a proposed exercise.

# 1 Introduction

## *1.1 Background*

An industrial control system (ICS) is a device or group of devices for managing and controlling other devices and systems. ICS is widely used in the Critical Infrastructure (such as electric power, city gas, and transportation) and the manufacturing industry that supports our daily lives, from those that are familiar to us. The conventional ICS has an isolated network as an independent system built with its hardware, software, communication protocol and has been considered resilient to cyber-attacks. However, in recent ICS, the adoption of COTS (Commercial Off The Shelf, such as Windows OS, Intel PC, and open source applications) devices is progressing due to versatility and budget. Also, DX (Digital Transformation) is being promoted for the purpose of plant operation support and operation optimization using IoT (Internet of Things) and AI (Artificial Intelligence) through open networks with the outside. As a result, ICS is exposed to the threat of cyber-attacks, and companies using ICS need to recognize not only the risk that has been assumed as the risk of plant shutdown but also cyber-attacks as one of the risks.

Cyber-attacks targeting ICS have been increasing in recent years. In 2010, malware called Stuxnet attacked Iran's uranium enrichment centrifuge [1]. Hackers caused power outages in Ukraine in 2015[2] and 2016[3]. In 2019, Norsk Hydros in Norway, the world's leading manufacturer of rolled aluminum products, was affected by cyber-attacks in plant operations [4]. In addition, the Information Security Agency (BSI) of the German Federal Government has published "Industrial Control System Security: Top 10 Threats and Countermeasures" [5].

In this way, cyber-attacks are a risk that threatens the safety of the plant, and companies must respond as a risk that threatens their business.

## *1.2 Response to a cyber incident*

### 1.2.1 Cyber-attacks targeting Industrial Control System

Fig. 1 shows a cybersecurity testbed and a virtual ICS network. The testbed (Fig. 1) heats water with a heater in the lower tank and circulates the heated water through a simple pipeline to the upper tank (using a pump). The testbed is also equipped with actual industrial control devices and is controlled automatically[6]. In this ICS network, OLE for Process Control (OPC) Servers collect and exchange process data and monitor them by using Supervisory Control And Data Acquisition (SCADA) function included in the OPC Servers. OPC Servers receive various parameters from the testbed's sensor, and SCADA gives instructions to the control equipment using the parameters' data.

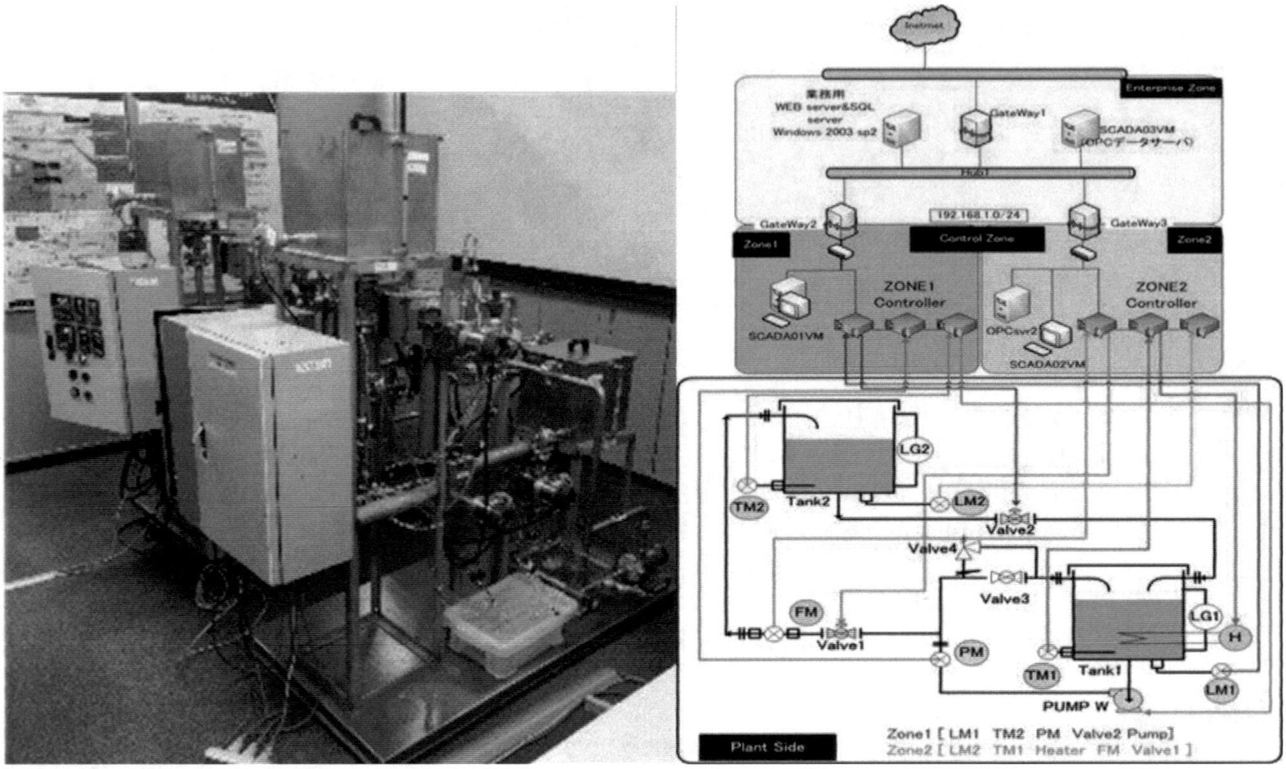


**Figure 1. Cyber Security Testbed and ICS network**

The authors have demonstrated cyber-attacks against the testbed, according to the steps of the anatomy of a cyber-attack. The attacking procedure is determined as follows.

1) Attack Step 1: Reconnoitering and enumeration

The attacker executes the Armitage to launch the Metasploit framework. Then, the attacker executes the Nmap, which scans a target network (ICS-2 Network 192.168.12.0/24). After completing the scan, the attacker gets target information.

2) Attack Step 2: Intrusion and advanced attacks

According to the functions of the Armitage, the attacker ascertains an attack method and vulnerability in the Metasploit framework. The attacker exploits payloads, which execute inappropriate code in the use of the vulnerability. After sending the payloads, the attacker obtains the administrator's permission for the OPC2 illegally.

3) Attack Step 3: Malware insertion

The attacker uploads a malware execution file to an OPC2 background process to execute a malware file. The configuration file of OPC2 was rewritten, and then the OPC2 was restarted. They were resulting from the attack, the valve controlled by the OPC2 shutdown.

4) Attack Step 4: Clean-up

The attacker deletes the malware file and the configuration file from OPC2 storage to not leave the cyber-attacks evidence and finish the cyber-attack.

As the valve is opened, the stage of the testbed changes. Usually, the testbed's state transition appears on the monitor of the operator (Fig. 2). The changing graph shows the liquid level of the upper tank. The liquid level is changing due to the malfunction of the valve. The operator can notice the testbed abnormality due to the unusual process deviations from the steady-state condition through the monitor. However, the attacker conceals the information about cyber-attack in order not to let anyone notice the cyber-attack. The testbed monitoring screen is concealed, as shown in Fig. 3, while the cyber-attack causes severe accidents at the testbed.

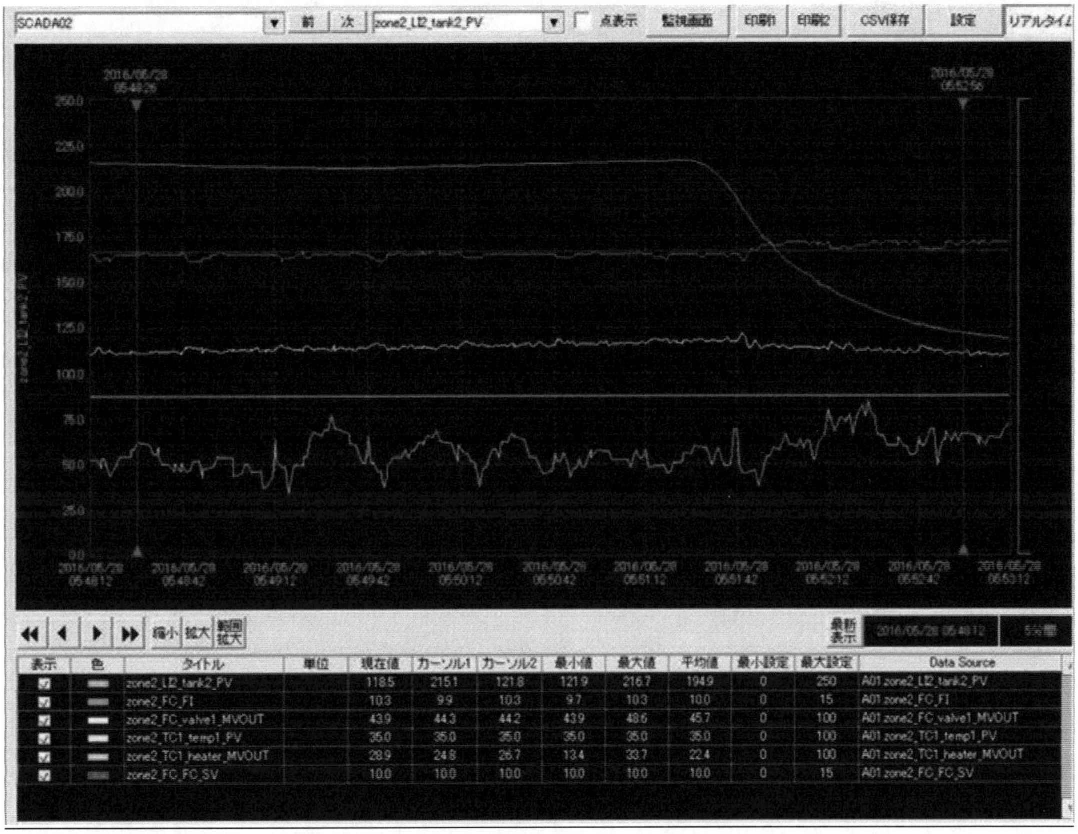


**Figure 2. HMI Screen Shot in Operation (Safety Incident)**

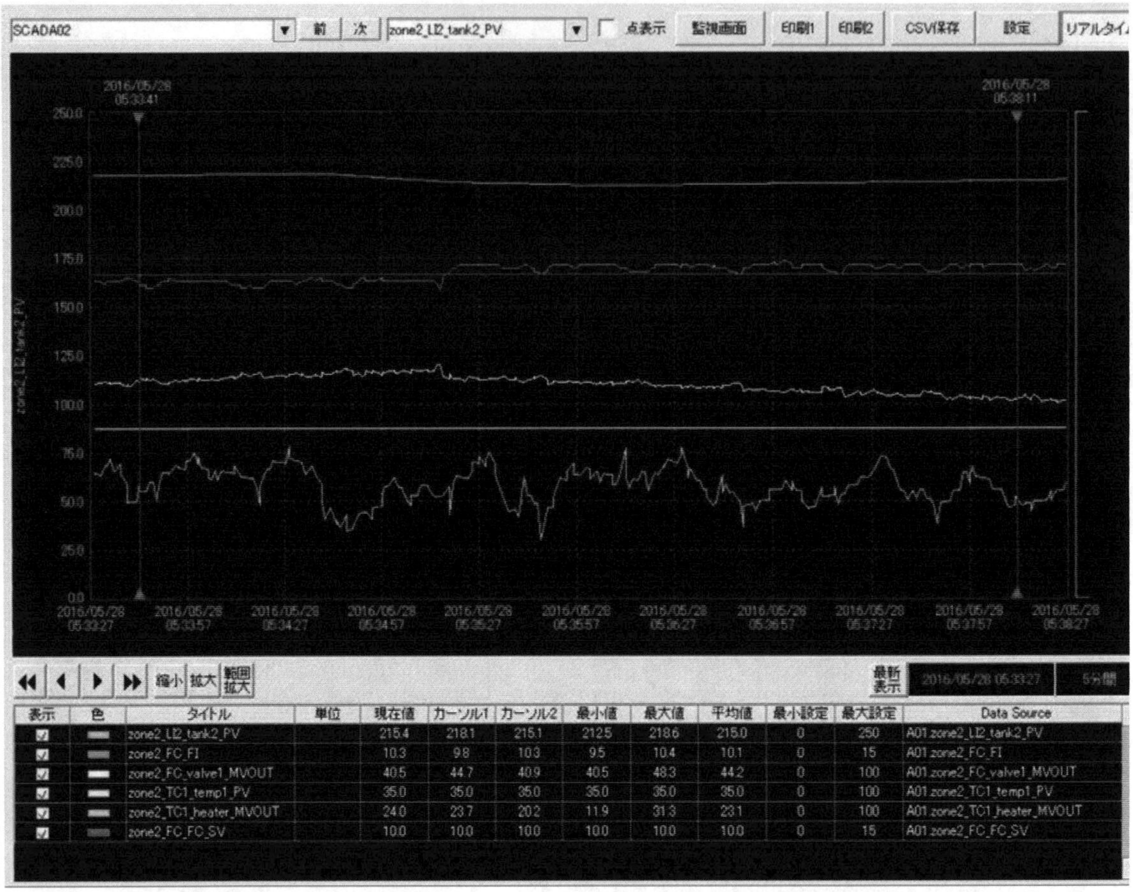


**Figure 3. HMI Screen Shot in Operation (concealed by the attacker)**

This demonstration shows a possibility of safety incidents caused by cyber-attacks to ICS.

### 1.2.2 Response to Cyber Incidents

The consequences of a cyber-attack depend on the characteristics of the plant. For example, an attacker hijacks a data server (DCS, OLE for Process Control server, etc.) that gives parameters and commands to control devices such as a PLC (Programmable Logic Controller) and gives the control devices malicious instructions. Such an attack can lead to abnormal operation and an incident, similar to a data server or control equipment failure or operator error with traditional equipment. That is, incidents caused by cyber-attacks in ICS systems are incidents that have already been considered in HAZOP scenarios with different initiating events. Operating companies have taken countermeasures to ensure the safety of the plant for many years. Companies are preparing countermeasures to ensure safety (Safety Response) under an incident at their plant. Human resource development is also carried out based on the countermeasures to respond when an incident occurs. In other words, even if an abnormality occurs due to a cyber-attack, it is possible to ensure the safety of the plant if the operation staff takes a proper Safety Response. However, problems peculiar to cyber-attacks, such as rewriting server configuration files and rewriting control device programs, may hinder the implemented safety-based Safety Response. Figure 4 shows the incident response structure that takes into account the cyber-attacks proposed here. If the plant is operating abnormally caused by the cause considered so far, it is possible to restore plant safety by performing a safety response, as shown in the left figure. However, there is a possibility that cyber-attacks will interfere with safety response (e.g., concealment of a plant's information on the monitor). Companies must consider where cyber-attacks interfere with safety response. Moreover, it is necessary to take the responsibility to eliminate the effect of cyber-attacks (security response) to take safety response as shown in the right figure. Therefore, if a cyber-attack causes a plant's abnormal operation, a response to rectify the situation is required in combination with a structure to add a security response to the safety response.

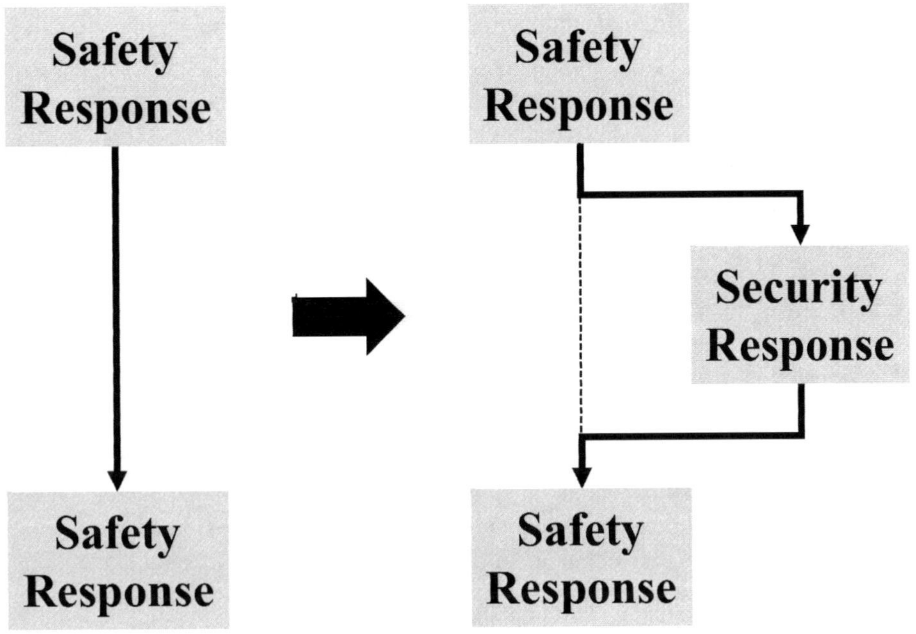


**Figure 4. Incident Response Structure[7]**

Responding to cyber-attacks targeting ICS systems, it is necessary to integrate a Safety Response to ensure the plant's safety and a Security Response to eliminate the effects of cyber-attacks. It is essential to create a plan that considers security incidents in addition to the existing safety response plan, and make it known to each operations staff.

### *1.3 Required organizational cooperation*

The right staff needs to take action at the right time. To respond to incidents competently, it is necessary to properly share the current situation and communicate it among the right staff.

When an incident occurs due to cyber-attacks at a plant, two responses, a Safety Response to ensure the plant's safety and a Security Response to remove the cyber-attack effects, proceed simultaneously. Since the Safety Response responds to ensure the plant's safety, the response's responsibility is on the OT side (operators working at the plant). Since the Security Response is a response to remove the effects of cyber-attacks, the response's responsibility is on the IT side (IT engineer). Also, considering a cyber-attack on ICS, the result of the cyber-attack appears in the plant, and the cause is the IT network. In other words, it is a structure in which the cause-effects occur in different places. Also, in the incidents that we have experienced so far, there was no problem in dealing with the cause, but in cyber incidents, the same incident may occur in all parts connected to the Internet. Therefore, information sharing is vital for appropriately responding to the above-mentioned new type of incident, "Simultaneous Multiple Accidents."

## 2 How to develop Security Personnel

### *2.1 Preparation required as a company*

・Formulation of ICS-BCP

Many companies have created a business continuity plan (BCP) in advance so that their business can continue even during natural disasters such as earthquakes and typhoons. It is important to develop a similar program for cyber incidents. It is necessary to analyze the risks in the event of a cyber incident and formulate an ICS-BCP that includes countermeasures for reducing, holding, avoiding, and mitigating the effects of the risks.

・Human Resource development

Even in unprecedented situations such as cyber incidents, companies must respond to continue their business. And it is the employees who respond to the incident. If the personnel's ability to respond to the incident is not sufficient, the company's damage will increase.
In other words, as a company, it is required to develop human resources who will respond.

### *2.2 Importance of virtual experience*

As mentioned in 2.1, companies must prepare a BCP in advance for cyber incidents and develop human resources to carry it out. However, many companies have no cyber-attacks experience with ICS, and it is not easy to imagine how cyber-attacks may affect companies.

A mechanism is needed to experience cyber incidents and discuss necessary countermeasures.

As an approach to that mechanism, the authors chose an exercise. Figure 5 shows the flow for developing a training exercise. By analyzing incidents in the real world, we can extract the causes of incidents and the commonality of cyber-attacks targeting ICS. By giving the extracted information rules and game characteristics, the exercise participants will learn systematically. Thus, participants experience cyber-attacks on ICS through exercises.

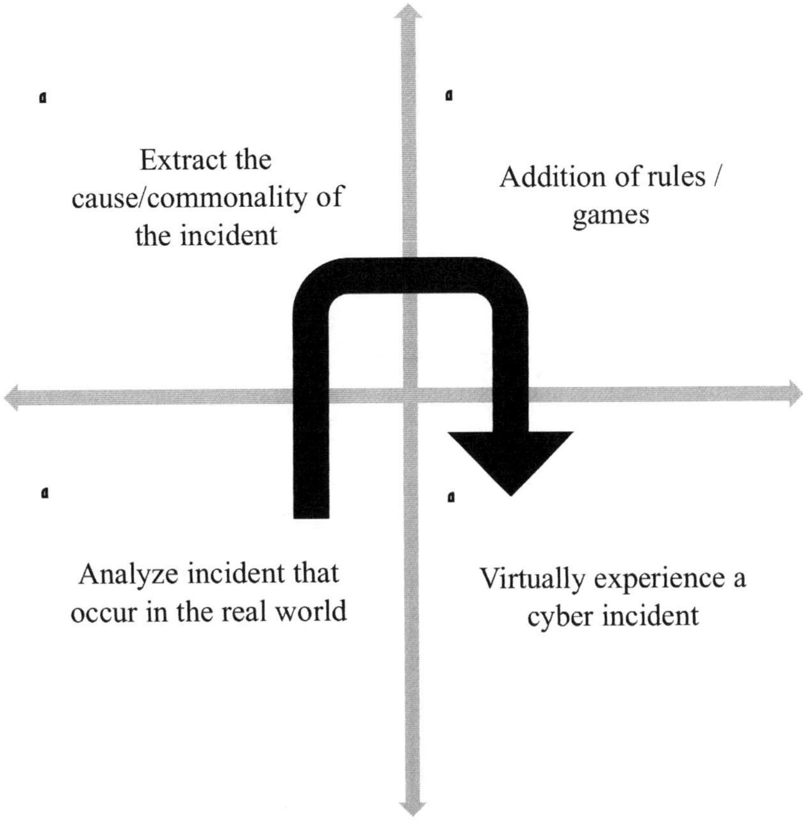


**Figure 5. Exercise development process**

Even staff who have never experienced a cyber incident can increase their experience with cyber incidents through exercises. By expanding this experience value, staff will more likely think and act even if people have a similar incident or an incident that people have never experienced before.

### *2.2 Human resource development method using virtual experience (exercise)*

By taking the following four steps, we aim to improve the incident response capability of the company.

1. Information input
   Many companies have no experience in responding to cyber incidents, so it is necessary to understand the impact of cyber incidents on companies and plants. It is also required for the parties concerned to understand that cyber-attacks on plants are problems on the IT side and OT side.

2. Create ICS-BCP from a management perspective
   Companies are required to generate stable profits. When an incident occurs, companies need to think about what to prioritize to keep the business running, when to recover, and alternatives if there is no recovery. Each company must create an ICS-BCP assuming a cyber incident and communicate it to employees.

3. ICS-BCP with added on-site perspective
   It is necessary to respond to incidents based on the ICS-BCP defined as a company created in step 2. However, this BCP only includes the perspective of how to respond as an organization.

   By checking the BCP from the viewpoint of the operational site, it can be made into an ICS-BCP that can be operated.

4. Share the idea of all layers
   Employees must understand ICS-BCP to respond even if a never-experienced-before incident occurs. Therefore, by conducting training centered on the planned ICS-BCP, we will deepen the understanding of employees and objectively evaluate the ICS-BCP.
   By repeating this training multiple times, it is possible to confirm omissions and deficiencies in BCP and increase ICS-BCP accuracy.

The activities mentioned above don't do only once. Each company can improve its resilience ability by making assumptions based on various preconditions (scenarios), rotating the PDCA cycle, and continuing discussions.

# 3 Exercises for improving resilience developed

Based on the recognition that not only hard skills but also soft skills are important for responding to cyber incidents, we have developed discussion-based ICS security exercises.

## *3.1 Theme of the exercise*

As mentioned above, it is very important to prepare a response plan to respond to an incident with hard and soft skills. Therefore, the purpose of the exercise is to understand the required actions to be taken by cyber-attacks and the importance of inter-organizational collaboration needed to take those actions.
The final goal is to be able to create an ICS-BCP using the knowledge gained from the exercises.

In order to make participants understand the purpose of the exercise, the exercise was divided into three phases according to the purpose of correspondence. (Fig.6)
When an incident occurs in a plant, whether it is a cyber incident or not, it is necessary to take measures to ensure the safety of the plant. Then, it is necessary to investigate the cause after ensuring the safety of the plant. If the cause is a cyber-attack, it is necessary to investigate the extent to which the impact of the cyber-attack has spread.
Initially, the safety workload is large, but the security workload gradually increases.
Then, there will be a response to confirm the extent of the impact of cyber-attacks not only on plant safety but also on business continuity.

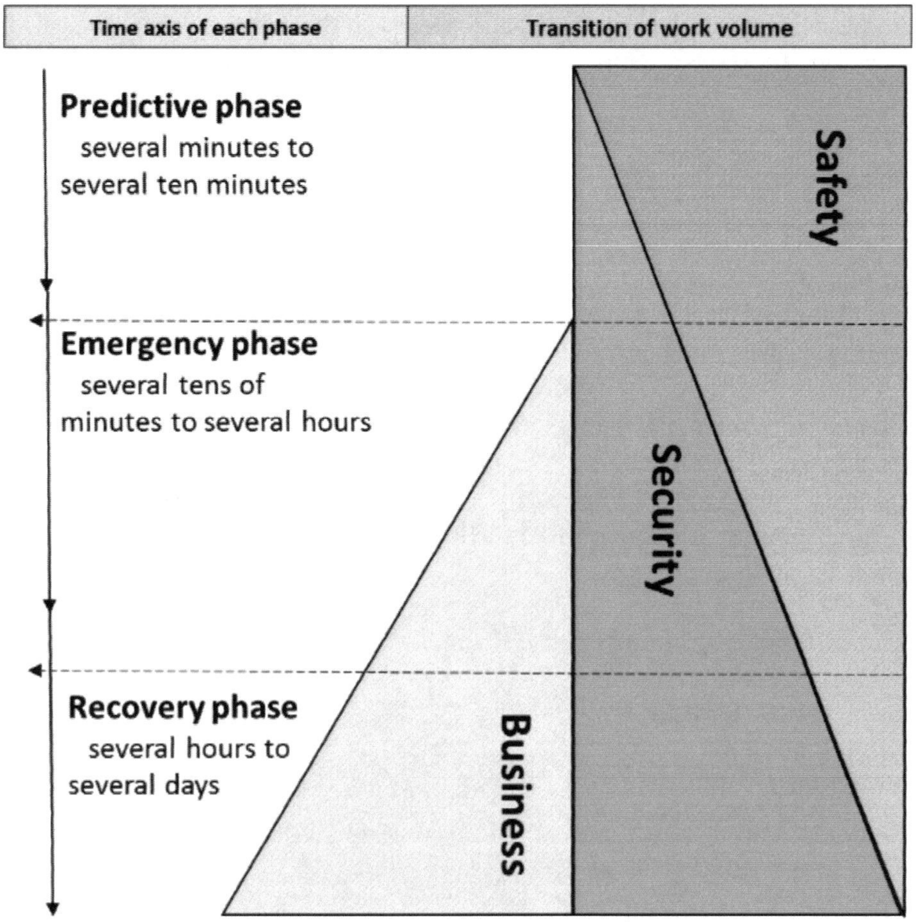


**Figure 6. Structure of Exercise with Workload Image among Safety, Security, and Business Continuity Activities[7]**

The authors visualized the state that this working volume change with the progress of time. (Fig6) Fig.6 expresses the work volume at a certain spot. The work volume is divided into safety, security, and business. The participant understands a change of this working volume through exercise.

1) Predictive Phase
The purpose of this phase is to ensure the safety of plants threatened by cyber-attacks. In this phase, the participants examine the safety response obstructed by a cyber-attack.
After detecting suspicious communication in a section that monitors the network, the exercise begins with an alert for the status of the liquid in the plant. In this phase, the participants must ensure the safety of the plant. Once safety is ensured, participants must investigate the abnormal operation. In this phase, participants can recognize how different causes can influence the safety response.

2) Emergency Phase
The purpose of this phase is to continue the business. The participants consider the security of the plant and how the exercise implicates a real cyber-attack. They must also take into account

how their business will be affected by the cyber-attack. In this phase, the participants consider practical measures taken to ensure business continuity throughout the company.
This phase begins after the abnormal operation of the plant reveals a cyber-attack. In this phase, participants can recognize that the entire company must be mobilized to treat a situation caused by a cyber-attack.

3) Recovery phase
The purpose of this phase is to restore own business to normal operation rapidly. The participants must find a way to nullify the influence of the cyber-attack so that the plant can return to normal operation. In addition, they must consider how business is affected by the temporary plant shut down. In this phase, the participants examine the start-up procedure of a plant crippled by a cyber-attack. In addition, the participants examine methods to prevent the recurrence of similar cyber-attacks.
This phase begins after the plant is stopped by a cyber-attack. In this phase, participants think about how to restart the plant. In this phase, participants can see how different causes affect the restoration of normal plant operations.

The following three types of training exercises have been developed.

- Card-Type Incident Response Exercise
- BCP Creation Exercise
- Simulation Exercise called "IMANE" (Incident MANagement Exercise)

### *3.2 Core scenarios for exercise*

The developed exercises are based on the same scenario with different methods.
The base scenario is required to be able to consider the ideal organizational cooperation and necessary soft/hard skills for the organization to respond quickly in the event of a cyber-attack through exercises. The created scenario was composed of three scenarios to think about the three aspects of safety, security, and business.

**Virtual Company:**
Participants from various companies, departments have different backgrounds. A common scenario is indispensable for the participants to share the exercise together. In addition, using a simple and realistic virtual company makes it easier to apply the awareness of the exercise to the business. Figure 7 is an organizational chart of the created virtual company. Practical prerequisites, such as the company's business profile and its organization's intercommunication rules, have been set to achieve the exercise's objectives. Participants can discuss how information should be shared and how an organization should respond to cyber-attacks. Figure 8 is the network structure of a virtual enterprise. The attacker will attack using the computer that was hijacked by the cyber-attack as a stepping stone. The attacker steps on the hijacked computer and moves to the computer he wants to attack. In other words, the effects of cyber-attacks are spreading not only to the attack target but also to the attack process. For mitigating the impacts of cyber-attacks, it is necessary to consider countermeasures within the range of the effects of cyber-attacks. By conducting exercises based on this network structure, participants can discuss the Security Response required to respond to cyber incidents.

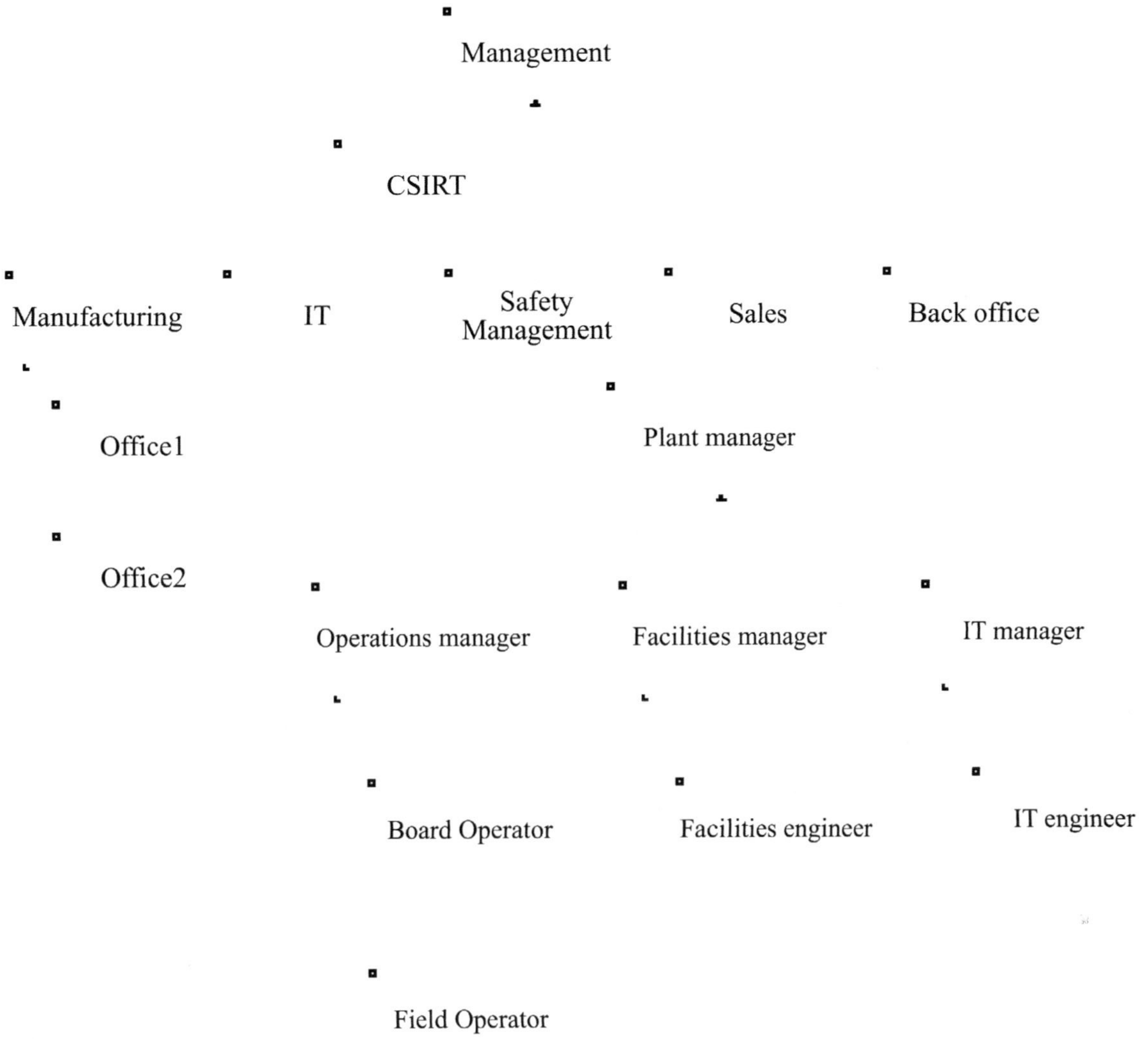


**Figure 7. The organization chart of a virtual company**

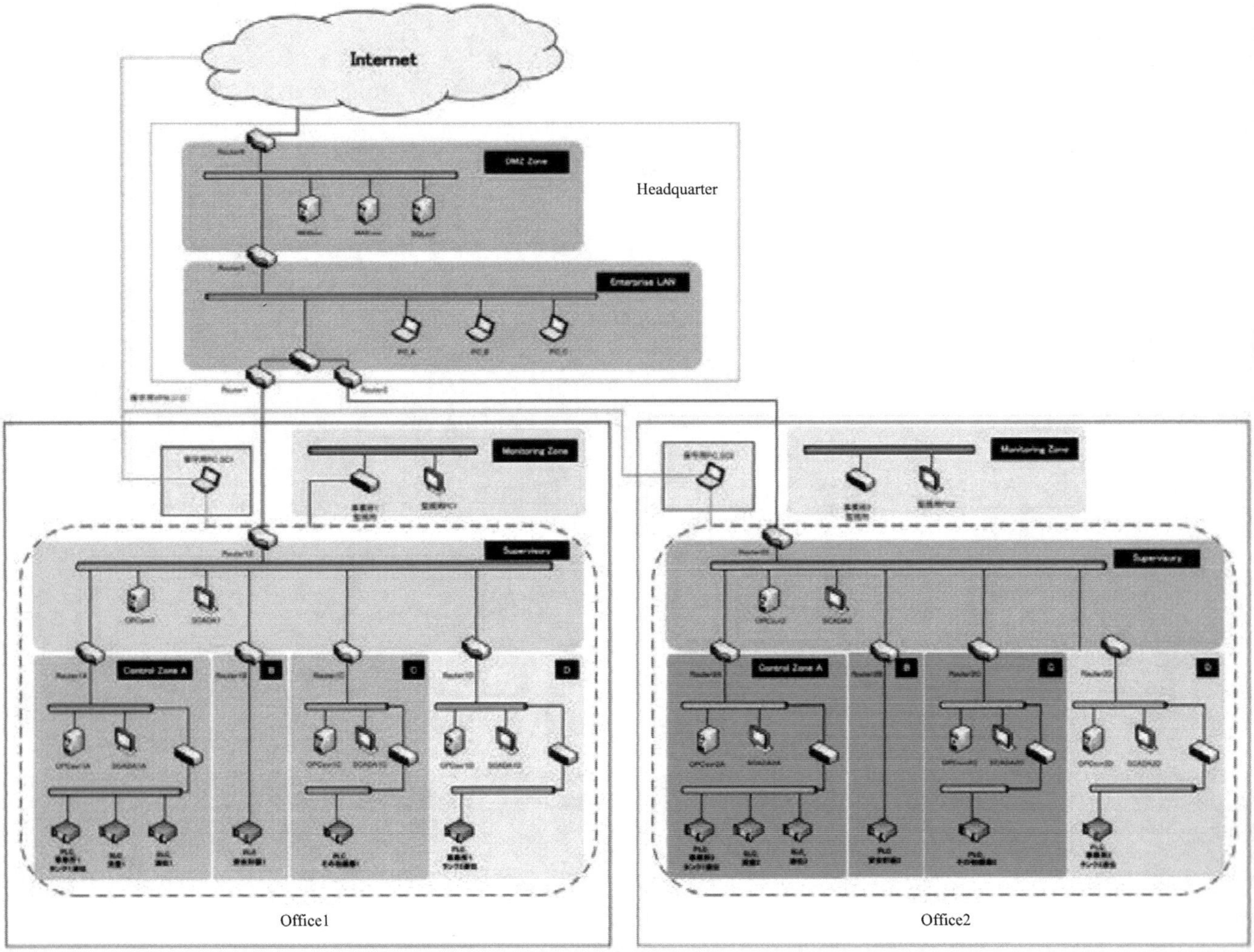


**Figure 8. The network structure of the virtual company**

**Plant Operation Status Scenario:**
A real-world plant consists of many devices, pipes, valves, and so on. If the plant is used as the subject of the exercise, the discussion may be complicated, and the purpose of the exercise may not be achieved. The plant used as the subject of the exercise should have a simple structure. By using a plant with a simple structure as the subject, it is possible to facilitate discussions focusing on the cyber incident that is the purpose of the exercise. In addition, it becomes easier to obtain meta-knowledge about cyber incidents, and it becomes easier to apply the knowledge gained through exercises to actual plants.

This scenario is based on our cybersecurity testbed. And we have introduced the control equipment used in the actual plant. This testbed aims to test and develop ICS security solutions. Based on the cyber-attack demonstration conducted as part of the research, the plant operation information caused by the cyber-attack is used for the exercise.

**Cyber-attack Scenario:**
The flow of cyber-attacks on plants is almost the same as cyber-attacks on IT systems.
To imagine how to respond to a cyber incident, it is important to deepen your cyber-attacks knowledge. There are many cyber-attacks methods, but attackers usually attack based on the cyber kill chain [8]. The cyber kill chain [9] classifies cyber-attacks into seven phases (Reconnaissance, Weaponization, Delivery, Exploitation, Installation, Command & Control, Actions on objectives). The idea is that appropriate measures can be taken in each phase, and cyber-attacks can be prevented if the attack does not proceed to the next phase.
This scenario is based on a cyber-attack demo conducted in our laboratory. It also includes the idea of a cyber kill chain. Throughout the exercises, participants can discuss the measures to protect the company by understanding cyber-attacks workflow and intention.

### *3.1 Card-Type Incident Response Exercise*

The first exercise is Card-Type Incident Response Exercise. This exercise aims to gain an awareness of what a cyberattack on a plant is and what actions will be required.
This exercise uses a prepared action card to create a workflow for responding to cyber-attacks. Participants will achieve the purpose of the exercise by considering the response to cyber incidents while understanding the given action card's contents.

#### 3.3.1 Exercise steps

These exercises consist of five steps: introduction, explanation of prerequisites, group work, presentations, and open discussions. Participants guess the company's situation from the given preconditions and create a workflow while thinking about the actions and information sharing necessary to make the plant safe and carry out business continuity.

**Introduction:**
The facilitator explains the purpose of the exercise and how to proceed. The general scenarios of the exercise are introduced to the participants. At this step, participants understand the importance of the exercise and, therefore, be motivated to apply themselves fully. By doing so, participants can increase the benefit of the exercise.

**Description of Prerequisites:**
Participants perform group work as a virtual company that owns a plant and provides products. The facilitator explains the preconditions about the virtual company at the beginning of the exercise. Prerequisites include the plant's status, the state of the IT network, and the company situation. Participants watch a video to understand the preconditions (In this phase, we use the videos we create to help participants understand the prerequisites.).

Participants envision the outcome based on the prerequisites.

**Group Work:**
The group task is to create a workflow by envisioning the outcome from the given information and considering the possible situation's response.
Originally, this exercise was conducted in group work in person. With COVID-19, we improved this exercise so that we could perform it using web conferencing.

In group work, the Excel file shown in Fig. 9 is used. Worksheets and action cards are pre-stored in this Excel file.

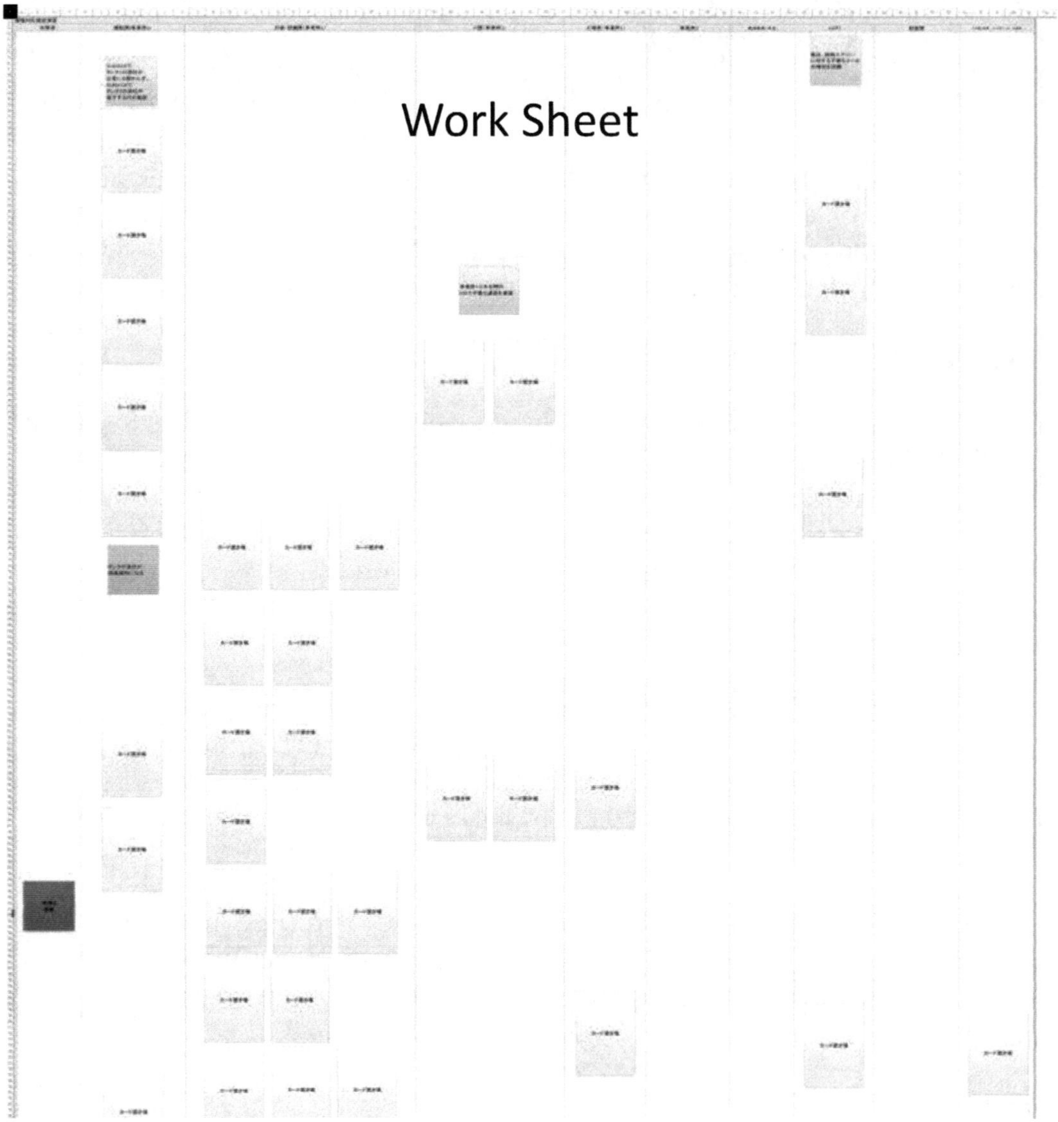


**Figure 9. Group worksheet (Excel Screen)**

The worksheet contains columns with the virtual company's departments, prerequisites, and action card storage roles. Action cards are classified into two types: normal cards and special cards (the upper part of Fig. 10 is the normal card and the lower part is the special card).

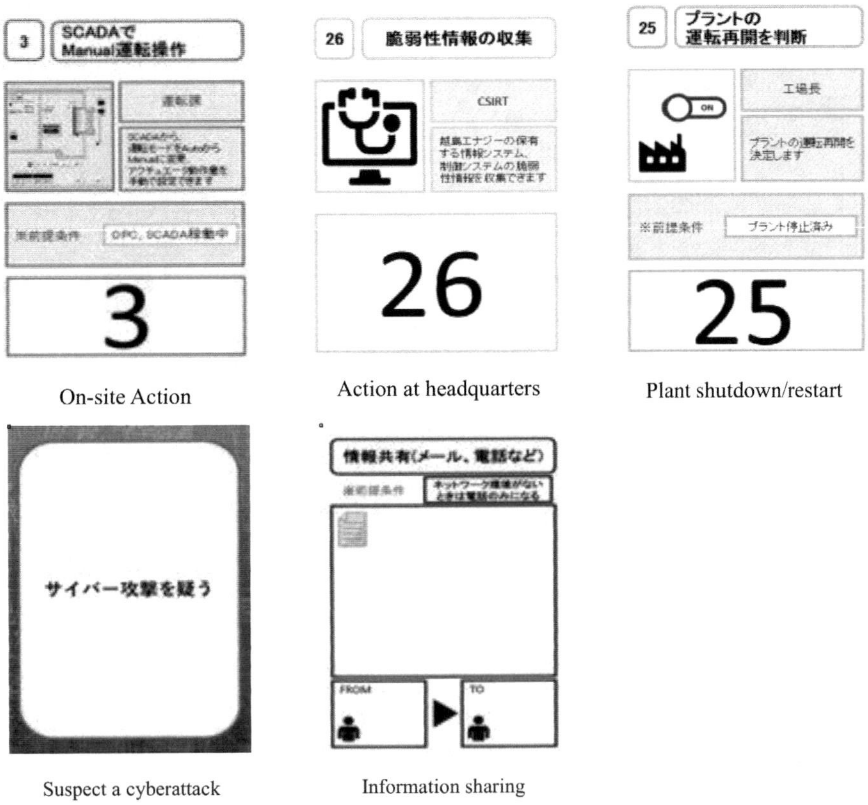


**Figure 10. Type of action cards**

The normal card is an incident response card, and there are a total of 38 cards of three types:

- "action at the site"
- "action at the head office"
- "shutdown/restart of the plant"

On the normal card, the result after taking action is written on the backside, and the participant thinks about the next action based on that information.

There are two types of special cards: "suspect/validation of cyber-attack" and "information sharing." The information-sharing card is used to share the information obtained by each department with other departments. In this exercise, participants discuss and strategize countermeasures from a bird's-eye-view, that is, without playing a specific role. However, in reality, information cannot be shared with others unless it is shared by the person who has the information. Participants use the information-sharing card to share the information held by each department. This makes it possible to express the flow of information sharing on the workflow. An important action in this exercise is to suspect/validate a cyberattack. No security response will be given unless the cause of the plant malfunction is suspected to be a cyber incident. There is information that can be a trigger to suspect that it is a cyber incident, and someone must judge that it is a cyber incident based on that information. By describing it on the workflow, participants identify the departments that will be the key persons in the cyber incident response. Participants will use these action cards to create a workflow, as shown in Figure 11.

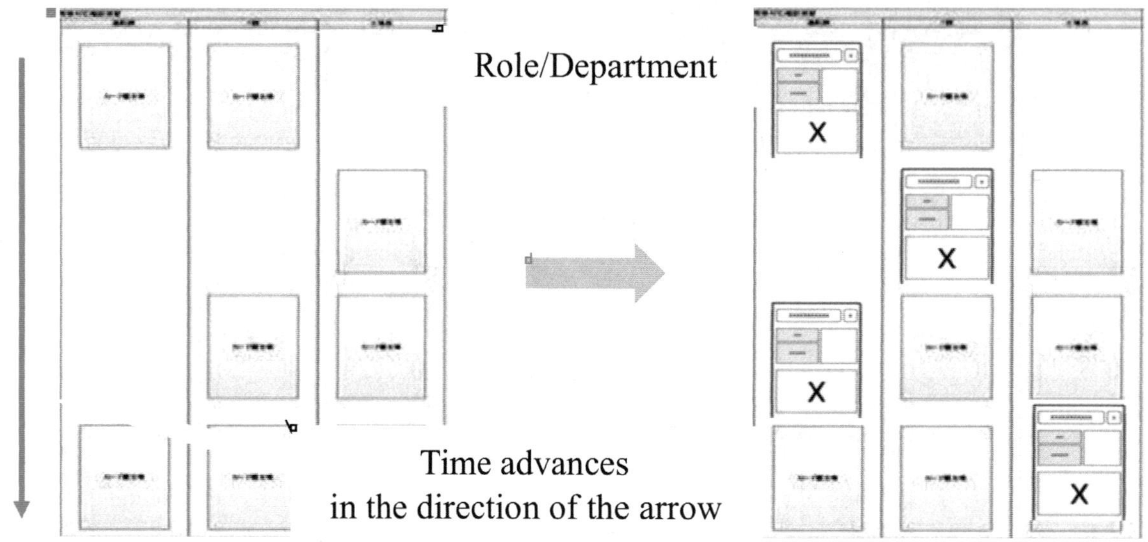


Work-flow(before group-work) Work-flow(after group-work)

**Figure 11. Workflow creation (image)**

**Presentation:**
For groups with a lot of work experience in plants, create workflows that focus on plant safety. For groups with a lot of work experience in the IT department, create workflows that mainly support IT systems. Workflow, which is the product of group work, depends on group members' background and experience. By sharing the workflow and the workflow creation process, the exercise participants confirm the similarities and differences between the ideas of other groups and their own group's ideas. This step can enhance the effectiveness of the exercise.

When each group makes a presentation, ask them to focus on the following three points.

- How did you think about "who (which department), when, why" and placed the action card?

- What did you imagine when you created the workflow from the information given?

- How did you feel through the exercises what kind of human resources, information sharing, and organization-related systems should respond to cyber incidents?

Participants can understand the ideas of other groups and gain new knowledge by listening to other groups' presentations and asking and answering questions.

**Open-Discussion:**
After presenting the group's plans, the facilitator will show an example of the exerciser's workflow. This workflow presented is the optimal solution that the exercise developer thinks. The facilitator will give the same presentation as the participants in the exercise. After the

facilitator's presentation is over, the ideal inter-organizational collaboration for cyber incidents, information sharing, and necessary human resources will be discussed.

Through the exercises, exercise participants will experience the response to cyber incidents and understand the impact of cyber-attacks on ICS. And it is possible to gain insights on ideal organizational cooperation and necessary soft skills and hard skills so that the organization can respond quickly in the event of a cyber-attack.

### *3.4 ICS-BCP Creation Exercise*

The second exercise is a BCP creation exercise. Participants create an incident response exercise based on a given scenario (such as a virtual business scenario). Participants will consider how to attack the plant from the attacker's point of view, consider actions to protect the company from cyber-attacks, and create an ICS-BCP. The purpose of the exercise is to understand the perspective of things needed to create an ICS-BCP.

#### 3.4.1 Exercise steps

This exercise consists of five steps: scenario explanation, group work, demonstration, exercise review, and open discussion. Participants in the exercise identify the risks that the company faces from the given prerequisites. Next, consider an example of a cyber-attack that may cause that risk. Then, think about how to respond to the incident caused by the cyber-attack and create an ideal ICS-BCP.

**Description of the Scenario:**
Virtual enterprise scenarios, plant scenarios, and cyberattack scenarios are required to create incident response exercises. In this exercise, we want participants to discuss how to create an ICS-BCP rather than how to create a scenario. Therefore, the scenario is provided by the facilitator in advance. In addition, since there id an infinite number of attack scenarios, some scenarios are presented, and participants select from those scenarios. In this step, those scenarios will be explained.

**Group-Work:**
This exercise also creates deliverables through group work.
The deliverables to be created are the worksheets and action cards required for the card-type incident response exercise.

- First, participants analyze the risks of a virtual company from a given virtual company and plant scenario.
- Next, participants create a cyber-attack scenario that causes the analyzed risk. Participants create a scenario of the impact on the company and the plant's operating status due to the created cyber-attack scenario.
- Finally, participants create an ideal workflow for the company to perform incident response exercises and create actions that use that workflow as action cards.

**Figure 12. An example of Team Discussion in Group Work**

**Demonstration:**
In this step, participants perform their own exercises on other groups. Exercises will be conducted for different groups using workflows and action cards created in group work. Through this step, participants confirmed that they could convey the message through the exercise (ideal response / inter-organizational collaboration) to other groups. Participants can also gain more cyber-incident experience by taking exercises in various scenarios.

**Group-work (Review of the exercise):**
There are many things that can be gained by doing exercises.
Participants review the exercises they have created based on the results of the exercises conducted by other groups. The group once again discusses whether the ideal workflow that they have considered is the optimal solution, what factors hinder the ideal workflow that they have created, and so on.

**Open-Discussion:**
Each group will share what they have reviewed their exercises and what they have learned through the exercises in this step. By listening to various people's opinions through this step, it is possible to discuss how to view the things necessary for creating an ICS-BCP.

### *3.5 IMANE (Simulation exercise)*

This exercise aims to confirm the effectiveness of the created ICS-BCP by simulating the incident response. This exercise is a simulation exercise that uses a personal computer to create a game based on a scenario, input chat, and respond to an incident by selecting a response.

#### 3.5.1 Exercise steps

This exercise consists of five steps: identifying stress factors such as no one to deal with or no tools to deal with, adding storylines, simulating incident response, reviewing the exercise, and open discussion. Participants will play the role given and aim for business continuity while responding to incidents as virtual company members.

**Identifying Stress Factors:**
The ICS-BCP created by ICS-BCP creation exercise is an ideal workflow. In other words, the plan is based on the premise that all the assumed actions can be performed. However, in reality, it may not be possible to perform actions that should be possible due to various factors, such as lack of resources such as people and tools required to perform actions. Participants will identify stress factors that can hinder their actions in operating the ICS-BCP they have created. Stress factors are the risks involved in conducting ICS-BCP, and it is important to have advanced preparations and alternatives to eliminate the stress factors.

**Adding storylines:**
The created ICS-BCP only identifies the actions and the shared information, instruction contents, and information on the operation transition of the plant due to the actions are hidden between the actions. Participants identify the hidden information and necessary information communication paths. Then, create a game scenario by incorporating the stress factors created in the previous step.

**Simulating Incident Response:**
This step simulates an incident response. Figure 13 shows the structure of this exercise. In previous exercises, exercise participants were required to create a workflow in groups without being assigned to a specific role. In this exercise, roles are assigned to each exercise participant, and incident response is performed by taking actions for each role. It is an exercise that is closer to the actual incident response.

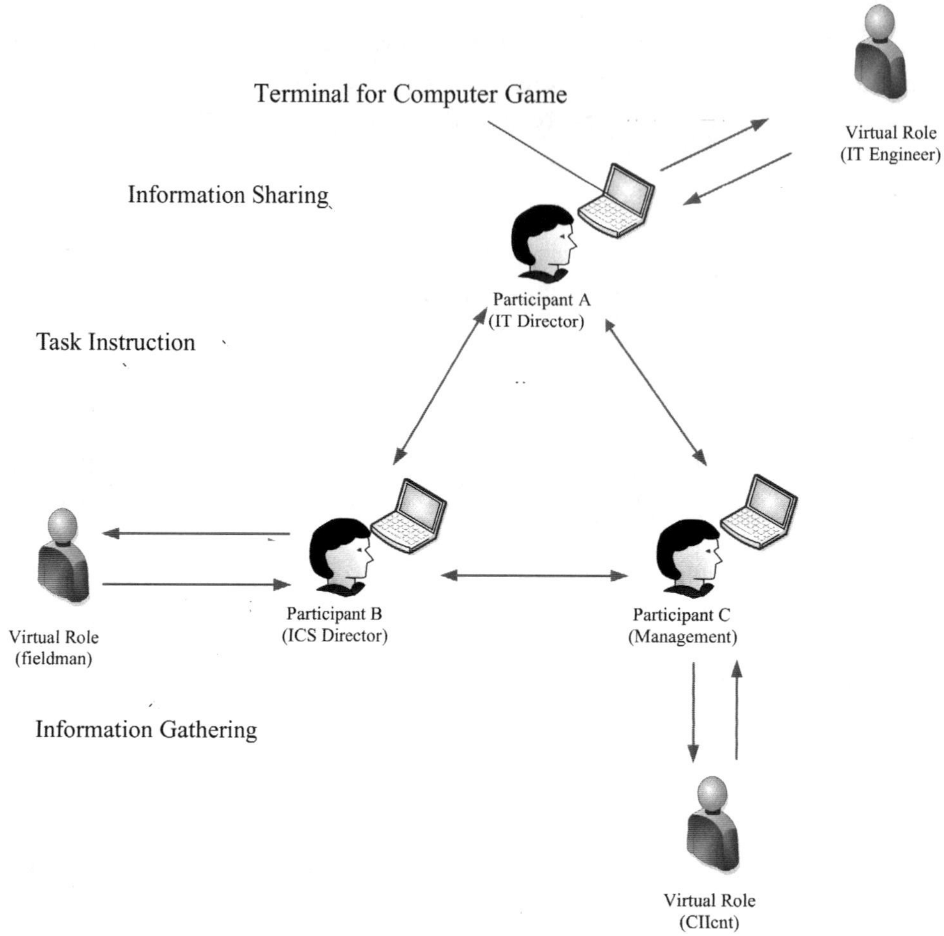


**Figure 13. IMANE structure**

Figure14 is a console screen operated by the participants. Participants share information with other players and take actions based on the information flowing in the tray. Each participant is required to understand the situation and take action based on the information displayed on their console screen. In addition, since the actions that can be performed for each role and the information is given are different, it is also required to share the information necessary for taking action to converge the situation with other players. Participants use the interface in Figure 14 to share information and select actions. The information displayed in the tray in Figure 15 is linked to the story created in the previous step, and the information displayed differs depending on the action performed so that the incident response can be made more realistic.

**Figure 14. The user interface of IMANE exercise**

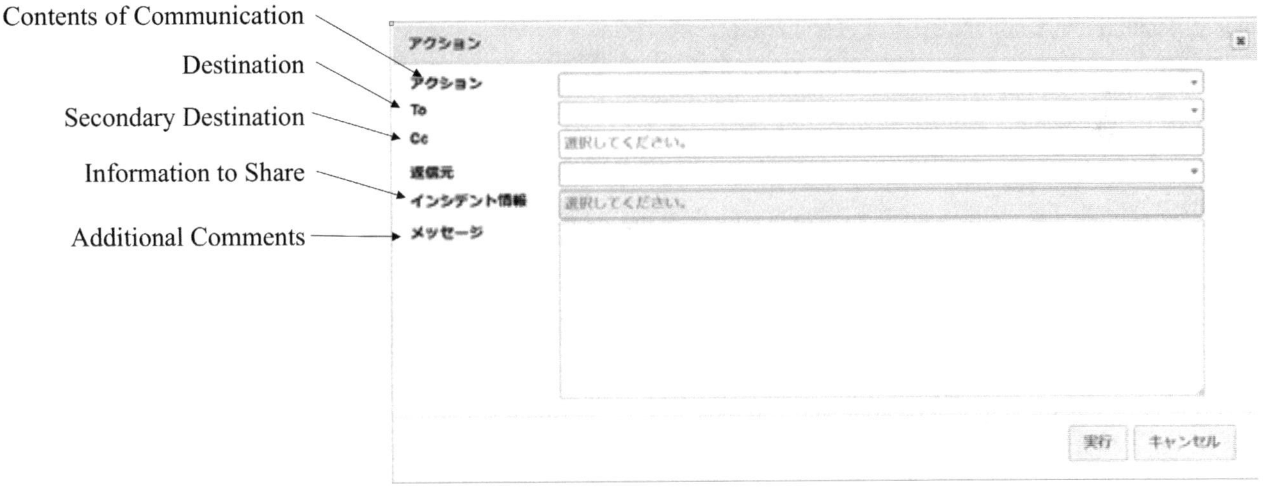


**Figure 15. The action user interface of IMANE Exercise**

**Review of the Exercise:**
After the end of this exercise, it can be output as a product of the exercise, as shown in Figure 16. This workflow includes the information given to each role, the actions taken, and the flow of information-sharing and participants can use this workflow to review the incident response after the exercise.

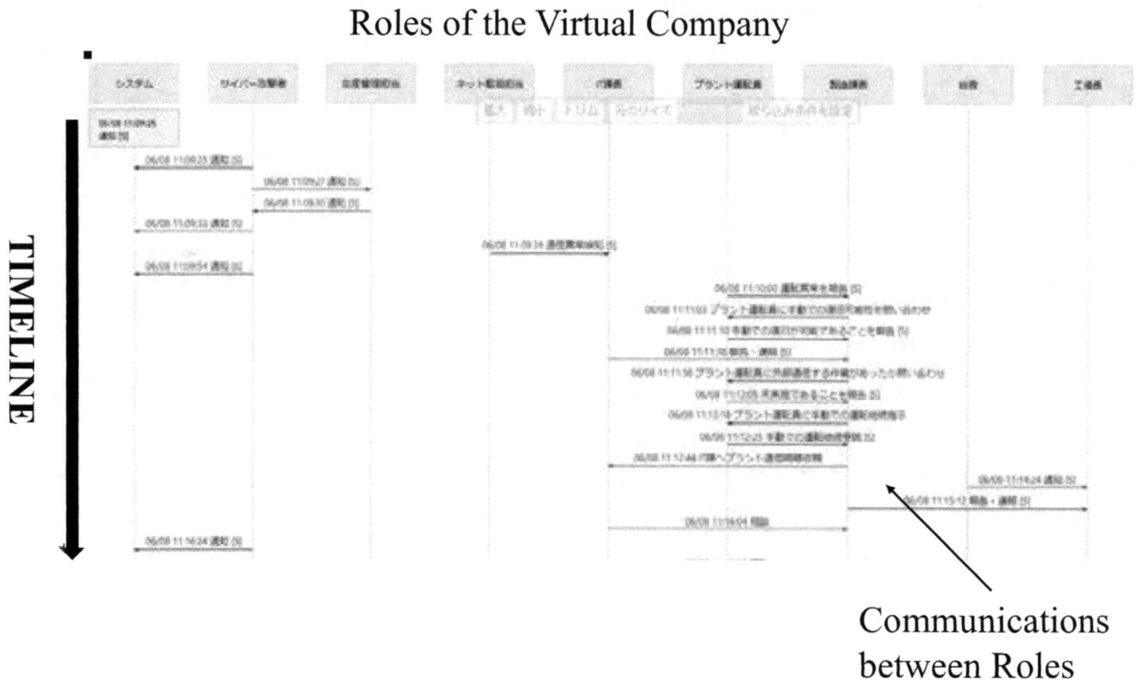


**Figure 16. Example deliverable of the exercise**

Using this deliverable, exercise participants can objectively review ICS-BCP by looking back on their response. It is possible to improve the accuracy of ICS-BCP by checking for missing or defective ICS-BCP.

**Open-Discussion:**
In this step, each participant shares the impressions gained through the exercise. We discuss the ideas needed to create an ICS-BCP and human resource development using this exercise.

By performing these exercises in various scenarios, the exercise participants can increase their imagination and resilience to cyber incidents.

## 4 Pilot Exercise

While improving the developed exercises step by step, many security and operation experts participated in the Nagoya Institute of Technology security workshops from 2018. To get an opinion to improve the exercises, we asked the participants to complete the questionnaire. The questionnaire was answered with a maximum of 5 points. In the exercise conducted on December 11, 2020, 28 people participated, and the following questionnaire results were obtained (Max point is 140). From the results of this questionnaire, it can be seen that the people who participated in the exercise understood the purpose of the exercise that we set and received high praise for the exercise.

Questionnaire content for Card-Type Incident Response Exercise and ICS-BCP Creation Exercise:

1. Was the goal set for the exercise, and was it possible to attain it?
2. Do you think you learned about organizational collaboration in the event of a cyber incident?
3. Do you think you learned about the organizations needed to respond to cyber incidents?
4. Was the information necessary to proceed with the exercise given?

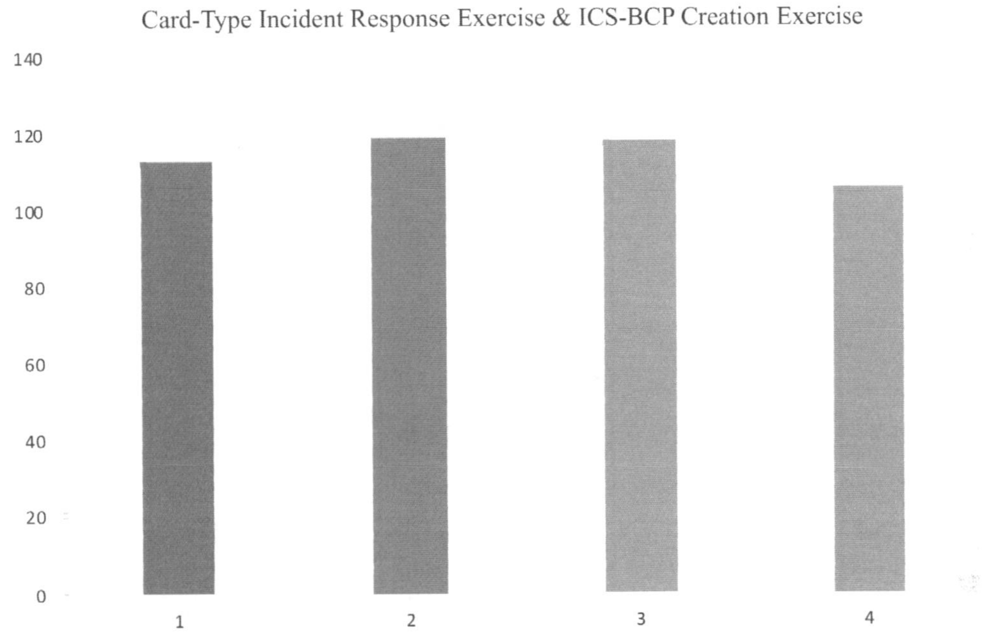


**Figure 17. Questionnaire results of Card-Type Incident Response Exercise & ICS-BCP Creation Exercise**

Questionnaire content for IMANE:

1. Was the goal set for the exercise, and was it possible to get it?
2. Do you think you learned about organizational collaboration in the event of a cyber incident?
3. Do you think you learned about the organizations needed to respond to cyber incidents?
4. Is it designed so that necessary information is communicated and given to each role?
5. Do you feel that you were able to experience information gathering and decision-making in an emergency?

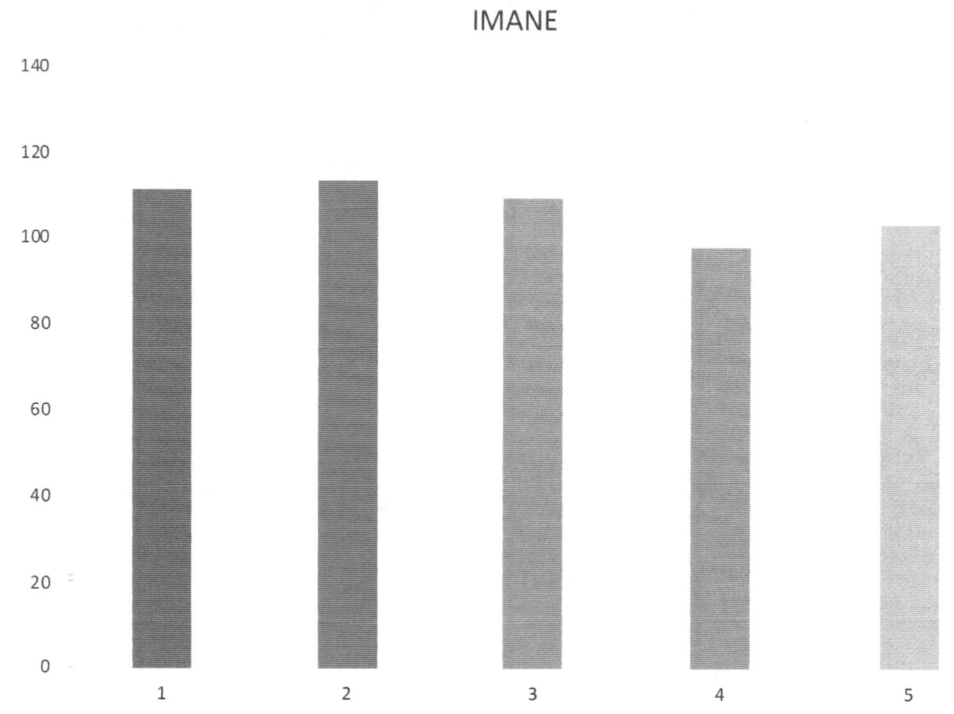


**Figure 18. Questionnaire results of IMANE**

In addition, the following was found by analyzing the deliverables of the exercise.

・ Plant safety first

Many workflows can keep the plant safe, even if the backgrounds of the participants in the exercises vary. Even if the cause of the plant abnormality is a cyber-attack, it is easy to imagine the visible situation, and it can be seen that appropriate measures can be taken.

・ No action has been taken to identify a cyber-attack

Analyzing the workflow, it is possible to select situation sharing and actions for performing Safety Response. However, information sharing for performing Security Response, such as identifying the range of influence of cyber-attacks is often not successful. This is thought to be due to the fact that they have never experienced cyber-attacks and lack the knowledge to consider actions. In addition, Safety Response is easy to imagine because it corresponds to the visible physical impact on the plant, but Security Response is difficult to imagine because it corresponds to the invisible network.

・Organization that is the center of the response

To respond to cyber incidents to ICS, it is necessary to fluidly change the organization that is the core of the response.

In the initial response, it is necessary to respond to plant operation abnormalities caused by cyber incidents. The person working at the plant takes action. After ensuring the plant's safety, it is necessary to eliminate the effects of cyber incidents. IT department takes this action. Finally, it is necessary to continue the business considering the impact of the cyber incident on the business.

Management considers actions for the business gains.
In this way, the work to be done changes over time, and it is necessary to respond while changing the organizational response that is the core of the response accordingly.

## 5 Concluding Remarks

In this paper, the authors presented the factors required of a company against the threat of cyber-attacks targeting ICS and explained the exercises developed to increase the company's resilience.

Through the workshop, we received positive opinions from the participants that they could gain various awareness of cyber-attacks. However, the effect of the exercise could not be confirmed quantitatively, and it could not be confirmed that the response-ability was improved.

In the future, we will conduct exercises for more people and obtain deliverables. Then, by analyzing the deliverables, we will consider the method of planning ICS-BCP considering cyber-attacks and the conditions necessary to confirm the validity of the planned ICS-BCP.

## 6 References

[1] Institute for Science and International Security, "Did Stuxnet Take Out 1,000 Centrifuges at the Natanz Enrichment Plant?", https://isis-online.org/isis-reports/detail/did-stuxnet-take-out-1000-centrifuges-at-the-Natanz-enrichment-plant/8 (accessed Jan. 25, 2021)

[2] ICS-CERT,
ICS Alert(IR-ALERT-H-16-056-01), "Cyber-Attack Against Ukrainian Critical Infrastructure", https://us-cert.cisa.gov/ics/alerts/IR-ALERT-H-16-056-01 (accessed Jan. 25, 2021)

[3] SANS Institute, ICS Defense Use Case No. 6, "Modular ICS Malware", https://ics.sans.org/media/E-ISAC_SANS_Ukraine_DUC_6.pdf (accessed Jan. 25, 2021)

[4] Norsk Hydro, "Cyber-attack on Hydro", https://www.hydro.com/en/media/on-the-agenda/cyber-attack/ (accessed Jan. 25, 2021)

[5] BSI, "Industrial Control System Security Top 10 Threats and Countermeasures2019", https://www.allianz-fuer-cybersicherheit.de/ACS/DE/_/downloads/BSI-CS/BSI-CS_005E.html#download=1 (accessed Jan.25, 2021)

[6] T. Aoyama, M. Koike, I. Koshijima, Y. Hashimoto, "A unified framework for safety and security assessment in critical infrastructures". In: Safety and Security Engineering V. WITPRESS LTD. (Sep 2013)

[7] Y. Ota, T. Aoyama, D. Nyambayar, I. Koshijima, "Cyber Incident Exercise for Safety Protection in Critical Infrastructure," Int J. Saf. Secur. Eng. 8,246-257 (2018)

[8] J. Sakhnini, H. Karimipour, A. Dehghantanha, R.M. Parizi, G. Srivastava, Security aspects of internet of things aided smart grids: a bibliometric survey. Internet Things (2019). https://doi.org/10.1016/j.iot.2019.100111

[9] A. Shostack, *Threat Modeling: Designing for Security* (Wiley, Indianapolis, In, 2014)

SPRING21
+17TH GCPS
A Joint AIChE and CCPS Meeting

# Hazard Identification during Process Related Maintenance and Operations Tasks

**T Michael O'Connor**
**Mary Kay O'Connor Process Safety Center**
**Texas A&M University, College Station, TX**
**mikeocon@gmail.com**

**Dr Zohra Salim**
**Mary Kay O'Connor Process Safety Center**
**Texas A&M University, College Station, TX**
**zohrahalim@tamu.edu**

Prepared for Presentation at
American Institute of Chemical Engineers
2021 Spring Meeting and 17th Global Congress on Process Safety
April 18 - 23, 2021

# Hazard Identification during Process Related Maintenance and Operations Tasks

T Michael O'Connor
Mary Kay O'Connor Process Safety Center
Texas A&M University
College Station, TX
mikeocon@gmail.com

Dr Zohra Halim
Mary Kay O'Connor Process Safety Center

**Keywords:** serious injury, fatality, chemical manufacturing, petroleum refining

## Abstract

Trends in process related fatalities are presented for chemical manufacturing and petroleum refining. The trend analysis revealed the more than half of fatalities occur in single fatality incidents. This observation led to study of the nature of these low fatality incidents. Three common tasks were identified that are associated with a large portion of process related fatalities and severe injuries. These 3 tasks are clearing, cleaning and purging lines and vessels; opening lines and vessels; and transferring fluids. The specific hazards encountered in incidents associated with these tasks are identified. The goal is to provide the basis to improve system design, training, procedures, job hazard analysis, PPE requirements and other factors based on hazards actually associated with major harm to individuals performing these tasks.

## Introduction

There are 3 primary components of this study:

- Trends in fatalities from 1984-2019 in chemical manufacturing (CM) and petroleum refining (PR)
- Tasks associated with severe injuries and fatalities in CM and PR
- Nature of serious injury incidents in CM only

Data is extracted from two US OSHA databases:

The first is the Fatality and Catastrophe Inspection Search (FCIS) database which includes all incidents with a fatality or 3 or more injuries. Since incidents without a fatality and less than 3 injuries are not included in the data injuries from this database are not included in the study. This database covers the years 1984 to the present whichallows long-term trending including the implementation of OSHA PSM in 1992. This analysis is limited to the states in OSHA Region 6. These states are TX, LA, AR, OK and NM. The data was accessed through OSHA's online query system.

The second database is the Severe Injury Reports (SIR) database. This database is limited to the 28 States with federal, not state, OSHA programs. This database includes all the Region 6 states. It covers the years 2015 to July 2020 and includes cases with hospitalization or amputation. This entire dataset can be downloaded as a spreadsheet.

For both databases the authors have interpreted the accident description to determine whether it is a process related incident and the task being undertaken. The serious injury database contains the hazardous substance, nature of injury, affected body part and event type. In some cases, the description is not sufficient to extract the needed data and these items are classified as unknown. Some incidents have been investigated by the Chemical Safety and Hazard Investigation Board (CSB) or other agencies. In those cases, information may have been extracted from those sources.

This paper focuses on both the Chemical Manufacturing (CM) and Petroleum Refining (PR) industry as determined from their NAICS and SIC codes. Both industries were analyzed for fatality trends from 1984-2020. The tasks associated with both severe injuries and fatalities were determined for both CM and PR. A more detailed analysis of selected tasks was performed only for CM. Detailed analysis of PR incidents are not yet completed.

## Scope of Incidents in Study

The following definitions were utilized to determine if an incident was process related and was to be included in the study:

Hazardous Material – Any substance the is toxic, reactive, flammable, asphyxiating or that presents a hazard to people, the environment, or property due to pressure or temperature.

Process Safety Incident – The sudden, unintended release of, or exposure to, a hazardous material, associated with an industrial process, that results in or might reasonably have resulted in death, injury, significant environmental or property damage, evacuation or sheltering in place.

These are generic definitions used by the MKOPSC. In this paper all of the incidents involve a fatality or injury since they are from the OSHA databases.

**Trends in Fatalities in Chemical Manufacturing and Petroleum Refining**

Long term trends with time are available from the Inspection Database for fatalities. Serious injuries are included in the database if there are 3 or more injuries or they occur with a fatality. However, incidents with one or two injuries are not included making any trending information questionable. The Serious Injury database only covers the years 2015 through 2020, making any trends of doubtful significance, so only fatalities were used for trending.

Plots are shown for CM and PR. Fatalities and fatal incidents over the years 1984 to 2020 are illustrated. There are a number of high fatality incidents in the latter 1980s. This is followed by a dramatic decrease in fatalities around the time of the implementation of OSHA PSM in 1992. It seems likely this decrease was due to both the increased focus on process safety due to the new regulation as well as a direct response to these high fatality incidents. The trends are somewhat similar for CM and PR with the exception of the BP Texas City explosion in 2005.

Figures 1 through 4 show the large number of incidents with a low number of fatalities. After 1992, for CM and PR, greater than 80% of fatalities occurred in single fatality incidents. This statistic in particular motivated a more detailed examination of these numerous incidents with mostly low numbers of fatalities as well as incidents with serious injuries.

In this plot (see Figure 1), the total height of the bars represents the total fatalities in a year. The colored bands making up the bars show individuals incidents and the number of fatalities in that event. The red arrow denotes the implementation of OSHA PSM in 1992. Clearly there is a sharp decrease in fatalities after that time. In addition, there are no incidents with more than 4 fatalities after PSM began.

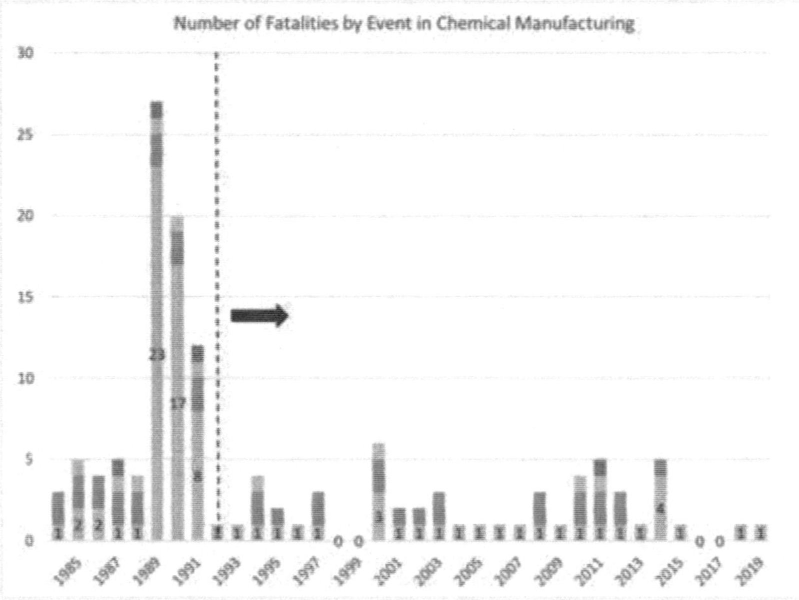


Figure 1 - Fatalities by Event in Chemical Manufacturing Region 6

This plot (see Figure 2) shows that not only the annual number of fatalities dropped in 1992 but also the number of fatal events.

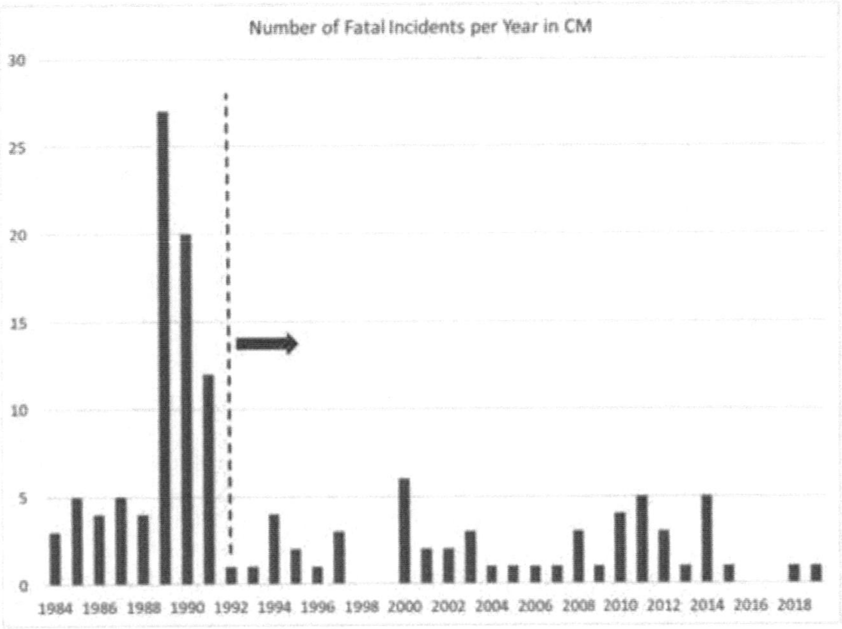


Figure 2 - Fatal Events in Chemical Manufacturing

Like CM there is a distinct reduction in fatalities in PR about the time of PSM implementation. This reduction appears to have occurred in 1994, a few years after PSM. There is another reduction about the time of the event at BP Texas City (see Figure 3). We theorize that this high-profile event may have re-focused the PR industry on process safety. This reduction has persisted for 14 years.

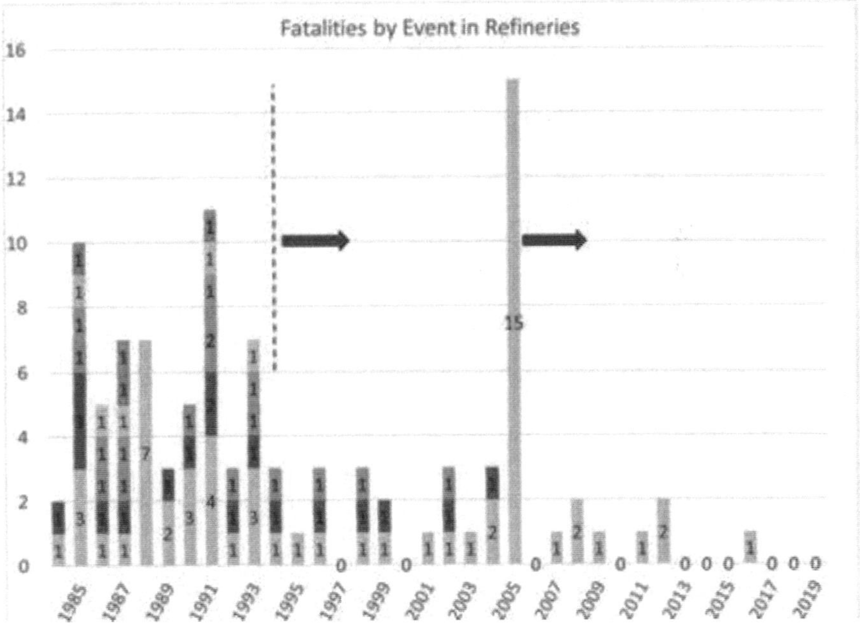


Figure 3 - Fatalities by Event in Petroleum Refining

It is clear that not only the number of fatalities was reduced around 1994 but also the number of events. Since 2002 there has been either zero or one fatal event per year (see Figure 4).

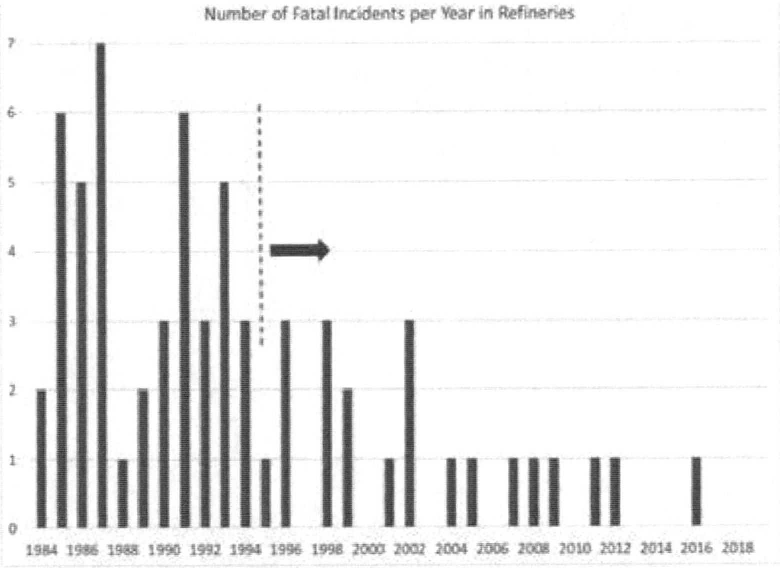


Figure 4 - Fatal Incidents in Petroleum Refining

Table 1

The fatality and fatal event annual rates are shown for CM and PR in Table 1. For CM there is a 50% reduction in fatal events per year after 1992. The fatalities per year dropped dramatically by a factor of almost 5. For PR there is a 60% reduction in fatal events in 1994 and an overall reduction of 75% after 2005. Fatalities in PR were reduced by nearly 50% in 1994 and a remarkable 10 to 1 after 2005. The relative magnitudes of the PR and CM rates are not comparable since they are for different numbers of facilities and employees.

| Event and Fatality Rates for CM and PR - Region 6 | | | |
|---|---|---|---|
| Chemical Manufacturing | | | |
| | 1984-1992 | 1993-2020 | |
| Fatal Events/yr | 4.1 | 2.0 | |
| Fatalities/yr | 10.0 | 2.2 | |
| Petroleum Refining | | | |
| | 1984-1994 | 1995-2005 | 2006-2020 |
| Fatal Events/yr | 4.0 | 1.6 | 1.0 |
| Fatalities/yr | 6.0 | 3.2 | 0.6 |

## Tasks Associated with Serious Injuries and Fatalities

As the descriptions of each incident were analyzed it was soon apparent that most of the incidents involved a task being performed by a small number of people and the consequences typically suffered by those workers. Both maintenance and operational activities are involved.

Table 2 lists the tasks for serious injury and fatal incidents, from both databases, in both CM and PR. The term "unrelated" is used to denote incidents that are not related to the activity of the victim. Most of these "unrelated" cases are mechanical failures, leaks or explosions and the victim is by chance in the vicinity.

Table 2

| Severe Injury and Fatal Incidents by Task | | | | | | | | | |
|---|---|---|---|---|---|---|---|---|---|
| | Chemical Manufacturing | | | | Petroleum Manufacturing | | | | |
| | Severe Injury Incidents | % | Fatal Incidents | % | Severe Injury Incidents | % | Fatal Incidents | % | % Averaged for Sort |
| Cleaning, clearing, purging lines and vessels | 35 | 16 | 16 | 20 | 10 | 29 | 11 | 11 | 19 |
| Opening lines and vessels | 47 | 21 | 8 | 10 | 3 | 9 | 15 | 15 | 14 |
| Transferring fluids | 43 | 20 | 8 | 10 | 3 | 9 | 7 | 7 | 11 |
| TOTAL % for 3 Tasks | | 57 | | 39 | | 47 | | 32 | |
| Unrelated* | 10 | 5 | 11 | 13 | | | 16 | 16 | 8 |
| Unknown | 20 | 9 | 3 | 4 | 2 | 6 | 8 | 8 | 7 |
| Maintenance | 18 | 8 | 3 | 4 | 5 | 15 | 4 | 4 | 8 |
| Operating | 6 | 3 | 6 | 7 | 3 | 9 | 4 | 4 | 6 |
| Starting equipment | 6 | 3 | 5 | 6 | 2 | 6 | 5 | 5 | 5 |
| Confined Space Entry | | | 7 | 9 | | | 8 | 8 | 4 |
| Mixing chemicals | 8 | 4 | 3 | 4 | | | 7 | 7 | 4 |
| Troubleshooting | 7 | 3 | | | 2 | 6 | 4 | 4 | 3 |
| Fall or step into hazard | 3 | 1 | | | 2 | 6 | 1 | 1 | 2 |
| Other | 5 | 2 | 4 | 5 | | | 1 | 1 | 2 |
| Sampling | 3 | 1 | 3 | 4 | 1 | 3 | | | 2 |
| Handling explosives | | | 3 | 4 | | | | | 1 |
| Lighting heater | | | | | | | 5 | 5 | 1 |
| Transferring solid | 7 | 3 | | | | | | | 1 |
| Reactor startup | | | 2 | 2 | | | | | 1 |
| Starting process | | | | | 1 | 3 | 2 | 2 | 1 |
| Hot Work | | | | | | | 3 | 3 | 1 |
| Testing | 2 | 1 | | | | | 1 | 1 | 0 |
| Inspecting | | | | | | | 1 | 1 | 0 |
| | | | | | | | | | |
| Total Incidents | 220 | 100 | 82 | 100 | 34 | 100 | 103 | 100 | 100 |

*Unrelated indicates the employee was not involved in a task telated to the event.

**Tasks Associated with Serious Injury and Fatal Incidents in Chemical Manufacturing.**

In both CM and PR, there are 3 activities that account for a large fraction of the fatalities. These tasks are:

- Cleaning, clearing, or purging lines and vessels
- Opening lines and vessels
- Transferring fluids and solids

These 3 types of tasks usually involve the intentional interruption of a pressure containing equipment component. That is of course the case with opening lines and vessels. For cleaning, clearing, or purging, it is necessary to open the pressure containment before proceeding. Transferring fluids often involves making or breaking connections as well as venting and draining lines and vessels

Cleaning, clearing, and purging by their nature involve removal of a substance, usually from the interior of the containment. Difficulties are encountered in numerous ways. These include:

- Substance is not what was expected
- Quantity of substance is greater than expected
- Unexpected pressure behind a blockage
- Blockages are propelled out of the containment by pressure
- Tools, covers and hatches are propelled by unexpected pressure
- Containment is over-pressurized when clearing an obstruction

Opening lines and vessels often involve problems similar to those above. These include:

- Unexpected pressure
- Unexpected substance
  Unexpected quantity

Transferring fluids involve many different situations based on the source of the fluid, receiving container, and type of transfer line or hose.

- Hoses often disconnect unexpectedly
- Hoses are subject to failure
- Disconnected hoses may contain an unexpected substance
- Hard pipes often contain vents and drain lines that are open
- Transfers may result in mixing incompatible substances
- Rail cars and tank trucks may have complex sets of valves for venting, draining, multiple compartments and other confusing situations
- Totes, drums, buckets, and other containers can be dropped or damaged, sometimes by forklifts
- Static electricity may be an issue in many of these operations

**Nature of the Serious Injury Incidents in CM**

The OSHA database, SIR, contains four factors that describe the incident and the resulting injuries. The four factors are the following:

- Type of Event
- Hazardous Substance
- Nature of Injury
- Affected Body Part

The results for all tasks in CM are shown in four tables. These four factors are the results of the incident and infer a limited amount about the causes. Hopefully information about the most common hazards will assist in focusing efforts for improved safety. Data for all incidents are included, not just those for the 3 most common tasks.

Table 3

| Chemical Manufacturing | | |
|---|---|---|
| Serious Injury - All Tasks | | |
| **Event Type** | **#** | **%** |
| Exposure through intact skin, eyes, or other exposed tissue | 82 | 37 |
| Contact with hot objects or substances | 74 | 33 |
| Inhalation of harmful substance-single episode | 21 | 9 |
| Ignition of vapors, gases, or liquids | 14 | 6 |
| Struck by dislodged flying object, particle | 9 | 4 |
| Explosion of pressure vessel, piping, or tire | 8 | 4 |
| Explosion, n.e.c. | 6 | 3 |
| Other | 9 | 4 |
| TOTAL | 223 | 100 |

**Event Type**

Two types of events account for 70% of all events in roughly equal frequency. These two are exposure of external tissue to a hazardous substance and contact with hot objects and surfaces. These are such broad categories that it does little to identify a particular type of event which deserves more attention than others. Two other event types also occur in a significant portion of incidents. These are inhalation of a substance and ignition of a substance. In planning for and executing a task, it is important to consider if any of these events could occur. This can lead to identifying procedures to avoid the event and to utilize proper PPE and emergency procedures.

Table 4

**Hazardous Substance**

Perhaps surprisingly, water is the most frequent hazardous substance involved in serious injuries. Of course, this is due in part to its widespread presence in many processes. Often the water is condensate and results in burns. Nearly as common are acids followed by resins. Resins are often associated with processes such as extrusion were contact burns are common. While the table shows a number of common substances there are an additional 40 substances that were only involved in one incident. In addition, there are 30 objects. In OSHA's classification system, objects, rather than a hazardous substance, are identified as the source of injury. These are usually cases where an object propelled by pressure or an explosion strikes the victim. These objects are typically shrapnel, pipe plugs, covers and hatches or hoses.

| Chemical Manufacturing | | |
|---|---|---|
| Serious Injuries - All tasks | | |
| Substances | # | % |
| **Water** | **27** | **12** |
| Unknown, not classified | 27 | 12 |
| **Acids** | **25** | **11** |
| **Resins** | **17** | **7** |
| Multiple | 12 | 5 |
| **Chlorine, chlorine bleach, compounds** | **9** | **4** |
| **Steam and vapors** | **9** | **4** |
| **Sodium and potassium hydroxide, potassium carbonate** | **8** | **4** |
| **Ammonia, anhydrous ammonia** | **6** | **3** |
| Chemicals, n.e.c. | 6 | 3 |
| **Phenol** | **4** | **2** |
| Liquids, unspecified | 4 | 2 |
| **Aromatics and hydrocarbon derivatives, except halogenated, n.e.c.** | **3** | **1** |
| Other Substances | 40 | 18 |
| Physical Objects* | 30 | 13 |
| TOTAL | 227 | 100 |
| * OSHA lists objects such as those propelled by pressure | | |

Table 5

**Nature of Injury**

The nature of injury generally reflects the type of event shown in the previous table. Thermal burns are naturally associated with contact with hot objects and substances and chemical burns with exposure of tissue. Traumatic injuries result from being struck by object and explosions. Poisoning, toxic, noxious, or allergenic reactions are typically due to inhalation. While there is little information in these databases about PPE it appears that it is often inadequate for the actual circumstances.

| Chemical Maufacturing | | |
|---|---|---|
| Serious Injuries - All Tasks | | |
| Nature of Injury | # | % |
| Thermal Burns | 93 | 40 |
| Chemical Burns | 79 | 34 |
| Traumatic | 26 | 11 |
| Poisoning, toxic, noxious, or allergenic effect | 26 | 11 |
| Other | 9 | 4 |
| TOTAL | 233 | 100 |

Table 6

**Affected Body Parts**

One third of the incidents involve multiple body parts. This may reflect the serious nature of the studied incidents. Severe injuries may be the result of releases of larger quantities of hazardous substance, higher pressures and other factors that lead to more widespread injuries. This contrasts with less serious injuries associated with more common cases involving days away from work. In a study of these injuries (Mannan), there were typically specific body parts injured. The most often injured body part was the face. The second most affected body part for severe injuries is body systems. This means internal systems implying inhalation or ingestion of a substance. The remaining body parts are more typical parts of face, extremities, trunk, and head. The adequacy of PPE may be a factor in the severity of these injuries.

| Chemical Manufacturing | | |
|---|---|---|
| Serious Injuries - All Tasks | | |
| Body Parts | # | % |
| Multiple body parts, n.e.c. | 77 | 33 |
| Body Systems | 28 | 12 |
| Nonclassifiable | 28 | 12 |
| Face | 25 | 11 |
| Lower Extremities | 25 | 11 |
| Upper Extremities | 24 | 10 |
| Trunk | 15 | 7 |
| Trunk & Head | 8 | 3 |
| TOTAL | 230 | 100 |

**Incidents with Detailed Investigations**

The CSB and the US Department of Labor investigated several of the CM incidents in this study. These incidents were Phillips 66 1989, Honeywell 2003 (3 incidents) and DuPont 2014. These incidents generally involve one of the 3 common tasks. Phillips 66 involved the removal of a plug in the settling leg of a reactor. This led to a release of isobutane resulting in 23 fatalities and more than 100 injuries. Honeywell had 3 incidents involving 3 different substances. The first was a chlorine release due to a mechanical failure. The second was an antimony pentachloride release which resulted in a fatality. The release was the result of removing a plug from a cylinder. The third release was hydrogen fluoride during the transfer of fluid using a venturi stick, resulting in hospitalization of one worker. The DuPont incident resulted in 4 fatalities during the removal of a hydrate plug, releasing mercaptan. These three investigation reports show that larger scale incidents may be the result of the 3 common tasks.

**Observations**

The trends in fatalities before and after the implementation of PSM in 1992 shows a remarkable improvement in fatalities and fatal events. Following the reduction in 1992, most fatalities occur in single fatality incidents. These trends are counter to what one might conclude by observing high profile incidents in the media. This is true in part because of incidents in storage, transportation and at user facilities which are not included in this study.

The nature of these incidents seems to straddle the line between process safety and occupational safety. It may be that there is a need to better coordinate the safety efforts of what are in many companies two distinct organizations. The same may be true of the educational system. For example at Texas A&M University, occupational safety is taught in the School of Public Health while process safety is addressed in the College of Engineering.

Based on the incident description, the vast majority of fatalities appear to occur with relatively small releases affecting workers in close proximity to the release. At least 80% of the events are associated with what seem to be routine maintenance and operations tasks. They do not appear to be the result of complex technical failures such as impaired control systems.

There has been great success in limiting fatalities since the introduction of OSHA PSM. In order to further reduce fatalities, it appears that additional focus on less complex small scale releases is warranted.

The nature of these incidents suggest that improvements may be needed in training for hazard recognition, procedures, and job safety analysis.

The incidents are often caused at least in part by the dynamic nature of the tasks being performed, where the operation is changing over time. Procedures are needed that recognize and adapt to these changes.

In many cases, there are only two physical barriers to protect the workers, the pressure containment and PPE. Since many of the tasks involve opening the containment, adequate PPE is especially important. PPE requirements should be based on all possible release scenarios.

## Recommendations for Industry and Academia

Process and Occupational Safety specialists should work together to understand and address small scale process incidents associated with common tasks. Some specific recommendations:

- Utilize this insight to better educate operations and maintenance personnel.
- Design process systems to minimize the need for hazardous tasks or reduce the severity of possible releases.
- Incorporate these findings in junior college process operator curriculum.
- Consider whether current procedures adequately cover the type of tasks most often associated with severe injuries and fatalities.

## Further Investigation by the Authors

This study focused on CM and PR. Fatalities were addressed only for OSHA Region 6. The other 9 regions should be studied to confirm the trends and patterns and increase the confidence in the results. The severe injury data covered the 22 federal OSHA states. The scope for severe injuries and fatalities should be expanded to Oil and Gas and other industries.

The descriptions of the incidents in this study were extracted by the authors. To identify and interpret the entire dataset, machine learning techniques should be applied.

Develop systematic methods for identifying hazards for the most common tasks which were identified in this study.

Investigate the need for and characteristics of procedures that would prevent or mitigate the consequences of incidents from the common tasks.

## References

https://www.osha.gov/data, This is the site for both data sources in this paper, "Fatality and Catastrophe Investigation Search" and "Severe Injury Reports".

Mannan etal, "Patterns and Trends in injuries due to chemicals based on OSHA occupational injuries and illness statistics", Journal of Hazardous Materials, 2008, p 349-356.

Chemical Safety and Hazard Investigation Board (CSB), Honeywell Incidents, Report No. 2003-13-I-LA, August 2005

Chemical Safety and Hazard Investigation Board (CSB), Toxic Chemical Release at the DuPont La Porte Chemical Facility, La Porte, Texas, Incident Date: November 15, 2014, No. 2015-01-I-TX

US Department of Labor, Phillips 66 Company's Houston Chemical Complex (HCC), Report to the President, April 1990.

# 618389 Incident Investigations to Enhance Safety Culture: Refinery Examples


**Abstract:**

The investigation of small incidents and near misses provides an opportunity for any company to change its safety culture at the grass roots level. This change is due in part to features unique to small incidents and near misses. The purpose of this paper is to outline the benefits of using small incidents and near misses to increase situational awareness and enhance safety culture. The paper will also include a refinery case study demonstrating how these small steps can make a big difference for process safety management.

**Scope:** The primary purpose of the investigation is to understand the underlying causes of the incident and to prevent it from happening over and over again. Events are not just human error or a result of a physical failure, each event is indicative of a fault in the system. The investigations are an important part of safety culture of company. They increase the awareness as well.

**Methods, procedures, Process:** The important elements of the event research are the timely research and report creation, the effective documentation of the findings and the sharing of the results. The method of the investigation is selected due to the potential risk of Incident/Near Miss/Safety Observation. In addition to the method, the assignment of team members is made by different levels of the organization depending on the level of potential risk assessment of the event.

**Results, Observations, Conclusions:** The current and potential evaluation of the impact of the incident to environment, human, reputation and to asset, the establishment of research teams, the appointment of root causes and actions, the automatic creation of the corrective and preventive actions or maintenance requests are made on specific investigation tab of the SAP (SAP EHSM) system. The publication of the accident research report are carried out by SAP EHSM as well.

In most of the applications Tier 1 and 2 Process Safety incidents are deeply investigated but sometimes Tier 3 and 4 incidents left behind in terms of investigation and actions completion speed. However, if you check all the Tier 3 and 4 incidents root causes comprehensively, you may see that they already call for the Tier 1 and 2 incidents. For example, you may find some corrosion reports as Tier 3 since it did not caused any Tier 1 or 2 incident. As you have investigation you will come up with some defaults due to crude oil quality or process technologies applied. Having such an investigation in the phase of Tier 3, you will be able to prevent such a big leakage. Moreover emphasizing on Tier 3 and 4 incidents investigation, gives the message that you are more proactive since you are not only dealing with the incidents only caused harm to people or asset but also potential risks. That will also help management to communicate to employees that, they are one-step ahead the incidents, which will lead employees to think potential risks in the workplace.

# 619133 Effective Implementation of Dhas – Trends One Year Past the NFPA Deadline

**Abstract:**

The deadline incorporated into NFPA 652 for existing facilities handling combustible dusts to complete a dust hazard analysis (DHA) has come and gone as of September 2020. How are companies doing with incorporating the findings of their DHAs into capital expense and operational expense budgets? Are authorities having jurisdiction (AHJs) even aware of the requirement? This presentation will review key findings from surveys of our clients and AHJs.

American Petroleum Institute
What Don't You Know about Your PSM Program?: How the Process Safety Site Assessment Program (PSSAP®) Can Help Sites Find and Address These Blind-Spots.
March 22, 2021

Often companies struggle to understand what process safety management areas need their focus. What specific topics am I neglecting? Which of my programs or procedures could use some help to be truly effective? How does my holistic program compare with my peers in the industry? The Process Safety Site Assessment Program (PSSAP®) is an industry-created, industry-led program that can help answer these questions.

The American Petroleum Institute (API) and the American Fuel and Petrochemical Manufacturers (AFPM) kicked-off the PSSAP program in 2012. Since then, API has administered over 130 of these assessments across the globe. These assessments cover the main focus areas of process safety management: Process Safety Leadership, Operating Practices, Mechanical Integrity (focused on fixed equipment), Safe Work Practices, Management of Change (MOC), Process Hazards Analysis (PHA), Facility Siting, Product Storage & Transfer, and Incident Learning. In addition to these traditional areas, the PSSAP program can also assess your Hydrofluoric Acid (HF) Alkylation/API RP-751 processes, as well as the Safeguards a site has in place to address risks, our newest offering.

But how can an assessment help my site? Won't I just receive a report? The PSSAP program is not a normal assessment. While our assessments are based on satisfying industry-created good practice protocols, an operator gets much more from our assessors than a report following their time on-site. API sends a full team of vetted, qualified assessors that are leaders in their field, with the average experience of one of our assessors being 40+ years. Our assessors spend time on site, and in addition to evaluating your programs for your blind spots, they also are constantly sharing the good-practices they have seen across the industry as part of their participation in this program. Not only does a site receive an assessment of their programs, they also can find out what the rest of the industry is doing to address these areas. Not only does a site get this through good-practice sharing from our assessors, but sites also get to see the benchmarking data from our previous assessments. With over 130 worldwide assessments, a site can statistically gauge how they compare with their peers. Sites also receive a detailed examination of the specific areas that could use more of their attention.

In addition to the different focus areas and our assessor pool, it is important to note that these assessments are scalable. Are you a small site? Is your operation too small for you to think such an assessment would be worthwhile? API created a shorter program for these sites, enabling them to access our industry-leading expert assessors so that all may benefit from their wealth of knowledge, continuing to drive refining, petrochemical and chemical operators toward our shared goal of zero process safety incidents.

We will explore our newest offerings in this forum, and will also show examples of what a site stands to gain from their participation in this industry-led and driven program.

# 619049 Bouncing Back from a Heart-Wrenching Process Safety Incident

**Abstract:**

During startup of Urea Fertilizer Plant after a brief shutdown, due to recurrence of a leakage, the field maintenance team was inspecting top head gasket of condenser on a 31-ft high platform when Carbamate solution splashed on the workers. They tried to escape via monkey ladder but fell down one after the other resulting in loss of life of one of the employees and hospitalization of other four.

Detailed incident investigation by a multi-disciplinary team highlighted number of gaps related to human elements, job handover/execution and crisis management. The incident occurred due to failure of multiple layers of protection including management of change, mechanical integrity, risk assessment, PTW, job supervision, fatigue element, over-occupancy, communication, training & development, emergency response, PPEs etc.

Interestingly, this happened in less than 3 years after successful compliance certification of Process Safety Management System but raw material unavailability and resulting financial crunch hugely jeopardized organization's process safety management system and safety culture.

The paper will cover this eye-opening case of huge PSM deterioration in very short time due to financial factors and how did the organization bounce back from that desperate situation. It will shed light on causes of incident, lessons learning and results achieved after revamp of the process safety management program. It will enable other participants to learn from story of our setbacks & revival, and implement similar cultural shift programs in their respective organizations.

# 618378 Butadiene Popcorn Polymer: Formation Mechanism and Inhibition

**Abstract:**

In olefin production, a porous crosslinked polymer like a “popcorn” forms occasionally in the apparatus due to olefin polymerization in the step of refining/recovery. Popcorn polymer also occurs during monomer recovery in synthetic rubber production. Numerous monomers can experience the formation of popcorn polymers, including diolefins (e.g. 1,3-butadiene, isoprene) and olefins (e.g. styrene, vinyl acetate, acrylic acid). Popcorn is a unique form of polymer and presents significant hazards to industrial processes. For example, popcorn polymer fouls and adheres to the apparatus/pipelines in the refining/recovery systems including distillation tower and heat exchanger. This often clogs the pipeline/apparatus leading to the decrease of efficiency. In more serious cases, excessive growth of popcorn polymer can generate significant mechanical stress to deform or rupture the pipeline/apparatus, causing release of olefin vapors which can lead to plant fire or explosions.

Popcorn polymers can occur in either liquid phase or gaseous phase. It is commonly formed in presence of rust, oxygen, and reactive monomer (e.g. butadiene). Popcorn polymer formation basically consists of two stages. The first stage is popcorn seed formation through a series of transformation steps. The second stage is rapid seed growth into large polymer lumps in presence of monomer in a self-accelerated propagation rate. The growth rate increases with time exponentially due to newly formed radical actives sites or seeds caused by popcorn fracture during growth.

Current industrial practice to remove popcorn polymer fouling includes temporal suspension and disassemble of the apparatus for mechanical cleaning. This reduces the run length causing significant economic loss. More effective practice is to use process inhibitors. Initial formation of popcorn seed is usually slow. An inhibitor that can scavenge peroxide is potentially effective. Some inhibitors have been historically reported to limit the seed formation. On the other hand, popcorn seed growth is difficult to inhibit once it is formed. Conventional inhibitors are generally not effective in preventing seed propagation. Popcorn seeds, when formed, are very resistant to mechanical cleaning. Small amount of seeds then grows into large popcorn polymers even in presence of conventional inhibitors. Hence, inhibiting the second popcorn seed propagation stage is challenging.

In this paper, popcorn formation mechanism and inhibition chemistry that has been historically used in the industry to limit the seed/polymer occurrence along with their supporting literature is reviewed. Importantly, new compounds are studied for inhibiting popcorn polymer formation and show good performance. As such, these compounds can potentially help address the challenges brought by popcorn polymer.

# 618379 Ethylene Decomposition from Heat of Compression in Small Volume Piping

**Abstract:**

Ethylene decomposition is a known process safety concern in high purity, high pressure ethylene services. Ethylene decomposition can be initiated by various mechanisms with one mechanism being the adiabatic compression of diatomic gases by high pressure ethylene.

This presentation will summarize a 2020 near miss event at the Beaumont Chemical Plant where a carbon blockage at the inlet to an ethylene product relief valve was identified during a relief valve maintenance task. As part of the investigation, it was determined that a relatively small volume of piping can be susceptible to an ethylene decomposition reaction in the presence of oxygen and nitrogen.

This presentation will provide a broad overview of ethylene decomposition fundamentals, an incident summary and suggest mitigation ideas to address this process safety risk.

# 619932 Furnace Reverse Steam Flow

**Abstract:**

This paper details the events that resulted in a failure of the steam piping between the steam drum and the convection section of a pyrolysis furnace during re-start of the furnace. The furnace had been shutdown due to an electrical outage, and was being restarted to decoke the furnace tubes. As the furnace was heating up to the decoking temperatures, the steam drum vent and the superheat vent were opened to vent steam. The superheater vent was slowly closed to increase the steam drum pressure as the heater temperatures increased, and the superheater vent was eventually closed to around 5%, fluctuating between 0-5% during the remainder of the startup. As the steam drum pressure continued to build to the point that it could be exported to the header, the piping between the steam drum and the firebox failed at an elbow, causing damage to the superheat coils, insulation on other piping. The superheated steam flow meter indicated a flow, but the investigation team determined the meter was incorrect and that there was either no flow or back flow going into the steam drum. The intent of this paper is to increase awareness on the hazards associated with reverse flow where the piping and equipment is not rated for the conditions associated with reverse flow, and not focus on the sequence of events related to this failure. Focusing solely on the incident overshadows the necessity to prevent reverse flow under any circumstance, known or not known to have occurred.

# 619124 Renewable Diesel As a Blend Component to Enhance Olefin Yields of Liquefied Waste Plastics Feedstocks

**Abstract:**

Demand for polymers and chemicals is increasing, as is their importance to society with the continuous development of advanced materials providing innovative solutions for the most demanding and everyday problems alike. Unfortunately this growth will contribute to an estimated 70% annual waste generation increase by 2050[1]. Steam cracking lies at the heart of meeting this demand and remains closely coupled with consuming fossil resources. In previous works, the suitability of renewable hydrocarbons as a drop-in steam cracker feedstock and the substantial environmental benefits that can be seen in the subsequent products have been presented [2,3]. To complement its commitment to reducing fossil oil dependency and related climate emissions, while offering a viable and scalable solution to address the plastic waste issue, Neste plans to refine liquefied plastic waste into high-quality drop-in raw material for the petrochemical industry. As part of this work it was surprisingly found that, due to their unique composition, renewable hydrocarbons have a synergistic effect when used as the major component in blends with liquefied waste plastic. In this paper the steam cracking performance of hydrocarbon streams originating from liquefied waste plastics is examined and beneficial effects of dilution with renewable hydrocarbons, as opposed to a fossil naphtha, are discussed.

**References**

*1. World Bank (2018), What a Waste 2.0: A Global Snapshot of Solid Waste Management to 2050.*

*2. Increasing Sustainability of the Steam Cracking Industry Through Drop-In Renewable Feedstock, Muñoz G., A.E., et al, 32nd. Ethylene Producers' Conference, 2020*

*3. Environmental impact assessments of innovative bio-based products, Task 1 of "Study on Support to R&I Policy in the Area of Bio-based Products and Services ", COWI A/S and Utrecht University EN December, 2018*

# 620756 The Proposed Ethylene Production NESHAP and Potential Complications with Reclaimed Water Use

**Abstract:**

Volatile Organic Compound (VOC) emissions from industrial cooling tower systems can occur due to corrosion or cracking of a heat exchanger's internal tubing material, allowing some process fluids to mix with or become entrained in the circulating cooling water. VOCs in the process fluids may subsequently be released from the cooling water into the atmosphere when the water is exposed to air. The Ethylene Production National Emission Standard for Hazardous Air Pollutants (NESHAP) includes heat exchange or cooling tower systems as emission sources at ethylene production facilities. Under this standard, an Ethylene Production Maximum Achievable Control Technology (EMACT) identifies work practices that specify monitoring and Leak, Detection and Repair (LDAR) to control potential heat exchange system VOC air emissions. In September 2019, the EPA proposed EMACT amendments for heat exchange systems that would further reduce potential air emissions. A leak action level is to be defined using the Modified El Paso Method for Determination of VOC Emissions from Water Sources. The resultant action level is equivalent to less than 1 ppmw of strippable VOC in the return cooling water. Since the monitoring point for this proposed EMACT is only after the heat exchanger(s) in the system, organics in makeup water, such as from a reclaimed/recycled water source, are not factored in as a background concentration and would be counted toward compliance. This paper presents an analysis of the impact this proposed EMACT may have on options to use reclaimed water sources for cooling water makeup at ethylene production facilities subject to this standard.

# 621082 Flare Monitoring and Data Management Challenges

**Abstract:**

Recently promulgated revisions to existing regulations have incorporated the flare monitoring requirements used by the refining sector and codified at 40 CFR Part 63, Subpart CC: National Emission Standards for Hazardous Air Pollutants (NESHAP) from Petroleum Refineries (MACT CC). The revised regulations include the Maximum Achievable Control Technology Standards (MACT) for Ethylene Production (EMACT), MACT for Organic Liquid Distribution (OLD MACT) as well as the Miscellaneous Organic NESHAP (MON). These rules require an increase in the use of continuous parametric monitoring systems (CPMS) on flares and an increased reliance on the management of the measured CPMS data for purposes of proper flare operation and ultimate compliance demonstration. This presentation will explore the new flare monitoring requirements and data management challenges experienced during the implementation of MACT CC in the refining sector. Specifically, considerations concerning:

- What is required to be measured on flares?
- How is the flare CPMS data quality assured (i.e., distinguishing between good vs. bad data)?
- How is the measured flare CPMS data validated, calculated, and averaged?
- What does the measured flare CPMS data indicate about flare operation?
- What notification, recordkeeping, and reporting requirements are required for measured flare CPMS?

# 619240 Digital approach for fouling predictive analysis on Steam turbines through operational data

**Abstract:**

Industrial Internet of Things (IIoT) applied to oil & gas equipment is nowadays a must-have-tool for engineering maintenance and operation teams to monitor and manage the installed fleet to ensure the highest availability and reliability. Operational data combined with subject matter expertise and smart analytics have consolidated capabilities for anomaly detection and early warning. Industrial market is more and more heading to production, maintenance and performance optimization horizons.

Deep domain expertise combined with physics-based analytics allows to meet this expectation.

Such analytics, based on design tool and OEM models (digital twin), let to highlight sub-optimal operation, related root cause and actionable recommendation to restore the expected design condition.

Fouling is one of the most common and relevant limitation for turbines and compressors operability producing performance deterioration, thrust bearing load increase and, depending on contaminants, possibility of corrosion cracking.

The paper highlights analytics techniques and capabilities to provide assessments related to operation, performances and maintenance that could be affected by machine fouling.

Steam turbine fouling is typically due to contaminants deposit and affected by steam purity, pressure and temperature in the different machine sections. Presence of deposits translates in a different geometry in the steam path producing an increase of pressures detectable through wheel chamber pressure transmitters. Main effects could be higher steam consumption, reduced power capability and reduced aerodynamic efficiency.

The analytic is based on a comparison between measured wheel chamber pressure and the expected one (in new and clean condition).

Furthermore, the wheel chamber pressure is proportional to the thrust bearing load: the analytic outputs also allow to estimate the stresses on the bearing with respect to margin to design allowable load.

# 619242 Design and Implementation of a Reliable and Cost-Effective Surge and Control System on Centrifugal Compressors

**Abstract:**

The proposed presentation will show the steps involved in the design and implementation of a reliable and cost-effective surge control system for centrifugal compressors. The paper highlights the basics of surge control system design, instrumentation, and guidelines for a controls retrofit and field testing. Surge and choke control as they pertain to the Cracked Gas Compressor, Propylene Compressor, and Ethylene Compressor will be addressed. Special attention will be given to process upsets that can cause surge and choke events in the compressors, and how a properly-designed control system can ameliorate the impacts caused by these upsets. Events such as a furnace shut-down can cause the Crack Gas Compressor to shut-down if the surge and performance control system is not adequately designed. A trip of the Propylene Compressor can cascade to the Ethylene Compressor and shut it down unless its Compressor Control System can use appropriate feed-forward signals to prevent that occurrence. Proper algorithms provide handles for the compressors to ride through upsets, rather than have overreactive control responses that lead to shut-downs or cause process instability.

Actual design and implementation of surge control systems will be illustrated with two successful field installations: 1) A turbomachinery surge control system for a gas fractionation plant with refrigeration applications done as an integrated control system, and 2) A surge control and loadsharing control system for two parallel mixed-refrigeration compressors in an LNG plant.

This tutorial content is meant to target rotating equipment, reliability and instrumentation engineers, and process engineers working in ethylene plants as well as other industries that make use of centrifugal and axial compressors. The material covered will be technical in nature, and it will be a good source of training material for new engineers, and a refresher for experienced engineers.

# Multi-Section Centrifugal Compressors:

# Compound vs. Back-to-Back Configurations

*Authors: Hannah Beachy, Elliott Group, Jeannette, PA*

*&*

*Jonathan Sutter, Elliott Group, Jeannette, PA*

## Biography

Hannah Beachy has a bachelor of science in Aerospace Engineering from Texas A&M University. She has worked for Elliott Group in various engineering roles since her graduation in 2011. Currently she works for the Engineered Products division creating design standards and tools.

Jonathan Sutter started with Elliott Group as a Project Engineer in 2010 and is currently an Application Engineer, a role he has fulfilled for six years. Jonathan received a BSME (2008) from The Ohio State University and is a licensed Professional Engineer in Ohio.

## Introduction

When considering a gas compressor within a process, often it is found that heat of compression exceeds the allowable temperature limits of the compressor materials (typically the elastomeric sub-components) or the allowable temperature limits of the gas itself. The standard solution to this issue is the utilization of a gas cooler. A multi-section, between-bearing centrifugal compressor solution offers a single casing that can be coupled with an intermediate cooler. However, there are two primary configurations for consideration: the compound compressor, and the back-to-back compressor. This paper aims to explore the differences between the two configurations from an internal standpoint, as well as the implications of these differences to the process engineer and/or purchaser.

After an explanation of the process flow path for the two configurations, comparison will be made on the basis of thrust resolution, main casing (shaft end) seal considerations, thermal gradients internal to the machine, internal leakage paths, and finally on process contamination. It is the authors' goal that this paper will lead to an understanding of the implications of a particular configuration.

It should be noted that multi-section, between-bearing centrifugal compressors can be offered with provisions for more than one inter-cooler, however the focus here is on the unit with a single inter-cooler (i.e. four process connections on the casing).

## Defining Configurations

For the purposes of this paper, "stage" will be used to define a single centrifugal impeller and its diffuser; a "section" will be used to define multiple centrifugal stages preceded by an inlet nozzle and terminated by a discharge nozzle.

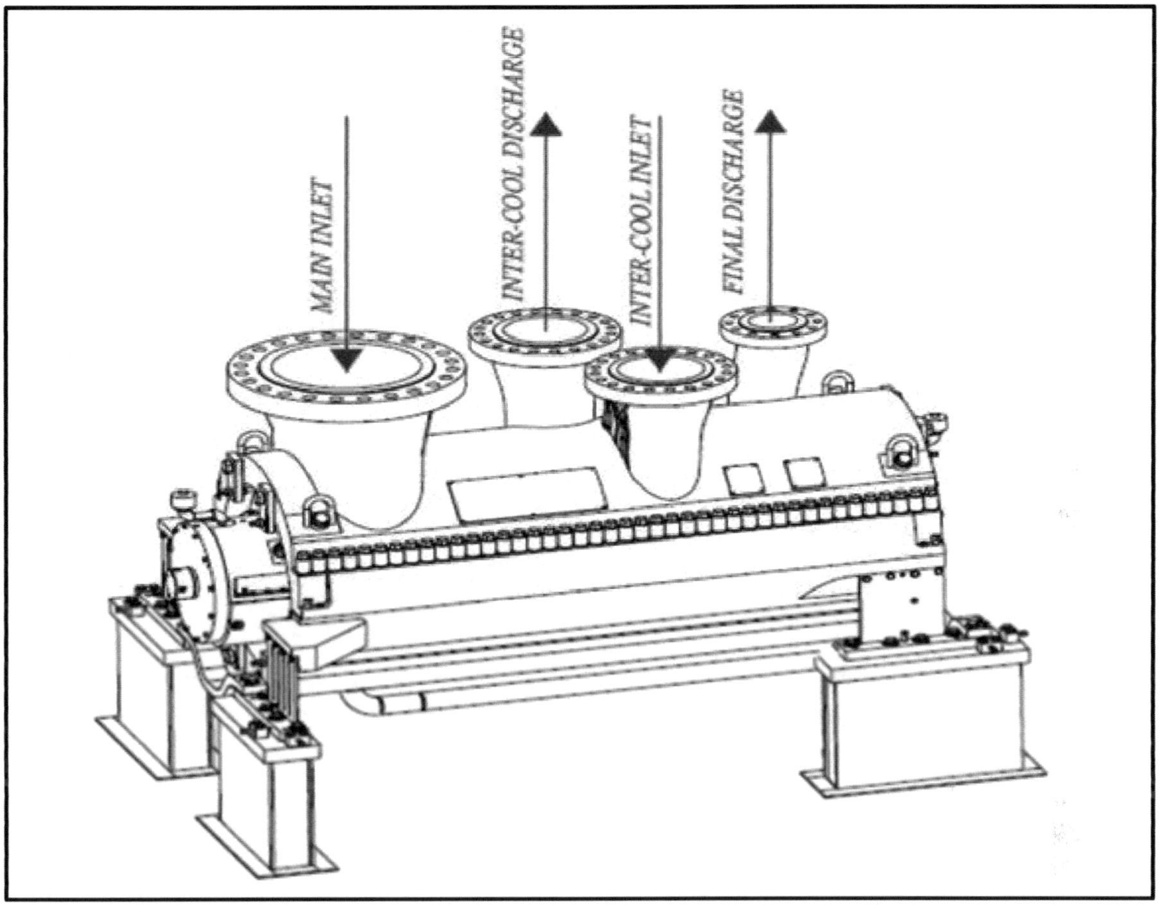


*Figure 1 An inter-cooled, between-bearing centrifugal compressor in the compound configuration.*

The most common inter-cooled configuration for a multi-section, between-bearing centrifugal compressor is the compound configuration (Figure 1). In this figure it can be seen that the process gas enters the compressor at the main inlet nozzle, goes through a section of compression, and exits at an intermediate (inter-cool) discharge nozzle. The gas is then routed to a process gas cooler (heat exchanger) before returning to the compressor at an intermediate (inter-cooled) inlet nozzle. Note that the inter-cooled discharge (terminating the first section of compression) is adjacent to the inter-cooled inlet (beginning the second section of compression) and the process gas physically flows in a series manner through the compressor. The stages of the first section are oriented in the same manner as the stages of the second section.

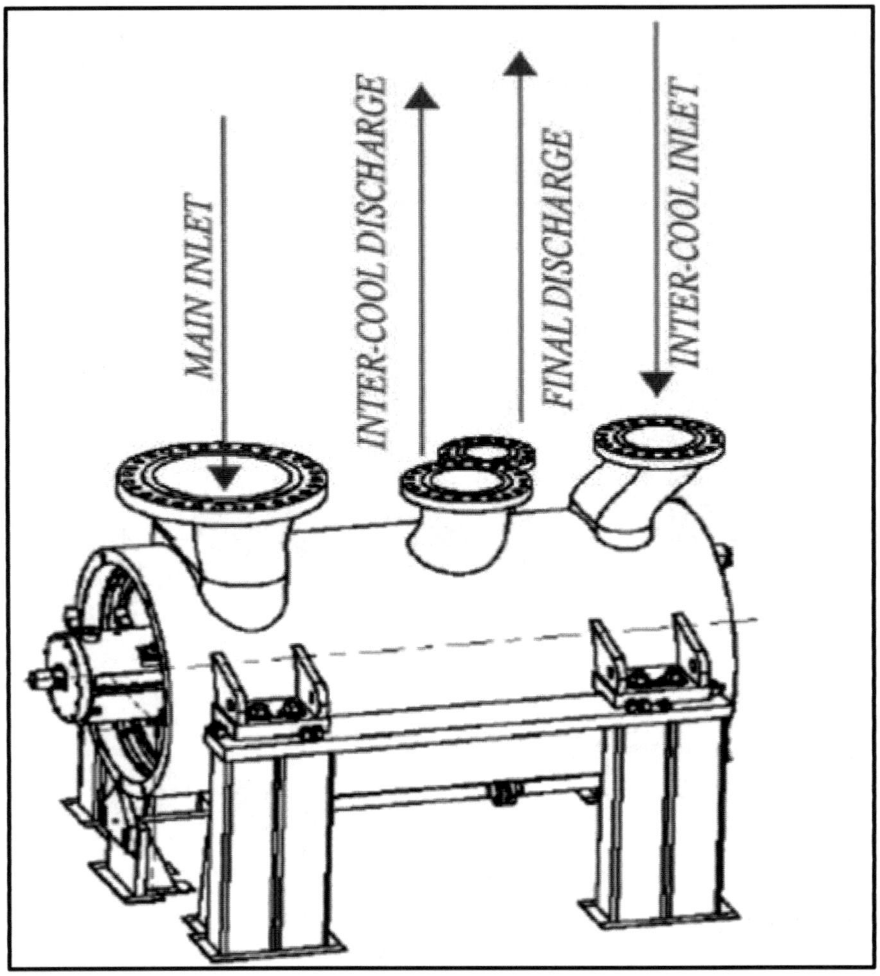


*Figure 2 An inter-cooled, between-between bearing centrifugal compressor in the back-to-back configuration.*

An alternative arrangement would be the back-to-back configuration (Figure 2). In this configuration, the process gas is introduced into the first section via a main inlet nozzle and discharges at an intermediate (inter-cooled) discharge nozzle. However, after the gas returns from the process gas cooler it is routed to the end of the compressor that is opposite the main inlet nozzle. The gas enters an intermediate (inter-cooled) inlet nozzle, goes through the second section of compression, and exits at the final discharge nozzle.

The final discharge nozzle is adjacent to the intermediate discharge nozzle, whereas the inlet nozzles (both main and inter-cooled) are located at the ends of the compressor body. In this manner the staging of the first section faces the opposite direction of the staging in the second section. For this reason, this particular configuration is known as "back-to-back". Additionally, it is worth noting that, due primarily to less favorable nozzle configurations, the back-to back unit typically has a longer axial length relative to the compound unit.

## Thrust Configurations

The differences in compressor configuration between compound compressors and back-to-back compressors impact the thrust analysis and magnitude. In general, the magnitude of the thrust will be significantly lower for a back-to-back configuration, but also more variable with more complex analysis. It is also much more common to have thrust reverse direction on a back-to-back compressor then on a compound compressor.

Thrust in the compressor has two primary internal sources: aerodynamic momentum changes and pressure differentials. Impellers, as pictured in Figure 3 impart momentum to the gas and experience a pressure differential from the eye of the impeller to the heal. Both these forces act in such a manner that the impeller will always thrust from the heal towards the eye. Every step in the shaft exposed to pressurized gas will also exert a force on the rotor. The magnitude will be determined by the pressure of the gas and the surface area of the step perpendicular to the axis of the rotor. The direction of the force will be along the axis of the rotor, towards the perpendicular surface, as show in Figure 4.

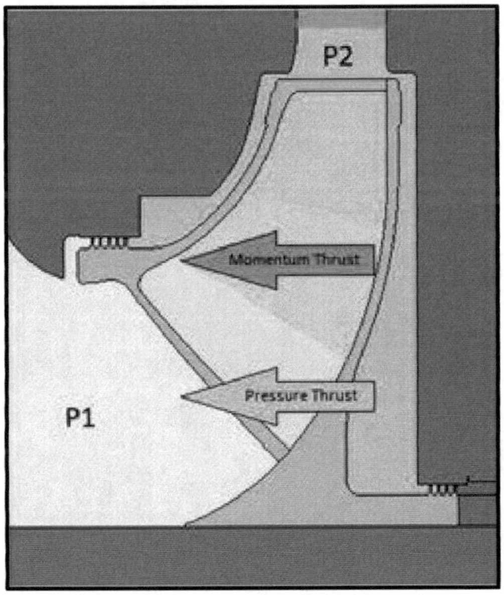


*Figure 3 Impeller illustrating pressure differential and thrust forces.*

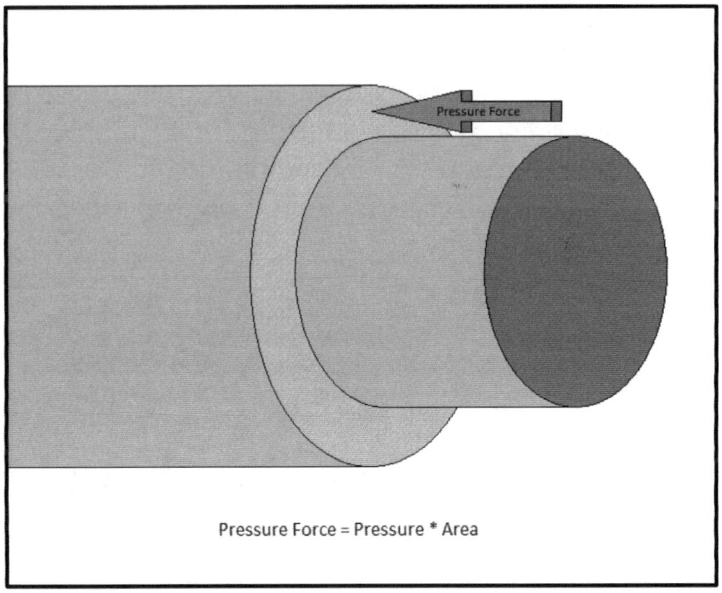


*Figure 4 Shaft section illustrating pressure force on faces of shaft steps.*

Figure 5 shows a compound compressor, indicating the primary sources of thrust. In the case of a compound compressor, all of the thrust from the impellers will be towards the main inlet. Some of the thrust from steps in the shaft will be towards the main inlet, and some will be towards the final discharge. For a typical compound compressor, the vast majority of the thrust will be towards the main inlet. In order to reduce the thrust loading, it is standard to add a balance piston seal after the final impeller, as shown in blue in

Figure 5. This seal causes a pressure drop between the last impeller and the main casing seal cavity. The differential pressure across this seal is higher on the inlet side and lower on the discharge side, causing a large thrust towards the discharge. The balance piston is usually sized such that the overall thrust is still towards the main inlet, but the magnitude is significantly reduced. This allows the selection of a smaller and cooler running thrust bearing.

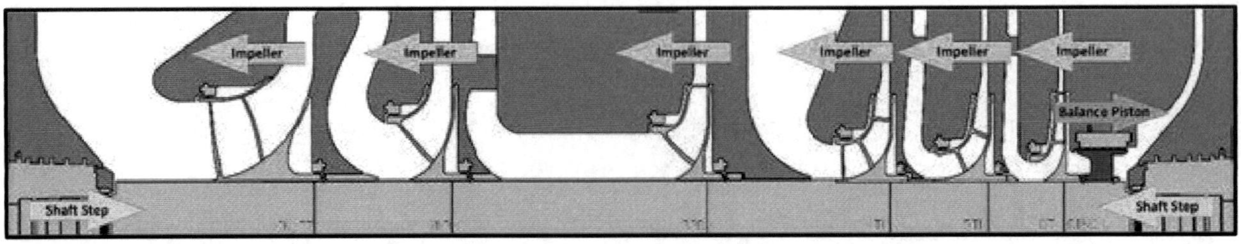


*Figure 5 Compound compressor showing internal sources of thrust with balance piston highlighted in blue.*

Figure 6 shows a back-to-back compressor, indicating the primary sources of thrust. In the case of a back-to-back compressor the impellers in section one will thrust towards the main inlet, and the impellers in the section two will thrust towards the intermediate inlet. Some of the thrust from steps in the shaft will be towards the main inlet, and some will be towards the intermediate inlet. In typical designs the resultant thrust from all shaft steps is towards the main inlet. The resultant thrust from all impellers depends on the balance between the two sections. If section one has a higher flow and/or pressure rise, the compressor will thrust towards the main inlet. If section two has a higher flow and/or pressure rise the compressor may thrust towards the intermediate inlet. For any given operating point the resultant thrust of all impellers will be less than for the same operating point with impellers in a compound configuration. A back-to-back configuration will typically have more variation in total thrust between operating points, including variation in the direction of thrust.

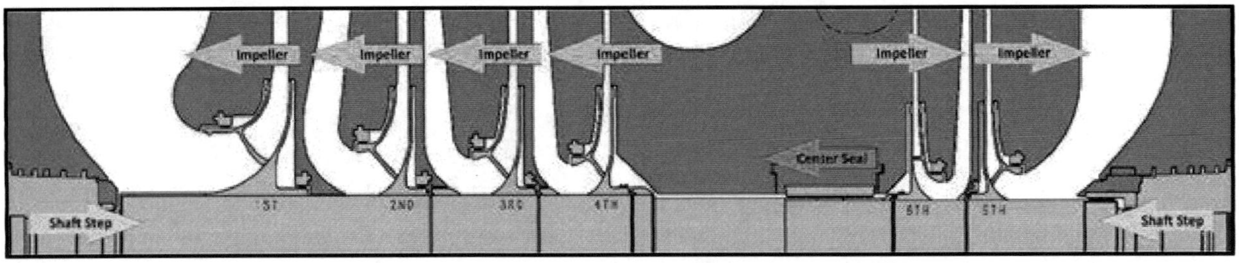


*Figure 6 Back-to-back compressor showing internal sources of thrust.*

## Main Casing Seals

The comparison of main casing seals on compound vs back-to-back compressors must consider two basic seal configurations: equalized and non-equalized. When equalized seals are implemented the main casing seals on both ends of the rotor experience nearly the same pressures. This is achieved through the addition of an equalizing seal and an equalizing line between the gas cavities on each end of the compressor. Non-equalized seals, as the name implies, do not see similar pressures at the main casing seals on both ends of the rotor.

Figure 7 shows a compound compressor with equalized seals. In the case of a compound compressor, the equalizing seal is also used to balance thrust, as previously discussed, and is usually referred to as the balance piston seal. The high pressure and temperature gas coming out of the final impeller flows across the labyrinth seal, into the seal cavity, through the balance piston line and into the main inlet of the machine. The cavity side of the balance piston seal is at a pressure only slightly higher than the main inlet of the machine.

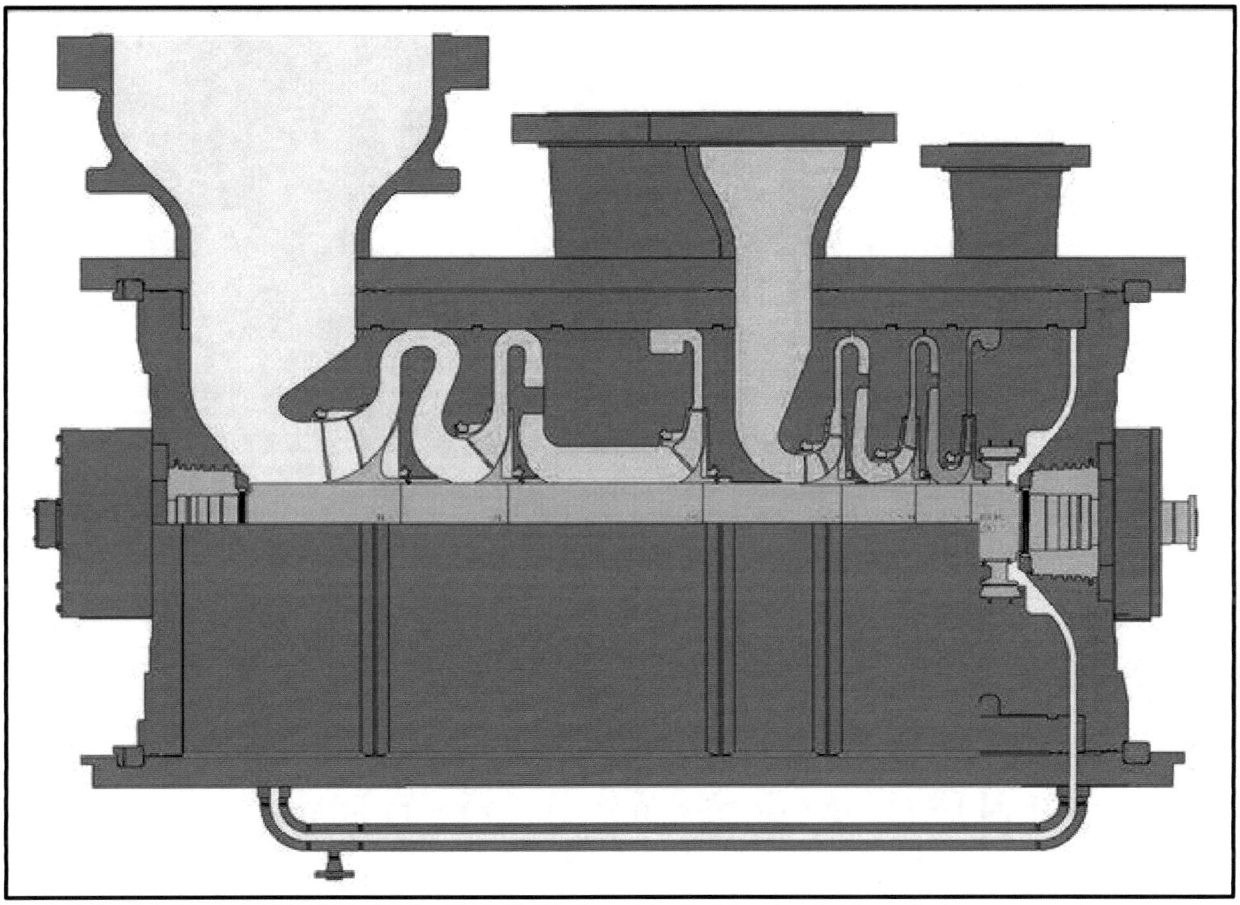

*Figure 7 Compound compressor with equalized seals. Darker color indicates higher pressure. For this configuration the seal cavities are both at pressures near the main inlet pressure.*

Figure 8 shows a back-to-back compressor with equalized seals. In this case, while the equalizing seal will have an impact on thrust, its primary purpose is not to balance the thrust. Typically, the equalizing seal implemented in a back-to-back configuration is much smaller than the balance piston for a compound compressor with the same process. Again, the seal cavity is at a pressure only slightly higher than the main inlet of the machine.

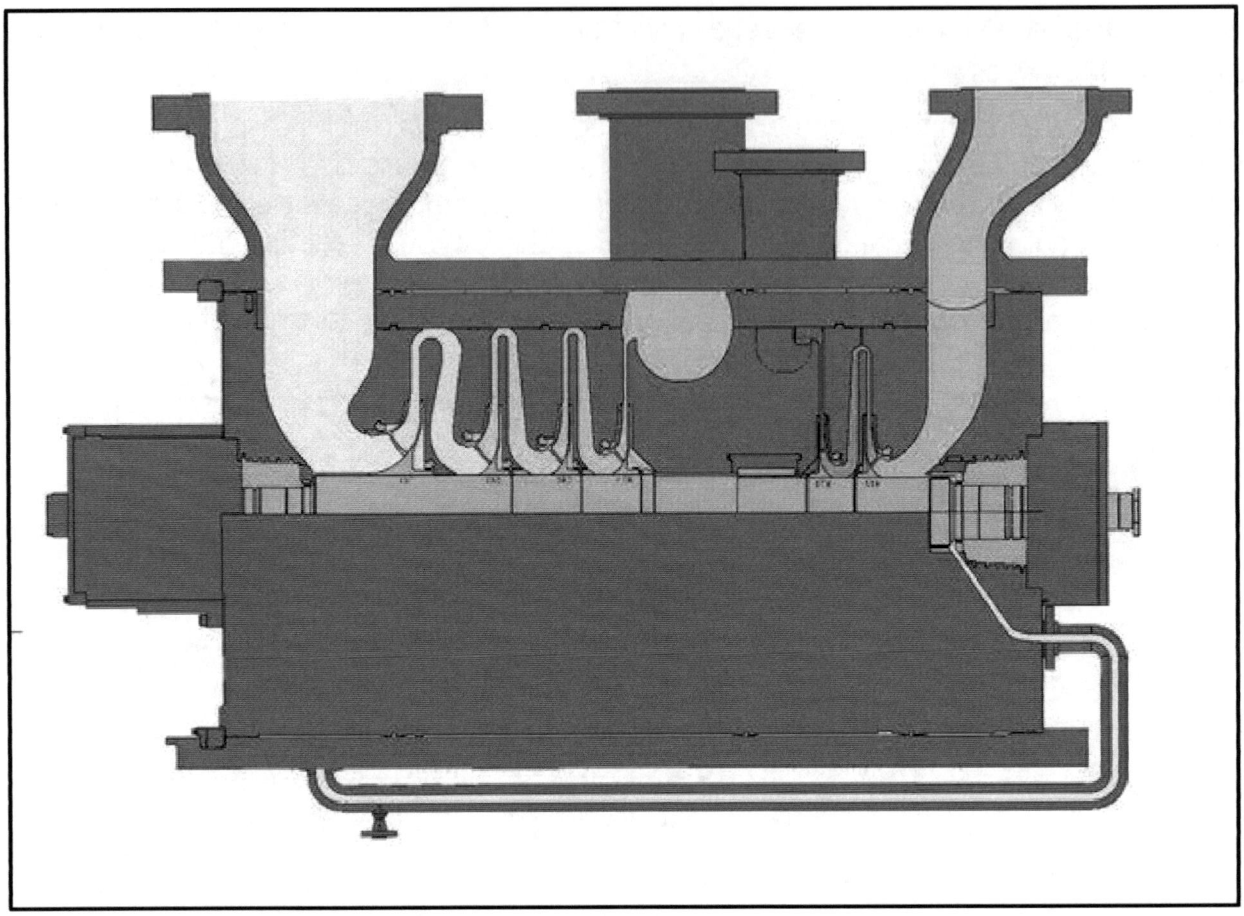

*Figure 8 Back-to-back compressor with equalized seals. Darker color indicates higher pressure. For this configuration both seal cavities experience pressures near the main inlet pressure*

Figure 9 shows a compound compressor with non-equalized seals. In most cases a compound compressor with non-equalized seals still has a balance piston and balance piston line. The balance piston line will not return to the main inlet as it does in an equalized configuration. Instead, it will return to some other location in the compressor, typically the intermediate discharge or inlet. The pressure in the seal cavity on the discharge side of the machine will be slightly higher than the pressure where the balance piston line is attached, and significantly higher than the main inlet pressure.

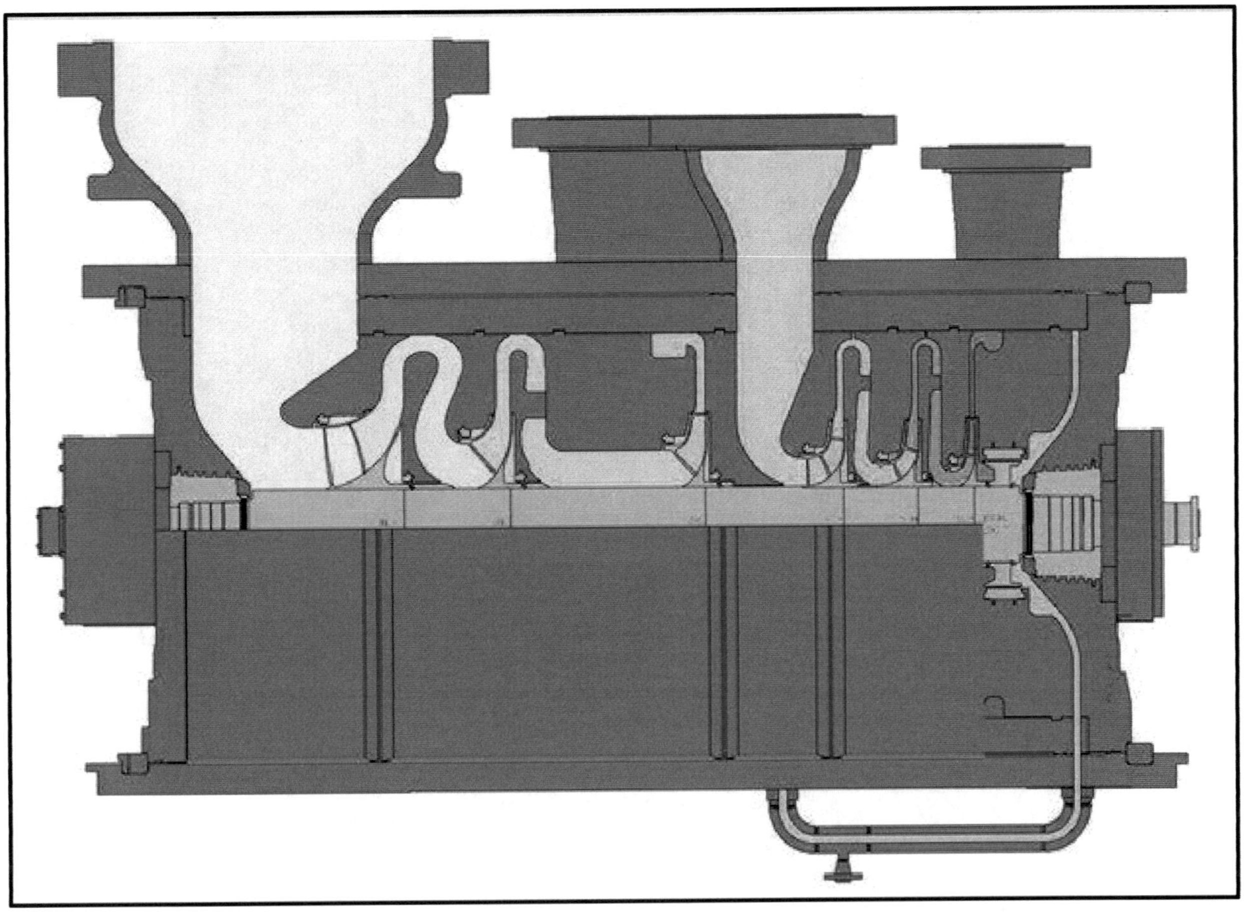

*Figure 9 Compound compressor with non-equalized seals. Darker color indicates higher pressure. For this configuration the seal on the main inlet side experiences the main inlet pressure. The seal cavity on the final discharge side experiences pressures near those of the return location, in this case the intermediate inlet.*

Figure 10 shows a back-to-back compressor with non-equalized seals. A non-equalized back-to-back compressor does not usually have any equalizing seal or equalizing line. The main casing seal on the main inlet side sees the inlet pressure to section one. The main casing seal on the intermediate inlet side sees the much higher inlet pressure to section two.

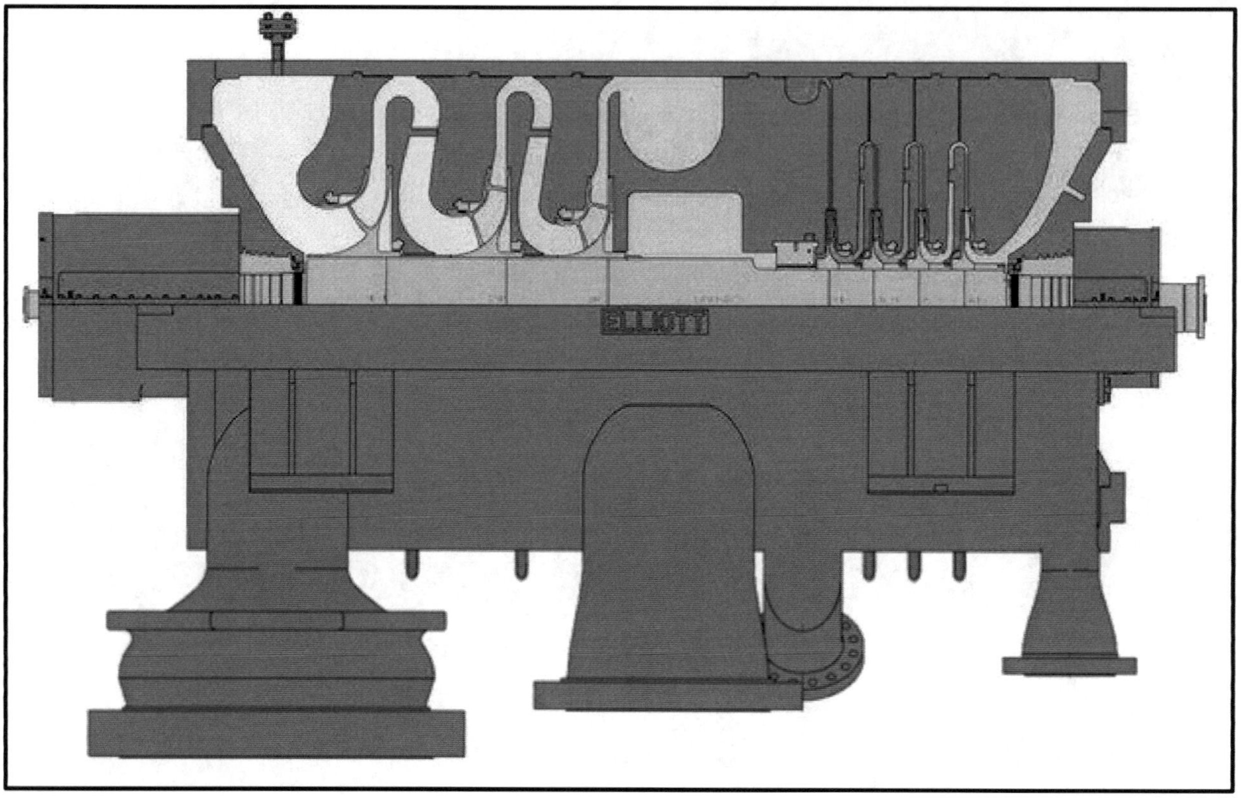


*Figure 10 Back-to-back compressor with non-equalized seals. Darker color indicates higher pressure. For this configuration the seal on the main inlet side experiences the main inlet pressure. The seal cavity on the intermediate inlet side experiences intermediate inlet pressure*

## Thermal Gradients

The physical arrangement of the two configurations creates another topic worth investigation: thermal gradients. More specifically, there are distinctly different thermal gradients found in the sectional dividing components of the compound and back-to-back configurations. These thermal gradients can have significant impact on the efficiency of the machine, an attribute both compressor manufacturers and purchasers are typically concerned about. For multi-sectional compressors, efficiency is presented in a sectional manner, and is usually measured at the flanges of the machine.

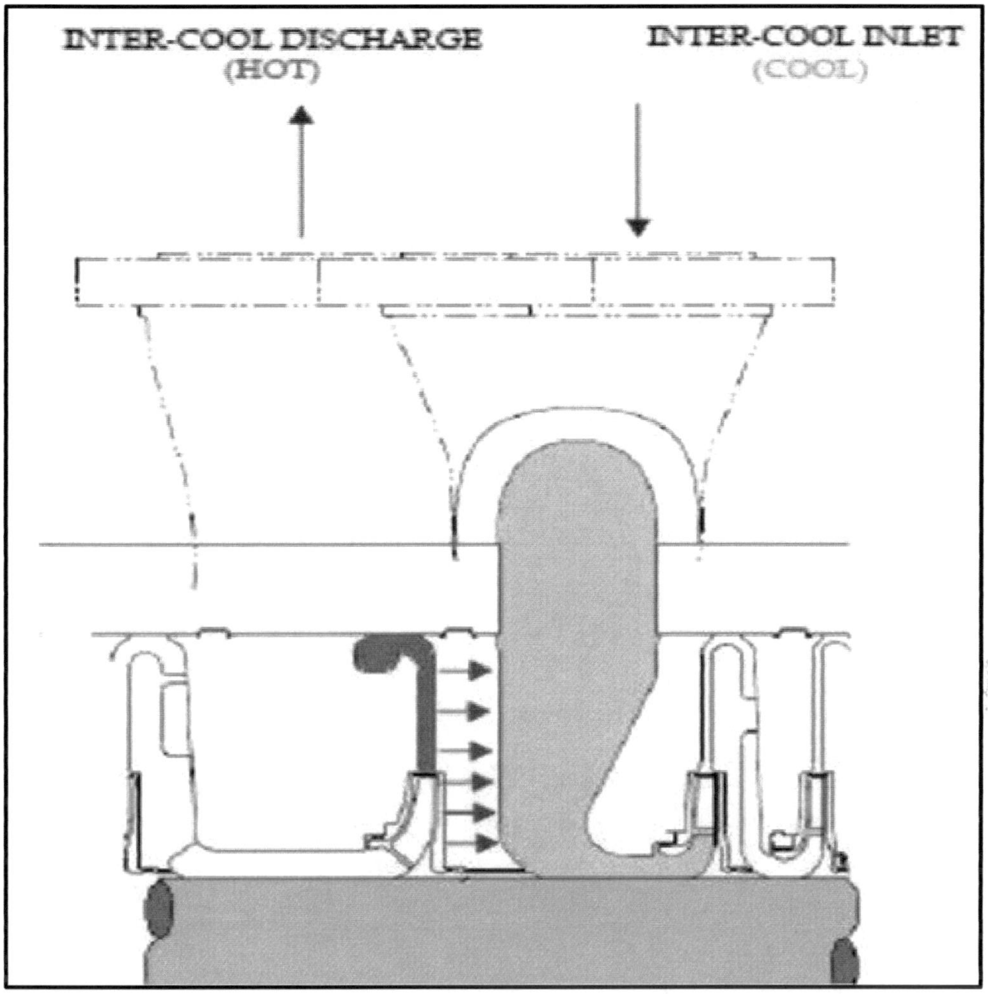


*Figure 11 In a compound configuration, heat (denoted by the purple arrows) is transferred from the hot gas in the inter-cool discharge to the cool gas in the inter-cool inlet.*

In the compound configuration, hot gas exiting the compressor via the discharge volute of section one (Figure 11) is in contact with the cool gas returning from the process gas cooler. The thermal gradient created across the volute wall is significant and causes heat to move from the discharge of the first section to the inlet of the second section. The gas is being cooled while traveling from the tip of the last impeller in section one to the discharge flange of section one, which in turn makes the sectional efficiency appear slightly better than that of a single section compressor at aero-dynamically identical conditions. The opposite situation occurs in the second section of the compound unit: cool gas returning from the process cooler is heated in between the inlet flange of section two and the eye of the first impeller. Before any work is done on the gas in the second section, the temperature has already risen from the initial temperature at the inlet flange of second section.

In the back-to-back configuration, the discharge volute of section one is abutted against the discharge volute of section two (Figure 12). In this configuration the heat transfer is much less as the temperatures are usually much closer. The amount of energy moving from section to section is smaller (note that the direction of energy transfer could be either way, depending on which section has a hotter discharge), and the heat is being exchanged between the discharge of section one and the discharge of section two. Recalling basic thermodynamics, it is more desirable to add heat to the discharge of the unit than to add the same amount of heat to the inlet of the unit. The back-to-back arrangement will require less power as compared to an areodynamically identical compound arrangement.

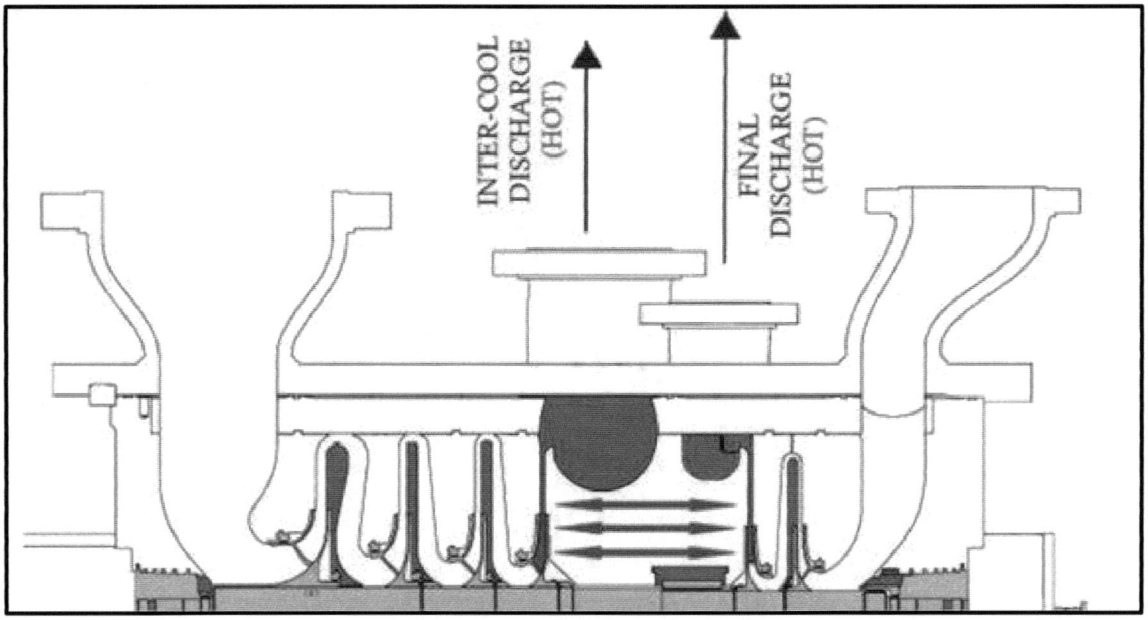


*Figure 12 In the back-to-back configuration, the direction of heat transfer is dependent on the temperature differential between the inter-cool discharge and final discharge. While heat transfer can occur in either direction (denoted by the purple arrows) and is dependent on operating conditions, the amount of heat transfer is much smaller relative to the compound configuration.*

In summary, for two identical compressors in the compound vs. back-to-back, the back-to-back would be expected to require slightly less power and show better efficiency for the same amount of compression.

## Internal Leakages

Compressor configuration has a significant impact on internal leak paths, and is important for both the designer and end-user to understand what leak-paths exist for compound vs back-to-back configurations. Internal to the compressor, gas leakage from one stage or section to another is usually reduced by the implementation of labyrinth seals. The teeth (or fins) on these seals function as a series of orifices, each reducing the pressure of the flow. The quantity of the flow will depend on the seal geometry and tooth count, pressure differential, gas composition and gas conditions. The estimated leakage across a labyrinth seal is usually quantified based on the Egli equations from the work of A. Egli.

The flow and leak paths for a compound compressor are shown in Figure 13. Both sections of the compressor will experience some recirculation at the eye and heal of each impeller. Most main casing seals have a buffer gas which leaks into the compressor. On the main inlet side, the buffer gas enters before the first impeller and mixes with the inlet gas flow. On the final discharge side the buffer gas will enter the cavity behind the balance piston, where it will mix with the gas discharging from the final impeller. This mixture will flow through the balance piston line and be added to the gas at the balance piston return location. On an equalized machine, the leakage will be added at the main inlet, prior to the first impeller. On a non-equalized machine, the leakage will be mixed with the flow at the intermediate discharge, intermediate inlet or wherever the balance piston line returns. Between the sections of a compound compressor there will be leakage across the shaft seal on the final impeller of the first section. The magnitude of the leakage will depend on seal configuration and pressure differential. The direction of the leakage will depend on whether the pressure rise in the final diffuser of section one is larger or smaller than the pressure loss in the process gas cooler and associated piping. Typically the flow will be from the second section into the first section.

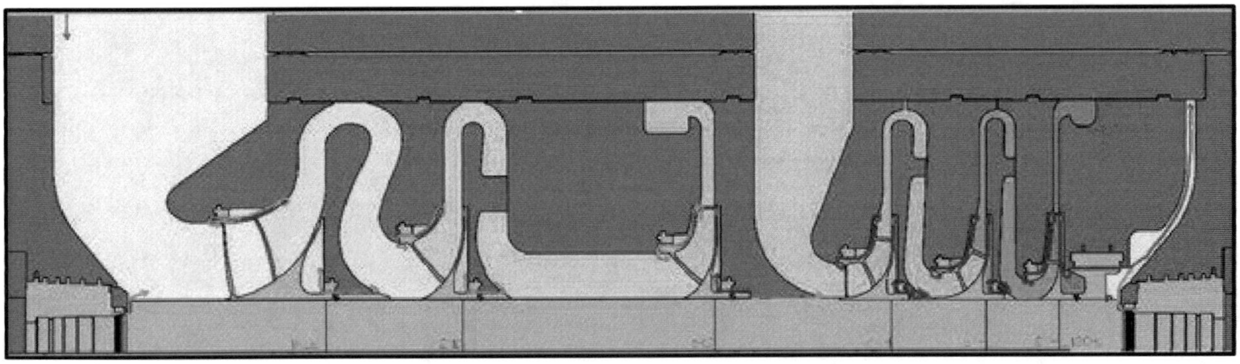

*Figure 13 Compound compressor illustrating leak paths. Leakage from the buffer is indicated by orange arrows. Leakage of section 1 gas is indicated by light blue arrows. Leakage of section 2 gas in indicated by dark blue arrows. Mix leakage from buffer and section 2 is indicated by purple arrows. Leakage across the final shaft seal in the first section may be from section1 to section 2, or from section 2 to section 1.*

The flow and leak paths for a back-to-back compressor are shown in Figure 14. As with the compound compressor, each stage will have some recirculation at the eye and heal of the impellers, and the buffer gas on the main inlet side will be added directly into the inlet gas flow. If the machine is non-equalized, the buffer gas on the intermediate inlet end of the machine will be added directly into the inlet gas flow for the second section. If the machine is equalized, the buffer gas will enter the cavity behind the equalizing seal, where it will be mixed with the leakage across the equalizing seal from the intermediate inlet. The mixture will then flow through the equalizing line to the main inlet where it will be added to the inlet gas prior to the first impeller. Between the sections of a back-to-back compressor there will be leakage across the center seal. The leakage will always be from the final discharge towards the intermediate discharge, and will be added to the flow exiting the first section of the machine.

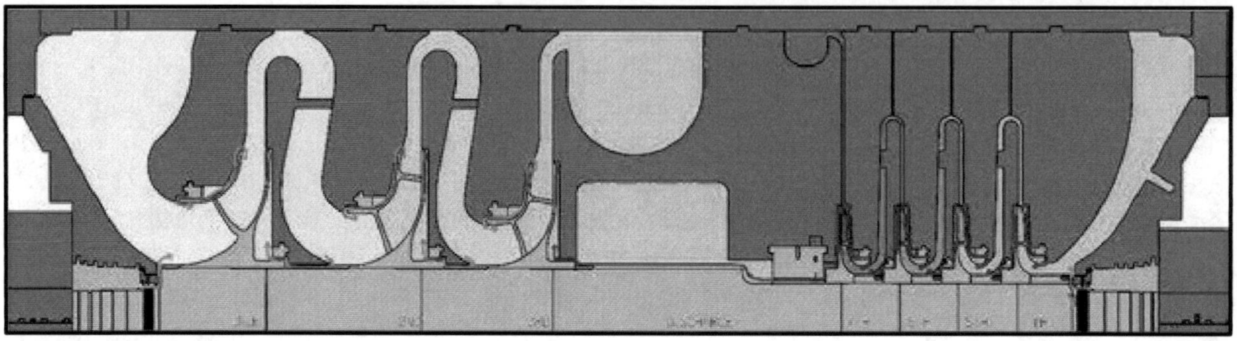

*Figure 14 Back-to-back compressor illustrating leak paths. Leakage from the buffer is indicated by orange arrows. Leakage of section 1 gas is indicated by light blue arrows. Leakage of section 2 gas in indicated by dark blue arrows. Leakage across the center seal will always be from section 2 to section 1.*

## Process Contamination

One final consideration of the two configurations may appeal to the process designer. Due to the internal leakage paths the back-to-back compressor offers an advantage in that the process gas within the lower pressure section never mixes with the process gas of the higher pressure section. The pressure at the discharge of section two will always be higher than the discharge pressure of section one, therefore the leakage will always be from section two towards section one (Figure 15).

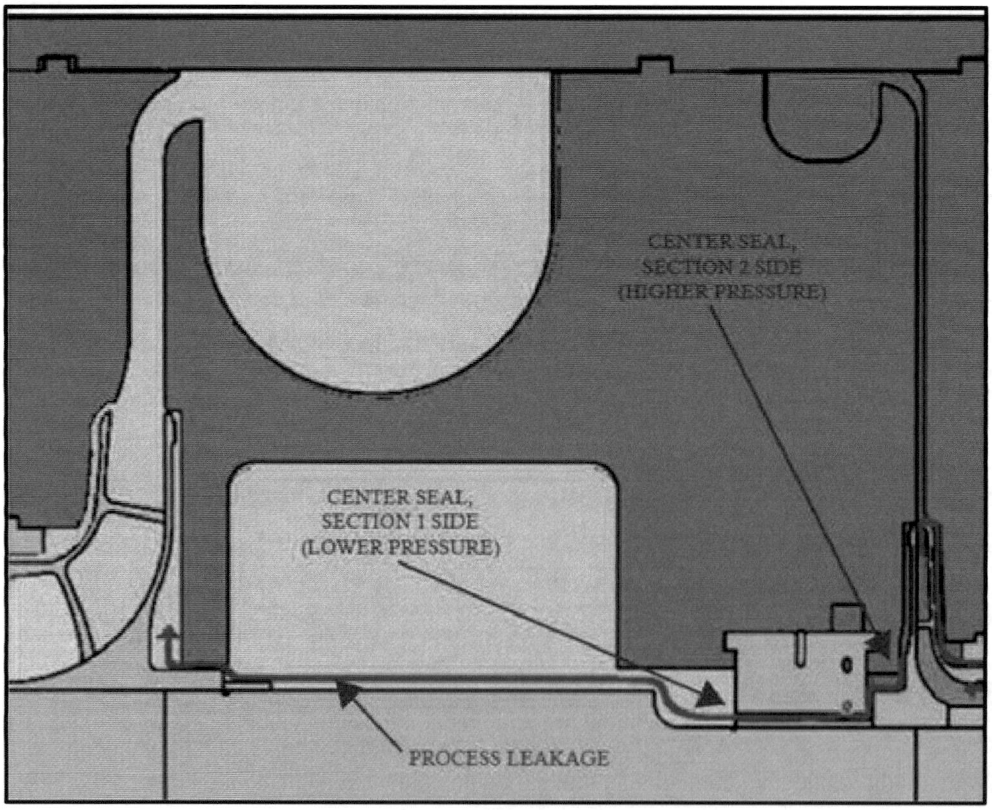

On the other hand, leakage direction in the compound configuration can vary during operation (Figure 16). To determine the direction of leakage, one must calculate pressures on either side of the heal seal on the last stage of the first section. It can roughly be assumed that 66.7% of the static pressure rise of a stage is achieved at the tip of the impeller (the remaining 33.3% of the static pressure rise being realized in the diffuser). Applying this generalization to the last stage of the first section (or more specifically, the heal seal of the last stage) and knowing the pressure loss from the inter-cool discharge to the inter-cool inlet allows for determination of leakage direction. It should be noted that (1) the process licensor and/or end-user rarely has the information necessary to determine the pressure rise of individual stages, and (2) knowing the leakage direction or quantity of leakage isn't as important as knowing *that there will be leakage in both directions during normal operation.* Therefore, the two sections will contaminate each other at various points of operation.

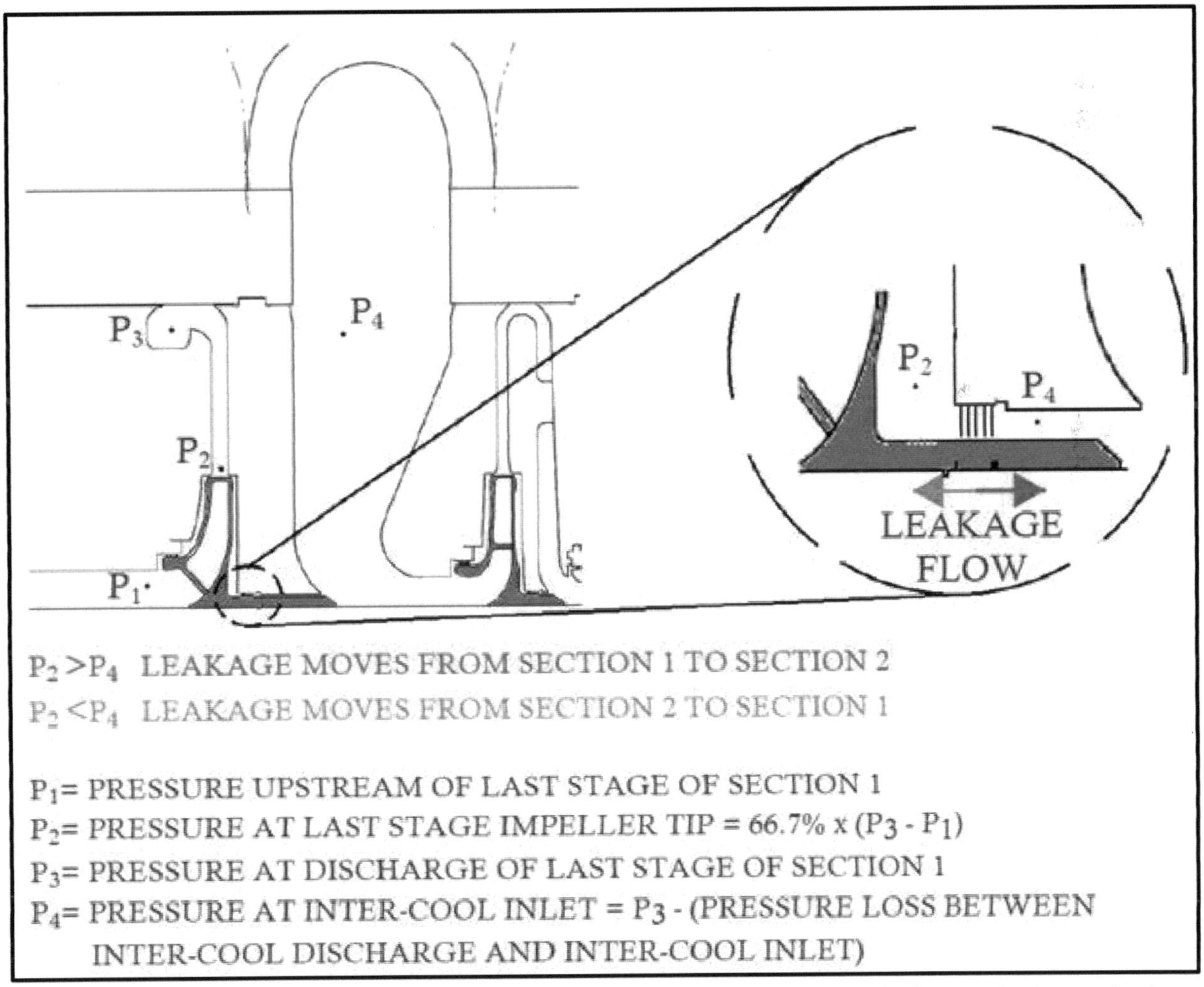


*Figure 16 Leakage (denoted by the blue line) in the back-to-back compressor will always be from section 2 to section 1.*

Within production of ethylene utilizing steam cracking, this leakage path is of particular importance in the last and second-to-last sections of compression. In this application, acid gas is removed from the process before the last section of compression. Contamination of the process gas in the final section of compression affects the performance of the demethanizor downstream of the final section of compression.

It is worth noting that compound compressors have been employed in this application but have historically been modified with a provision for a high pressure buffer between the two sections. While this provision is effective in preventing contamination, the compressor manufacturer prefers the back-to-back configuration for simplicity, and the plant operator has one less auxiliary system to monitor and maintain.

## Conclusions

Both the compound and back-to-back configurations are viable solutions to creating a multi-section inter-cooled machine, but the differences between these configurations must be understood in order to correctly select and implement the machine.

The compound configuration presents several advantages, particularly in the initial purchase. For the same aerodynamic process and components, the compound configuration will typically have better rotor dynamic performance. This is due to its generally shorter bearing span, and to the absence of a heavy center seal near the mid span. The relatively shorter bearing span, enabled by favorable nozzle configurations, also means the compound configuration will typically come at a lower initial cost. Although the thrust for the compound configuration is generally higher, its magnitude is typically less variable, and the thrust analysis much less complex.

The back-to-back machine may be more difficult and expensive in initial manufacture, but provides some advantages when it comes to implementation within the plant and process. For the same aerodynamic process and components, a back-to-back configuration will usually provide slightly lower power requirements and higher efficiencies due to more favorable thermal gradients between the sections. It also will generally have lower thrust magnitudes resulting in a smaller, cooler thrust bearing. Finally, the risk of contaminating the process gas in the second section with process gas in the first section during normal operation is eliminated for a back-to-back machine.

Knowledge of the basic differences between compound and back-to-back configurations should allow the process engineer or purchaser to evaluate the advantages and disadvantages of each and the implications to their application.

# 618530 Development and Installation of SIL2 Torquemeter for Functional Safety Process Control of Steam Turbine Driven Compressor

**Abstract:**

**Abstract for SIL2 Torquemeter for Functional Safety Process Control of Steam Turbine driven Compressor**

**Introduction**
Use of a torquemeter to accurately measure shaft horsepower independent of fuel, pressure, temperature, and flow variations is known to many oil and gas plant turbomachinery users worldwide. Used to inform the end user of turbomachinery train location specific degradation, the basic condition monitoring torquemeter system is a useful plant tool for performance trending and diagnostics. The large fuel use of steam turbines in ethylene manufacture makes plant efficiency extremely important to bottom line considerations.
For end users of torquemeter devices, it has always been possible for them to be used as process control devices, since accurate delivered power measurement helps run close to performance limits, maximizing plant throughput and therefore revenue. The single measurement aspect of the torquemeter system does not provide for a functionally safe system in process control terms.
This paper outlines the successful development of a process system, using novel technology, which can be applied as a retrofit or new build system.
**Summary**
For one ethylene plant operated by Total, their existing torquemeter was successfully monitoring shaft horsepower and was set to react to an increase in demand from the compressor by ensuring the steam turbine power delivery was within acceptable limits for the compressor rotor rating. Any potential over run in power from the steam turbine was prevented by linking the torquemeter reading to the steam demand valve, which closed on the first sign of potential over torque. Since this process loop was safety critical, after a customer functional safety review a requirement was identified for a two out of three (2oo3) redundant voting system of torque measurement, so that in the unlikely event of a torquemeter failure, the manufacturing process could continue safely. The Safety Integrity Level required for plant compliance from the customer was SIL 2. Accuracy during normal running was to be maintained at +/-1% of reading at normal torque and operating speed. Single failure mode (plant still running) had to maintain accuracy at +/-5% of reading at normal torque and speed. To move from a single reading torquemeter to a 2oo3 voting functional safety torquemeter required the integration of three pairs of measuring coils, detecting three independent values, which could be processed in three independent SIL Modules to provide 3 completely independent values for comparison by the customers data acquisition system and associated emergency shutdown system (ESD). This paper will explain how proof of concept study was carried out using a small R&D developed torquemeter; how the recorded data showing optimal alignment, separation and signal strengths led to the final design; how the end-user was involved in all stages of the development; and how the final design was installed. It will also summarize some of the SIL concepts and work carried out to achieve the required level 2 product.

## AUTHOR INDEX